Bäcklund and Darboux Transformations

This book describes the remarkable connections that exist between the classical differential geometry of surfaces and modern soliton theory. The authors explore the extensive body of literature from the nineteenth and early twentieth centuries by such eminent geometers as Bianchi, Darboux, Bäcklund, and Eisenhart on transformations of privileged classes of surfaces which leave key geometric properties unchanged. Prominent amongst these are Bäcklund-Darboux transformations with their remarkable associated nonlinear superposition principles and importance in soliton theory. It is with these transformations and the links they afford between the classical differential geometry of surfaces and the nonlinear equations of soliton theory that the present text is concerned. In this geometric context, solitonic equations arise out of the Gauss-Mainardi-Codazzi equations for various types of surfaces that admit invariance under Bäcklund-Darboux transformations.

This text is appropriate for use at a higher undergraduate or graduate level for applied mathematicians or mathematical physicists.

Professor Colin Rogers currently holds a Chair in Applied Mathematics at the University of New South Wales. He has written some 150 research papers in international research journals and is the co-author of the monograph *Bäcklund Transformations and Their Applications* (1982).

Dr. Wolfgang K. Schief presently holds the position of Queen Elizabeth II, ARC Research Fellow, at the University of New South Wales. Dr. Schief has written more than 60 research papers in international research journals.

Cambridge Texts in Applied Mathematics

Maximum and Minimum Principles
M.J. Sewell

Solitons
P.G. Drazin and R.S. Johnson

The Kinematics of Mixing
J.M. Ottino

Introduction to Numerical Linear Algebra and Optimisation
Philippe G. Ciarlet

Integral Equations
David Porter and David S.G. Stirling

Perturbation Methods
E.J. Hinch

The Thermomechanics of Plasticity and Fracture
Gerard A. Maugin

Boundary Integral and Singularity Methods for Linearized Viscous Flow
C. Pozrikidis

Nonlinear Wave Processes in Acoustics
K. Naugolnykh and L. Ostrovsky

Nonlinear Systems
P.G. Drazin

Stability, Instability and Chaos
Paul Glendinning

Applied Analysis of the Navier–Stokes Equations
C.R. Doering and J.D. Gibbon

Viscous Flow
H. Ockendon and J.R. Ockendon

Scaling, Self-Similarity and Intermediate Asymptotics
G.I. Barenblatt

A First Course in the Numerical Analysis of Differential Equations
Arieh Iserles

Complex Variables: Introduction and Applications
Mark J. Ablowitz and Athanassios S. Fokas

Mathematical Models in the Applied Sciences
A.C. Fowler

Thinking About Ordinary Differential Equations
Robert E. O'Malley

A Modern Introduction to the Mathematical Theory of Water Waves
R.S. Johnson

Rarefied Gas Dynamics
Carlo Cercignani

Symmetry Methods for Differential Equations
Peter E. Hydon

High Speed Flow
C.J. Chapman

Wave Motion
J. Billingham and A.C. King

An Introduction to Magnetohydrodynamics
P.A. Davidson

Linear Elastic Waves
John G. Harris

Infinite Dimensional Dynamical Systems
James C. Robinson

Introduction to Symmetry Analysis
Brian J. Cantwell

Vorticity and Incompressible Flow
Andrew J. Majda and Andrea L. Bertozzi

Bäcklund and Darboux Transformations
Geometry and Modern Applications in Soliton Theory

C. ROGERS
Professor of Applied Mathematics
The University of New South Wales

W.K. SCHIEF
Queen Elizabeth II, ARC Research Fellow
The University of New South Wales

CAMBRIDGE UNIVERSITY PRESS
Cambridge, New York, Melbourne, Madrid, Cape Town, Singapore,
São Paulo, Delhi, Dubai, Tokyo, Mexico City

Cambridge University Press
The Edinburgh Building, Cambridge CB2 8RU, UK

Published in the United States of America by Cambridge University Press, New York

www.cambridge.org
Information on this title: www.cambridge.org/9780521012881

First published 2002

A catalogue record for this publication is available from the British Library

Library of Congress Cataloguing in Publication data

Rogers, C.

Bäcklund and Darboux transformations : geometry and modern applications in
soliton theory / C. Rogers, W.K. Schief.

p. cm. – (Cambridge texts in applied mathematics)

Includes bibliographical references and index.

ISBN 0-521-81331-X – ISBN 0-521-01288-0 (pb.)

1. Solitons. 2. Bäcklund transformations. 3. Darboux transformations.
I. Schief, W.K. (Wolfgang Karl) 1964– II. Title. III. Series.

QC174.26 W28 2002

530.124 – dc21 2001043453

ISBN 978-0-521-81331-0 Hardback
ISBN 978-0-521-01288-1 Paperback

Dedicated to the memory of
Professor David Crighton, FRS

Für Christel und Wolfgang

Contents

Preface xv

Acknowledgements xvii

General Introduction and Outline **1**

1 Pseudospherical Surfaces and the Classical Bäcklund Transformation. The Bianchi System **17**

1.1 The Gauss-Weingarten Equations for Hyperbolic Surfaces. Pseudospherical Surfaces. The Sine-Gordon Equation 18

1.2 The Classical Bäcklund Transformation for the Sine-Gordon Equation 22

1.3 Bianchi's Permutability Theorem. Generation of Multi-Soliton Solutions 28

- 1.3.1 Bianchi's Permutability Theorem 28
- 1.3.2 Physical Applications 30

1.4 Pseudospherical Soliton Surfaces. Breathers 31

- 1.4.1 The Pseudosphere 32
- 1.4.2 A Pseudospherical Helicoid 35
- 1.4.3 Two-Soliton Surfaces 37
- 1.4.4 Breathers 38
- 1.4.5 Stationary Breather Surfaces 39

1.5 Parallel Surfaces. Induced Bäcklund Transformation for a Class of Weingarten Surfaces 41

- 1.5.1 Surfaces of Constant Mean Curvature. A Theorem of Bonnet 42
- 1.5.2 An Induced Bäcklund Transformation 43

1.6 The Bianchi System. Its Auto-Bäcklund Transformation 45

1.6.1 Hyperbolic Surfaces. Spherical Representation 46
1.6.2 A Bäcklund Transformation for Hyperbolic Surfaces 49
1.6.3 The Bianchi System 53

2 The Motion of Curves and Surfaces. Soliton Connections 60
2.1 Motions of Curves of Constant Torsion or Curvature. The Sine-Gordon Connection 61
2.1.1 A Motion of an Inextensible Curve of Constant Torsion 62
2.1.2 A Motion of an Inextensible Curve of Constant Curvature 63
2.2 A 2×2 Linear Representation for the Sine-Gordon Equation 64
2.3 The Motion of Pseudospherical Surfaces. A Weingarten System and Its Bäcklund Transformation 68
2.3.1 A Continuum Limit of an Anharmonic Lattice Model 71
2.3.2 A Weingarten System 71
2.3.3 Bäcklund Transformations 73
2.4 The mKdV Equation. Moving Curve and Soliton Surface Representations. A Solitonic Weingarten System 80
2.4.1 The mKdV Equation 80
2.4.2 Motion of a Dini Surface 82
2.4.3 A Triply Orthogonal Weingarten System 85

3 Tzitzeica Surfaces. Conjugate Nets and the Toda Lattice Scheme 88
3.1 Tzitzeica Surfaces. Link to an Integrable Gasdynamics System 89
3.1.1 The Tzitzeica and Affinsphären Equations 89
3.1.2 The Affinsphären Equation in a Gasdynamics Context 95
3.2 Construction of Tzitzeica Surfaces. An Induced Bäcklund Transformation 101
3.3 Laplace-Darboux Transformations. The Two-Dimensional Toda Lattice. Conjugate Nets 109
3.3.1 Laplace-Darboux Transformations 110
3.3.2 Iteration of Laplace-Darboux Transformations. The Two-Dimensional Toda Lattice 111
3.3.3 The Two-Dimensional Toda Lattice: Its Linear Representation and Bäcklund Transformation 113
3.3.4 Conjugate Nets 117

4 Hasimoto Surfaces and the Nonlinear Schrödinger Equation. Geometry and Associated Soliton Equations 119
4.1 Binormal Motion and the Nonlinear Schrödinger Equation. The Heisenberg Spin Equation 120

4.1.1 A Single Soliton NLS Surface 122
4.1.2 Geometric Properties 124
4.1.3 The Heisenberg Spin Equation 128
4.2 The Pohlmeyer-Lund-Regge Model. SIT and SRS Connections. Compatibility with the NLS Equation 129
4.2.1 The Pohlmeyer-Lund-Regge Model 130
4.2.2 The SIT Connection 132
4.2.3 The SRS Connection 134
4.2.4 Compatibility of the Maxwell-Bloch System with the NLS Equation 135
4.3 Geometry of the NLS Equation. The Auto-Bäcklund Transformation 137
4.3.1 The Nonlinear Schrödinger Equation 142
4.3.2 The Auto-Bäcklund Transformation 146

5 Isothermic Surfaces. The Calapso and Zoomeron Equations 152
5.1 The Gauss-Mainardi-Codazzi Equations for Isothermic Surfaces. The Calapso Equation. Dual Isothermic Surfaces 152
5.2 The Geometry of Isothermic Surfaces in $\mathbb{R}^{n+2}$ 156
5.2.1 Conjugate and Orthogonal Coordinates 157
5.2.2 Isothermic Surfaces 159
5.2.3 Specialisations and Generalisations 160
5.3 The Vector Calapso System. Its Scalar Lax Pair 162
5.3.1 The Vector Calapso System 162
5.3.2 A Scalar Lax Pair 164
5.3.3 Reductions 166
5.4 The Fundamental Transformation 167
5.4.1 Parallel Nets. The Combescure Transformation 167
5.4.2 The Radial Transformation 168
5.4.3 The Fundamental Transformation 169
5.5 A Bäcklund Transformation for Isothermic Surfaces 171
5.5.1 The Fundamental Transformation for Conjugate Coordinates 171
5.5.2 The Ribaucour Transformation 173
5.5.3 A Bäcklund Transformation for Isothermic Surfaces 175
5.6 Permutability Theorems and Their Geometric Implications 178
5.6.1 A Permutability Theorem for Conjugate Nets. Planarity 178
5.6.2 A Permutability Theorem for Orthogonal Conjugate Nets. Cyclicity 181
5.6.3 A Permutability Theorem for Isothermic Surfaces. Constant Cross-Ratio 184

5.7 An Explicit Permutability Theorem for the Vector Calapso System 187
5.7.1 The Ribaucour-Moutard Connection 187
5.7.2 A Permutability Theorem 189
5.8 Particular Isothermic Surfaces. One-Soliton Surfaces and Cyclides 191
5.8.1 One-Soliton Isothermic Surfaces 191
5.8.2 A Class of Solutions Generated by the Moutard Transformation 192
5.8.3 Dupin Cyclides 198

6 General Aspects of Soliton Surfaces. Role of Gauge and Reciprocal Transformations 204
6.1 The AKNS 2×2 Spectral System 205
6.1.1 The Position Vector of Pseudospherical Surfaces 205
6.1.2 The *su*(2) Linear Representation and Its Associated Soliton Surfaces. The AKNS Case $r = -\bar{q}$ 209
6.2 NLS Eigenfunction Hierarchies. Geometric Properties. The Miura Transformation 216
6.2.1 Soliton Surface Position Vectors as Solutions of Eigenfunction Equations 217
6.2.2 The Serret-Frenet Equations and the NLS Hierarchy 220
6.3 Reciprocal Transformations. Loop Solitons 222
6.3.1 Reciprocal Transformations and the Loop Soliton Equation 222
6.3.2 Loop Solitons 225
6.4 The Dym, mKdV, and KdV Hierarchies. Connections 229
6.4.1 Invariance under Reciprocal Transformations. A Class of Planar Curve Motions 230
6.4.2 The Dym, mKdV, and KdV Hierarchies 233
6.4.3 A Permutability Theorem 235
6.4.4 A Geometric Derivation of the mKdV Hierarchy 237
6.5 The Binormal Motion of Curves of Constant Curvature. Extended Dym Surfaces 240
6.5.1 Curves of Constant Curvature 242
6.5.2 Extended Dym Surfaces. The *su*(2) Linear Representation 246
6.5.3 A CC-Ideal Formulation 249
6.5.4 A Matrix Darboux Transformation. A Bäcklund Transformation for the Extended Dym and m^2KdV Equations 252
6.5.5 Soliton Surfaces 255

6.6 The Binormal Motion of Curves of Constant Torsion. The Extended Sine-Gordon System 258
- 6.6.1 The Extended Sine-Gordon System 259
- 6.6.2 Fundamental Forms. An *su*(2) Linear Representation 260
- 6.6.3 A Bäcklund Transformation 262
- 6.6.4 An Analogue of the Bianchi Transformation. Dual Surfaces 263

7 Bäcklund Transformation and Darboux Matrix Connections 266

7.1 The Connection for Pseudospherical and Nonlinear Schrödinger Surfaces 267
- 7.1.1 Pseudospherical Surfaces 267
- 7.1.2 NLS Surfaces 271

7.2 Darboux Matrix and Induced Bäcklund Transformations for the AKNS System. The Constant Length Property 277
- 7.2.1 An Elementary Matrix Darboux Transformation 277
- 7.2.2 Invariance of a *su*(2) Constraint 280
- 7.2.3 The AKNS Class $r = -\bar{q}$ and Its Elementary Bäcklund Transformation 282
- 7.2.4 The Constant Length Property 285

7.3 Iteration of Matrix Darboux Transformations. Generic Permutability Theorems 287
- 7.3.1 Iteration of Matrix Darboux Transformations 288
- 7.3.2 Generic Permutability Theorems 292

8 Bianchi and Ernst Systems. Bäcklund Transformations and Permutability Theorems 297

8.1 Bianchi Surfaces. Application of the Sym-Tafel Formula 298

8.2 Matrix Darboux Transformations for Non-Isospectral Linear Representations 300

8.3 Invariance of the *su*(2) Constraint. A Distance Property 302

8.4 The Ernst Equation of General Relativity 303
- 8.4.1 Linear Representations 305
- 8.4.2 The Dual 'Ernst Equation' 306

8.5 The Ehlers and Matzner-Misner Transformations 309

8.6 The Neugebauer and Harrison Bäcklund Transformations 311

8.7 A Matrix Darboux Transformation for the Ernst Equation 319

8.8 A Permutability Theorem for the Ernst Equation and Its Dual. A Classical Bianchi Connection 324

9 Projective-Minimal and Isothermal-Asymptotic Surfaces 329

9.1 Analogues of the Gauss-Mainardi-Codazzi Equations in Projective Differential Geometry 330

9.2 Projective-Minimal, Godeaux-Rozet, and Demoulin Surfaces 333
9.3 Linear Representations 335
9.3.1 The Wilczynski Tetrahedral and a 4×4 Linear Representation 336
9.3.2 The Plücker Correspondence and a 6×6 Linear Representation 337
9.4 The Demoulin System as a Periodic Toda Lattice 341
9.5 A Bäcklund Transformation for Projective-Minimal Surfaces 343
9.5.1 Invariance of the $so(3, 3)$ Linear Representation 345
9.5.2 Invariance of the $sl(4)$ Linear Representation 350
9.6 One-Soliton Demoulin Surfaces 353
9.7 Isothermal-Asymptotic Surfaces. The Stationary mNVN Equation 357
9.7.1 The Stationary mNVN Equation 358
9.7.2 The Stationary NVN Equation 360
9.8 A Bäcklund Transformation for Isothermal-Asymptotic Surfaces 365
9.8.1 An Invariance of the mNVN Equation 365
9.8.2 An Invariance of the NVN Equation and a Bäcklund Transformation for Isothermal-Asymptotic Surfaces 367

Appendix A **The $su(2)$–$so(3)$ Isomorphism** **371**

Appendix B **CC-Ideals** **374**

Appendix C **Biographies** **380**

Bibliography and Author Index 383

Subject Index 403

Preface

'Only connect'.
E.M. Forster, *Howards End*

The deep connections that exist between the classical differential geometry of surfaces and modern soliton theory are by now well established. Thus, Bäcklund transformations, together with Darboux-type transformations in the form of the Levy transformation and the so-called Fundamental Transformation of differential geometry, have proved to be important tools in the generation of solutions to the nonlinear equations of soliton theory. Eisenhart, in the preface to his monograph *Transformations of Surfaces* published in 1922, asserted that

> During the past twenty-five years many of the advances in differential geometry of surfaces in euclidean space have had to do with transformations of surfaces of a given type into surfaces of the same type.

Thus, distinguished geometers such as Bianchi, Calapso, Darboux, Demoulin, Guichard, Jonas, Ribaucour, and Weingarten all conducted detailed investigations into various privileged classes of surfaces that admit such transformations.

It is with the class of surfaces that admit invariance under Bäcklund-Darboux transformations that the present monograph is concerned. Invariance under a Bäcklund transformation turns out to be a generic property of all solitonic equations. In the geometric context of this monograph, solitonic equations are seen to arise out of the nonlinear Gauss-Mainardi-Codazzi equations for various types of surfaces that admit invariance under Bäcklund-Darboux transformations. The linear Gauss-Weingarten equations for such surfaces provide, on injection of a Bäcklund parameter, linear representations for the underlying nonlinear soliton equations.

Accordingly, Bäcklund-Darboux transformations with their origin in the nineteenth century provide a natural bridge between classical differential geometry and modern soliton theory. Pseudospherical surfaces, surfaces of constant mean curvature, Bianchi and isothermic surfaces are amongst those shown in the classical literature to admit Bäcklund transformations and associated nonlinear superposition principles known as permutability theorems. The latter provide purely algebraic algorithms for the iterative generation of solutions of the solitonic equations linked with such classes of surfaces.

Here, our aim has been to provide a monograph describing, through the medium of Bäcklund-Darboux transformations, the many remarkable connections between results of classical differential geometry of the nineteenth and early twentieth centuries and soliton theory of modern times. The level of treatment is very much that of the classical works of Darboux and Bianchi to which this monograph owes an enormous debt. This is to be regarded as an introductory text for practitioners in soliton theory who wish to become acquainted with underlying geometric aspects of the subject. It is appropriate for use as a upper level undergraduate or graduate-level text for applied mathematicians or mathematical physicists. Indeed, it has grown out of a course on the geometry of soliton theory given over several years at the University of New South Wales.

Acknowledgements

The authors are, above all, indebted to the late Professor David Crighton, FRS, of Cambridge University who supported and encouraged this endeavour from the outset. They also wish to express their gratitude to the Center for Dynamical Systems and Nonlinear Studies, Georgia Institute of Technology, where this project was initiated. The support of the Australian Research Council is also gratefully acknowledged.

General Introduction and Outline

The foundations of the differential geometry of curves and surfaces were laid in the early part of the nineteenth century with the monumental works of Monge (1746–1818) and Gauss (1777–1855). Monge's major contributions were collected in his *Applications de l'Analyse à la Géometrie* published in 1807. The 1850 edition of that work is of particular value in that it includes an annotation by Liouville (1809–1882) detailing additional contributions to the subject by such luminaries as Frenet (1816–1888), Serret (1819–1885), Bertrand (1822–1900) and Saint-Venant (1796–1886), whose work in geometry was motivated by his interest in elasticity. Gauss' treatise on the geometry of surfaces, instigated by a geodetic study sponsored by the Elector of Hanover, was the *Disquisitiones Generales Circa Superficies Curvas* published in 1828. Therein, Gauss set down the system of equations that bears his name and which time has shown to be fundamental to the analysis of surfaces. Indeed, this Gauss system and the symmetries that it admits for privileged classes of surfaces underpin the remarkable connection between classical differential geometry and modern soliton theory to be the subject of this monograph.

The origins of soliton theory are likewise to be found in the early part of the nineteenth century. Thus, it was in 1834 that the Scottish engineer John Scott Russell recorded the first sighting, along a canal near Edinburgh, of the solitary hump-shaped wave to be rediscovered in 1965 in the context of the celebrated Fermi-Pasta-Ulam problem by Kruskal and Zabusky and termed a *soliton*. Scott Russell observed that his so-called *great wave of translation* proceeded with a speed proportional to its height. In a vivid account of water tank experiments set up to reproduce this large amplitude surface phenomenon, and described in a report to the British Association in 1844, there is also depicted the creation of two such waves. However, the limited duration of Scott Russell's experiments apparently did not allow him to observe the dramatic interaction properties of these waves in their entirety. Moreover, at that time, neither the nonlinear evolution equation descriptive of their propagation nor the analytic means to predict their interaction properties were to hand.

It was in 1895 that two Dutch mathematicians, Korteweg and de Vries, derived the nonlinear wave equation which now bears their name and adopts the canonical form

$$u_t + u_{xxx} + 6uu_x = 0. \tag{0.1}$$

This models long wave propagation in a rectangular channel and provides, through a simple travelling wave solution, a theoretical confirmation of the existence of the controversial solitary wave observed some sixty years earlier by Scott Russell on the Union canal. However, it is less well-known that what is now called the Korteweg–de Vries (KdV) equation had, in fact, been set down earlier by Boussinesq in his memoir of 1877 entitled *Essai sur la Théorie des Eaux Courantes*. Indeed, a pair of equations equivalent to the KdV equation (0.1) appeared as early as 1871 in two papers by Boussinesq devoted to wave propagation in rectangular channels.

The KdV equation was to be rediscovered in the mid-twentieth century by Gardner and Morikawa in 1960 in an analysis of the transmission of hydromagnetic waves. It has since been shown to be a canonical model for a rich diversity of large amplitude wave systems arising in the theory of solids, liquids and gases.

The advent of modern soliton theory was heralded in 1965 by the rediscovery of the KdV equation in the context of the celebrated Fermi-Pasta-Ulam problem. Thus, in a pioneering study by Kruskal and Zabusky, the KdV equation was obtained as a continuum limit of an anharmonic lattice model with cubic nonlinearity. The existence of solitary waves in this nonlinear model which possess the remarkable property that they preserve both their amplitude and speed subsequent upon interaction was revealed via a computational study. The term *soliton* was coined to describe such waves which had originally been observed in a hydrodynamic context by Scott Russell. However, the problem of obtaining an analytical expression descriptive of the interaction of solitons still remained.

It turns out that, remarkably, a generic method for the description of soliton interaction has its roots in a type of transformation originally introduced by Bäcklund in the nineteenth century to generate pseudospherical surfaces, that is, surfaces of constant negative Gaussian curvature $\mathcal{K} = -1/\rho^2$. The study of such surfaces goes back at least to Edmond Bour in 1862, who generated the celebrated sine-Gordon equation

$$\omega_{uv} = \frac{1}{\rho^2} \sin \omega \tag{0.2}$$

from the Gauss-Mainardi-Codazzi system for pseudospherical surfaces parametrised in terms of asymptotic coordinates. The sine-Gordon equation was subsequently rederived independently by both Bonnet in 1867 and Enneper in 1868 in a similar manner.

A purely geometric construction for pseudospherical surfaces was reformulated in mathematical terms as a transformation by Bianchi in 1879. In 1882, Bäcklund published details of his celebrated transformation $\mathbb{B}_\sigma$ which allows the iterative construction of pseudospherical surfaces. In 1883, Lie presented the decomposition $\mathbb{B}_\sigma = \mathbb{L}_\sigma^{-1}\mathbb{B}_{\pi/2}\mathbb{L}_\sigma$ which shows that the Bäcklund transformation $\mathbb{B}_\sigma$, in fact, represents a conjugation of Lie transformations $\mathbb{L}_\sigma$, $\mathbb{L}_\sigma^{-1}$ with the parameter-independent Bianchi transformation $\mathbb{B}_{\pi/2}$. Thus, the Lie transformations serve to intrude the key parameter σ into the original Bianchi transformation.

In 1892, under the title *Sulla Trasformazione di Bäcklund per le Superficie Pseudosferiche*, in a masterly breakthrough, Bianchi demonstrated that the Bäcklund transformation $\mathbb{B}_\sigma$ admits a commutativity property $\mathbb{B}_{\sigma_2}\mathbb{B}_{\sigma_1} = \mathbb{B}_{\sigma_1}\mathbb{B}_{\sigma_2}$ a consequence of which is a nonlinear superposition principle embodied in what is termed a *permutability theorem*. The evidence that Bianchi's permutability theorem has important application in nonlinear physics had to await the work of Seeger et al. in 1953 on crystal dislocations. Therein, in the context of Frenkel and Kontorova's dislocation theory of 1938, the superposition of so-called *eigenmotions* was obtained via the classical permutability theorem. Indeed, the interaction of what today is called a breather with a kink-type dislocation was both described analytically by means of the permutability theorem and displayed graphically. The typical solitonic features to be later discovered numerically in 1965 for the KdV equation, namely, preservation of velocity and shape following interaction, as well as the concomitant phase shift, were all derived by means of the permutability theorem for the sine-Gordon equation in this remarkable paper.

In 1958, Skyrme derived a higher-dimensional sine-Gordon equation in a nonlinear theory of particle interaction, while in 1965 the same equation was set down by Josephson in his seminal study of the tunnelling phenomenon in superconductivity for which he was later to gain the Nobel Prize. In 1967, Lamb derived the classical sine-Gordon equation in an analysis of the propagation of ultrashort light pulses. Lamb, aware of the earlier work of Seeger et al., exploited the permutability theorem associated with the Bäcklund transformation to generate an analytic expression for pulse decomposition corresponding to the two-soliton solution. Later, in 1971, he used the permutability theorem to analyse the decomposition of $2N\pi$ light pulses into N stable 2π pulses. The experimental evidence for such a decomposition phenomenon had been provided by Gibbs and Slusher in 1970, who recorded the decomposition of a 6π pulse into three 2π pulses in a Rb vapour. In the same year, Scott had noted how the permutability theorem may also be exploited in the study of long Josephson junctions.

In 1973, Wahlquist and Estabrook demonstrated that the KdV equation, like the sine-Gordon equation, admits invariance under a Bäcklund-type transformation and moreover possesses an associated permutability theorem. The novel pulse interaction properties observed by Zabusky and Kruskal in their original numerical study of the KdV equation are captured analytically in the multi-soliton solutions generated by iterative application of this permutability theorem.

In 1974, a Bäcklund transformation for the nonlinear Schrödinger (NLS) equation

$$iq_t + q_{xx} + \nu q^2|q| = 0 \tag{0.3}$$

was constructed by Lamb via a classical method developed by Clairin in 1910. A nonlinear superposition principle may again be constructed by means of the Bäcklund transformation. The NLS equation has important applications in fibre optics. It seems to have been first set down independently by Kelley and Talanov in 1965 in studies of the self-focusing of optical beams in nonlinear Kerr media. Subsequently, in 1968, Zakharov derived the NLS equation in a study of deep water gravity waves. Hasimoto, in 1971, obtained the same equation in an approximation to the hydrodynamical motion of a thin isolated vortex filament. Implicit was a geometric derivation of the NLS equation wherein it is associated with a motion of an inextensible curve in $\mathbb{R}^3$. This association of an integrable equation with the spatial motion of an inextensible curve will arise naturally in our study of the geometry of solitons.

Thus, by 1974, the Bäcklund transformations for the canonical soliton equations (0.1)–(0.3) were all in place and in that year a National Science Foundation meeting was convened at Vanderbilt University in the USA to assess the status and potential role of Bäcklund transformations in soliton theory. In 1973, the celebrated generalised ZS-AKNS spectral system had been introduced by Ablowitz et al. A broad spectrum of 1+1-dimensional nonlinear evolution equations amenable to the Inverse Scattering Transform (IST) can be encapsulated as compatibility conditions for this ZS-AKNS system. The latter was exploited by Chen to derive auto-Bäcklund transformations for (0.1)–(0.3) in an elegant manner.

The linear structure of the ZS-AKNS system permits the application in soliton theory of another important class of transformations with their origin in the nineteenth century, namely, Darboux transformations. The latter arose in a study by Darboux in 1882 of Sturm-Liouville problems. However, they are but a special case of transformations due to Moutard and introduced earlier in 1878 in connection with the sequential reduction of linear hyperbolic equations to canonical form. Iterated Darboux transformations were constructed by Crum in 1955 in connection with related Sturm-Liouville problems. In 1975, the Crum transformation was taken up by Wadati et al. and used to generate multi-soliton solutions of integrable equations associated with the ZS-AKNS system. In geometric terms, these iterated versions of Darboux transformations occur in the classical theory of surfaces as Levy sequences as described in Eisenhart's *Transformations of Surfaces*.

In 1976, Lund and Regge, en route to the celebrated solitonic system which bears their name, made the crucial observation that the ZS-AKNS system for the sine-Gordon equation is nothing but a 2×2 representation of the classical Gauss-Weingarten system for pseudospherical surfaces. This connection was made independently in the same year by Pohlmeyer.

Thus, by 1976, it was clear that Bäcklund and Darboux transformations, with their origins in the classical differential geometry of surfaces, have deep connections with soliton theory. The aim of the present monograph is to bring together these strands and to give an account not only of their historical connections, but also of modern advances. It builds upon the complementary earlier monograph by Rogers and Shadwick (1982), which presented a non-geometric account of Bäcklund transformations and their applications in soliton theory and continuum mechanics. The geometric viewpoint in this monograph is inspired in many respects by the work of Antoni Sym published in 1981 under the title *Soliton Theory is Surface Theory*. It is the exploration of this theme that, in part, motivated the present work.

Chapter 1 presents an account of the connection between the classical Bäcklund transformation and its variants and modern soliton theory. It opens with the derivation of a classical nonlinear system due to Bianchi which embodies the Gauss-Mainardi-Codazzi equations for hyperbolic surfaces described in asymptotic coordinates. Specialisation to pseudospherical surfaces produces the celebrated sine-Gordon equation. There follows, in Section 1.2, a description of the geometric procedure for the construction of pseudospherical surfaces along with the derivation of the induced auto-Bäcklund transformation for the sine-Gordon equation. In Section 1.3, Bianchi's permutability theorem is derived via this Bäcklund transformation, and a lattice is introduced whereby multi-soliton solutions may be generated in a purely algebraic manner. Pseudospherical surfaces corresponding to one- and two-soliton solutions of the sine-Gordon equation are constructed in Section 1.4. Thus, the stationary single soliton solution is seen to correspond to the pseudosphere, while the non-stationary soliton leads to the Dini surface, namely the helicoid generated by simultaneous rotation and translation of Huygen's tractrix. The two-soliton solution is obtained via the permutability theorem, and pseudospherical surfaces corresponding to entrapped periodic solutions known as breathers are presented. In Section 1.5, it is shown that the Bäcklund transformation for surfaces parallel to pseudospherical surfaces may be induced in a straightforward manner. This extends the action of the classical Bäcklund transformation to a class of Weingarten surfaces. The chapter concludes with a treatment of another important class of surfaces which have a solitonic connection, namely that which bears the name of Bianchi. This class is determined by the system of equations

$$
\begin{gathered}
a_v + \frac{1}{2}\frac{\rho_v}{\rho}a - \frac{1}{2}\frac{\rho_u}{\rho}b\cos\omega = 0, \\
b_u + \frac{1}{2}\frac{\rho_u}{\rho}b - \frac{1}{2}\frac{\rho_v}{\rho}a\cos\omega = 0, \\
\omega_{uv} + \frac{1}{2}\left(\frac{\rho_u}{\rho}\frac{b}{a}\sin\omega\right)_u + \frac{1}{2}\left(\frac{\rho_v}{\rho}\frac{a}{b}\sin\omega\right)_v - ab\sin\omega = 0, \\
\rho_{uv} = 0,
\end{gathered}
\tag{0.4}
$$

where $\mathcal{K} = -1/\rho^2$ is the Gaussian curvature and u, v are asymptotic coordinates. In 1890, Bianchi presented a purely geometric construction for such hyperbolic surfaces. The determining constraint $\rho_{uv} = 0$ was retrieved one hundred years later by Levi and Sym (1990) in their search for the subclass of hyperbolic surfaces which possess an associated integrable Gauss-Mainardi-Codazzi system. Their procedure was based on the intrusion by Lie group methods of a spectral parameter into a 2×2 linear representation of the Gauss-Weingarten system for hyperbolic surfaces. In Section 1.6, a spherical representation is used to show that the Bianchi system (0.4) is, in fact, equivalent to the nonlinear sigma-type model

$$
\begin{gathered}
(\rho N N_u)_v + (\rho N N_v)_u = 0, \quad N^2 = \mathbb{1}, \quad N^\dagger = N \\
\rho_{uv} = 0.
\end{gathered}
\tag{0.5}
$$

Thus, this important system of modern soliton theory has its origin in classical differential geometry. Indeed, a vector version of (0.5) is implicit in the work of Bianchi.

An elliptic variant of the Bianchi system is shown to deliver the well-known Ernst equation of general relativity, namely

$$
\mathcal{E}_{z\bar{z}} + \frac{1}{2}\frac{\rho_{\bar{z}}}{\rho}\mathcal{E}_z + \frac{1}{2}\frac{\rho_z}{\rho}\mathcal{E}_{\bar{z}} = \frac{\mathcal{E}_z\mathcal{E}_{\bar{z}}}{\Re(\mathcal{E})}, \quad \rho_{z\bar{z}} = 0. \tag{0.6}
$$

To conclude, a Bäcklund transformation that connects hyperbolic surfaces is constructed in a geometric manner. This is then specialised to provide an invariance which admits the constraint associated with the Bianchi system. The resulting Bäcklund transformation is then applied to a degenerate seed Bianchi surface to generate a one-soliton Bianchi surface.

Chapter 2 is concerned with how certain motions of privileged curves and surfaces can lead to solitonic equations. Thus, in Section 2.1, the classical sine-Gordon equation is arrived at by consideration of motions of an inextensible curve of constant curvature or torsion. In the latter case, the curve sweeps out a pseudospherical surface. In Section 2.2, the AKNS spectral problem for the sine-Gordon equation is derived via the $so(3)$–$su(2)$ isomorphism applied to its 3×3 Gauss-Weingarten representation. In Section 2.3, the discussion turns to privileged motions of pseudospherical surfaces which are associated with soliton equations said to be compatible with, or symmetries of, the sine-Gordon equation. Particular classes of motion of pseudospherical surfaces are considered. One is linked to a continuum version of an anharmonic lattice model which incorporates the important modified Korteweg-de Vries (mKdV) equation

$$\omega_t + \omega_{xxx} + 6\omega^2\omega_x = 0. \tag{0.7}$$

This mKdV equation, like the KdV equation (0.1) to which it is connected by the Miura transformation, is of considerable physical importance and arises, in particular, in plasma physics in the theory of the propagation of Alfvén waves.

Another important motion of pseudospherical surfaces, purely normal in character, is shown to produce a classical system due to Weingarten and Bianchi which may be found in Eisenhart's *A Treatise on the Differential Geometry of Curves and Surfaces* in connection with triply orthogonal systems of surfaces wherein one constituent family is pseudospherical. This system adopts the form

$$\begin{gathered}
\theta_{xyt} - \theta_x\theta_{yt}\cot\theta + \theta_y\theta_{xt}\tan\theta = 0,\\
\left(\frac{\theta_{xt}}{\cos\theta}\right)_x - \frac{1}{\rho}\left(\frac{1}{\rho}\sin\theta\right)_t - \frac{\theta_y\theta_{yt}}{\sin\theta} = 0,\\
\left(\frac{\theta_{yt}}{\sin\theta}\right)_y + \frac{1}{\rho}\left(\frac{1}{\rho}\cos\theta\right)_t + \frac{\theta_x\theta_{xt}}{\cos\theta} = 0,\\
\theta_{xx} - \theta_{yy} = \frac{1}{\rho^2}\sin\theta\cos\theta.
\end{gathered} \tag{0.8}$$

Bäcklund transformations for both the continuum lattice model and the above system are then shown to be induced by gauge transformations acting on an

AKNS representation. To conclude this chapter, in Section 2.4, the mKdV equation is generated via the motion of an inextensible curve of zero torsion. The motion of solitonic Dini surfaces is then investigated and triply orthogonal Weingarten systems of surfaces are thereby constructed.

In Chapter 3, the discussion turns to the classical surfaces of Tzitzeica which, like pseudospherical surfaces, emerge as having an underlying soliton connection. It was in the first decade of the twentieth century that the Romanian geometer Tzitzeica investigated the class of surfaces which is associated with the important nonlinear hyperbolic equation

$$(\ln h)_{\alpha\beta} = h - h^{-2}, \tag{0.9}$$

to be rediscovered some seventy years later in a solitonic context. In fact, the study by Tzitzeica of the surfaces associated with this equation may be said to have initiated the important subject of affine geometry. Therein, the Tzitzeica equation (0.9) describes the so-called affinsphären.

In Section 3.1, the class of surfaces Σ determined by the so-called Tzitzeica condition $\mathcal{K} = -c^2 d^4$, $c = \text{const}$ is introduced, wherein d is the distance from the origin to the tangent plane to Σ at a generic point. The linear representation of the Tzitzeica equation as originally set down by Tzitzeica is rederived and its dual is then used as a route to another important avatar of (0.9), namely the affinsphären equation

$$\left(\frac{R_u}{R^2 v^2}\right)_u = \left(\frac{R R_v}{v^2}\right)_v \tag{0.10}$$

as obtained by the German geometer Jonas in 1953. This integrable equation is then shown to arise naturally in a Lagrangian description of an anisentropic gasdynamics system for a certain three-parameter class of constitutive laws. In Section 3.2, a Bäcklund transformation for the construction of suites of Tzitzeica surfaces is derived in a geometric manner, and its connection with the classical Moutard transformation of 1878 is elucidated. The action of the Bäcklund transformation on the trivial seed solution $h = 1$ of the Tzitzeica equation (0.9) is then used to generate an affinsphäre with rotational symmetry. Tzitzeica surfaces corresponding to one- and two-soliton solutions of (0.9) are then constructed. In particular, a Tzitzeica surface corresponding to a breather solution is displayed.

It turns out that the Tzitzeica equation is embedded in another classical system which surprisingly has an even longer history. This solitonic system has become known as the two-dimensional Toda lattice model

$$(\ln h_n)_{uv} = -h_{n+1} + 2h_n - h_{n-1}, \quad n \in \mathbb{Z}. \tag{0.11}$$

This nonlinear differential-difference scheme, to be rediscovered almost a century later in modern soliton theory, is actually to be found in a treatise of Darboux published in 1887. There, it was derived in the iteration of what have become known as Laplace-Darboux transformations. The latter, like the contemporary Moutard transformation, arose in connection with the iterative reduction of linear hyperbolic equations to canonical form. They have interesting application to the theory of conjugate nets in the classical differential geometry of surfaces. This aspect of Laplace-Darboux transformations is described at length in Eisenhart's *Transformations of Surfaces*. Here, in Section 3.3, the notion of a Laplace-Darboux transformation is introduced along with key associated invariants. It is shown how application of a Laplace-Darboux transformation leads to the Toda lattice scheme (0.11). The Tzitzeica equation is then generated as a particular periodic Toda lattice. An invariance of the general two-dimensional Toda lattice model is presented which, in particular, preserves periodicity. It is then shown how Laplace-Darboux transformations may be applied iteratively to produce a suite of surfaces on which the parametric lines constitute conjugate nets.

In Chapter 4, we focus upon the NLS equation (0.3). The latter seems to have escaped the attention of the geometers of the nineteenth century even though it has a simple geometric origin in the evolution of an inextensible curve moving through space with speed $v = \kappa \boldsymbol{b}$, where κ is its curvature and $\boldsymbol{b}$ its binormal. In Section 4.1, the NLS equation is derived in a geometric manner, and soliton surfaces corresponding to single soliton and breather solutions are presented along with general geometric properties and the connection to the Heisenberg spin equation

$$\boldsymbol{S}_t = \boldsymbol{S} \times \boldsymbol{S}_{ss}, \quad \boldsymbol{S}^2 = 1, \tag{0.12}$$

where t is time and s is arc length. In Section 4.2, a solitonic system linked to the NLS equation, namely the Pohlmeyer-Lund-Regge model,

$$\begin{aligned} \theta_{\xi\xi} - \theta_{\eta\eta} - \epsilon^2 \cos\theta \sin\theta + \left(\phi_\xi^2 - \phi_\eta^2\right) \cos\theta \ \operatorname{cosec}^3 \theta = 0, \\ \left(\phi_\xi \cot^2 \theta\right)_\xi = \left(\phi_\eta \cot^2 \theta\right)_\eta \end{aligned} \tag{0.13}$$

is also derived in a geometric manner. This system arises in the study of relativistic vortices. It is shown to be related, in turn, to the sharpline self-induced transparency (SIT) system

$$\begin{aligned} \chi_{tx} &= \sin\chi + \nu_t \nu_x \tan\chi, \\ \nu_{tx} &= -\nu_x \chi_t \cot\chi - \nu_t \chi_x (\cos\chi \sin\chi)^{-1} \end{aligned} \tag{0.14}$$

which stems from the unpumped Maxwell-Bloch system

$$\begin{gathered} E_x = P, \quad P_t = EN, \\ N_t = -\frac{1}{2}(\bar{E}P + E\bar{P}), \quad N^2 + P\bar{P} = 1. \end{gathered} \tag{0.15}$$

In the above, E and $P = e^{iv} \sin\chi$ denote, in turn, the slowly varying amplitudes of the electric field and polarisation, while $N = \cos\chi$ is the atomic inversion. The unpumped Maxwell-Bloch system is likewise shown to be linked to the stimulated Raman scattering (SRS) system

$$A_{1X} = -SA_2, \quad A_{2X} = \bar{S}A_1, \quad S_T = A_1\bar{A}_2, \tag{0.16}$$

where A_1, A_2 are the electric field amplitudes of the pump and Stokes waves, respectively. The connection between the SIT and SRS systems and the NLS equation is then established via the compatibility of the latter with the unpumped Maxwell-Bloch system. Thus, an appropriate time evolution of the eigenfunction pair in the AKNS representation for the NLS equation produces the system (0.15). In geometric terms, this unpumped Maxwell-Bloch system arises out of certain motions of Hasimoto surfaces in the same way as the mKdV equation or Weingarten system come from appropriate compatible motions of pseudospherical surfaces. In Section 4.3, the NLS equation is derived in an alternative manner via a geometric formulation originally developed in a kinematic analysis of certain hydrodynamical motions by Marris and Passman in 1969. The auto-Bäcklund transformation for the NLS equation is derived in this representation at the level of the generation of Hasimoto surfaces. Spatially periodic solutions of 'smoke-ring' type are thereby generated.

Chapter 5 is concerned with yet another classical class of surfaces which have a soliton connection, namely isothermic surfaces. These surfaces seem to have their origin in work by Lamé in 1837 motivated by problems in heat conduction. An important subclass of isothermic surfaces were subsequently investigated in a paper by Bonnet in 1867. These Bonnet surfaces admit non-trivial families of isometries which leave invariant the principal curvatures κ_1 and κ_2 and, accordingly, both the Gaussian curvature $\mathcal{K} = \kappa_1\kappa_2$ and mean curvature $\mathcal{M} = (\kappa_1 + \kappa_2)/2$. In Section 5.1, the Gauss-Mainardi-Codazzi system associated with isothermic surfaces parametrised in curvature coordinates is set down, namely

$$\begin{gathered} \theta_{xx} + \theta_{yy} + \kappa_1\kappa_2 e^{2\theta} = 0, \\ \kappa_{1y} + (\kappa_1 - \kappa_2)\theta_y = 0, \qquad \kappa_{2x} + (\kappa_2 - \kappa_1)\theta_x = 0 \end{gathered} \tag{0.17}$$

and a reduction originally obtained by Calapso in 1903 is made to the single fourth-order nonlinear equation

$$\left(\frac{z_{xy}}{z}\right)_{xx} + \left(\frac{z_{xy}}{z}\right)_{yy} + (z^2)_{xy} = 0. \tag{0.18}$$

In Section 5.2, the notion of isothermic surfaces in $\mathbb{R}^{n+2}$ is introduced, and an integrable generalisation of the classical isothermic system (0.17) is derived. In Section 5.3, a vector Calapso system which represents an extension of (0.18) for isothermic surfaces in $\mathbb{R}^{n+2}$ is constructed along with a Lax pair. Connection is made with the stationary Davey-Stewartson II equation which is shown to describe isothermic surfaces in $\mathbb{R}^4$. Likewise, the Davey-Stewartson III equation may be associated with isothermic surfaces in Minkowski space $\mathbb{M}^4$. Section 5.4 turns to classical results on the transformation of conjugate nets which have importance in soliton theory. It is shown how the conjugate net equation is invariant under the so-called Fundamental Transformation which, in turn, can be decomposed into a Combescure and two radial transformations. In Section 5.5, this classical result for $\mathbb{R}^3$ is shown to be actually valid in spaces of arbitrary dimension. The classical notion of the Ribaucour transformation for curvature nets is likewise generalised. These results are then used to obtain a Bäcklund transformation for isothermic surfaces in $\mathbb{R}^{n+2}$. In Section 5.6, a permutability theorem for the Fundamental Transformation for conjugate nets in $\mathbb{R}^{n+2}$ is constructed, and various important geometric implications are recorded. Thus, planarity, cyclicity and constant cross-ratio properties of the Bianchi quadrilateral are established. In Section 5.7, a Bäcklund transformation and associated compact permutability theorem are obtained for the vector Calapso system associated with isothermic surfaces in $\mathbb{R}^{n+2}$. Section 5.8 is devoted to the construction of explicit solutions of the classical isothermic system and Calapso equation via Darboux-Ribaucour and Moutard transformations. Thus, one-soliton isothermic surfaces are generated along with a 'lump' solution of the Calapso equation and integrable surfaces in the form of Dupin cyclides. Localised solutions of the zoomeron equation are seen to be related via a Lie point symmetry to important dromion solutions of the Davey-Stewartson III equation.

Chapter 6 introduces the key Sym-Tafel formula for the construction of soliton surfaces associated with an $su(2)$ linear representation. In Section 6.1, its use is illustrated in the construction of pseudospherical surfaces via the AKNS representation for the sine-Gordon equation. The more general result for the AKNS class $r = -\bar{q}$ is then established. This incorporates the NLS hierarchy

$$\begin{pmatrix} q \\ -\bar{q} \end{pmatrix}_t = i A_N L^N \begin{pmatrix} q \\ \bar{q} \end{pmatrix}, \quad L = i \begin{pmatrix} -\partial_x - \frac{1}{2} q \partial_x^{-1} \bar{q} & \frac{1}{2} q \partial_x^{-1} q \\ -\frac{1}{2} \bar{q} \partial_x^{-1} \bar{q} & \partial_x + \frac{1}{2} \bar{q} \partial_x^{-1} q \end{pmatrix}, \tag{0.19}$$

where L is a recursion operator and the A_N are constants. It is shown that the higher order members of the NLS hierarchy are compatible with the canonical NLS equation. Thus, in geometric terms, they are associated with privileged motions of Hasimoto surfaces.

In Section 6.2, it is shown that the generic position vector to the soliton surface associated with the NLS hierarchy (0.19) itself satisfies an associated integrable system, namely the potential NLS eigenfunction hierarchy. A gauge transformation applied to the AKNS representation for (0.19) is used to generate the NLS eigenfunction hierarchy. A geometric interpretation follows for a Miura-type transformation which links the NLS hierarchy and this associated eigenfunction hierarchy. In Section 6.3, the notion of reciprocal transformation is introduced and used to generate the loop soliton equation

$$X_T = \pm \left(\frac{X_Z}{\sqrt{1 + X_Z^2}} \right)_{ZZ} \tag{0.20}$$

which is linked to the mKdV equation by a combination of reciprocal and gauge transformations. The presence of loop solitons in the complex NLS hierarchy is then established. Loop solitons are seen to be naturally associated with the generation of soliton surfaces. In Section 6.4, the invariance under a reciprocal transformation of the Dym equation

$$\rho_t = \rho^{-1}(\rho^{-1})_{xxx} \tag{0.21}$$

and, more generally, of the Dym hierarchy

$$\rho_t = \rho^{-1}(-D^3 r I r)^n \rho \rho_x, \quad n = 1, 2, \ldots \tag{0.22}$$

is recorded, where $D\phi = \phi_x$, $I\phi := \int_x^\infty \phi(\sigma, t) d\sigma$, and $r = \rho^{-1}$. This reciprocal invariance, when appropriately conjugated with a Galilean transformation, induces the spatial part of the auto-Bäcklund transformation generic to the KdV hierarchy

$$u_t = K^n u_x, \quad n = 1, 2, \ldots, \quad K = \frac{\partial^2}{\partial x^2} - 4u + 2u_x \int_x^\infty dx. \tag{0.23}$$

The permutability theorem for the potential KdV equation is then derived from the Bäcklund transformation. An interpretation in terms of the nonlinear extrapolation ϵ-algorithm of numerical analysis is given. There follows a purely geometric derivation of the mKdV hierarchy in terms of the planar motion of a curve. In Section 6.5, there is a return to the geometric formulation used previously to generate the NLS equation. Here, it is exploited to provide a natural geometric derivation of integrable extended versions of the Dym and sine-Gordon equations in terms of the binormal motion of inextensible curves of constant curvature and torsion, respectively. Thus, an extended Dym equation

$$\tau_b = \left[\frac{1}{\kappa}\left(\frac{1}{\tau^{1/2}}\right)_{ss} - \tau^{3/2} + \kappa\left(\frac{1}{\tau^{1/2}}\right)\right]_s \tag{0.24}$$

is generated via the motion of an inextensible curve of constant curvature κ. This travels with velocity $\boldsymbol{v} = \tau^{-1/2}\boldsymbol{b}$. On the other hand, the motion of an inextensible curve of constant torsion τ leads to an integrable extended sine-Gordon system, namely

$$\begin{aligned} \omega_{sb} - \tau(\cos\omega \tanh\phi)_b &= \sin\omega\cosh\phi, \\ \phi_s &= \tau \sin\omega, \end{aligned} \tag{0.25}$$

where $\theta_s = \kappa$ and $\theta_b = \tau^{-1}\sinh\phi$. In this case, the curve has velocity $v = -(\theta_b/2\tau)\boldsymbol{b}$. Invariance under a reciprocal transformation is used to establish the existence of a parallel dual soliton surface associated with each soliton surface of the extended Dym equation. Auto-Bäcklund transformations for the extended Dym equation and sine-Gordon system are constructed. These are then used to generate novel soliton surfaces. To conclude, an analogue of Bianchi's classical transformation for pseudospherical surfaces is set down.

In Chapter 7, the important connection between Bäcklund transformations and matrix Darboux transformations is established. In Section 7.1, the Sym-Tafel formula for the generic position vector of soliton surfaces is applied to show that the original Bäcklund transformation for the construction of pseudospherical surfaces provides a prototype for a matrix version of a classical Darboux transformation. It is established that the Bäcklund transformation for the construction of NLS soliton surfaces can likewise be represented as a matrix Darboux transformation which acts on the underlying $su(2)$ representation. In Section 7.2, an elementary matrix Darboux transformation is constructed which leaves invariant the AKNS representation for the NLS hierarchy. The auto-Bäcklund transformation for the NLS equation and its hierarchy is then induced. A generic geometric property of these Bäcklund transformations is established, namely that they preserve distance between corresponding points. Indeed, this constant length property, evident in the classical auto-Bäcklund transformation for the generation of pseudospherical surfaces, is seen to extend to soliton surfaces linked to the AKNS class $r = -\bar{q}$. Section 7.3 deals with the iteration of elementary matrix Darboux transformations and a pivotal commutativity property is established. In geometric terms, it is shown that repetition of matrix Darboux transformations generates a suite of surfaces whose neighbouring members possess the constant length property. To conclude, iterated matrix Darboux transformations are exploited to construct a permutability theorem generic to the AKNS class $r = -\bar{q}$ of soliton equations. This represents a generalisation of Bianchi's classical permutability theorem for the sine-Gordon equation. The role of permutability theorems in general in the integrable discretisation of soliton equations and surfaces is a matter of current research.

In Chapter 8, the discussion turns to the geometric properties of important soliton systems which admit non-isospectral linear representations. Both the classical Bianchi system (0.4) and the elliptic counterpart encapsulated in (0.6) fall into this category. In Section 8.1, the generic position vector to Bianchi surfaces as well as their associated fundamental forms are retrieved via a non-isospectral variant of the Sym-Tafel formula. In Section 8.2, a generalised elementary matrix Darboux transformation valid for a wide class of non-isospectral Lax pairs is presented. Then, in Section 8.3, a distance property is recorded for Bäcklund transformations at the surface level. In Section 8.4, it is recalled that the unit normal $\boldsymbol{N}$ to Bianchi surfaces of total curvature $\mathcal{K} = -1/\rho^2$ with $\rho_{uv} = 0$ obeys the vectorial equation

$$(\rho \boldsymbol{N} \times \boldsymbol{N}_u)_v + (\rho \boldsymbol{N} \times \boldsymbol{N}_v)_u = \boldsymbol{0}, \tag{0.26}$$

which, on appropriate parametrisation, produces the complex equation

$$\mathcal{E}_{uv} + \frac{1}{2}\frac{\rho_v}{\rho}\mathcal{E}_u + \frac{1}{2}\frac{\rho_u}{\rho}\mathcal{E}_v = \frac{2\mathcal{E}_u\mathcal{E}_v\bar{\mathcal{E}}}{|\mathcal{E}|^2 + 1}. \tag{0.27}$$

The elliptic analogue of the latter, namely

$$\xi_{z\bar{z}} + \frac{1}{2}\frac{\rho_{\bar{z}}}{\rho}\xi_z + \frac{1}{2}\frac{\rho_z}{\rho}\xi_{\bar{z}} = \frac{2\xi_z\xi_{\bar{z}}\bar{\xi}}{|\xi|^2 - 1}, \tag{0.28}$$

where $\rho_{z\bar{z}} = 0$, describes surfaces of Bianchi type in three-dimensional Minkowski space. Introduction of the Ernst potential $\mathcal{E} = (1-\xi)/(1+\xi)$ into (0.28) produces the Ernst equation of general relativity as set down in (0.6). With this important geometric interpretation of the Ernst equation in mind, we proceed to give an account of its Bäcklund and Darboux transformations. These can be used to construct important solutions of Einstein's vacuum equations. As a preliminary, Neugebauer's original non-isospectral linear representation for the Ernst equation is described. This is next set in a broader context leading naturally to the dual Ernst equation which, like the original Ernst equation, governs stationary, axi-symmetric gravitational fields in a vacuum. The solutions of this dual Ernst equation are then shown to be related to those of the Ernst equation by a contact transformation. In Section 8.5, the latter is conjugated with a pair of additional Möbius invariances known in the literature as the Ehlers and Matzner-Misner transformations, respectively. The importance in general relativity of the resultant so-called Geroch transformations is then discussed.

It was in 1978 that Harrison first derived a Bäcklund transformation for the Ernst equation. Independently, in 1979, Neugebauer constructed another Bäcklund transformation which subsequently has been shown to be a basic building block for all other Bäcklund transformations admitted by the Ernst equation. Section 8.6 opens with a description of the seminal Neugebauer transformation couched in terms of pseudopotentials. It is shown that it incorporates both the Ehlers and Matzner-Misner transformations. The composition laws for Neugebauer transformations are then set down, and the mechanism for their iteration is explained. The Harrison transformation is then derived as a conjugation of Neugebauer transformations embodied in a commutation theorem. It is recalled that a single application of the Harrison transformation produces the Schwarzschild solution on the Papapetrou background, while N applications with seed solution the Kerr black hole metric leads to a nonlinear superposition of N Kerr-NUT fields. In Section 8.7, a matrix Darboux transformation for a generalised Ernst system and its specialisations to the Ernst equation and its dual are presented. In Section 8.8, successive application of two such transformations is shown to lead to permutability theorems for the Ernst equation and its dual. To conclude, it is demonstrated that the celebrated Harrison transformation for the Ernst equation is the direct analogue of the classical Bäcklund transformation for the Bianchi system (0.4).

Chapter 9 describes developments in soliton theory which are linked to the geometry of projective-minimal and isothermal-asymptotic surfaces. In Section 9.1, the analogues of the Gauss-Weingarten and Gauss-Mainardi-Codazzi equations are set down for surfaces in projective space $\mathbb{P}^3$, and certain projective invariants are recorded. In Section 9.2, the requirement of invariance of the projective Gauss-Mainardi-Codazzi equations under a simple Lie point symmetry is shown to lead to a specialisation which may be identified as the Euler-Lagrange equations associated with projective-minimal surfaces. This suggests that projective-minimal surfaces which arise as extrema of a projective area functional constitute another class of integrable surfaces. Indeed, this turns out to be the case. Godeaux-Rozet, Demoulin and Tzitzeica surfaces are then extracted as particular projective-minimal surfaces. Godeaux-Rozet and Demoulin surfaces have been relatively unstudied in soliton theory in comparison with those associated with the name of Tzitzeica. Nonetheless, the Demoulin surfaces, in particular, have a literature going back at least to 1933 and provide through their associated Gauss-Mainardi-Codazzi equations an important integrable extension of (0.9), namely

$$
\begin{aligned}
(\ln h)_{xy} &= h - \frac{1}{hk},\\
(\ln k)_{xy} &= k - \frac{1}{hk}.
\end{aligned}
\tag{0.29}
$$

In Section 9.3, a 4×4 linear representation for projective-minimal surfaces based on the Wilczynski moving tetrahedral is introduced. The classical Plücker correspondence is then adopted to derive a 6×6 linear representation. The geometric significance of the Plücker correspondence becomes apparent in this context in Section 9.4 when Godeaux sequences of surfaces in $\mathbb{P}^5$ are introduced. Godeaux sequences of period 6 are then seen to lead to the Demoulin system which, in turn, is connected to the two-dimensional Toda lattice. In Section 9.5, a Bäcklund transformation for projective-minimal surfaces is constructed by imposition of suitable constraints on the Fundamental Transformation. In Section 9.6, a Bäcklund transformation for the Demoulin system (0.29) is recorded and then used to generate a one-soliton Demoulin surface via action on the seed solution $h = k = 1$. In Section 9.7, the discussion turns to isothermal-asymptotic surfaces. The Gauss-Mainardi-Codazzi equations underlying these surfaces turn out to be integrable. Indeed, a Gauss-Weingarten system for isothermal-asymptotic surfaces is here shown to be connected to the standard linear representation for the stationary modified Nizhnik-Veselov-Novikov (mNVN) equation. Links between the stationary mNVN and NVN equations are established. In Section 9.8, the connection between their linear representations is exploited, in concert with the Lelieuvre formulae, to construct a Bäcklund transformation for isothermal-asymptotic surfaces.

1

Pseudospherical Surfaces and the Classical Bäcklund Transformation. The Bianchi System

The explicit study of surfaces of constant negative total curvature goes back to the work of Minding [261] in 1838. Thus, in that year, Minding's theorem established the important result that these surfaces are isometric, that is, points on two such surfaces can be placed in one-to-one correspondence in a way that the metric is preserved. Beltrami [28] subsequently gave the term pseudospherical to these surfaces and made important connections with Lobachevski's non-Euclidean geometry.

It was Bour [54], in 1862, who seems to have first set down what is now termed the sine-Gordon equation arising out of the compatibility conditions for the Gauss equations for pseudospherical surfaces expressed in asymptotic coordinates.

In 1879, Bianchi [31] in his habilitation thesis presented, in mathematical terms, a geometric construction for pseudospherical surfaces. This result was extended by Bäcklund [21] in 1883 to incorporate a key parameter which allows the iterative construction of such pseudospherical surfaces. The Bäcklund transformation was subsequently shown by Bianchi [32], in 1885, to be associated with an elegant invariance of the sine-Gordon equation. This invariance has become known as the Bäcklund transformation for the sine-Gordon equation. It includes an earlier parameter-independent result of Darboux [94]. The Bäcklund transformation has important applications in soliton theory. Indeed, it appears that the property of invariance under Bäcklund and associated Darboux transformations as originated in [92] is enjoyed by all soliton equations. The contribution of Bianchi and Darboux to the geometry of surfaces and, in particular, the role of Bäcklund transformations preserving certain geometric properties have been discussed by Chern [77] and Sym et al. in [385]. It is with Bäcklund and Darboux transformations, their geometric origins and their application in modern soliton theory that we shall be concerned in the present monograph.

1.1 The Gauss-Weingarten Equations for Hyperbolic Surfaces. Pseudospherical Surfaces. The Sine-Gordon Equation

Here, the study of pseudospherical surfaces is set in the broader context of hyperbolic surfaces via a nonlinear system due to Bianchi [37]. The background is that of basic classical differential geometry of curves and surfaces to be found in such standard works as do Carmo [108] or Struick [352]. The latter work is a rich source of material on the history of the subject.

Let $\boldsymbol{r} = \boldsymbol{r}(u, v)$ denote the position vector of a generic point P on a surface Σ in $\mathbb{R}^3$. Then, the vectors $\boldsymbol{r}_u$ and $\boldsymbol{r}_v$ are tangential to Σ at P and, at such points at which they are linearly independent,

$$\boldsymbol{N} = \frac{\boldsymbol{r}_u \times \boldsymbol{r}_v}{|\boldsymbol{r}_u \times \boldsymbol{r}_v|} \tag{1.1}$$

determines the unit normal to Σ. The 1st and 2nd fundamental forms of Σ are defined by

$$\begin{aligned} \mathrm{I} &= \quad d\boldsymbol{r} \cdot d\boldsymbol{r} \;\;= E\,du^2 + 2F\,du dv + G\,dv^2, \\ \mathrm{II} &= -d\boldsymbol{r} \cdot d\boldsymbol{N} = e\,du^2 + 2f\,du dv + g\,dv^2, \end{aligned} \tag{1.2}$$

where

$$\begin{gathered} E = \boldsymbol{r}_u \cdot \boldsymbol{r}_u, \quad F = \boldsymbol{r}_u \cdot \boldsymbol{r}_v, \quad G = \boldsymbol{r}_v \cdot \boldsymbol{r}_v, \\ e = -\boldsymbol{r}_u \cdot \boldsymbol{N}_u = \boldsymbol{r}_{uu} \cdot \boldsymbol{N}, \quad g = -\boldsymbol{r}_v \cdot \boldsymbol{N}_v = \boldsymbol{r}_{vv} \cdot \boldsymbol{N}. \\ f = -\boldsymbol{r}_u \cdot \boldsymbol{N}_v = -\boldsymbol{r}_v \cdot \boldsymbol{N}_u = \boldsymbol{r}_{uv} \cdot \boldsymbol{N}. \end{gathered} \tag{1.3}$$

An important classical result due to Bonnet [53] states that the sextuplet $\{E, F, G;\ e, f, g\}$ determines the surface Σ up to its position in space.

The Gauss equations associated with Σ are [352]

$$\begin{aligned} \boldsymbol{r}_{uu} &= \Gamma^1_{11}\boldsymbol{r}_u + \Gamma^2_{11}\boldsymbol{r}_v + e\boldsymbol{N}, \\ \boldsymbol{r}_{uv} &= \Gamma^1_{12}\boldsymbol{r}_u + \Gamma^2_{12}\boldsymbol{r}_v + f\boldsymbol{N}, \\ \boldsymbol{r}_{vv} &= \Gamma^1_{22}\boldsymbol{r}_u + \Gamma^2_{22}\boldsymbol{r}_v + g\boldsymbol{N}, \end{aligned} \tag{1.4}$$

while the Weingarten equations comprise

$$\begin{aligned} \boldsymbol{N}_u &= \frac{fF - eG}{H^2}\boldsymbol{r}_u + \frac{eF - fE}{H^2}\boldsymbol{r}_v, \\ \boldsymbol{N}_v &= \frac{gF - fG}{H^2}\boldsymbol{r}_u + \frac{fF - gE}{H^2}\boldsymbol{r}_v, \end{aligned} \tag{1.5}$$

where

$$H^2 = |\boldsymbol{r}_u \times \boldsymbol{r}_v|^2 = EG - F^2. \tag{1.6}$$

The Γ^i_{jk} in (1.4) are the usual Christoffel symbols given by the relations

$$\Gamma^i_{jk} = \frac{g^{il}}{2}(g_{jl,k} + g_{kl,j} - g_{jk,l}), \tag{1.7}$$

where, with $x^1 = u$, $x^2 = v$,

$$\mathrm{I} = g_{jk} dx^j dx^k, \tag{1.8}$$

and

$$g^{jk} g_{kl} = \delta^j_l. \tag{1.9}$$

In the above, the Einstein convention of summation over repeated indices has been adopted.

The compatibility conditions $(\boldsymbol{r}_{uu})_v = (\boldsymbol{r}_{uv})_u$ and $(\boldsymbol{r}_{uv})_v = (\boldsymbol{r}_{vv})_u$ applied to the *linear* Gauss system (1.4) produce the *nonlinear* Mainardi-Codazzi system

$$\begin{aligned} \left(\frac{e}{H}\right)_v - \left(\frac{f}{H}\right)_u + \frac{e}{H}\Gamma^2_{22} - 2\frac{f}{H}\Gamma^2_{12} + \frac{g}{H}\Gamma^2_{11} &= 0, \\ \left(\frac{g}{H}\right)_u - \left(\frac{f}{H}\right)_v + \frac{e}{H}\Gamma^1_{22} - 2\frac{f}{H}\Gamma^1_{12} + \frac{g}{H}\Gamma^1_{11} &= 0 \end{aligned} \tag{1.10}$$

or, equivalently,

$$\begin{aligned} e_v - f_u &= e\Gamma^1_{12} + f\left(\Gamma^2_{12} - \Gamma^1_{11}\right) - g\Gamma^2_{11}, \\ f_v - g_u &= e\Gamma^1_{22} + f\left(\Gamma^2_{22} - \Gamma^1_{12}\right) - g\Gamma^2_{12}, \end{aligned} \tag{1.11}$$

augmented by the 'Theorema egregium' of Gauss. The latter provides an expression for the *Gaussian (total) curvature*

$$\mathcal{K} = \frac{eg - f^2}{EG - F^2} \tag{1.12}$$

in terms of E, F, G alone according to, in Liouville's representation,

$$\mathcal{K} = \frac{1}{H}\left[\left(\frac{H}{E}\Gamma^2_{11}\right)_v - \left(\frac{H}{E}\Gamma^2_{12}\right)_u\right]. \tag{1.13}$$

In physical terms, the 'Theorema egregium' implies that the total curvature of a surface Σ is invariant under bending without stretching.

If the total curvature of Σ is negative, that is, if Σ is a hyperbolic surface, then the *asymptotic lines* on Σ may be taken as parametric curves. Then $e = g = 0$ and the Mainardi-Codazzi equations (1.10) reduce to,

$$\left(\frac{f}{H}\right)_u + 2\Gamma^2_{12}\frac{f}{H} = 0, \quad \left(\frac{f}{H}\right)_v + 2\Gamma^1_{12}\frac{f}{H} = 0 \tag{1.14}$$

while

$$\mathcal{K} = -\frac{f^2}{H^2} =: -\frac{1}{\rho^2} \tag{1.15}$$

and

$$\Gamma^1_{12} = \frac{GE_v - FG_u}{2H^2}, \tag{1.16}$$

$$\Gamma^2_{12} = \frac{EG_u - FE_v}{2H^2}. \tag{1.17}$$

The angle ω between the parametric lines is such that

$$\cos\omega = \frac{F}{\sqrt{EG}}, \quad \sin\omega = \frac{H}{\sqrt{EG}} \tag{1.18}$$

and since $E, G > 0$, we may take, without loss of generality,

$$E = \rho^2 a^2, \quad G = \rho^2 b^2, \tag{1.19}$$

whence I and II reduce to

$$\begin{aligned} \mathrm{I} &= \rho^2(a^2du^2 + 2ab\cos\omega\, du dv + b^2dv^2), \\ \mathrm{II} &= 2\rho ab\sin\omega\, du dv. \end{aligned} \tag{1.20}$$

The Mainardi-Codazzi equations (1.11) now show that

$$a_v + \frac{1}{2}\frac{\rho_v}{\rho}a - \frac{1}{2}\frac{\rho_u}{\rho}b\cos\omega = 0, \tag{1.21}$$

$$b_u + \frac{1}{2}\frac{\rho_u}{\rho}b - \frac{1}{2}\frac{\rho_v}{\rho}a\cos\omega = 0, \tag{1.22}$$

while the representation (1.13) for the total curvature yields

$$\omega_{uv} + \frac{1}{2}\left(\frac{\rho_u}{\rho}\frac{b}{a}\sin\omega\right)_u + \frac{1}{2}\left(\frac{\rho_v}{\rho}\frac{a}{b}\sin\omega\right)_v - ab\sin\omega = 0. \tag{1.23}$$

The nonlinear system of Gauss-Mainardi-Codazzi equations (1.21)–(1.23) was originally set down by Bianchi (see [37]). Its importance in soliton theory has been noted by Cenkl [74] and subsequently by Levi and Sym [234]. It will be returned to later in that connection subject to an additional constraint, namely $\rho_{uv} = 0$. The system then becomes solitonic.

In the particular case when $\mathcal{K} = -1/\rho^2 < 0$ is a constant, Σ is termed a *pseudospherical* surface. The Mainardi-Codazzi equations (1.21), (1.22) then yield $a = a(u)$, $b = b(v)$. If Σ is now parametrised by arc length along asymptotic lines (corresponding to the transformation $du \to du' = \sqrt{E(u)}\,du, dv \to dv' = \sqrt{G(v)}\,dv$), then the fundamental forms become, on dropping the primes,

$$\begin{aligned} \mathrm{I} &= du^2 + 2\cos\omega\, du dv + dv^2, \\ \mathrm{II} &= \frac{2}{\rho}\sin\omega\, du dv, \end{aligned} \tag{1.24}$$

while (1.23) reduces to the celebrated *sine-Gordon* equation

$$\boxed{\omega_{uv} = \frac{1}{\rho^2}\sin\omega.} \tag{1.25}$$

The associated Gauss equations yield

$$\begin{aligned} \boldsymbol{r}_{uu} &= \omega_u \cot\omega\, \boldsymbol{r}_u - \omega_u \operatorname{cosec}\omega\, \boldsymbol{r}_v, \\ \boldsymbol{r}_{uv} &= \frac{1}{\rho}\sin\omega \boldsymbol{N}, \\ \boldsymbol{r}_{vv} &= -\omega_v \operatorname{cosec}\omega\, \boldsymbol{r}_u + \omega_v \cot\omega\, \boldsymbol{r}_v, \end{aligned} \tag{1.26}$$

while those of Weingarten give

$$\begin{aligned} \boldsymbol{N}_u &= \frac{1}{\rho}\cot\omega\, \boldsymbol{r}_u - \frac{1}{\rho}\operatorname{cosec}\omega\, \boldsymbol{r}_v, \\ \boldsymbol{N}_v &= -\frac{1}{\rho}\operatorname{cosec}\omega\, \boldsymbol{r}_u + \frac{1}{\rho}\cot\omega\, \boldsymbol{r}_v. \end{aligned} \tag{1.27}$$

In the twentieth century, the sine-Gordon equation has been shown, remarkably, to arise in a diversity of areas of physical interest (see [311]). It was the work of Seeger et al. [201, 345, 346] that first demonstrated how the classical Bäcklund transformation for this equation has important application in the theory of crystal dislocations. Indeed, in [345], within the context of Frenkel's and Kontorova's dislocation theory, the superposition of so-called 'eigenmotions' was obtained by means of the classical Bäcklund transformation. The interaction of what today are called breathers with kink-type dislocations was both described analytically and displayed graphically. The typical solitonic features to be subsequently discovered by Zabusky and Kruskal [389] in 1965 for the Korteweg-de Vries equation, namely preservation of velocity and shape following interaction, as well as the concomitant phase shift, were all recorded for the sine-Gordon equation in this remarkable paper of 1953.[1] Connections between the geometry of pseudospherical surfaces and other solitonic equations have been later investigated in [26, 78, 79, 141, 190, 292, 294, 321, 363].

Lamb [223] and Barnard [23] showed that the nonlinear superposition principle associated with the Bäcklund transformation for the sine-Gordon equation has application in the theory of ultrashort optical pulse propagation. In particular, solitonic decomposition phenomena observed experimentally in *Rb* vapour by Gibbs and Slusher [150] were thereby reproduced theoretically. In addition, the classical Bäcklund transformation has also found application in the theory of long Josephson junctions [344].

The preceding provides an historical motivation, both with regard to theory and application, for beginning our study of Bäcklund transformations with the classical result for the sine-Gordon equation. It will be seen that this Bäcklund transformation, in fact, corresponds to a conjugation of invariant transformations due to Bianchi and Lie. The Lie symmetry serves to intrude a key *Bäcklund parameter* into the Bianchi transformation which enables its iteration and the generation thereby of what are, in physical terms, multi-soliton solutions. Therein, the Bäcklund parameters have an important physical interpretation.

1.2 The Classical Bäcklund Transformation for the Sine-Gordon Equation

Underlying the original Bäcklund transformation for the sine-Gordon equation is a simple geometric construction for pseudospherical surfaces. Thus, if a point P is taken on an initial pseudospherical surface Σ and a line segment PP' of constant length and tangential to Σ at P is constructed in a manner dictated by a Bäcklund transformation as described below, then the locus of the points P' as P traces out Σ is another pseudospherical surface Σ' with the same total curvature as Σ. The procedure may be iterated to generate a sequence of pseudospherical surfaces with the same total curvature as the original seed surface Σ.

[1] "Man sieht ... daß beim Durchdringen von Wellengruppe und Versetzung weder die Energie noch die Geschwindigkeit beider geändert wird. Es tritt lediglich eine Verschiebung des Versetzungsmittelpunktes ... und des Schwerpunktes der Wellengruppe ... auf" [345, p 189].

Let Σ be a pseudospherical surface with total curvature $\mathcal{K} = -1/\rho^2$ and with generic position vector $\boldsymbol{r} = \boldsymbol{r}(u, v)$, where u, v correspond to the parametrisation by arc length along asymptotic lines. In this parametrisation, $\boldsymbol{r}_u, \boldsymbol{r}_v$ and $\boldsymbol{N}$ are all unit vectors, but $\boldsymbol{r}_u$ and $\boldsymbol{r}_v$ are not orthogonal. Accordingly, it proves convenient to introduce an orthonormal triad $\{\boldsymbol{A}, \boldsymbol{B}, \boldsymbol{C}\}$, where

$$\begin{aligned} \boldsymbol{A} = \boldsymbol{r}_u, \quad \boldsymbol{B} &= -\boldsymbol{r}_u \times \boldsymbol{N} = -\boldsymbol{r}_u \times \frac{(\boldsymbol{r}_u \times \boldsymbol{r}_v)}{\sin\omega}, \quad \boldsymbol{C} = \boldsymbol{N} \\ &= \operatorname{cosec}\omega\, \boldsymbol{r}_v - \cot\omega\, \boldsymbol{r}_u. \end{aligned} \tag{1.28}$$

The Gauss-Weingarten equations (1.26), (1.27) can now be used to obtain expressions for the derivatives of $\boldsymbol{A}$, $\boldsymbol{B}$ and $\boldsymbol{C}$ with respect to u and v, namely

$$\begin{aligned} \begin{pmatrix} \boldsymbol{A} \\ \boldsymbol{B} \\ \boldsymbol{C} \end{pmatrix}_u &= \begin{pmatrix} 0 & -\omega_u & 0 \\ \omega_u & 0 & 1/\rho \\ 0 & -1/\rho & 0 \end{pmatrix} \begin{pmatrix} \boldsymbol{A} \\ \boldsymbol{B} \\ \boldsymbol{C} \end{pmatrix}, \\ \begin{pmatrix} \boldsymbol{A} \\ \boldsymbol{B} \\ \boldsymbol{C} \end{pmatrix}_v &= \begin{pmatrix} 0 & 0 & (1/\rho)\sin\omega \\ 0 & 0 & -(1/\rho)\cos\omega \\ -(1/\rho)\sin\omega & (1/\rho)\cos\omega & 0 \end{pmatrix} \begin{pmatrix} \boldsymbol{A} \\ \boldsymbol{B} \\ \boldsymbol{C} \end{pmatrix}. \end{aligned} \tag{1.29}$$

This linear system is compatible if and only if ω satisfies the sine-Gordon equation (1.25).

A new pseudospherical surface Σ' with position vector $\boldsymbol{r}'$ is now sought in the form

$$\boldsymbol{r}' = \boldsymbol{r} + L\cos\phi\boldsymbol{A} + L\sin\phi\boldsymbol{B}, \tag{1.30}$$

where $L = |\boldsymbol{r}' - \boldsymbol{r}|$ is constant. Here, $\phi(u, v)$ is to be constrained by the requirement that on Σ', as on Σ, the coordinates u, v correspond to parametrisation along asymptotic lines. A necessary condition for this to be the case is that Σ' have a 1st fundamental form of the type $(1.24)_1$. In particular, this requires that

$$\boldsymbol{r}'_u \cdot \boldsymbol{r}'_u = 1, \quad \boldsymbol{r}'_v \cdot \boldsymbol{r}'_v = 1, \tag{1.31}$$

where, on use of (1.30) and the relations (1.29), we have

$$\begin{aligned} \boldsymbol{r}'_u &= [1 - L(\phi_u - \omega_u)\sin\phi]\boldsymbol{A} + L(\phi_u - \omega_u)\cos\phi\boldsymbol{B} + \frac{L}{\rho}\sin\phi\boldsymbol{C}, \\ \boldsymbol{r}'_v &= (\cos\omega - L\phi_v\sin\phi)\boldsymbol{A} + (\sin\omega + L\phi_v\cos\phi)\boldsymbol{B} + \frac{L}{\rho}\sin(\omega - \phi)\boldsymbol{C}. \end{aligned} \tag{1.32}$$

The conditions (1.31) now yield, in turn,

$$\phi_u = \omega_u + \frac{1}{L}\left(1 \pm \sqrt{1 - \frac{L^2}{\rho^2}}\right) \sin\phi \tag{1.33}$$

and

$$\phi_v = \frac{1}{L}\left(1 \mp \sqrt{1 - \frac{L^2}{\rho^2}}\right) \sin(\phi - \omega). \tag{1.34}$$

Accordingly, if we set

$$\beta = \frac{\rho}{L}\left(1 \pm \sqrt{1 - \frac{L^2}{\rho^2}}\right) = \frac{L}{\rho}\left(1 \mp \sqrt{1 - \frac{L^2}{\rho^2}}\right)^{-1}, \tag{1.35}$$

then the relations (1.33), (1.34), deliver the necessary requirements

$$\phi_u = \omega_u + \frac{\beta}{\rho} \sin\phi, \tag{1.36}$$

$$\phi_v = \frac{1}{\beta\rho} \sin(\phi - \omega) \tag{1.37}$$

on the angle ϕ in order that Σ' be a pseudospherical surface parametrised by arc length along asymptotic lines. In fact, the pair of equations, (1.36), (1.37), is sufficient in this regard. Moreover, these equations are compatible modulo the sine-Gordon equation (1.25).

On use of (1.36), (1.37), the expressions (1.32) become

$$\boldsymbol{r}'_u = \left(1 - \frac{L}{\rho}\beta \sin^2\phi\right)\boldsymbol{A} + \frac{L}{\rho}\beta \sin\phi \cos\phi \boldsymbol{B} + \frac{L}{\rho} \sin\phi \boldsymbol{C}, \tag{1.38}$$

$$\begin{aligned} \boldsymbol{r}'_v &= \left[\cos\omega - \frac{L}{\rho\beta} \sin\phi \sin(\phi - \omega)\right]\boldsymbol{A} \\ &\quad + \left[\sin\omega + \frac{L}{\rho\beta} \cos\phi \sin(\phi - \omega)\right]\boldsymbol{B} - \frac{L}{\rho} \sin(\phi - \omega)\boldsymbol{C}, \end{aligned} \tag{1.39}$$

so that $\boldsymbol{r}'_u \cdot \boldsymbol{r}'_v = \cos(2\phi - \omega)$ and the 1st fundamental form of Σ' becomes

$$\mathrm{I}' = du^2 + 2\cos(2\phi - \omega)\, du dv + dv^2. \tag{1.40}$$

Furthermore, the unit normal $\boldsymbol{N}'$ to Σ' is given by

$$\boldsymbol{N}' = \frac{\boldsymbol{r}'_u \times \boldsymbol{r}'_v}{|\boldsymbol{r}'_u \times \boldsymbol{r}'_v|} = -\frac{L}{\rho}\sin\phi\boldsymbol{A} + \frac{L}{\rho}\cos\phi\boldsymbol{B} + \left(1 - \frac{L\beta}{\rho}\right)\boldsymbol{C}, \quad (1.41)$$

whence, on use of (1.30), it is seen that $(\boldsymbol{r}' - \boldsymbol{r}) \cdot \boldsymbol{N}' = 0$. Accordingly, the vector $\boldsymbol{r}' - \boldsymbol{r}$ joining corresponding points on Σ and Σ' is tangential to Σ'. It is recalled that it is tangential to Σ by construction. Moreover,

$$\boldsymbol{N}'_u = -\frac{L\beta}{\rho^2}\sin\phi\cos\phi\boldsymbol{A} + \left(\frac{L\beta}{\rho^2}\cos^2\phi - \frac{1}{\rho}\right)\boldsymbol{B} + \frac{L}{\rho^2}\cos\phi\boldsymbol{C}, \quad (1.42)$$

$$\begin{aligned}\boldsymbol{N}'_v = &\left[\frac{L}{2\rho^2\beta}\sin(\omega - 2\phi) + \frac{1}{\rho}\left(1 - \frac{L}{2\rho\beta}\right)\sin\omega\right]\boldsymbol{A} \\ &+ \left[\frac{L}{2\rho^2\beta}\cos(\omega - 2\phi) - \frac{1}{\rho}\left(1 - \frac{L}{2\rho\beta}\right)\cos\omega\right]\boldsymbol{B} \quad (1.43) \\ &- \frac{L}{\rho^2}\cos(\omega - \phi)\boldsymbol{C},\end{aligned}$$

whence

$$\boldsymbol{r}'_u \cdot \boldsymbol{N}'_u = 0, \quad \boldsymbol{r}'_u \cdot \boldsymbol{N}'_v = \boldsymbol{r}'_v \cdot \boldsymbol{N}'_u = -\frac{1}{\rho}\sin(2\phi - \omega), \quad \boldsymbol{r}'_v \cdot \boldsymbol{N}'_v = 0.$$

The 2nd fundamental form for Σ' is

$$\mathrm{II}' = \frac{2}{\rho}\sin(2\phi - \omega)\,du\,dv \quad (1.44)$$

and this together with I' as given by (1.40) shows that Σ' is a pseudospherical surface parametrised by arc length along asymptotic lines. The angle between the asymptotic lines on Σ' is given by

$$\omega' = 2\phi - \omega, \quad (1.45)$$

where ω' plays the same role in relation to Σ' as is played by ω in relation to Σ. In particular, ω' must satisfy the sine-Gordon equation

$$\omega'_{uv} = \frac{1}{\rho^2}\sin\omega'. \quad (1.46)$$

Use of the relation (1.45) to eliminate ϕ in (1.36) and (1.37) now yields

$$\boxed{\begin{aligned}\left(\frac{\omega'-\omega}{2}\right)_u &= \frac{\beta}{\rho}\sin\left(\frac{\omega'+\omega}{2}\right)\\ \left(\frac{\omega'+\omega}{2}\right)_v &= \frac{1}{\beta\rho}\sin\left(\frac{\omega'-\omega}{2}\right).\end{aligned}} \quad \mathbb{B}_\beta \tag{1.47}$$

This is the standard form of the Bäcklund transformation which links the sine-Gordon equations (1.25) and (1.46).

It is noted that, under $\mathbb{B}_\beta$,

$$N' \cdot N = 1 - \frac{L\beta}{\rho} = \text{const}, \tag{1.48}$$

that is, the tangent planes at corresponding points on Σ and Σ' meet at a constant angle ζ where $\beta = \tan(\zeta/2)$. In Bianchi's original geometric construction, of which the Bäcklund result is an extension,

$$L = \rho, \quad \beta = 1 \tag{1.49}$$

so that these tangent planes are orthogonal. Bäcklund's relaxation of the orthogonality requirement allows the key parameter β to be inserted into the Bianchi transformation. In fact, the Bäcklund transformation $\mathbb{B}_\beta$ may be viewed as a composition of a Bianchi transformation with a simple Lie group invariance. Thus, the sine-Gordon equation (1.25) is invariant under the scaling

$$u^* = \beta u, \quad v^* = \frac{v}{\beta}, \qquad \beta \neq 0 \tag{1.50}$$

so that, any solution $\omega = \omega(u, v)$ generates a one-parameter class of solutions $\omega^*(u^*, v^*) = \omega(\beta u, v/\beta)$.[2] Lie observed that conjugation of the invariance (1.50) with the original Bianchi transformation

$$\begin{aligned}\left(\frac{\omega'-\omega}{2}\right)_{u^*} &= \frac{1}{\rho}\sin\left(\frac{\omega'+\omega}{2}\right),\\ \left(\frac{\omega'+\omega}{2}\right)_{v^*} &= \frac{1}{\rho}\sin\left(\frac{\omega'-\omega}{2}\right)\end{aligned} \tag{1.51}$$

produces the Bäcklund transformation (1.47).

[2] Importantly, this Lie point invariance also inserts the Bäcklund parameter β into the 'linear representation' (1.29) and delivers a one-parameter family of pseudospherical surfaces associated with a given solution ω of the sine-Gordon equation.

In terms of the construction of pseudospherical surfaces, the Bäcklund transformation corresponds to the following result: let $\boldsymbol{r}$ be the coordinate vector of the pseudospherical surface Σ corresponding to a solution ω of the sine-Gordon equation (1.25). Let ω' denote the Bäcklund transform of ω via $\mathbb{B}_\beta$. Then, the coordinate vector $\boldsymbol{r}'$ of the pseudospherical surface Σ' corresponding to ω' is given by

$$\boldsymbol{r}' = \boldsymbol{r} + \frac{L}{\sin\omega}\left[\sin\left(\frac{\omega-\omega'}{2}\right)\boldsymbol{r}_u + \sin\left(\frac{\omega+\omega'}{2}\right)\boldsymbol{r}_v\right], \tag{1.52}$$

where $L = \rho \sin\zeta$.

1.2.0.1 Key Observations

- The *nonlinear* sine-Gordon equation (1.25) is derived as the compatibility condition for the *linear* Gauss equations (1.26).
- The Bäcklund transformation $\mathbb{B}_\beta$ given by (1.47) acts on the sine-Gordon equation (1.25) and leaves it invariant. Indeed, the action of $\mathbb{B}_\beta$ is restricted to (1.25) in that (1.47) is a valid system for ω' if and only if (1.25) holds: otherwise the compatibility condition $\omega'_{uv} = \omega'_{vu}$ is not satisfied.
- $\mathbb{B}_\beta$ contains a parameter $\beta = \tan(\zeta/2)$ injected into the underlying Bianchi transformation by a Lie group invariance.
- At the *linear* level, the Bäcklund transformation is represented by (1.52) and acts on the Gauss system (1.26) associated with pseudospherical surfaces parametrised by arc length along asymptotic lines. The transformation (1.52) acting on the underlying linear representation (1.26) induces the Bäcklund transformation $\mathbb{B}_\beta$ operating at the *nonlinear* level.

In that $\mathbb{B}_\beta$ represents a correspondence between solutions of the same equation, it is commonly termed an *auto-Bäcklund* transformation.

In the next section, a nonlinear superposition principle associated with the auto-Bäcklund transformation $\mathbb{B}_\beta$ will be derived whereby, in particular, multi-soliton solutions of the nonlinear sine-Gordon equation (1.25) may be generated by *purely algebraic procedures*. The algorithmic nature of the latter makes them well-suited to implementation by symbolic computation packages. Such nonlinear superposition principles are generically associated with the auto-Bäcklund transformations admitted by solitonic equations.

Exercises

1. Establish the relations (1.33), (1.34) governing the angle ϕ.
2. Derive the expression (1.52) descriptive of the action of the Bäcklund transformation $\mathbb{B}_\beta$ at the pseudospherical surface level.

1.3 Bianchi's Permutability Theorem. Generation of Multi-Soliton Solutions

Next, we turn to the application of the auto-Bäcklund transformation (1.47) to construct multi-soliton solutions of the sine-Gordon equation.

Let us start with the seed 'vacuum' solution $\omega = 0$ of (1.25). The Bäcklund transformation (1.47) shows that a second, but nontrivial, solution ω' of (1.46) may be constructed by integration of the pair of first-order equations

$$\begin{aligned} \omega'_u &= \frac{2\beta}{\rho} \sin\left(\frac{\omega'}{2}\right), \\ \omega'_v &= \frac{2}{\beta\rho} \sin\left(\frac{\omega'}{2}\right), \end{aligned} \tag{1.53}$$

leading to the new *single soliton* solution

$$\omega' = 4\tan^{-1}\left[\exp\left(\frac{\beta}{\rho}u + \frac{1}{\beta\rho}v + \alpha\right)\right], \tag{1.54}$$

where α is an arbitrary constant of integration. It should be noted that, here, it is the quantities

$$\begin{aligned} \omega'_u &= \frac{2\beta}{\rho} \operatorname{sech}\left(\frac{\beta}{\rho}u + \frac{1}{\beta\rho}v + \alpha\right), \\ \omega'_v &= \frac{2}{\beta\rho} \operatorname{sech}\left(\frac{\beta}{\rho}u + \frac{1}{\beta\rho}v + \alpha\right), \end{aligned} \tag{1.55}$$

which have the characteristic hump shape associated with a soliton.

Remarkably, analytic expressions for multi-soliton solutions which encapsulate their nonlinear interaction may now be obtained by an entirely algebraic procedure. This is a consequence of an elegant nonlinear superposition principle derived from the auto-Bäcklund transformation $\mathbb{B}_\beta$ and originally set down by Bianchi [35] in 1892. It is described in his monumental work [37] and is now known as:

1.3.1 Bianchi's Permutability Theorem

Suppose ω is a seed solution of the sine-Gordon equation (1.25) and that ω_1 and ω_2 are the Bäcklund transforms of ω via $\mathbb{B}_{\beta_1}$ and $\mathbb{B}_{\beta_2}$, that is, $\omega_1 = \mathbb{B}_{\beta_1}(\omega)$, $\omega_2 = \mathbb{B}_{\beta_2}(\omega)$. Let $\omega_{12} = \mathbb{B}_{\beta_2}(\omega_1)$ and $\omega_{21} = \mathbb{B}_{\beta_1}(\omega_2)$. The situation may be represented schematically by a *Bianchi diagram* as given in Figure 1.1.

It is natural to enquire if there are any circumstances under which the commutative condition $\omega_{12} = \omega_{21}$ applies. To investigate this matter, we set down the

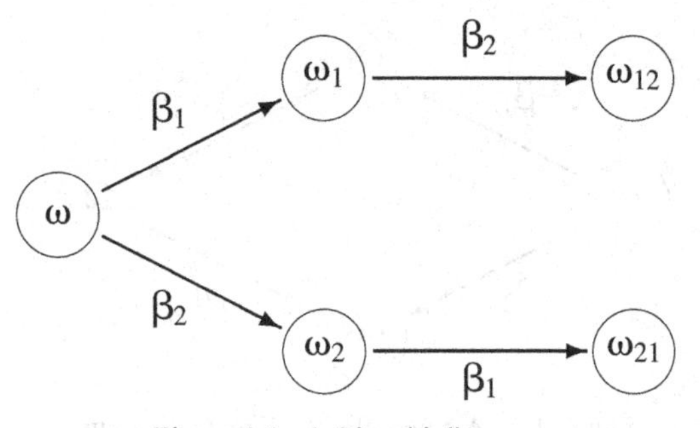

Figure 1.1. A Bianchi diagram.

u-parts of the Bäcklund transformations associated with the Bianchi diagram. Thus,

$$\omega_{1,u} = \omega_u + \frac{2\beta_1}{\rho}\sin\left(\frac{\omega_1+\omega}{2}\right), \tag{1.56}$$

$$\omega_{2,u} = \omega_u + \frac{2\beta_2}{\rho}\sin\left(\frac{\omega_2+\omega}{2}\right), \tag{1.57}$$

$$\omega_{12,u} = \omega_{1,u} + \frac{2\beta_2}{\rho}\sin\left(\frac{\omega_{12}+\omega_1}{2}\right), \tag{1.58}$$

$$\omega_{21,u} = \omega_{2,u} + \frac{2\beta_1}{\rho}\sin\left(\frac{\omega_{21}+\omega_2}{2}\right). \tag{1.59}$$

If we now put

$$\omega_{12} = \omega_{21} = \Omega, \tag{1.60}$$

then the operations (1.56) – (1.57) + (1.58) – (1.59) yield

$$0 = \frac{2}{\rho}\left[\beta_1\left\{\sin\left(\frac{\omega_1+\omega}{2}\right) - \sin\left(\frac{\Omega+\omega_2}{2}\right)\right\}\right. \tag{1.61}$$

$$\left.+\,\beta_2\left\{\sin\left(\frac{\Omega+\omega_1}{2}\right) - \sin\left(\frac{\omega_2+\omega}{2}\right)\right\}\right], \tag{1.62}$$

whence

$$\tan\left(\frac{\Omega-\omega}{4}\right) = \frac{\beta_2+\beta_1}{\beta_2-\beta_1}\tan\left(\frac{\omega_2-\omega_1}{4}\right).$$

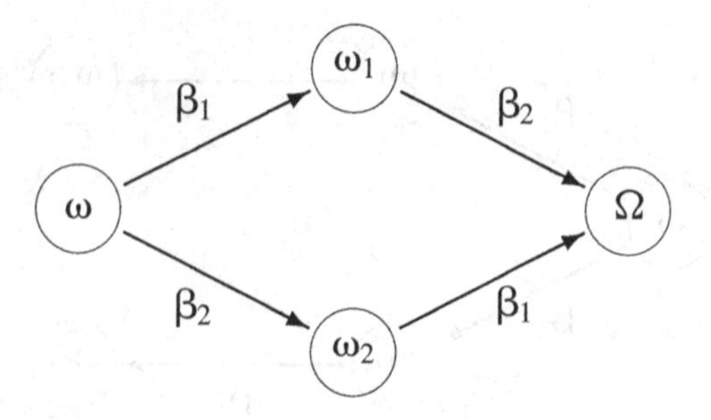

Figure 1.2. A commutative Bianchi diagram.

Accordingly, if the commutativity condition (1.60) holds, then it is necessary that

$$\boxed{\Omega = \omega + 4\tan^{-1}\left[\frac{\beta_2+\beta_1}{\beta_2-\beta_1}\tan\left(\frac{\omega_2-\omega_1}{4}\right)\right].} \tag{1.63}$$

If this expression for Ω is substituted back into (1.58) and (1.59) in place of ω_{12} and ω_{21}, then these equations may be seen to be satisfied modulo (1.56) and (1.57). Moreover, the corresponding relations in the v-part of the Bäcklund transformation are also satisfied by the expression (1.63). These considerations allow closure of the Bianchi diagram as indicated in Figure 1.2. The relation (1.63) represents a nonlinear superposition principle known as a *permutability theorem* which acts on the solution set $\{\omega, \omega_1, \omega_2\}$ to produce a new solution Ω.

The commutative property now allows a *Bianchi lattice* to be constructed corresponding to iterated application of the permutability theorem. N-soliton solutions of the sine-Gordon equation may be thereby generated by purely algebraic procedures. These represent a nonlinear superposition of N single soliton solutions (1.54) with Bäcklund parameters $\beta = \beta_1, \ldots, \beta_N$. Thus, at each application of the Bäcklund transformation, a new Bäcklund parameter β_i is introduced and an ith order soliton generated. The procedure is indicated via a Bianchi lattice in Figure 1.3.

1.3.2 Physical Applications

Seeger et al. [345] exploited the permutability theorem for the sine-Gordon equation to investigate interaction properties of kink and breather-type solutions in connection with a crystal dislocation model. Later, this procedure was adapted by Lamb [223] and subsequently by Barnard [23] in an analysis

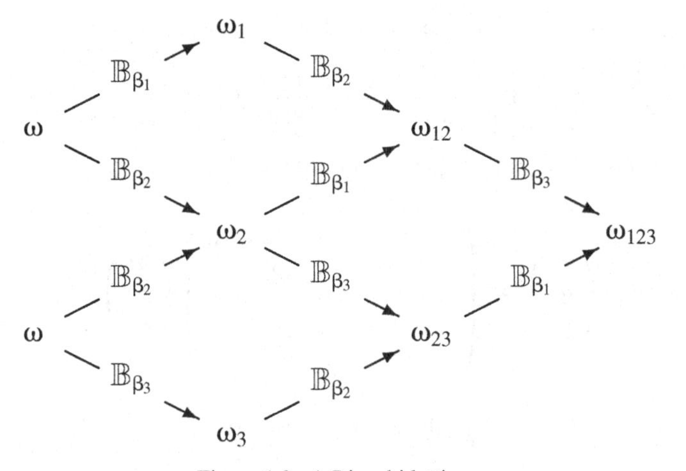

Figure 1.3. A Bianchi lattice.

of the propagation of ultrashort optical pulses in a resonant medium. Therein, analytic expressions for '$2N\pi$' light pulses were obtained via the nonlinear superposition principle. These $2N\pi$ pulses exhibit the distinctive property that they ultimately decompose into N stable 2π pulses. Experimental evidence for this phenomenon is provided, in particular, by the work of Gibbs and Slusher [150] which describes the decomposition of a 6π pulse into three 2π pulses in *Rb* vapour. An account of such decomposition in ultrashort pulse propagation is presented in the companion monograph by Rogers and Shadwick [311].

1.4 Pseudospherical Soliton Surfaces. Breathers

Here, the Bäcklund transformation is used in its linear version (1.52) to construct pseudospherical surfaces corresponding to soliton solutions of the sine-Gordon equation. In this context, it proves more convenient to parametrise the pseudospherical surfaces in terms of curvature coordinates

$$x = u + v, \quad y = u - v. \tag{1.64}$$

If we set $\omega = 2\theta$, then the 1st and 2nd fundamental forms (1.24) become

$$\mathrm{I} = \cos^2\theta\, dx^2 + \sin^2\theta\, dy^2, \tag{1.65}$$

$$\mathrm{II} = \frac{1}{\rho}\sin\theta\cos\theta\,(dx^2 - dy^2). \tag{1.66}$$

Thus, an orthonormal triad can be introduced according to

$$\boldsymbol{A}^* = \frac{\boldsymbol{r}_x}{\cos\theta}, \quad \boldsymbol{B}^* = \frac{\boldsymbol{r}_y}{\sin\theta}, \quad \boldsymbol{C}^* = \boldsymbol{N}, \tag{1.67}$$

and the Gauss-Weingarten equations (1.26) and (1.27) then yield

$$\begin{pmatrix} \boldsymbol{A}^* \\ \boldsymbol{B}^* \\ \boldsymbol{C}^* \end{pmatrix}_x = \begin{pmatrix} 0 & \theta_y & \frac{1}{\rho}\sin\theta \\ -\theta_y & 0 & 0 \\ -\frac{1}{\rho}\sin\theta & 0 & 0 \end{pmatrix} \begin{pmatrix} \boldsymbol{A}^* \\ \boldsymbol{B}^* \\ \boldsymbol{C}^* \end{pmatrix}, \tag{1.68}$$

$$\begin{pmatrix} \boldsymbol{A}^* \\ \boldsymbol{B}^* \\ \boldsymbol{C}^* \end{pmatrix}_y = \begin{pmatrix} 0 & \theta_x & 0 \\ -\theta_x & 0 & -\frac{1}{\rho}\cos\theta \\ 0 & \frac{1}{\rho}\cos\theta & 0 \end{pmatrix} \begin{pmatrix} \boldsymbol{A}^* \\ \boldsymbol{B}^* \\ \boldsymbol{C}^* \end{pmatrix}. \tag{1.69}$$

This linear system in $\{\boldsymbol{A}^*, \boldsymbol{B}^*, \boldsymbol{C}^*\}$ is compatible if and only if

$$\theta_{xx} - \theta_{yy} = \frac{1}{\rho^2}\sin\theta\cos\theta. \tag{1.70}$$

This version of the sine-Gordon equation in curvature coordinates is the most common in physical applications. Therein, x usually denotes a spatial variable and y time. In that context, it is usual to call (1.70) the $1+1$-dimensional sine-Gordon equation.[3]

1.4.1 The Pseudosphere

Here, it is shown that the *stationary* one-soliton solution of (1.70), namely

$$\theta = 2\tan^{-1}\left[\exp\left(\frac{x}{\rho} + \alpha\right)\right], \tag{1.71}$$

as obtained by setting $u = v = x/2, \beta = 1$ in (1.54), corresponds to a pseudospherical surface of revolution known as the Beltrami *pseudosphere* [352].

To establish the connection between the stationary single soliton solution (1.71) and the pseudosphere, it is recalled that the position vector $\boldsymbol{r}$ of the surface of revolution generated by the rotation of a plane curve $z = \phi(r)$ about

[3] In that it contains one spatial and one temporal independent variable.

the z-axis is given by

$$\boldsymbol{r} = \begin{pmatrix} r\cos\eta \\ r\sin\eta \\ \phi(r) \end{pmatrix}. \tag{1.72}$$

Here, the circles $r =$ const are the parallels and the curves $\eta =$ const are known as the meridians. The 1st and 2nd fundamental forms associated with the surface (1.72) are given by

$$\begin{aligned} \mathrm{I} &= [1+\phi'(r)^2]dr^2 + r^2 d\eta^2, \\ \mathrm{II} &= \frac{\phi''(r)dr^2}{\sqrt{1+\phi'(r)^2}} + \frac{r\phi'(r)d\eta^2}{\sqrt{1+\phi'(r)^2}}. \end{aligned} \tag{1.73}$$

Thus, $F = f = 0$ so that the coordinate lines $r =$ const, $\eta =$ const, namely the parallels and meridians, respectively, are lines of curvature on the surface of revolution. If we write

$$\mathrm{I} = d\xi^2 + r^2 d\eta^2, \tag{1.74}$$

where

$$d\xi = \sqrt{1+\phi'(r)^2}\, dr, \quad r = r(\xi) \tag{1.75}$$

then Gauss' theorem (1.13) shows that the total curvature is given by

$$\mathcal{K} = -\frac{1}{r}\frac{d^2 r}{d\xi^2}, \tag{1.76}$$

whence the general pseudospherical surface of revolution with $\mathcal{K} = -1/\rho^2$ adopts the form

$$r = c_1 \cosh\frac{\xi}{\rho} + c_2 \sinh\frac{\xi}{\rho}, \tag{1.77}$$

where ρ is constant. In particular, in the case $c_1 = c_2 = c$ corresponding to so-called parabolic pseudospherical surfaces of revolution, the meridians are given by

$$r = ce^{\xi/\rho}, \tag{1.78}$$

while

$$z = \phi(r) = \int \sqrt{1-(c^2/\rho^2)e^{2\xi/\rho}}\, d\xi. \tag{1.79}$$

The substitution

$$\sin\psi = \frac{c}{\rho} e^{\xi/\rho}$$

in (1.79) yields

$$z = \rho\left(\cos\psi + \ln\left|\tan\frac{\psi}{2}\right|\right), \tag{1.80}$$

whence

$$dz = \cot\psi \, dr$$

so that ψ is the angle that the tangent to the meridian makes with the z-axis. The distance $d = r \operatorname{cosec}\psi$ from a generic point on the meridian to the z-axis measured along the tangent is seen to be ρ and so is a constant. A curve with this property is called a *tractrix*. Hence, the parabolic pseudospherical surface of revolution is generated by the rotation about the z-axis of a tractrix. This surface is known as the pseudosphere.

To determine the solution of the sine-Gordon equation (1.70) corresponding to the pseudosphere, the latter must be parametrised according to (1.65) and (1.66). In terms of ψ and η, the position vector of the pseudosphere is given by

$$\boldsymbol{r} = \begin{pmatrix} \rho\sin\psi\cos\eta \\ \rho\sin\psi\sin\eta \\ \rho\left(\cos\psi + \ln\left|\tan\frac{\psi}{2}\right|\right) \end{pmatrix}, \tag{1.81}$$

whence

$$\begin{aligned} \mathrm{I} &= \rho^2\cot^2\psi \, d\psi^2 + \rho^2\sin^2\psi \, d\eta^2, \\ \mathrm{II} &= \rho\cot\psi \, d\psi^2 - \rho\sin\psi\cos\psi \, d\eta^2. \end{aligned} \tag{1.82}$$

If we now introduce x and y according to

$$dx = \rho \operatorname{cosec}\psi \, d\psi, \quad y = \rho\eta \tag{1.83}$$

then I and II in (1.82) adopt the forms (1.65) and (1.66), respectively, with $\theta = \psi$. Integration of (1.83) produces the one-soliton solution (1.71).

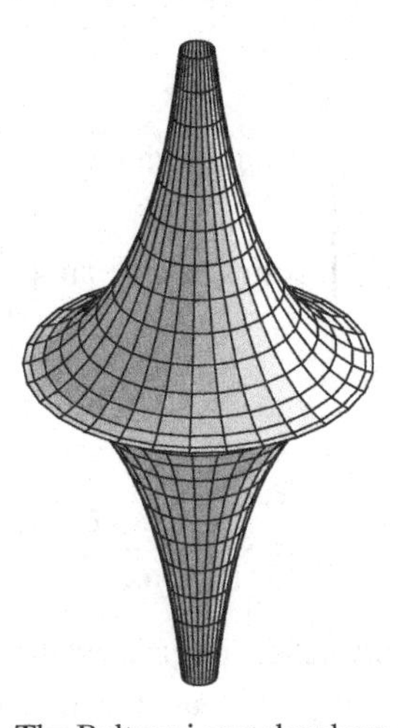

Figure 1.4. The Beltrami pseudosphere ($\zeta = \pi/2$).

In terms of the lines of curvature parameters x and y, the position vector of the pseudosphere is[4]

$$\boldsymbol{r}(x, y) = \begin{pmatrix} \rho \operatorname{sech}\left(\dfrac{x}{\rho} + \alpha\right) \cos\left(\dfrac{y}{\rho}\right) \\ \rho \operatorname{sech}\left(\dfrac{x}{\rho} + \alpha\right) \sin\left(\dfrac{y}{\rho}\right) \\ \rho\left[\dfrac{x}{\rho} + \alpha - \tanh\left(\dfrac{x}{\rho} + \alpha\right)\right] \end{pmatrix}. \tag{1.84}$$

Here, the coordinate lines $x = \text{const}$ and $y = \text{const}$ are parallels and meridians, respectively. A pseudosphere plotted using the coordinate vector (1.84) is displayed in Figure 1.4.

1.4.2 A Pseudospherical Helicoid

The surface generated by a curve which is rotated about an axis and simultaneously translated parallel to that axis in such a way that the ratio of the velocity of translation to the velocity of rotation is constant, is known as an *helicoid*. In particular, an helicoid generated by the tractrix is pseudospherical and is known as a *Dini* surface. In terms of the parameters x, y as above, its coordinate vector

[4] A change in the value of the constant of integration α is merely equivalent to a change in the origin. Accordingly, in the sequel, it will be set to be zero.

is given by

$$\boldsymbol{r}(x, y) = \begin{pmatrix} \rho \sin \zeta \operatorname{sech} \chi \cos \left(\dfrac{y}{\rho}\right) \\ \rho \sin \zeta \operatorname{sech} \chi \sin \left(\dfrac{y}{\rho}\right) \\ x - \rho \sin \zeta \tanh \chi \end{pmatrix}, \tag{1.85}$$

where

$$\chi = \frac{x - y \cos \zeta}{\rho \sin \zeta} \tag{1.86}$$

and ζ is a constant. The pseudosphere is retrieved in the case $\zeta = \pi/2$. Here, the tangent length of the tractrix is $\rho \sin \zeta$ and the helicoidal parameter associated with the relative rates of translation and rotation of the generating tractrix is $\rho \cos \zeta$. The corresponding solution of the 1+1-dimensional sine-Gordon equation (1.70) is the moving one-soliton solution given by (1.54) rewritten in terms of curvature coordinates and with $\beta = \tan(\zeta/2)$, namely

$$\begin{aligned} \theta = \frac{\omega}{2} &= 2 \arctan\left[\exp\left\{\frac{1}{2\rho}\left(\beta + \frac{1}{\beta}\right)x + \frac{1}{2\rho}\left(\beta - \frac{1}{\beta}\right)y\right\}\right] \\ &= 2 \arctan \exp \chi. \end{aligned} \tag{1.87}$$

A Dini surface with the coordinate vector (1.85) is plotted in Figure 1.5.

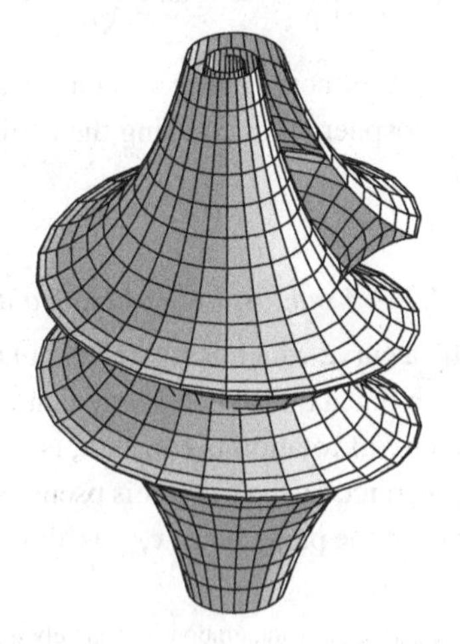

Figure 1.5. A surface of Dini.

1.4.3 Two-Soliton Surfaces

The Bäcklund transformation (1.47) for the 1+1-dimensional sine-Gordon equation (1.70) in terms of curvature coordinates x, y reads

$$\left.\begin{aligned} \theta'_x - \theta_x &= \frac{1}{\rho \sin\zeta}(\sin\theta' \cos\theta - \cos\zeta \cos\theta' \sin\theta) \\ \theta'_y - \theta_y &= \frac{1}{\rho \sin\zeta}(\cos\theta' \sin\theta - \cos\zeta \sin\theta' \cos\theta) \end{aligned}\right\} \mathbb{B}_\beta, \tag{1.88}$$

where $\beta = \tan(\zeta/2)$. Moreover, if $\boldsymbol{r}$ is the coordinate vector of the pseudospherical surface corresponding to the solution θ of (1.70), then the coordinate vector of the new pseudospherical surface corresponding to $\theta' = \mathbb{B}_\beta(\theta)$ is given by

$$\boldsymbol{r}' = \boldsymbol{r} + L\left[\frac{\cos\theta'}{\cos\theta}\boldsymbol{r}_x - \frac{\sin\theta'}{\sin\theta}\boldsymbol{r}_y\right], \tag{1.89}$$

where $L = \rho \sin\zeta$.

The nonlinear superposition principle (1.63) yields

$$\tan\left(\frac{\theta_{12} - \theta_0}{2}\right) = \frac{\sin\left(\frac{\zeta_2 + \zeta_1}{2}\right)\tan\left(\frac{\theta_2 - \theta_1}{2}\right)}{\sin\left(\frac{\zeta_2 - \zeta_1}{2}\right)}, \tag{1.90}$$

where θ_0 is a seed solution, $\theta_1 = \mathbb{B}_{\beta_1}(\theta_0)$, $\theta_2 = \mathbb{B}_{\beta_2}(\theta_0)$ and $\theta_{12} = \mathbb{B}_{\beta_2}(\theta_1) = \mathbb{B}_{\beta_1}(\theta_2)$. In particular, if the vacuum solution $\theta_0 = 0$ is taken as seed, then

$$\theta_i = 2\arctan(\exp\chi_i), \quad i = 1, 2, \tag{1.91}$$

where

$$\chi_i = \frac{1}{\rho \sin\zeta_i}(x - y\cos\zeta_i), \quad \zeta_1 \neq \zeta_2 \tag{1.92}$$

and the relation (1.90) produces, on use of the invariance $\theta \to -\theta$ of the sine-Gordon equation (1.70), the *two-soliton* solution

$$\Theta^{\pm} = \pm 2\arctan\left[\frac{\sin\left(\frac{\zeta_2 + \zeta_1}{2}\right)\sinh\left(\frac{\chi_1 - \chi_2}{2}\right)}{\sin\left(\frac{\zeta_2 - \zeta_1}{2}\right)\cosh\left(\frac{\chi_1 + \chi_2}{2}\right)}\right]. \tag{1.93}$$

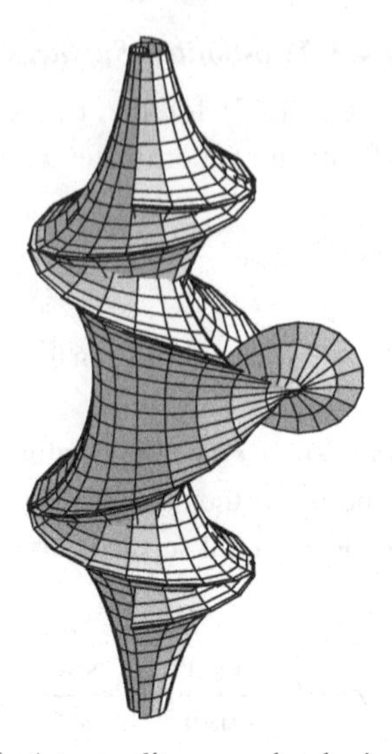

Figure 1.6. A two-soliton pseudospherical surface.

The coordinate vector $\boldsymbol{r}'$ of the pseudospherical surface corresponding to the solution (1.93) of (1.70) may now be obtained by substitution of $\theta = \theta_1$, $\zeta = \zeta_2, \theta' = \Theta^{\pm}$ into (1.89). A pseudospherical surface corresponding to a two-soliton solution is displayed in Figure 1.6.

1.4.4 Breathers

There exists an important subclass of entrapped periodic two-soliton solutions known as *breathers*. Here, an analytic expression for the breather solution is obtained via the permutability theorem, and associated pseudospherical surfaces are constructed.

In terms of the Bäcklund parameters $\beta_i = \tan(\zeta_i/2)$, the two-soliton solution Θ^+ given by (1.93) becomes

$$\Theta^+ = 2\tan^{-1}\left[\frac{\beta_2+\beta_1}{\beta_2-\beta_1}\frac{\sinh\left(\dfrac{\chi_1-\chi_2}{2}\right)}{\cosh\left(\dfrac{\chi_1+\chi_2}{2}\right)}\right] \tag{1.94}$$

with constituent single soliton solutions (1.91), where

$$\chi_i = \frac{1}{2\beta_i\rho}\left[\left(1+\beta_i^2\right)x - \left(1-\beta_i^2\right)y\right]. \tag{1.95}$$

To get a periodic solution, complex-conjugate Bäcklund parameters $\beta_1 = c + id$,

$\beta_2 = c - id$ are introduced, whence (1.93) yields

$$\Theta^+ = 2\arctan\left[\frac{c\sin\left(\dfrac{d}{2\rho(c^2+d^2)}\xi\right)}{d\cosh\left(\dfrac{c}{2\rho(c^2+d^2)}\eta\right)}\right] \tag{1.96}$$

with $\xi = [1-(c^2+d^2)]x - [1+(c^2+d^2)]y$ and $\eta = [1+(c^2+d^2)]x - [1-(c^2+d^2)]y$. Hence, Θ^+ is real and periodic in the variable ξ.

If we require that $|\beta_1| = 1$ so that $c^2+d^2 = 1$, then a solution which is periodic in y is obtained, namely

$$\Theta^+ = -2\arctan\left[\frac{c\sin(dy/\rho)}{d\cosh(cx/\rho)}\right]. \tag{1.97}$$

This is known as the *stationary* breather since it is not translated as y evolves.

1.4.5 Stationary Breather Surfaces

It is recalled that θ_{12} may be generated either as $\mathbb{B}_{\beta_2}(\theta_1)$ or as $\mathbb{B}_{\beta_1}(\theta_2)$, whence the expression (1.89) admits the symmetric representation

$$\begin{aligned}\boldsymbol{r}_{12} = \frac{1}{2}\Bigg[&\boldsymbol{r}_1 + \boldsymbol{r}_2 + \rho\cos\theta_{12}\left(\sin\zeta_2\frac{\boldsymbol{r}_{1,x}}{\cos\theta_1} + \sin\zeta_1\frac{\boldsymbol{r}_{2,x}}{\cos\theta_2}\right)\\ &+ \rho\sin\theta_{12}\left(\sin\zeta_2\frac{\boldsymbol{r}_{1,y}}{\sin\theta_1} + \sin\zeta_1\frac{\boldsymbol{r}_{2,y}}{\sin\theta_2}\right)\Bigg].\end{aligned} \tag{1.98}$$

For stationary breather solutions,

$$\begin{gathered}\sin\zeta_1 = \sin\zeta_2 = \frac{1}{c},\\ \chi_2 = \frac{1}{2\rho}(cx - idy) = \bar{\chi}_1, \quad \boldsymbol{r}_2 = \bar{\boldsymbol{r}}_1, \quad \theta_2 = \bar{\theta}_1.\end{gathered} \tag{1.99}$$

Hence, on use of (1.98) with $\boldsymbol{r}_1$, $\boldsymbol{r}_2$ given by (1.85) where $\zeta = \zeta_1$ and ζ_2 respectively, the pseudospherical surface corresponding to the stationary breather solution (1.97) is seen to be real with position vector given by, on setting $\rho = 1$,

$$\begin{aligned}\boldsymbol{r}_{\text{breather}} = &\begin{pmatrix}0\\0\\x\end{pmatrix} + \frac{2d}{c}\frac{\sin(dy)\cosh(cx)}{d^2\cosh^2(cx) + c^2\sin^2(dy)}\begin{pmatrix}\sin y\\ -\cos y\\ 0\end{pmatrix}\\ &+ \frac{2d^2}{c}\frac{\cosh(cx)}{d^2\cosh^2(cx) + c^2\sin^2(dy)}\begin{pmatrix}\cos y\cos(dy)\\ \sin y\cos(dy)\\ -\sinh(cx)\end{pmatrix},\end{aligned} \tag{1.100}$$

where $c = \sqrt{1 - d^2}$. It is readily verified that the lines of curvature $y = \text{const}$ are planar and, accordingly, the above pseudospherical surfaces constitute *Enneper* surfaces. The latter have been studied in detail by Steuerwald [351].

To every rational number d between zero and unity, there corresponds a pseudospherical stationary breather surface which is periodic in the y-parameter. If we write $d = p/q$, where p and q are co-prime integers with $p < q$, then the period of the breather solution is $2\pi q/p$. Stationary breather pseudospherical surfaces corresponding to various values of the parameter d are displayed in Figure 1.7.

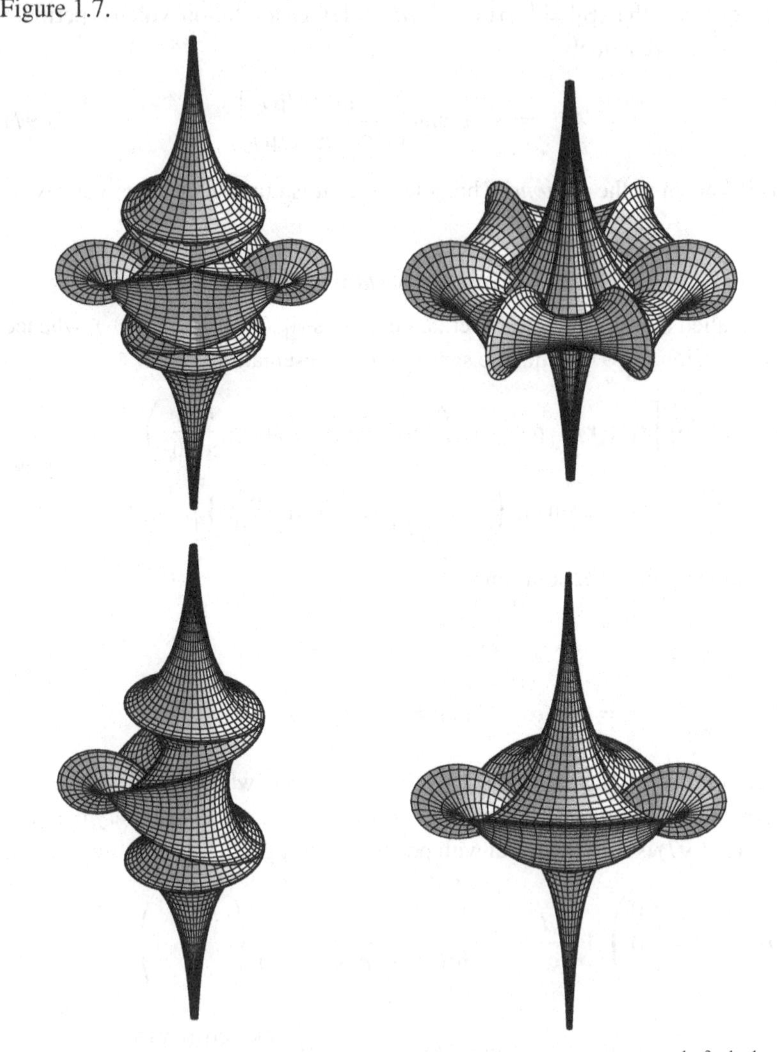

Figure 1.7. Stationary breather pseudospherical surfaces [181]: $\frac{p}{q} = \frac{1}{4}, \frac{3}{4}, \frac{1}{5}, \frac{1}{2}$.

By repeated application of (1.89), one can obtain the position vector of the pseudospherical surfaces associated with N-soliton solutions of the sine-Gordon equation as generated via iteration of the permutability theorem.

Exercises

1. Establish the Bäcklund relations (1.88) and (1.89).
2. Plot the time evolution of Θ_x^+ for the two-soliton solution (1.93) over a range which depicts both the nonlinear interaction and eventual decomposition into two separate single soliton solutions.

1.5 Parallel Surfaces. Induced Bäcklund Transformation for a Class of Weingarten Surfaces

Given a surface Σ with the generic position vector $\boldsymbol{r}$, a surface $\tilde{\Sigma}$ with current point

$$\tilde{\boldsymbol{r}} = \boldsymbol{r} + c\boldsymbol{N}, \quad c \text{ constant} \tag{1.101}$$

is said to be *parallel* to Σ. Here, c represents the constant distance along the normal between Σ and $\tilde{\Sigma}$. If lines of curvature $x = \text{const}$, $y = \text{const}$ on Σ are taken as parametric curves, then (1.101) implies that

$$\begin{aligned} \tilde{\boldsymbol{r}}_x &= \boldsymbol{r}_x + c\boldsymbol{N}_x = \boldsymbol{r}_x - \frac{ce}{E}\boldsymbol{r}_x = (1 - c\kappa_1)\boldsymbol{r}_x, \\ \tilde{\boldsymbol{r}}_y &= \boldsymbol{r}_y + c\boldsymbol{N}_y = \boldsymbol{r}_y - \frac{cg}{G}\boldsymbol{r}_y = (1 - c\kappa_2)\boldsymbol{r}_y, \end{aligned} \tag{1.102}$$

where κ_1 and κ_2 are the *principal curvatures* at the generic point $\boldsymbol{r}$ on Σ.[5] Thus,

$$\begin{gathered} \tilde{E} = (1 - c\kappa_1)^2 E, \quad \tilde{F} = 0, \quad \tilde{G} = (1 - c\kappa_2)^2 G, \\ \tilde{H}^2 = (1 - c\kappa_1)^2 (1 - c\kappa_2)^2, \end{gathered} \tag{1.103}$$

while

$$\tilde{\boldsymbol{N}} = \frac{\tilde{\boldsymbol{r}}_x \times \tilde{\boldsymbol{r}}_y}{\tilde{H}} = \epsilon \boldsymbol{N}, \tag{1.104}$$

where $\epsilon = \pm 1$, depending on whether $(1 - c\kappa_1)(1 - c\kappa_2)$ is positive or negative. Moreover,

$$\tilde{e} = \epsilon(1 - c\kappa_1)e, \quad \tilde{f} = 0, \quad \tilde{g} = \epsilon(1 - c\kappa_2)g. \tag{1.105}$$

[5] In terms of the principal curvatures κ_1, κ_2 the Gaussian curvature $\mathcal{K}$ and mean curvature $\mathcal{M}$ are given respectively by $\mathcal{K} = \kappa_1\kappa_2$, $\mathcal{M} = \frac{\kappa_1 + \kappa_2}{2}$.

The parallel surface $\tilde{\Sigma}$ is seen to be parametrised along lines of curvature which correspond to those on the original surface Σ. Tangents at corresponding points are parallel.

The principal curvatures $\tilde{\kappa}_1, \tilde{\kappa}_2$ on $\tilde{\Sigma}$ are

$$\tilde{\kappa}_1 = \frac{\tilde{e}}{\tilde{E}} = \frac{\epsilon\kappa_1}{(1-c\kappa_1)}, \quad \tilde{\kappa}_2 = \frac{\tilde{g}}{\tilde{G}} = \frac{\epsilon\kappa_2}{(1-c\kappa_2)}, \tag{1.106}$$

so that the mean curvature $\tilde{\mathcal{M}}$ and total curvature $\tilde{\mathcal{K}}$ of the parallel surface $\tilde{\Sigma}$ are given by

$$\tilde{\mathcal{M}} = \frac{1}{2}(\tilde{\kappa}_1 + \tilde{\kappa}_2) = \frac{\epsilon(\mathcal{M} - c\mathcal{K})}{1 - 2c\mathcal{M} + c^2\mathcal{K}}, \tag{1.107}$$

$$\tilde{\mathcal{K}} = \tilde{\kappa}_1\tilde{\kappa}_2 = \frac{\mathcal{K}}{1 - 2c\mathcal{M} + c^2\mathcal{K}}. \tag{1.108}$$

1.5.1 *Surfaces of Constant Mean Curvature. A Theorem of Bonnet*

If $\Sigma = \tilde{\Sigma}|_{c=0}$ is pseudospherical with $\mathcal{K} = -1/\rho^2$, then the mean and total curvature of the parallel surfaces $\tilde{\Sigma}$ are given parametrically in terms of $\mathcal{M}$ via

$$\tilde{\mathcal{M}} = \frac{\epsilon(\mathcal{M} + c/\rho^2)}{1 - 2c\mathcal{M} - c^2/\rho^2}, \quad \tilde{\mathcal{K}} = \frac{-1}{\rho^2(1 - 2c\mathcal{M} - c^2/\rho^2)}, \tag{1.109}$$

whence, we obtain the relation

$$(\rho^2 + c^2)\tilde{\mathcal{K}} + 2c\epsilon\tilde{\mathcal{M}} = -1, \tag{1.110}$$

or, equivalently, in terms of the principal curvatures,

$$\left[\tilde{\kappa}_1 + \frac{c\epsilon}{\rho^2 + c^2}\right]\left[\tilde{\kappa}_2 + \frac{c\epsilon}{\rho^2 + c^2}\right] = \frac{-\rho^2}{(\rho^2 + c^2)^2}. \tag{1.111}$$

Thus, the surfaces $\tilde{\Sigma}$ parallel to a given pseudospherical surface Σ are particular *Weingarten* surfaces, that is, their mean and total curvatures $\tilde{\mathcal{M}}$ and $\tilde{\mathcal{K}}$, or, equivalently, their principal curvatures $\tilde{\kappa}_1$ and $\tilde{\kappa}_2$, are functionally related.[6]

If the surface Σ is of constant positive Gaussian curvature

$$\mathcal{K} = \frac{1}{\rho^2} \tag{1.112}$$

[6] Connections between Weingarten surfaces and certain solitonic equations are discussed in [75].

then the Gauss equation reduces to an integrable elliptic sinh-Gordon equation. The mean curvature of the parallel surfaces is given by

$$\tilde{\mathcal{M}} = \frac{\epsilon(\mathcal{M} - c/\rho^2)}{1 - 2c\mathcal{M} + c^2/\rho^2} \tag{1.113}$$

which is independent of $\mathcal{M}$ if and only if

$$c = \pm\rho\,. \tag{1.114}$$

In this case, $\tilde{\mathcal{M}}$ is constant and we can formulate the following theorem due to Bonnet.

Theorem 1. *A surface of constant positive Gaussian curvature* $\mathcal{K} = 1/\rho^2$ *admits two parallel surfaces of constant mean curvature* $\pm 1/(2\rho)$. *The distance to the surface of positive Gaussian curvature is* ρ.

In conclusion, it is remarked that the parameter c may be regarded as being associated with a constant speed normal motion of a base pseudospherical surface $\tilde{\Sigma}|_{c=0}$ which evolves into a member of the associated Weingarten class (1.110). In this evolution, the lines of curvature, which are invariant under the normal displacement (1.101), generate mutually orthogonal surfaces which are, in turn, orthogonal to the parallel surfaces $\tilde{\Sigma}$.

1.5.2 An Induced Bäcklund Transformation

If a Bäcklund transformation is known for a class of surfaces Σ, then a Bäcklund transformation is naturally induced for the class of parallel surfaces. In particular, a Bäcklund transformation is readily obtained for Weingarten surfaces governed by a relation of the type (1.111). The result is a consequence of the Bäcklund transformation (1.89), that is

$$\boldsymbol{r}' = \boldsymbol{r} + \rho(\boldsymbol{N} \times \boldsymbol{N}'), \tag{1.115}$$

where $\boldsymbol{N}'$ is given by (1.41) whence, in terms of curvature coordinates,

$$\begin{aligned} \boldsymbol{N}' &= \frac{L}{\rho \sin 2\theta}[-\cos(\theta' - \theta) + \cos(\theta' + \theta)]\boldsymbol{r}_x \\ &\quad + \frac{L}{\rho \sin 2\theta}[-\cos(\theta' - \theta) - \cos(\theta' + \theta)]\boldsymbol{r}_y + \left(1 - \frac{L\beta}{\rho}\right)\boldsymbol{N} \\ &= \frac{L}{\rho}\left(\frac{\sin\theta'}{\cos\theta}\boldsymbol{r}_x - \frac{\cos\theta'}{\sin\theta}\boldsymbol{r}_y\right) + \left(1 - \frac{L\beta}{\rho}\right)\boldsymbol{N}. \end{aligned} \tag{1.116}$$

Parallel surfaces $\tilde{\Sigma}$ and $\tilde{\Sigma}'$ to Σ and Σ', respectively, are given by

$$\tilde{\boldsymbol{r}} = \boldsymbol{r} + c\boldsymbol{N}, \;\; \tilde{\boldsymbol{r}}' = \boldsymbol{r}' + c'\boldsymbol{N}', \tag{1.117}$$

whence

$$\begin{aligned}
\tilde{\boldsymbol{r}}' &= \tilde{\boldsymbol{r}} + \rho(\tilde{\boldsymbol{N}} \times \tilde{\boldsymbol{N}}') + c'\tilde{\boldsymbol{N}}' - c\boldsymbol{N} \\
&= \tilde{\boldsymbol{r}} + L\left[\frac{\cos\theta'}{\cos\theta}\boldsymbol{r}_x - \frac{\sin\theta'}{\sin\theta}\boldsymbol{r}_y\right] \\
&\quad + \frac{c'L}{\rho}\left[\frac{\sin\theta'}{\cos\theta}\boldsymbol{r}_x - \frac{\cos\theta'}{\sin\theta}\boldsymbol{r}_y\right] + \left[c'\left(1 - \frac{L\beta}{\rho}\right) - c\right]\boldsymbol{N} \\
&= \tilde{\boldsymbol{r}} + L\left[\frac{\cos\theta' + (c'/\rho)\sin\theta'}{\cos\theta - (c'/\rho)\sin\theta}\tilde{\boldsymbol{r}}_x - \frac{\sin\theta' + (c'/\rho)\cos\theta'}{\sin\theta + (c'/\rho)\cos\theta}\tilde{\boldsymbol{r}}_y\right] \\
&\quad + \epsilon\left[c'\left(1 - \frac{L\beta}{\rho}\right) - c\right]\tilde{\boldsymbol{N}}.
\end{aligned} \tag{1.118}$$

This provides a Bäcklund transformation acting on the class of Weingarten surfaces $\tilde{\Sigma}$ with mean and total curvatures $\tilde{\mathcal{M}}$, $\tilde{\mathcal{K}}$ related by (1.110). Since

$$\mathcal{K}' = -\frac{1}{\rho^2}, \;\; \mathcal{M}' = -\frac{1}{\rho}\cot 2\theta', \tag{1.119}$$

it is seen that, under this Bäcklund transformation,

$$\begin{aligned}
\tilde{\mathcal{K}}' &= \frac{-1}{\rho^2(\rho^2 - c'^2 + 2c'\rho\cot 2\theta')}, \\
\tilde{\mathcal{M}}' &= \frac{\epsilon(c' - \rho\cot 2\theta')}{\rho^2 - c'^2 + 2c'\rho\cot 2\theta'}.
\end{aligned} \tag{1.120}$$

It is recalled that in (1.118), θ and θ' are related by the Bäcklund transformation $\mathbb{B}_\beta$ as given by (1.88). Moreover, (1.118) yields

$$|\tilde{\boldsymbol{r}}' - \tilde{\boldsymbol{r}}|^2 = L^2 + c'^2 - 2cc'\cos\zeta + c^2 \tag{1.121}$$

so that $\tilde{\boldsymbol{r}}' - \tilde{\boldsymbol{r}}$ is of constant length.

The Weingarten surfaces characterised by the relation

$$(c^2 \pm \rho^2)\mathcal{K} + 2c\epsilon\mathcal{M} = -1 \tag{1.122}$$

have here been shown to be solitonic in that they are parallel to surfaces of constant Gaussian curvature. It proves that if one assumes that the quantities c

and ρ in (1.122) are 'harmonic' with respect to a certain coordinate system, then the associated 'generalised Weingarten surfaces' [329, 332, 333] are likewise integrable. In the following section, we shall discuss the particular case $c = 0$, $\rho_{uv} = 0$, that is

$$\left(\frac{1}{\sqrt{\mathcal{K}}}\right)_{xy} = 0, \tag{1.123}$$

in terms of asymptotic coordinates. The corresponding surfaces are classical and known as Bianchi surfaces. Moreover, Bobenko [40] has established the integrability of 'inverse harmonic mean curvature surfaces' given by

$$(\partial_x^2 + \partial_y^2)\left(\frac{1}{\mathcal{M}}\right) = 0 \tag{1.124}$$

associated with $c = \pm\rho$, $(\partial_x^2 + \partial_y^2)\rho = 0$ and the minus sign in (1.122), where x and y denote 'conformal' coordinates (cf. Chapter 5). These surfaces generalise the constant mean curvature surfaces mentioned in the previous subsection.

1.6 The Bianchi System. Its Auto-Bäcklund Transformation

In Section 1.1, it was shown that hyperbolic surfaces parametrised in terms of asymptotic coordinates are governed by the system

$$\begin{aligned} a_v + \frac{1}{2}\frac{\rho_v}{\rho}a - \frac{1}{2}\frac{\rho_u}{\rho}b\cos\omega &= 0, \\ b_u + \frac{1}{2}\frac{\rho_u}{\rho}b - \frac{1}{2}\frac{\rho_v}{\rho}a\cos\omega &= 0, \\ \omega_{uv} + \frac{1}{2}\left(\frac{\rho_u}{\rho}\frac{b}{a}\sin\omega\right)_u + \frac{1}{2}\left(\frac{\rho_v}{\rho}\frac{a}{b}\sin\omega\right)_v - ab\sin\omega &= 0. \end{aligned} \tag{1.125}$$

The corresponding fundamental forms read

$$\begin{aligned} \mathrm{I} &= \rho^2(a^2du^2 + 2ab\cos\omega\, dudv + b^2dv^2), \\ \mathrm{II} &= 2\rho ab\sin\omega\, dudv \end{aligned} \tag{1.126}$$

so that the Gaussian curvature takes the form

$$\mathcal{K} = -\frac{1}{\rho^2} < 0. \tag{1.127}$$

In 1890, Bianchi [33] presented a purely geometric construction of hyperbolic surfaces subject to the constraint

$$\rho_{uv} = 0. \tag{1.128}$$

These surfaces have been termed *Bianchi surfaces*. Here, we derive a Bäcklund transformation for hyperbolic surfaces without restriction as to the Gaussian curvature. This Bäcklund transformation may be specialised to deliver classical Bäcklund transformations associated, in turn, with pseudospherical, Tzitzeica and Bianchi surfaces. As a preliminary, we derive the Mainardi-Codazzi equations associated with the spherical representation of hyperbolic surfaces. Interestingly, it will be shown that the normal of these surfaces, regarded as a function of u and v, obeys a vector equation which is well-known in soliton theory.

1.6.1 Hyperbolic Surfaces. Spherical Representation

For a given surface $\Sigma : \boldsymbol{r} = \boldsymbol{r}(u, v)$, the normal $\boldsymbol{N} = \boldsymbol{N}(u, v)$ generates, as u and v vary, a coordinate system on the unit sphere $\boldsymbol{N}^2 = 1$. This coordinate system is called the *spherical* or *Gauss representation* of the surface. The corresponding 1st fundamental form of the sphere reads

$$d\boldsymbol{N}^2 = \mathcal{E}du^2 + 2\mathcal{F}dudv + \mathcal{G}dv^2, \tag{1.129}$$

where

$$\mathcal{E} = \boldsymbol{N}_u^2, \quad \mathcal{F} = \boldsymbol{N}_u \cdot \boldsymbol{N}_v, \quad \mathcal{G} = \boldsymbol{N}_v^2. \tag{1.130}$$

The metric of the spherical representation may be expressed in terms of the fundamental forms of Σ. In fact, in terms of asymptotic coordinates, we find that

$$E = \rho^2\mathcal{E}, \quad F = -\rho^2\mathcal{F}, \quad G = \rho^2\mathcal{G} \tag{1.131}$$

for a hyperbolic surface with Gaussian curvature

$$\mathcal{K} = -\frac{1}{\rho^2}. \tag{1.132}$$

Accordingly, the Weingarten equations for the hyperbolic surface may be brought into the form

$$\boldsymbol{r}_u = \frac{\rho}{\mathcal{H}}(\mathcal{F}\boldsymbol{N}_u - \mathcal{E}\boldsymbol{N}_v), \quad \boldsymbol{r}_v = \frac{\rho}{\mathcal{H}}(\mathcal{F}\boldsymbol{N}_v - \mathcal{G}\boldsymbol{N}_u) \tag{1.133}$$

with $\mathcal{H}^2 = \mathcal{E}\mathcal{G} - \mathcal{F}^2$, and the Mainardi-Codazzi equations reduce to

$$\frac{\rho_u}{\rho} = -2\tilde{\Gamma}_{12}^2, \quad \frac{\rho_v}{\rho} = -2\tilde{\Gamma}_{12}^1, \tag{1.134}$$

where $\tilde{\Gamma}^k_{ij}$ are the Christoffel symbols associated with the metric of the spherical representation. Moreover, compatibility of (1.134) requires that

$$\frac{\partial}{\partial u}\tilde{\Gamma}^1_{12} = \frac{\partial}{\partial v}\tilde{\Gamma}^2_{12}. \tag{1.135}$$

Conversely, a coordinate system on the unit sphere represented by some vector-valued function $\boldsymbol{N} = \boldsymbol{N}(u, v)$, $\boldsymbol{N}^2 = 1$ may be identified as the spherical representation of the asymptotic lines on a hyperbolic surface if and only if the corresponding coefficients $\mathcal{E}$, $\mathcal{F}$ and $\mathcal{G}$ obey the relation (1.135). The proof of this theorem is by construction. Thus, the relation (1.135) guarantees the existence of a function ρ satisfying (1.134) which implies, in turn, the compatibility of the system (1.133). It is then readily verified that $\boldsymbol{N}$ is indeed the normal of the surface defined by $\boldsymbol{r}$. It is noted that the surface vector $\boldsymbol{r}$ is defined only up to homothetic transformations, that is, scalings and translations, since, in turn, ρ is given only up to a multiplicative factor and $\boldsymbol{r}$ up to an additive constant of integration.

The above result may be re-formulated. Thus, the fundamental forms may be written, in terms of the normal $\boldsymbol{N}$ of a hyperbolic surface, as

$$\begin{aligned} \mathrm{I} &= \rho^2\left(\boldsymbol{N}_u^2 du^2 - 2\boldsymbol{N}_u \cdot \boldsymbol{N}_v du dv + \boldsymbol{N}_v^2 dv^2\right), \\ \mathrm{II} &= 2\rho(\boldsymbol{N}_u \times \boldsymbol{N}_v) \cdot \boldsymbol{N} du dv = \pm 2\rho\,|\boldsymbol{N}_u \times \boldsymbol{N}_v| du dv, \end{aligned} \tag{1.136}$$

by virtue of

$$\mathcal{K} = -\frac{1}{\rho^2} = -\frac{f^2}{EG - F^2} = -\frac{f^2}{\rho^4\left[\boldsymbol{N}_u^2\boldsymbol{N}_v^2 - (\boldsymbol{N}_u \cdot \boldsymbol{N}_v)^2\right]}. \tag{1.137}$$

Now, the Gauss-Weingarten equations imply that $\boldsymbol{N}$ satisfies the hyperbolic equation

$$\boldsymbol{N}_{uv} + \frac{1}{2}\frac{\rho_v}{\rho}\boldsymbol{N}_u + \frac{1}{2}\frac{\rho_u}{\rho}\boldsymbol{N}_v + \mathcal{F}\boldsymbol{N} = \boldsymbol{0}. \tag{1.138}$$

Hence, on taking the cross-product with $\boldsymbol{N}$, we obtain the vector equation

$$(\rho\boldsymbol{N} \times \boldsymbol{N}_u)_v + (\rho\boldsymbol{N} \times \boldsymbol{N}_v)_u = \boldsymbol{0} \tag{1.139}$$

which is the integrability condition for (1.133) written in the form

$$\boldsymbol{r}_u = \rho\boldsymbol{N}_u \times \boldsymbol{N}, \quad \boldsymbol{r}_v = -\rho\boldsymbol{N}_v \times \boldsymbol{N}. \tag{1.140}$$

The latter relations are commonly referred to as the *Lelieuvre formulae* (see [118]).

Conversely, any solution of the vector equation (1.139) with $\boldsymbol{N}^2 = 1$ guarantees the existence of a vector-valued function $\boldsymbol{r}$ satisfying (1.140). This, in turn, implies that

$$\boldsymbol{r}_u \cdot \boldsymbol{N} = 0, \quad \boldsymbol{r}_v \cdot \boldsymbol{N} = 0, \quad \boldsymbol{r}_u \cdot \boldsymbol{N}_u = 0, \quad \boldsymbol{r}_v \cdot \boldsymbol{N}_v = 0, \tag{1.141}$$

whence $\boldsymbol{r}$ may be regarded as the position vector of a surface parametrised in terms of asymptotic coordinates and with corresponding 1st and 2nd fundamental forms (1.136). In fact, one may directly verify that the Gauss-Mainardi-Codazzi equations (1.125) are satisfied modulo (1.139). We conclude that the vector equation (1.139) or, equivalently,

$$\begin{gathered}\varepsilon_{uv} + \frac{1}{2}\frac{\rho_v}{\rho}\varepsilon_u + \frac{1}{2}\frac{\rho_u}{\rho}\varepsilon_v = 2\frac{\varepsilon_u\varepsilon_v\bar{\varepsilon}}{|\varepsilon|^2+1}, \\ \boldsymbol{N} = \frac{1}{|\varepsilon|^2+1}\begin{pmatrix} \varepsilon + \bar{\varepsilon} \\ -i(\varepsilon - \bar{\varepsilon}) \\ |\varepsilon|^2 - 1 \end{pmatrix},\end{gathered} \tag{1.142}$$

is but another manifestation of the Gauss-Mainardi-Codazzi equations (1.125). In particular, the classical Bianchi system is equivalent to $(1.142)_1$ supplemented by $\rho_{uv} = 0$.

Interestingly, the vector equation (1.139) has an important meaning in soliton theory. Thus, if we introduce the matrix-valued function

$$N = \boldsymbol{N} \cdot \boldsymbol{\sigma}, \tag{1.143}$$

where $\boldsymbol{\sigma} = (\sigma_1, \sigma_2, \sigma_3)^{\mathrm{T}}$ and the σ_i are the usual Pauli matrices defined by

$$\sigma_1 = \begin{pmatrix} 0 & 1 \\ 1 & 0 \end{pmatrix}, \quad \sigma_2 = \begin{pmatrix} 0 & -i \\ i & 0 \end{pmatrix}, \quad \sigma_3 = \begin{pmatrix} 1 & 0 \\ 0 & -1 \end{pmatrix}, \tag{1.144}$$

then an equivalent form of (1.139) is given by

$$(\rho N N_u)_v + (\rho N N_v)_u = 0, \quad N^2 = \mathbb{1}, \quad N^\dagger = N. \tag{1.145}$$

This constitutes an extension of the *nonlinear sigma model* [394]. Here, the underlying Lie group is $O(3)$. It is remarked that a nonlinear sigma-type model based on the Lie group $O(2, 1)$ is given by

$$(\rho S S_u)_v + (\rho S S_v)_u = 0, \quad \rho_{uv} = 0, \quad S^2 = -\mathbb{1}, \quad \bar{S} = S. \tag{1.146}$$

If the matrix S is parametrised according to

$$S = \frac{1}{\varepsilon + \bar{\varepsilon}}\begin{pmatrix} i(\varepsilon - \bar{\varepsilon}) & -2\varepsilon\bar{\varepsilon} \\ 2 & -i(\varepsilon - \bar{\varepsilon}) \end{pmatrix} \tag{1.147}$$

and u, v are taken as the complex-conjugate variables z, $\bar{z}$, then we retrieve Ernst's equation of general relativity [121]

$$\varepsilon_{z\bar{z}} + \frac{1}{2}\frac{\rho_{\bar{z}}}{\rho}\varepsilon_z + \frac{1}{2}\frac{\rho_z}{\rho}\varepsilon_{\bar{z}} = \frac{\varepsilon_z \varepsilon_{\bar{z}}}{\Re(\varepsilon)}, \quad \rho_{z\bar{z}} = 0. \tag{1.148}$$

Accordingly, this Ernst equation may be regarded as an elliptic counterpart of the Bianchi system.

It is noted that 2+1-dimensional versions of the Bianchi system, as well as of the cognate Ernst equation of general relativity, may be constructed by setting these nonlinear sigma-type models in the yet more general context of so-called LKR systems [324].

1.6.2 A Bäcklund Transformation for Hyperbolic Surfaces

Here, we restrict our attention to the construction of a Bäcklund transformation that obeys the tangency condition

$$\boldsymbol{r}' = \boldsymbol{r} + p\boldsymbol{r}_u + q\boldsymbol{r}_v, \tag{1.149}$$

that is, we require that the line segment which connects corresponding points on the surfaces Σ and Σ' be tangential to Σ. If we also assume that Σ' like Σ is parametrised in terms of asymptotic coordinates u, v then the conditions

$$e' = \boldsymbol{r}'_{uu} \cdot \boldsymbol{N}' = 0, \quad g' = \boldsymbol{r}'_{vv} \cdot \boldsymbol{N}' = 0 \tag{1.150}$$

constitute two nonlinear differential equations of second order for the functions p and q. Any solution of these equations gives rise to a transformation of the form (1.149) between hyperbolic surfaces Σ and Σ'.

The situation changes dramatically if one demands that the difference vector $\boldsymbol{r}' - \boldsymbol{r}$ be also tangential to the second surface Σ'. It is then convenient to introduce an orthonormal triad consisting of the normal $\boldsymbol{N}$ and unit vectors which are tangential to the lines of curvature on Σ. It is readily verified that the directions of the lines of curvature are given by

$$\frac{\boldsymbol{r}_u}{a} \pm \frac{\boldsymbol{r}_v}{b} \tag{1.151}$$

or, equivalently,

$$\frac{\boldsymbol{N}_u}{a} \pm \frac{\boldsymbol{N}_v}{b}. \tag{1.152}$$

Thus, following Bianchi [33], we choose the orthonormal basis of tangent vectors

$$\boldsymbol{V} = \frac{1}{2\sin\frac{\Omega}{2}}\left(\frac{\boldsymbol{N}_u}{a} - \frac{\boldsymbol{N}_v}{b}\right), \quad \boldsymbol{W} = \frac{1}{2\cos\frac{\Omega}{2}}\left(\frac{\boldsymbol{N}_u}{a} + \frac{\boldsymbol{N}_v}{b}\right), \tag{1.153}$$

where

$$\Omega = \omega + \pi. \tag{1.154}$$

It is noted that Bianchi's considerations were based on the spherical representation of the surface Σ and its 1st fundamental form

$$d\boldsymbol{N}^2 = a^2du^2 + 2ab\cos\Omega\, du\, dv + b^2dv^2 \tag{1.155}$$

so that Ω denotes the angle between the coordinate lines on the sphere swept out by $\boldsymbol{N} = \boldsymbol{N}(u, v)$.

In terms of the orthonormal triad $\{\boldsymbol{V}, \boldsymbol{W}, \boldsymbol{N}\}$, the Gauss-Weingarten equations become

$$\begin{aligned}
\begin{pmatrix}\boldsymbol{V}\\ \boldsymbol{W}\\ \boldsymbol{N}\end{pmatrix}_u &= \begin{pmatrix} 0 & -\Omega_1 & -a\sin\frac{\Omega}{2} \\ \Omega_1 & 0 & -a\cos\frac{\Omega}{2} \\ a\sin\frac{\Omega}{2} & a\cos\frac{\Omega}{2} & 0 \end{pmatrix}\begin{pmatrix}\boldsymbol{V}\\ \boldsymbol{W}\\ \boldsymbol{N}\end{pmatrix}, \\
\begin{pmatrix}\boldsymbol{V}\\ \boldsymbol{W}\\ \boldsymbol{N}\end{pmatrix}_v &= \begin{pmatrix} 0 & \Omega_2 & b\sin\frac{\Omega}{2} \\ -\Omega_2 & 0 & -b\cos\frac{\Omega}{2} \\ -b\sin\frac{\Omega}{2} & b\cos\frac{\Omega}{2} & 0 \end{pmatrix}\begin{pmatrix}\boldsymbol{V}\\ \boldsymbol{W}\\ \boldsymbol{N}\end{pmatrix},
\end{aligned} \tag{1.156}$$

where Ω_1 and Ω_2 are defined by

$$\Omega_1 = \frac{1}{2}\Omega_u - \frac{1}{2}\frac{a}{b}\frac{\rho_v}{\rho}\sin\Omega, \quad \Omega_2 = \frac{1}{2}\Omega_v - \frac{1}{2}\frac{b}{a}\frac{\rho_u}{\rho}\sin\Omega. \tag{1.157}$$

These are compatible if and only if a, b, ρ and $\omega = \Omega - \pi$ satisfy the Gauss-Mainardi-Codazzi equations (1.125). The system (1.156) is readily derived by use of the Weingarten equations in the form

$$\begin{aligned} \boldsymbol{r}_u &= -\rho a \cos\frac{\Omega}{2}\,\boldsymbol{V} + \rho a \sin\frac{\Omega}{2}\,\boldsymbol{W}, \\ \boldsymbol{r}_v &= \rho b \cos\frac{\Omega}{2}\,\boldsymbol{V} + \rho b \sin\frac{\Omega}{2}\,\boldsymbol{W}. \end{aligned} \tag{1.158}$$

Accordingly, the position vector $\boldsymbol{r}$ to Σ is obtained by integration of (1.158) once $\boldsymbol{V}$ and $\boldsymbol{W}$ are known.

A Bäcklund transformation for hyperbolic surfaces is readily derived. Thus, since both surfaces are assumed to be parametrised in terms of the asymptotic coordinates u, v, the generic position vectors $\boldsymbol{r}$ and $\boldsymbol{r}'$ satisfy the Lelieuvre formulae (1.140), namely

$$\begin{aligned} \boldsymbol{r}_u &= \boldsymbol{\nu}_u \times \boldsymbol{\nu}, \quad & \boldsymbol{r}_v &= -\boldsymbol{\nu}_v \times \boldsymbol{\nu}, \\ \boldsymbol{r}'_u &= \boldsymbol{\nu}'_u \times \boldsymbol{\nu}', \quad & \boldsymbol{r}'_v &= -\boldsymbol{\nu}'_v \times \boldsymbol{\nu}', \end{aligned} \tag{1.159}$$

where

$$\boldsymbol{\nu} = \sqrt{\rho}\boldsymbol{N}, \quad \boldsymbol{\nu}' = \sqrt{\rho'}\boldsymbol{N}'. \tag{1.160}$$

In these variables, the governing equations for $\boldsymbol{\nu}$ and $\boldsymbol{\nu}'$ (1.138) take the form

$$\boldsymbol{\nu}_{uv} = \Lambda\boldsymbol{\nu}, \quad \boldsymbol{\nu}'_{uv} = \Lambda'\boldsymbol{\nu}' \tag{1.161}$$

and the Gaussian curvatures read

$$\mathcal{K} = -\frac{1}{|\boldsymbol{\nu}|^4}, \quad \mathcal{K}' = -\frac{1}{|\boldsymbol{\nu}'|^4}. \tag{1.162}$$

The requirement that $\boldsymbol{r}' - \boldsymbol{r}$ be tangential to both Σ and Σ' implies that

$$\boldsymbol{r}' - \boldsymbol{r} = m\boldsymbol{\nu}' \times \boldsymbol{\nu}. \tag{1.163}$$

Insertion of $\boldsymbol{r}'$ as given by (1.163) into $(1.159)_3$ now yields

$$\boldsymbol{\nu}'_u \times (\boldsymbol{\nu}' - m\boldsymbol{\nu}) + \boldsymbol{\nu}_u \times (m\boldsymbol{\nu}' - \boldsymbol{\nu}) - m_u\boldsymbol{\nu}' \times \boldsymbol{\nu} = \boldsymbol{0} \tag{1.164}$$

so that the component in the direction $\boldsymbol{\nu}' - m\boldsymbol{\nu}$ delivers

$$(m^2 - 1)(\boldsymbol{\nu} \times \boldsymbol{\nu}_u) \cdot \boldsymbol{\nu}' = 0. \tag{1.165}$$

Similarly, the remaining relation $(1.159)_4$ gives rise to

$$(m^2 - 1)(\boldsymbol{\nu} \times \boldsymbol{\nu}_v) \cdot \boldsymbol{\nu}' = 0. \tag{1.166}$$

Accordingly, if we assume that the vectors $\boldsymbol{\nu}_u, \boldsymbol{\nu}_v$ and $\boldsymbol{\nu}$ are linearly independent and $\Sigma' \neq \Sigma$, then $m^2 = 1$. Without loss of generality, we may therefore set

$$m = 1 \tag{1.167}$$

which, in turn, reduces (1.164) and its counterpart to

$$(\boldsymbol{\nu}'_u + \boldsymbol{\nu}_u) \times (\boldsymbol{\nu}' - \boldsymbol{\nu}) = \mathbf{0}, \quad (\boldsymbol{\nu}'_v - \boldsymbol{\nu}_v) \times (\boldsymbol{\nu}' + \boldsymbol{\nu}) = \mathbf{0}. \tag{1.168}$$

The preceding necessary conditions may be brought into the form

$$\boldsymbol{\nu}'_u + \boldsymbol{\nu}_u = k(\boldsymbol{\nu}' - \boldsymbol{\nu}), \quad \boldsymbol{\nu}'_v - \boldsymbol{\nu}_v = l(\boldsymbol{\nu}' + \boldsymbol{\nu}), \tag{1.169}$$

where k, l are as yet unspecified functions. Differentiation of $(1.169)_1$ with respect to v and $(1.169)_2$ with respect to u and evaluation modulo (1.161) produces the constraints

$$\begin{aligned}(\Lambda' - kl - k_v)\boldsymbol{\nu}' + (\Lambda - kl + k_v)\boldsymbol{\nu} &= \mathbf{0},\\ (\Lambda' - kl - l_u)\boldsymbol{\nu}' - (\Lambda - kl + l_u)\boldsymbol{\nu} &= \mathbf{0}.\end{aligned} \tag{1.170}$$

Thus, if it is assumed that $\boldsymbol{\nu}$ and $\boldsymbol{\nu}'$ are non-parallel, then

$$\Lambda = -k_v + kl, \quad \Lambda' = k_v + kl, \quad l_u = k_v. \tag{1.171}$$

The latter relation may be satisfied identically by introducing a potential ψ according to

$$k = -(\ln \psi)_u, \quad l = -(\ln \psi)_v \tag{1.172}$$

so that the two remaining relations yield

$$\psi_{uv} = \Lambda\psi, \quad \Lambda' = \Lambda - 2(\ln \psi)_{uv}, \tag{1.173}$$

and the system (1.169) adopts the form

$$(\psi\boldsymbol{\nu}')_u = -\psi\boldsymbol{\nu}_u + \psi_u\boldsymbol{\nu}, \quad (\psi\boldsymbol{\nu}')_v = \psi\boldsymbol{\nu}_v - \psi_v\boldsymbol{\nu}. \tag{1.174}$$

The relations (1.173), (1.174) determine the classical *Moutard transformation* to be discussed in connection with Tzitzeica surfaces in Chapter 3. In the present context, if $\boldsymbol{r}$ is the position vector of a hyperbolic surface Σ given in terms of asymptotic coordinates and $\boldsymbol{\nu}$ is the corresponding scaled normal satisfying the *Moutard equation* $(1.161)_1$, then the system (1.174) is compatible if ψ is a solution of the Moutard equation $(1.173)_1$. The position vector $\boldsymbol{r}'$ as given by (1.163) with $m = 1$ then represents a second hyperbolic surface Σ' which is again parametrised in terms of asymptotic coordinates u, v. Moreover, $\boldsymbol{r}' - \boldsymbol{r}$ is tangential to both Σ and Σ'.

1.6.3 The Bianchi System

The Gauss-Mainardi-Codazzi equations (1.125) constitute an *underdetermined* system for the functions ω, a, b and ρ. Thus, it is natural to supplement this system by constraints which are invariant under the Bäcklund transformation derived in the previous subsection. Here, we require the Gaussian curvature $\mathcal{K}$ to be invariant under the Bäcklund transformation, that is

$$\mathcal{K}' = \mathcal{K} = -\frac{1}{\rho^2}. \tag{1.175}$$

Upon insertion of the relations (1.160) with $\rho' = \rho$ and the parametrisation

$$\boldsymbol{N}' = \cos\sigma \boldsymbol{N} + \sin\sigma(\cos\theta \boldsymbol{V} + \sin\theta \boldsymbol{W}), \tag{1.176}$$

where σ denotes the angle between the normals $\boldsymbol{N}$ and $\boldsymbol{N}'$, into the system (1.168), we obtain

$$\begin{aligned} \theta_u &= \frac{1}{2}\Omega_u + a\tan\frac{\sigma}{2}\cos\left(\theta + \frac{\Omega}{2}\right) - \frac{1}{2}\frac{a}{b}\frac{\rho_v}{\rho}\sin\Omega, \\ \theta_v &= -\frac{1}{2}\Omega_v - b\cot\frac{\sigma}{2}\cos\left(\theta - \frac{\Omega}{2}\right) + \frac{1}{2}\frac{b}{a}\frac{\rho_u}{\rho}\sin\Omega \end{aligned} \tag{1.177}$$

and the relations

$$\sigma_u = -\frac{\rho_u}{\rho}\tan\frac{\sigma}{2}, \quad \sigma_v = \frac{\rho_v}{\rho}\cot\frac{\sigma}{2} \tag{1.178}$$

which are compatible if and only if

$$\rho_{uv} = 0. \tag{1.179}$$

In this way, we retrieve Bianchi's classical Bäcklund transformation for hyperbolic surfaces with Gaussian curvature

$$\mathcal{K} = -\frac{1}{\rho^2}, \qquad \rho = U(u) + V(v). \tag{1.180}$$

These are known as Bianchi surfaces.

To make the Bäcklund transformation for Bianchi surfaces somewhat more explicit, it is noted that the system (1.178) may be integrated to give

$$\mu = \tan\frac{\sigma}{2} = \pm\sqrt{\frac{V(v)-k}{U(u)+k}}, \tag{1.181}$$

where k is a constant of integration. On the other hand, it is well-known that Frobenius systems of the form (1.177), that is

$$\theta_{u^i} = f_i + \sin\theta g_i + \cos\theta h_i \tag{1.182}$$

with $(u^1, u^2) = (u, v)$, are equivalent to Riccati equations and hence may be linearised. Indeed, on setting

$$\theta = 2\arctan\frac{\phi^1}{\phi^2}, \tag{1.183}$$

the general solution of (1.177) may be expressed in terms of solutions of the linear system

$$\begin{aligned}\phi_u &= \left[-\mu a \sin\frac{\Omega}{2}X_1 + \mu a\cos\frac{\Omega}{2}X_2 + \frac{1}{2}\left(\Omega_u - \frac{a}{b}\frac{\rho_v}{\rho}\sin\Omega\right)X_3\right]\phi,\\ \phi_v &= \left[-\mu^{-1} b\sin\frac{\Omega}{2}X_1 - \mu^{-1} b\cos\frac{\Omega}{2}X_2 + \frac{1}{2}\left(-\Omega_v + \frac{b}{a}\frac{\rho_v}{\rho}\sin\Omega\right)X_3\right]\phi,\end{aligned} \tag{1.184}$$

where

$$\phi = \begin{pmatrix}\phi^1\\ \phi^2\end{pmatrix} \tag{1.185}$$

and

$$X_1 = \frac{1}{2}\begin{pmatrix}1 & 0\\ 0 & -1\end{pmatrix}, \quad X_2 = \frac{1}{2}\begin{pmatrix}0 & 1\\ 1 & 0\end{pmatrix}, \quad X_3 = \frac{1}{2}\begin{pmatrix}0 & 1\\ -1 & 0\end{pmatrix}. \tag{1.186}$$

In the terminology of soliton theory, the linear system (1.184) represents a 'non-isospectral' Lax pair for the Bianchi system (1.125), (1.128) since we may regard μ as a non-constant 'spectral parameter'. In fact, as in the case of pseudospherical surfaces, it constitutes precisely the $su(2)$ version (cf. Section 2.2) of the Gauss-Weingarten equations (1.156) if $k \to \infty$. A gauge-equivalent form of this Lax pair has been used by Levi and Sym [234] to obtain a matrix Darboux transformation for the Bianchi system.

The Bäcklund transformation (1.163) or, equivalently,

$$\boldsymbol{r}' = \boldsymbol{r} - \left(\frac{2\mu}{1+\mu^2}\right)\frac{1}{\sin\Omega}\left[\cos\left(\theta - \frac{\Omega}{2}\right)\frac{\boldsymbol{r}_u}{a} + \cos\left(\theta + \frac{\Omega}{2}\right)\frac{\boldsymbol{r}_v}{b}\right] \tag{1.187}$$

is now applied to the simplest (degenerate) Bianchi surfaces. The first application is due to Bianchi who considered the hyperbolic paraboloid

$$z = \frac{1}{2}(x^2 - y^2) \tag{1.188}$$

as seed surface. In terms of asymptotic coordinates, the corresponding position vector reads

$$x = \frac{1}{\sqrt{2}}(u+v), \quad y = \frac{1}{\sqrt{2}}(u-v), \quad z = uv, \tag{1.189}$$

and the Gaussian curvature is indeed of the required form, namely

$$\mathcal{K} = -\frac{1}{(u^2+v^2+1)^2}. \tag{1.190}$$

It is then readily shown that a particular solution of (1.177) leads to the second surface Σ' given by

$$\begin{gathered} x' = \frac{1}{\sqrt{2}}\left(3u - v\frac{1-3u^2}{1+u^2}\right), \quad y' = \frac{1}{\sqrt{2}}\left(3u + v\frac{1-3u^2}{1+u^2}\right), \\ z' = \frac{uv}{\sqrt{2}}\left(\frac{3-u^2}{1+u^2}\right). \end{gathered} \tag{1.191}$$

A particular clipping of this surface is depicted in Figure 1.8.

The second application of Bianchi's transformation is associated with a degenerate seed surface for which the coordinate lines are parallel, that is $\omega = 0$. We make the natural choice

$$\Omega = \pi, \quad a = \alpha(\rho)\rho_u, \quad b = \beta(\rho)\rho_v \tag{1.192}$$

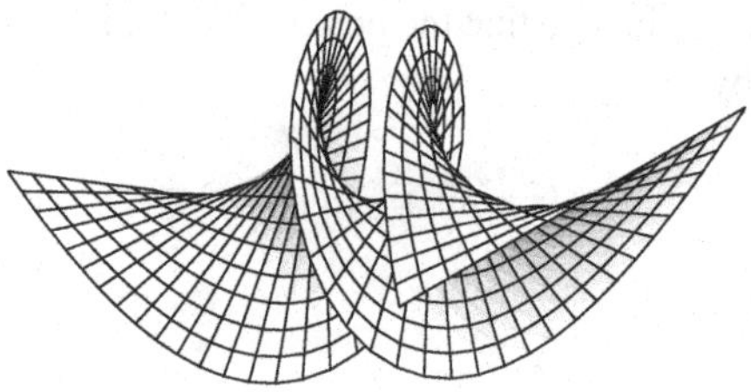

Figure 1.8. A Bianchi surface generated from a hyperbolic paraboloid.

which reduces the Gauss-Mainardi-Codazzi equations to

$$\alpha_\rho + \frac{1}{2}\frac{\alpha}{\rho} - \frac{1}{2}\frac{\beta}{\rho} = 0, \quad \beta_\rho + \frac{1}{2}\frac{\beta}{\rho} - \frac{1}{2}\frac{\alpha}{\rho} = 0, \tag{1.193}$$

the general solution of which is

$$\alpha = c_1 + \frac{c_2}{\rho}, \quad \beta = c_1 - \frac{c_2}{\rho}. \tag{1.194}$$

Comparison of the Frobenius systems (1.177), namely

$$\begin{aligned} \frac{\theta_u}{\sin\theta} &= -\left(c_1 + \frac{c_2}{\rho}\right)\rho_u \tan\frac{\sigma}{2} \\ \frac{\theta_v}{\sin\theta} &= -\left(c_1 - \frac{c_2}{\rho}\right)\rho_v \cot\frac{\sigma}{2} \end{aligned} \tag{1.195}$$

with (1.178) then shows that

$$\tan\frac{\theta}{2} = \exp\gamma, \qquad \gamma = c_2\sigma - c_1\rho\sin\sigma + c_3, \tag{1.196}$$

whence

$$\sin\theta = \frac{1}{\cosh\gamma}, \qquad \cos\theta = -\tanh\gamma. \tag{1.197}$$

On the other hand, the Gauss-Weingarten equations (1.156) are readily integrated to give

$$\boldsymbol{V} = \begin{pmatrix} \cos\delta \\ \sin\delta \\ 0 \end{pmatrix}, \quad \boldsymbol{W} = \begin{pmatrix} 0 \\ 0 \\ 1 \end{pmatrix}, \quad \boldsymbol{N} = \begin{pmatrix} \sin\delta \\ -\cos\delta \\ 0 \end{pmatrix} \tag{1.198}$$

with $\delta = c_1\rho\cos\sigma + c_2\ln\rho$ so that the position vector $\boldsymbol{r}$ of the degenerate seed surface Σ is given by

$$\boldsymbol{r} = \left(\tfrac{1}{2}c_1\rho^2 + c_2\rho\cos\sigma\right)\boldsymbol{W}. \tag{1.199}$$

If we write

$$\boldsymbol{r}' = \boldsymbol{r} + \rho\sin\sigma(\sin\theta\,\boldsymbol{V} - \cos\theta\,\boldsymbol{W}) \tag{1.200}$$

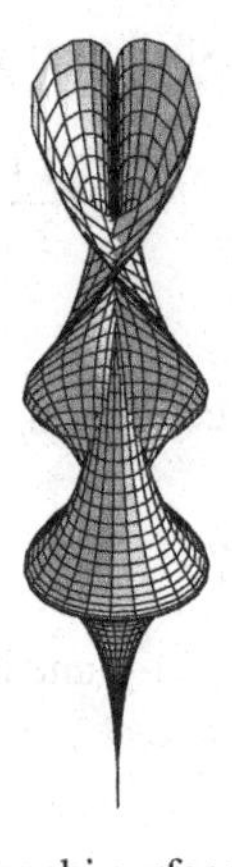

Figure 1.9. A Bianchi surface for $c_2 = c_3 = 0$.

for the generic position vector of the second surface Σ', then

$$\boldsymbol{r}' = \begin{pmatrix} \rho \dfrac{\sin\sigma}{\cosh\gamma} \cos\delta \\ \rho \dfrac{\sin\sigma}{\cosh\gamma} \sin\delta \\ \frac{1}{2}c_1\rho^2 + c_2\rho\cos\sigma + \rho\sin\sigma\tanh\gamma \end{pmatrix}. \tag{1.201}$$

It is seen that, for $c_2 = 0$, the position vector $\boldsymbol{r}'$ is periodic in σ. A typical surface of this type is shown in Figure 1.9. Here, the coordinate lines are $\rho = \text{const}$ and $\sigma = \text{const}$.

One-soliton Bianchi surfaces have been constructed and displayed graphically in [290].

Exercises

1. (a) Derive the 1st fundamental form

$$dN^2 = \mathcal{E}du^2 + 2\mathcal{F}dudv + \mathcal{G}dv^2$$

of the spherical representation of a surface, where

$$\begin{aligned} \mathcal{E} &= H^{-2}(e^2G - 2efF + f^2E), \\ \mathcal{F} &= H^{-2}(efG - (eg + f^2)F + fgE), \\ \mathcal{G} &= H^{-2}(f^2G - 2fgF + g^2E). \end{aligned}$$

(b) Show that the normal $\boldsymbol{N}$ satisfies the hyperbolic equation

$$\boldsymbol{N}_{uv} + \frac{1}{2}\frac{\rho_v}{\rho}\boldsymbol{N}_u + \frac{1}{2}\frac{\rho_u}{\rho}\boldsymbol{N}_v + \mathcal{F}\boldsymbol{N} = 0$$

in asymptotic coordinates.

(c) Show that the line segment connecting the north pole of the unit sphere

$$x^2 + y^2 + (z-1)^2 = 1$$

and a generic point on the (x, y)-plane labelled by $\varepsilon = x_0 + i y_0$ intersects the sphere at

$$\begin{pmatrix} x \\ y \\ z-1 \end{pmatrix} = \frac{1}{|\varepsilon|^2+1}\begin{pmatrix} \varepsilon + \bar{\varepsilon} \\ -i(\varepsilon - \bar{\varepsilon}) \\ |\varepsilon|^2 - 1 \end{pmatrix}.$$

Thus, the parametrisation $(1.142)_2$ of the normal $\boldsymbol{N}$ represents the *stereographic projection* of the unit sphere onto the complex plane.

2. (a) Show that in asymptotic coordinates the vectors

$$\frac{\boldsymbol{r}_u}{a} \pm \frac{\boldsymbol{r}_v}{b}, \quad \frac{\boldsymbol{N}_u}{a} \pm \frac{\boldsymbol{N}_v}{b}$$

are tangential to the lines of curvature.

(b) Derive the linear system (1.156) for the orthonormal triad $\{\boldsymbol{V}, \boldsymbol{W}, \boldsymbol{N}\}$.

3. (a) Show that the nonlinear equation

$$\dot{\theta} = f(t) + g(t)\sin\theta + h(t)\cos\theta$$

transforms into the *Riccati* equation

$$\dot{y} = \tfrac{1}{2}f(t)(1+y^2) + g(t)y + \tfrac{1}{2}h(t)(1-y^2)$$

on use of the transformation

$$\theta = 2\arctan y.$$

(b) Verify that the general solution of the above Riccati equation is given by

$$y = \frac{\phi^1}{\phi^2},$$

where $\phi = (\phi^1 \ \phi^2)^{\mathsf{T}}$ is the general solution of the *linear* system

$$\dot{\phi} = [g(t)X_1 + h(t)X_2 + f(t)X_3]\phi$$

and the matrices X_i are defined in (1.186).

4. (a) Show that for the hyperbolic paraboloid (1.189), the solution of (1.177) is given by

$$\tan\left(\frac{\theta}{2} + \frac{\omega}{4}\right) = \frac{\sqrt{u^2 + v^2 + 1}}{u(c\sqrt{v^2 + 1} - v)},$$

with the choice

$$\mu = \frac{\sqrt{v^2 + 1}}{u}.$$

(b) Derive the components (1.191) of the surface vector $\boldsymbol{r}'$ for $c = 0$.

2

The Motion of Curves and Surfaces. Soliton Connections

The geometric link between soliton theory and the motion of inextensible curves may be said to have its origin in an analysis by Da Rios [97] in 1906 on the spatial evolution of an isolated vortex filament in an unbounded, inviscid liquid. Therein, Da Rios, who studied under Levi-Civita at the University of Padua, invoked what is now known as the localized induction approximation to derive a pair of coupled nonlinear equations governing the time evolution of the curvature and torsion of the vortex filament. The importance of the Da Rios results was realised by Levi-Civita and were subsequently collected and extended by him in a survey published in 1932 [235]. However, it was not until 1965 that the Da Rios equations were rediscovered by Betchov [30]. In 1972, Hasimoto [162], motivated by the earlier geometric study of Betchov and preceded by experimental work on a distorted vortex ring by Kambe and Takao [189], showed that the Da Rios equations may be combined to produce the celebrated nonlinear Schrödinger equation of soliton theory.

Lamb [224], later in 1977, linked the spatial motion of certain curves with the sine-Gordon, modified Korteweg-de Vries and nonlinear Schrödinger equations. Lakshmanan et al. [222], in turn, derived the Heisenberg spin chain equation via the spatial motion of a curve. In recent times, Doliwa and Santini [112] established a connection between the motion of inextensible curves and solitonic systems via an embedding in a space of constant curvature. Literature on the integrable motion of curves is catalogued in that work.

The study of triply orthogonal systems of surfaces has a long history going back to Lamé [226]. The nonlinear equations descriptive of triply orthogonal systems containing a family of pseudospherical surfaces were discussed by Weingarten, Bianchi, and Darboux (see Eisenhart [118]). This Weingarten system may be generated by consideration of a normal motion of a pseudospherical surface parametrised in terms of curvature coordinates. Its auto-Bäcklund transformation is readily induced from the classical Bäcklund transformation for the sine-Gordon equation. In fact, it has recently been shown that the classical Lamé system descriptive of triply orthogonal systems of surfaces in general is amenable to the inverse scattering transform [391]. Thus, the Lamé system, originally set down in 1840, seems to be the oldest solitonic system to be found in classical differential geometry. Indeed, the existence of Bäcklund transformations for the lamé system indicative of its integrable nature were certainly known to Darboux.

The connection between the motion of curves and surfaces and modern soliton theory is the concern of the present chapter.

2.1 Motions of Curves of Constant Torsion or Curvature. The Sine-Gordon Connection

In Hasimoto's pioneering paper and in the later study by Lamb [224], the motion of soliton curves was developed in terms of a complex quantity incorporating both curvature and torsion. Here, it proves convenient to work in terms of the usual orthonormal triad $\{\boldsymbol{t}, \boldsymbol{n}, \boldsymbol{b}\}$. In this section, the sine-Gordon equation is derived via the motion of an inextensible curve of constant torsion which, in turn, sweeps out a pseudospherical surface.

If $\boldsymbol{r} = \boldsymbol{r}(s, t)$ is the position vector of a curve C moving in space, then the unit tangent, principal normal and binormal vectors vary along C according to the well-known Serret-Frenet relations

$$\begin{aligned} \boldsymbol{t}_s &= \kappa \boldsymbol{n}, \\ \boldsymbol{n}_s &= \tau \boldsymbol{b} - \kappa \boldsymbol{t}, \\ \boldsymbol{b}_s &= -\tau \boldsymbol{n}, \end{aligned} \tag{2.1}$$

where s measures arc length along C, κ is its curvature and τ its torsion. In the present moving curve context, the time t enters into the system (2.1) as a parameter.

The general temporal evolution in which the $\{\boldsymbol{t}, \boldsymbol{n}, \boldsymbol{b}\}$ triad remains orthonormal adopts the form

$$\begin{aligned} \boldsymbol{t}_t &= \alpha \boldsymbol{n} + \beta \boldsymbol{b}, \\ \boldsymbol{n}_t &= -\alpha \boldsymbol{t} + \gamma \boldsymbol{b}, \\ \boldsymbol{b}_t &= -\beta \boldsymbol{t} - \gamma \boldsymbol{n}. \end{aligned} \tag{2.2}$$

Here, it is required that the arc length and time derivatives commute. This implies inextensibility of C. Accordingly, the compatibility conditions $\boldsymbol{t}_{st} = \boldsymbol{t}_{ts}$, etc., applied to the systems (2.1), (2.2) yield

$$\begin{aligned} \alpha_s &= \kappa_t + \beta\tau, \\ \beta_s &= \kappa\gamma - \tau\alpha, \\ \gamma_s &= \tau_t - \kappa\beta. \end{aligned} \tag{2.3}$$

2.1.1 A Motion of an Inextensible Curve of Constant Torsion

A consequence of the system (2.3) is that

$$(\alpha^2 + \beta^2 + \gamma^2)_s = 2(\alpha\kappa_t + \gamma\tau_t). \tag{2.4}$$

In particular, if $\alpha = 0$ and $\tau_t = 0$, then $\beta^2 + \gamma^2 = \delta^2(t)$ and we may set

$$\beta = \delta(t)\sin\sigma, \quad \gamma = \delta(t)\cos\sigma,$$

so that the compatibility conditions (2.3) reduce to

$$\sigma_{st} = -\delta(t)\tau(s)\sin\sigma, \tag{2.5}$$

where

$$\kappa = \sigma_s. \tag{2.6}$$

If we put $\tau = 1/\rho = \tau_0$ and $\delta = -1/\rho$, then (2.5) becomes the sine-Gordon equation (1.25) under the correspondence $\{\sigma, s, t\} \leftrightarrow \{\omega, u, v\}$. Moreover, the relations (2.1), (2.2) now adopt the form

$$\begin{pmatrix} \boldsymbol{t} \\ \boldsymbol{n} \\ \boldsymbol{b} \end{pmatrix}_s = \begin{pmatrix} 0 & \sigma_s & 0 \\ -\sigma_s & 0 & 1/\rho \\ 0 & -1/\rho & 0 \end{pmatrix} \begin{pmatrix} \boldsymbol{t} \\ \boldsymbol{n} \\ \boldsymbol{b} \end{pmatrix}, \tag{2.7}$$

$$\begin{pmatrix} \boldsymbol{t} \\ \boldsymbol{n} \\ \boldsymbol{b} \end{pmatrix}_t = \begin{pmatrix} 0 & 0 & (-1/\rho)\sin\sigma \\ 0 & 0 & (-1/\rho)\cos\sigma \\ (1/\rho)\sin\sigma & (1/\rho)\cos\sigma & 0 \end{pmatrix} \begin{pmatrix} \boldsymbol{t} \\ \boldsymbol{n} \\ \boldsymbol{b} \end{pmatrix}. \tag{2.8}$$

The compatibility condition for this system produces the sine-Gordon equation.

In geometric terms, the motion of a curve of constant torsion and with curvature $\kappa = \omega_u$ has now been linked with the sine-Gordon equation (1.25). It is natural to ask if there is a connection between this moving curve derivation of the sine-Gordon equation and the classical derivation associated with pseudospherical surfaces. In fact, a comparison of the two linear representations (1.29) and (2.7), (2.8) makes it apparent that the two systems are indeed equivalent under the correspondences[1]

$$\{\omega, u, v\} \leftrightarrow \{-\sigma, s, t\}, \quad \{\boldsymbol{A}, \boldsymbol{B}, \boldsymbol{C}\} \leftrightarrow \{\boldsymbol{t}, \boldsymbol{n}, \boldsymbol{b}\}. \tag{2.9}$$

Thus, the constant-torsion curve associated with the sine-Gordon equation will trace out a pseudospherical surface as it moves and, at each instant, will be an asymptotic line on the surface. This is consistent with the classical Beltrami-Enneper theorem which states that asymptotic lines on a hyperbolic surface of Gaussian curvature $-1/\rho^2$ have torsion of magnitude $1/\rho$ [352].

The correspondence (2.9) allows us to obtain an expression for the velocity $\boldsymbol{r}_t$ of the moving curve of constant torsion in Lamb's formulation. Thus, $(1.28)_2$ shows that

$$\boldsymbol{r}_v = \cos\omega \boldsymbol{A} + \sin\omega \boldsymbol{B},$$

whence

$$\boldsymbol{v} = \boldsymbol{r}_t = \cos\sigma \boldsymbol{t} - \sin\sigma \boldsymbol{n}. \tag{2.10}$$

The moving constant torsion curve corresponding to a particular solution of the sine-Gordon equation may be obtained via the pseudospherical surface associated with that solution. Thus, we take the position vector of the pseudospherical surface in asymptotic coordinates and animate one of the parametric lines with respect to the other parameter. If this is done for the two-soliton solution, then we obtain a curve with two localised loops which, as time elapses, move along the curve, interact and ultimately pass through each other unchanged but undergoing a phase shift in position due to the nonlinear interaction.

2.1.2 A Motion of an Inextensible Curve of Constant Curvature

An alternative specialisation in (2.4) is suggested, namely that with $\gamma = 0$ and $\kappa_t = 0$. The system (2.3) then admits the solution

$$\begin{aligned} \alpha &= -\frac{\cos\sigma}{\rho}, \quad \beta = \frac{\sin\sigma}{\rho}, \\ \kappa &= \frac{1}{\rho}, \quad \tau = \sigma_s, \end{aligned} \tag{2.11}$$

[1] It should be noted that the sine-Gordon equation (1.25) is invariant under $\omega \leftrightarrow -\omega$.

where σ is a solution of the sine-Gordon equation

$$\sigma_{st} = \frac{1}{\rho^2} \sin \sigma. \tag{2.12}$$

The underlying linear representation with compatibility condition (2.12) is, in this instance

$$\begin{pmatrix} \boldsymbol{t} \\ \boldsymbol{n} \\ \boldsymbol{b} \end{pmatrix}_s = \begin{pmatrix} 0 & 1/\rho & 0 \\ -1/\rho & 0 & \sigma_s \\ 0 & -\sigma_s & 0 \end{pmatrix} \begin{pmatrix} \boldsymbol{t} \\ \boldsymbol{n} \\ \boldsymbol{b} \end{pmatrix}, \tag{2.13}$$

$$\begin{pmatrix} \boldsymbol{t} \\ \boldsymbol{n} \\ \boldsymbol{b} \end{pmatrix}_t = \begin{pmatrix} 0 & (-1/\rho)\cos\sigma & (1/\rho)\sin\sigma \\ (1/\rho)\cos\sigma & 0 & 0 \\ (-1/\rho)\sin\sigma & 0 & 0 \end{pmatrix} \begin{pmatrix} \boldsymbol{t} \\ \boldsymbol{n} \\ \boldsymbol{b} \end{pmatrix}. \tag{2.14}$$

This system may be set in correspondence with the linear representation (1.29) via

$$\{\omega, u, v\} \leftrightarrow \{\sigma, s, t\}, \quad \{\boldsymbol{A}, \boldsymbol{B}, \boldsymbol{C}\} \leftrightarrow \{\boldsymbol{b}, \boldsymbol{n}, -\boldsymbol{t}\}. \tag{2.15}$$

However, here, $\boldsymbol{t} \cdot \boldsymbol{N} = \boldsymbol{t} \cdot \boldsymbol{C} = -1 \neq 0$ so that the moving curve of constant curvature does not lie on and hence does not sweep out the associated pseudospherical surface.

2.2 A 2 × 2 Linear Representation for the Sine-Gordon Equation

It has been seen that the sine-Gordon equation may be derived as the compatibility condition for the 3×3 linear system (1.29) generated by the Gauss-Weingarten equations associated with pseudospherical surfaces or, alternatively, for the 3×3 linear system (2.7), (2.8) descriptive of the evolution of a $\{\boldsymbol{t}, \boldsymbol{n}, \boldsymbol{b}\}$ triad for a curve of constant torsion. These two representations have been shown to be equivalent.

The sine-Gordon equation may also be generated as the compatibility condition for a linear 2×2 system. It is interesting to record that such a linear representation is implicit in the work of Loewner [238] published in 1952 on the application of infinitesimal Bäcklund transformations to the hodograph equations of gasdynamics. Indeed, a modern reinterpretation of Loewner's work has revealed deep connections with soliton theory [210, 211].

In 1973, Ablowitz et al. [1] introduced the canonical AKNS 2×2 linear system which, in particular, includes a representation for the classical sine-Gordon equation. This 2×2 linear representation is connected here to the 3×3

linear representation (1.29) in terms of the orthonormal triad $\{\boldsymbol{A}, \boldsymbol{B}, \boldsymbol{C}\}$ and associated with the Gauss-Weingarten equations descriptive of pseudospherical surfaces parametrised in terms of asymptotic coordinates.

We start with the observation that the system (1.29) is equivalent to the system

$$\begin{aligned} \Psi_u &= S\Psi, \\ \Psi_v &= T\Psi, \end{aligned} \tag{2.16}$$

where

$$\begin{aligned} S &= \begin{pmatrix} 0 & -\omega_u & 0 \\ \omega_u & 0 & 1/\rho \\ 0 & -1/\rho & 0 \end{pmatrix}, \\ T &= \begin{pmatrix} 0 & 0 & (1/\rho)\sin\omega \\ 0 & 0 & (-1/\rho)\cos\omega \\ (-1/\rho)\sin\omega & (1/\rho)\cos\omega & 0 \end{pmatrix} \end{aligned} \tag{2.17}$$

and Ψ is the matrix given by

$$\Psi = \begin{pmatrix} A_1 & A_2 & A_3 \\ B_1 & B_2 & B_3 \\ C_1 & C_2 & C_3 \end{pmatrix} \tag{2.18}$$

with A_i, B_i and C_i the components, in turn, of the unit vectors $\boldsymbol{A}, \boldsymbol{B}, \boldsymbol{C}$. The fact that the triad $(\boldsymbol{A}, \boldsymbol{B}, \boldsymbol{C})$ is orthonormal and right-handed implies that the matrix Ψ is both special (i.e., $\det \Psi = 1$) and orthogonal (i.e., $\Psi^{\mathrm{T}}\Psi = \mathbb{1}$.) The set of such matrices forms a (Lie) group under multiplication known as $SO(3)$ (special orthogonal matrices). It constitutes a representation of the group of rotations of three-dimensional Euclidean space [343].

The compatibility condition for the linear system (2.16) is

$$S_v - T_u + [S, T] = 0 \tag{2.19}$$

where $[S, T] = ST - TS$ denotes the commutator of S and T. With S and T given by (2.17), the condition (2.19) is equivalent to the sine-Gordon equation (1.25).

To a Lie group corresponds a Lie algebra consisting of a vector space V equipped with a Lie bracket $[\cdot,\cdot] : V \times V \to V$ that is bilinear, anti-commutative and which satisfies the Jacobi Identity, namely

$$[a,[b,c]] + [b,[c,a]] + [c,[a,b]] = 0, \quad \forall\, a,b,c \in V. \tag{2.20}$$

The Lie algebra of $SO(3)$ may be identified with the real vector space with basis $\{L_1, L_2, L_3\}$ where the matrices L_i, $i = 1, 2, 3$ are given by

$$L_1 = \begin{pmatrix} 0 & 0 & 0 \\ 0 & 0 & -1 \\ 0 & 1 & 0 \end{pmatrix}, \quad L_2 = \begin{pmatrix} 0 & 0 & 1 \\ 0 & 0 & 0 \\ -1 & 0 & 0 \end{pmatrix}, \quad L_3 = \begin{pmatrix} 0 & -1 & 0 \\ 1 & 0 & 0 \\ 0 & 0 & 0 \end{pmatrix} \tag{2.21}$$

and the Lie bracket is the commutator operator. The matrices L_i satisfy the commutation relations

$$[L_1, L_2] = L_3, \quad [L_2, L_3] = L_1, \quad [L_3, L_1] = L_2, \tag{2.22}$$

which characterise the Lie algebra of $SO(3)$. This Lie algebra is denoted by $so(3)$. Here, $S, T \in so(3)$ with

$$S = \omega_u L_3 - \frac{1}{\rho} L_1, \tag{2.23}$$

$$T = \frac{1}{\rho} \cos\omega\, L_1 + \frac{1}{\rho} \sin\omega\, L_2. \tag{2.24}$$

The connection between the 3×3 matrix $\{\boldsymbol{A}, \boldsymbol{B}, \boldsymbol{C}\}$ linear representation and the 2×2 AKNS linear representation may now be made if we exploit the fact that $so(3)$ admits a 2×2 representation in terms of the Pauli matrices

$$\sigma_1 = \begin{pmatrix} 0 & 1 \\ 1 & 0 \end{pmatrix}, \quad \sigma_2 = \begin{pmatrix} 0 & -i \\ i & 0 \end{pmatrix}, \quad \sigma_3 = \begin{pmatrix} 1 & 0 \\ 0 & -1 \end{pmatrix}. \tag{2.25}$$

The matrices σ_i, $i = 1, 2, 3$ satisfy the commutation relations

$$[\sigma_1, \sigma_2] = 2i\sigma_3, \quad [\sigma_2, \sigma_3] = 2i\sigma_1, \quad [\sigma_3, \sigma_1] = 2i\sigma_2 \tag{2.26}$$

and comparison of the latter with the commutation relations (2.22) shows that the Lie algebra generated under commutation by the triad $\{e_1, e_2, e_3\}$, where

$$e_k = \frac{\sigma_k}{2i}, \tag{2.27}$$

is isomorphic to that generated by $\{L_1, L_2, L_3\}$ under the correspondence $e_k \leftrightarrow L_k$. This allows us to construct a 2×2 linear representation for the sine-Gordon equation (1.25) from the 3×3 linear representation. Thus, if $S = s_1 L_1 + s_2 L_2 + s_3 L_3$ and $T = t_1 L_1 + t_2 L_2 + t_3 L_3$ are elements of $so(3)$ such that the compatibility condition (2.19) holds, then the matrices

$$P = s_1 e_1 + s_2 e_2 + s_3 e_3, \quad Q = t_1 e_1 + t_2 e_2 + t_3 e_3$$

satisfy

$$P_v - Q_u + [P, Q] = 0. \tag{2.28}$$

Hence, the sine-Gordon equation (1.25) may indeed be generated as the compatibility condition for a 2×2 linear representation, namely

$$\begin{aligned} \Phi_u &= P\Phi, \\ \Phi_v &= Q\Phi, \end{aligned} \tag{2.29}$$

where

$$\begin{aligned} P &= \omega_u e_3 - \frac{1}{\rho} e_1 = \frac{i}{2}\begin{pmatrix} -\omega_u & 1/\rho \\ 1/\rho & \omega_u \end{pmatrix}, \\ Q &= \frac{1}{\rho}\cos\omega\, e_1 + \frac{1}{\rho}\sin\omega\, e_2 = -\frac{i}{2\rho}\begin{pmatrix} 0 & e^{-i\omega} \\ e^{i\omega} & 0 \end{pmatrix}. \end{aligned} \tag{2.30}$$

Introduction of the *gauge* transformation $\tilde{\Phi} = G\Phi$, where

$$G = \begin{pmatrix} 1 & 1 \\ i & -i \end{pmatrix},$$

now takes the linear representation determined by (2.29), (2.30) to the gauge equivalent system

$$\tilde{\Phi}_u = GPG^{-1}\tilde{\Phi} = \begin{pmatrix} i/2\rho & -\omega_u/2 \\ \omega_u/2 & -i/2\rho \end{pmatrix}\tilde{\Phi},$$

$$\tilde{\Phi}_v = GQG^{-1}\tilde{\Phi} = -\frac{i}{2\rho}\begin{pmatrix} \cos\omega & \sin\omega \\ \sin\omega & -\cos\omega \end{pmatrix}\tilde{\Phi}.$$

Use of the invariance

$$\tilde{\omega} = -\omega, \quad \tilde{u} = \frac{u}{\lambda}, \quad \tilde{v} = \lambda v \tag{2.31}$$

of (1.25) now injects the real 'spectral' parameter λ into the above to produce the standard AKNS 2×2 linear representation for the sine-Gordon equation (1.25), viz

$$\tilde{\Phi}_{\tilde{u}} = \frac{1}{2}\left[\begin{pmatrix} 0 & \tilde{\omega}_{\tilde{u}} \\ -\tilde{\omega}_{\tilde{u}} & 0 \end{pmatrix} + i\frac{\lambda}{\rho}\begin{pmatrix} 1 & 0 \\ 0 & -1 \end{pmatrix}\right]\tilde{\Phi},$$
$$\tilde{\Phi}_{\tilde{v}} = \frac{i}{2\lambda\rho}\begin{pmatrix} -\cos\tilde{\omega} & \sin\tilde{\omega} \\ \sin\tilde{\omega} & \cos\tilde{\omega} \end{pmatrix}\tilde{\Phi}. \tag{2.32}$$

2.3 The Motion of Pseudospherical Surfaces. A Weingarten System and Its Bäcklund Transformation

The connection between the sine-Gordon equation and the geometry of stationary pseudospherical surfaces or, alternatively, the motion of curves of constant torsion on such surfaces has now been established. In this section, certain motions of pseudospherical surfaces are investigated. One type of motion is seen to produce a continuum version of an anharmonic lattice model which, in a reduction, yields the well-known modified Korteweg-de Vries equation (cf. [257]). Another type of motion leads to a classical system corresponding to a subclass of the Lamé equations descriptive of triply orthogonal surfaces [118]. This consideration of the motion of surfaces leads naturally to the important idea of *compatible* integrable systems.

The motion of a pseudospherical surface $\Sigma : \boldsymbol{r} = \boldsymbol{r}(u, v, t)$ parametrised in asymptotic coordinates u, v is investigated. At each instant t, the total curvature $\mathcal{K} = \mathcal{K}(t) < 0$ is constant and negative on Σ. The orthonormal basis $\{\boldsymbol{A}, \boldsymbol{B}, \boldsymbol{C}\}$ as introduced in (1.28) is used and so the Gauss-Weingarten equations are given by (1.29) wherein now, however, $\rho = \rho(t)$ and $\omega = \omega(u, v, t)$. The general time evolution which maintains the orthonormality of the triad $\{\boldsymbol{A}, \boldsymbol{B}, \boldsymbol{C}\}$ is adjoined, namely

$$\begin{pmatrix} \boldsymbol{A} \\ \boldsymbol{B} \\ \boldsymbol{C} \end{pmatrix}_t = \begin{pmatrix} 0 & a & b \\ -a & 0 & c \\ -b & -c & 0 \end{pmatrix}\begin{pmatrix} \boldsymbol{A} \\ \boldsymbol{B} \\ \boldsymbol{C} \end{pmatrix}, \tag{2.33}$$

where a, b and c are real functions of u, v and t. The linear system (1.29), in which time t enters only parametrically, encapsulates the information that the surface Σ is pseudospherical and parametrised by arc length along asymptotic lines. To construct a time evolution in which these properties are preserved, it is required that (2.33) be compatible with the system (1.29). This imposes the conditions $\boldsymbol{A}_{ut} = \boldsymbol{A}_{tu}, \boldsymbol{A}_{vt} = \boldsymbol{A}_{tv}$, etc. which, in turn, produce the following linear non-homogeneous system for a, b and c:

$$\begin{pmatrix} a \\ b \\ c \end{pmatrix}_u = \begin{pmatrix} 0 & 1/\rho & 0 \\ -1/\rho & 0 & -\omega_u \\ 0 & \omega_u & 0 \end{pmatrix} \begin{pmatrix} a \\ b \\ c \end{pmatrix} + \begin{pmatrix} -\omega_u \\ 0 \\ 1/\rho \end{pmatrix}_t, \tag{2.34}$$

$$\begin{pmatrix} a \\ b \\ c \end{pmatrix}_v = \begin{pmatrix} 0 & -(1/\rho)\cos\omega & -(1/\rho)\sin\omega \\ (1/\rho)\cos\omega & 0 & 0 \\ (1/\rho)\sin\omega & 0 & 0 \end{pmatrix} \begin{pmatrix} a \\ b \\ c \end{pmatrix} + \begin{pmatrix} 0 \\ (1/\rho)\sin\omega \\ -(1/\rho)\cos\omega \end{pmatrix}_t. \tag{2.35}$$

This system is compatible modulo the sine-Gordon equation (1.25).

Thus far, the motion of the surface Σ has been specified via the evolution of the frame field $\{\boldsymbol{A}, \boldsymbol{B}, \boldsymbol{C}\}$. In what follows, it also proves convenient to work in terms of the velocity $\boldsymbol{r}_t$ of Σ. If $\boldsymbol{r}_t$ has the representation

$$\boldsymbol{r}_t = l\boldsymbol{A} + m\boldsymbol{B} + n\boldsymbol{C} \tag{2.36}$$

then

$$\begin{aligned} \boldsymbol{A}_t &= \boldsymbol{r}_{ut} = \boldsymbol{r}_{tu} = (l\boldsymbol{A} + m\boldsymbol{B} + n\boldsymbol{C})_u \\ &= (l_u + m\omega_u)\boldsymbol{A} + (-l\omega_u + m_u - n/\rho)\boldsymbol{B} + (m/\rho + n_u)\boldsymbol{C}, \\ \boldsymbol{B}_t &= [\omega_t - n/\rho - (l_u + m\omega_u)\cot\omega + l_v \operatorname{cosec}\omega]\boldsymbol{A} \\ &\quad + [(-\omega_t + l\omega_u - m_u + 2n/\rho)\cot\omega + m_v \operatorname{cosec}\omega]\boldsymbol{B} \\ &\quad + [l/\rho - (n_u + 2m/\rho)\cot\omega + n_v \operatorname{cosec}\omega]\boldsymbol{C}, \end{aligned}$$

while $\boldsymbol{C}_t = \boldsymbol{A}_t \times \boldsymbol{B} + \boldsymbol{A} \times \boldsymbol{B}_t$. The requirement that $\{\boldsymbol{A}, \boldsymbol{B}, \boldsymbol{C}\}$ remains an orthonormal triad provides the necessary conditions

$$l_u + m\omega_u = 0, \tag{2.37}$$

$$\omega_t - l\omega_u + m_u - 2n/\rho - m_v \sec\omega = 0, \tag{2.38}$$

$$l_v \cos\omega + m_v \sin\omega = 0, \tag{2.39}$$

in which case,

$$a = -l\omega_u + m_u - n/\rho, \tag{2.40}$$

$$b = n_u + m/\rho, \tag{2.41}$$

$$c = -(n_u + 2m/\rho)\cot\omega + l/\rho + n_v \operatorname{cosec}\omega. \tag{2.42}$$

Rearrangement of the constraints (2.37)–(2.39) shows that, in order to obtain a valid motion, it is required to determine l such that

$$l_{uv} = l_v \omega_u \cot \omega + l_u \frac{\sin \omega}{\rho^2 \omega_u}, \tag{2.43}$$

while the quantities m and n are then given by

$$m = -l_u / \omega_u \tag{2.44}$$

and

$$n = \frac{\rho}{2}(\omega_t - l\omega_u + m_u - m_v \sec \omega), \tag{2.45}$$

respectively.

Residual constraints on the motion are obtained by insertion of $\{a, b, c\}$ as given by (2.40)–(2.42) into (2.34), (2.35). These are determined to be

$$n_{uu} = (n_u \cot \omega - n_v \operatorname{cosec} \omega)\omega_u + \frac{n}{\rho^2} + \frac{2}{\rho}(m\omega_u \cot \omega - m_u), \tag{2.46}$$

$$n_{vv} = (n_v \cot \omega - n_u \operatorname{cosec} \omega)\omega_v + \frac{n}{\rho^2} + \frac{2 \sec \omega}{\rho}(m_v - m\omega_v \cot \omega), \tag{2.47}$$

$$n_{uv} = \frac{n}{\rho^2} \cos \omega + \left(\frac{1}{\rho}\right)_t \sin \omega. \tag{2.48}$$

Thus, a viable evolution (2.36) requires solution of the system (2.43)–(2.48).

If a set $\{a, b, c\}$ and thereby the motion of the triad $\{\boldsymbol{A}, \boldsymbol{B}, \boldsymbol{C}\}$ is known, then (2.40)–(2.42) provide a linear inhomogeneous system for $\{l, m, n\}$ to determine $v = \boldsymbol{r}_t$. For instance, one class of solutions has $\dot{\rho} = 0$ and

$$\{a, b, c\} = \{\zeta(t)\rho\, \omega_u,\ \delta(t) \sin \omega,\ -\zeta(t) - \delta(t) \cos \omega\} \tag{2.49}$$

together with the auxiliary linear evolutionary condition

$$\omega_t = \rho[\delta(t)\omega_v - \zeta(t)\omega_u], \tag{2.50}$$

where $\delta(t)$, $\zeta(t)$ are arbitrary functions of t. A motion consistent with the solution set (2.49) is given by

$$\{l, m, n\} = \{\rho(\delta \cos \omega - \zeta),\ \rho\delta \sin \omega,\ 0\}. \tag{2.51}$$

This corresponds to a sliding motion of Σ in which there is no normal propagation.

2.3.1 A Continuum Limit of an Anharmonic Lattice Model

Another possible motion $\{l, m, n\}$ is given by

$$\{l, m, n\} = \left\{\rho\left[\frac{\omega_u^2}{4} + \delta\cos\omega - \zeta\right], \rho\left[\delta\sin\omega - \frac{\omega_{uu}}{2}\right], \omega_u\right\}, \tag{2.52}$$

where

$$\begin{aligned} \omega_t &= \frac{\rho}{2}\omega_{uuu} + \frac{\rho}{4}\omega_u^3 + \left(\frac{3}{2\rho} - \rho\zeta\right)\omega_u + \delta\rho\,\omega_v, \\ \omega_{uv} &= \frac{1}{\rho^2}\sin\omega \end{aligned} \tag{2.53}$$

and $\delta = \delta(t)$, $\zeta = \zeta(t)$ are arbitrary while $\dot{\rho} = 0$ so that the total curvature $\mathcal{K}$ remains constant on Σ throughout the motion. The set $\{a, b, c\}$ is obtained by substitution of (2.52) into (2.40)–(2.42). The corresponding linear triad system consisting of (1.29) augmented by the time evolution (2.33) has compatibility condition the coupled nonlinear system (2.53). Elimination of ω_v therein yields

$$\omega_{ut} = \frac{\rho}{2}\omega_{uuuu} + \frac{3}{4}\rho\,\omega_u^2\omega_{uu} + \left(\frac{3}{2\rho} - \rho\zeta\right)\omega_{uu} + \frac{\delta}{\rho}\sin\omega. \tag{2.54}$$

This solitonic equation with $\delta \neq 0$ and $\zeta = 3/(2\rho^2)$ was originally derived by Konno et al. [203] as the continuum limit of a model of wave propagation in an anharmonic lattice.

The specialisation $\delta = 0$, $\zeta = 3/(2\rho^2)$ in (2.54) produces the *modified Korteweg-de Vries* (mKdV) equation in $\omega' = \omega_u$, namely

$$\omega'_{t'} + \omega'_{u'u'u'} + 6\omega'^2\omega'_{u'} = 0, \tag{2.55}$$

where $u' = u/2$, $t' = -t$ and $\rho = 16$. The mKdV equation has important applications, in particular, in the analysis of nonlinear Alfvén waves in a collisionless plasma [188]. Its connection with acoustic wave propagation in anharmonic lattices has been described by Zabusky [388].

2.3.2 A Weingarten System

In a purely normal motion in which the Gaussian curvature $\mathcal{K} = -1/\rho^2$ of Σ is now allowed to evolve in time, the relations (2.37)–(2.39) yield

$$\{l, m, n\} = \{0, 0, \rho\,\omega_t/2\}. \tag{2.56}$$

In this case, substitution into the normal component system (2.46)–(2.48) produces

$$\begin{aligned} \omega_{uut} &= \omega_u \omega_{ut} \cot \omega - \omega_u \omega_{vt} \operatorname{cosec} \omega + \frac{1}{\rho^2} \omega_t, \\ \omega_{vvt} &= \omega_v \omega_{vt} \cot \omega - \omega_v \omega_{ut} \operatorname{cosec} \omega + \frac{1}{\rho^2} \omega_t, \\ \omega_{uv} &= \frac{1}{\rho^2} \sin \omega. \end{aligned} \tag{2.57}$$

The normal motion (2.56) has

$$\boldsymbol{r}_t = \rho \theta_t \boldsymbol{N}, \qquad (\theta = \omega/2) \tag{2.58}$$

and, in terms of curvature coordinates $x = u + v,\ y = u - v$, the system (2.57) is equivalent to

$$\boxed{\begin{aligned} &\theta_{xyt} - \theta_x \theta_{yt} \cot \theta + \theta_y \theta_{xt} \tan \theta = 0, \\ &\left(\frac{\theta_{xt}}{\cos \theta} \right)_x - \frac{1}{\rho} \left(\frac{1}{\rho} \sin \theta \right)_t - \frac{1}{\sin \theta} \theta_y \theta_{yt} = 0, \\ &\left(\frac{\theta_{yt}}{\sin \theta} \right)_y + \frac{1}{\rho} \left(\frac{1}{\rho} \cos \theta \right)_t + \frac{1}{\cos \theta} \theta_x \theta_{xt} = 0, \\ &\theta_{xx} - \theta_{yy} = \frac{1}{\rho^2} \sin \theta \cos \theta. \end{aligned}} \tag{2.59}$$

This system appears in Eisenhart [119] in connection with a special class of triply orthogonal surfaces. It was studied extensively by Weingarten in the case ρ-constant, by Bianchi and Darboux. In particular, it was shown by Darboux that the general solution of the system depends upon five arbitrary functions of a single variable (see also [182]). It should be noted that, in the system (2.59), each of $(2.59)_2$ and $(2.59)_3$ is a consequence of $(2.59)_4$ and the other. Consequently, one of $(2.59)_2$ or $(2.59)_3$ is redundant and may be discarded.

In the above motion, since Σ undergoes a purely normal propagation, it follows that the surfaces swept out by the parametric lines on Σ as t evolves will be orthogonal to Σ at all times. However, lines of curvature on a surface are mutually orthogonal, so that the family of surfaces swept out by the x-parametric lines will be orthogonal to that swept out by the y-parametric lines. Thus, the system consisting of these two families, augmented by the family whose members are the pseudospherical surfaces formed at each time t, constitute a triply orthogonal system of surfaces. This explains the appearance of what we here term the Weingarten system, namely (2.59), which describes a triply orthogonal system of surfaces in which the members of one family are pseudospherical. Triply orthogonal systems associated with multi-soliton solutions of the Weingarten system may be readily constructed. These are generated by a nonlinear superposition principle of the type (1.63) associated with an auto-Bäcklund transformation of the Weingarten system as derived in the following subsection.

2.3.3 *Bäcklund Transformations*

Given the position vector of a generic point on a moving pseudospherical surface Σ, a Bäcklund transformation may be applied at each instant t to generate a new pseudospherical surface. Here, we seek to do this in such a way that constraints on the motion are preserved. In particular, auto-Bäcklund transformations are constructed which, in turn, maintain the restrictions on the motion associated with the Weingarten and anharmonic lattice systems. These auto-Bäcklund transformations generically admit (1.63) as a nonlinear superposition principle for the generation of solutions. They are induced by invariance of 2×2 linear representations under a suitable gauge transformation.

Thus, to the 2×2 linear system (2.29), with P and Q given by (2.30), we adjoin a 2×2 time evolution corresponding to the representation (2.33). We consider the linear representation

$$\begin{aligned} \Phi_u &= P(\omega)\Phi, \\ \Phi_v &= Q(\omega)\Phi, \\ \Phi_t &= R(\omega)\Phi, \end{aligned} \tag{2.60}$$

where, since

$$\begin{pmatrix} 0 & a & b \\ -a & 0 & c \\ -b & -c & 0 \end{pmatrix} = -aL_3 + bL_2 - cL_1,$$

under the correspondence $e_k \leftrightarrow L_k$ we obtain

$$R(\omega) = \frac{1}{2}\begin{pmatrix} ia & -b+ic \\ b+ic & -ia \end{pmatrix} \tag{2.61}$$

with $a = a(\omega)$, $b = b(\omega)$ and $c = c(\omega)$.

Introduction of the gauge transformation $\Phi' = H\Phi$, into (2.60) yields

$$\begin{aligned}\Phi'_u &= P'\Phi', \\ \Phi'_v &= Q'\Phi', \\ \Phi'_t &= R'\Phi',\end{aligned} \tag{2.62}$$

where

$$\begin{gathered}P' = (H_u + HP)H^{-1}, \quad Q' = (H_v + HQ)H^{-1}, \\ R' = (H_t + HR)H^{-1}.\end{gathered} \tag{2.63}$$

An orthonormal triad $\{\boldsymbol{A}', \boldsymbol{B}', \boldsymbol{C}'\}$, where $\boldsymbol{A}' = \boldsymbol{r}'_u$, $\boldsymbol{B}' = -\boldsymbol{A}' \times \boldsymbol{N}'$ and $\boldsymbol{C}' = \boldsymbol{N}'$ are given, in turn, by (1.38), (1.39) and (1.41), is now introduced on the Bäcklund transform Σ' of the surface Σ. If Ψ is the matrix with $\boldsymbol{A}$, $\boldsymbol{B}$ and $\boldsymbol{C}$ as rows and Ψ' is the corresponding primed matrix then, on use of (1.52), it is readily shown that

$$\Psi' = \Lambda\Psi, \tag{2.64}$$

where the transformation matrix Λ is given by

$$\Lambda = \begin{pmatrix} 1 - \dfrac{L}{\rho}\beta\sin^2\phi & \dfrac{L\beta}{\rho}\sin\phi\cos\phi & \dfrac{L}{\rho}\sin\phi \\ \dfrac{L}{\rho}\beta\sin\phi\cos\phi & 1 - \dfrac{L\beta}{\rho}\cos^2\phi & -\dfrac{L}{\rho}\cos\phi \\ -\dfrac{L}{\rho}\sin\phi & \dfrac{L}{\rho}\cos\phi & 1 - L\dfrac{\beta}{\rho} \end{pmatrix}, \tag{2.65}$$

with $\phi = (\omega + \omega')/2$.

Since $\Psi, \Psi' \in SO(3)$, it follows that $\Lambda \in SO(3)$, that is,

$$\Lambda^{\mathsf{T}}\Lambda = \mathbb{1}, \quad \det\Lambda = 1. \tag{2.66}$$

In geometric terms, these relations are just a consequence of the observation that any two right-handed orthonormal triads are related by a rotation.

Use of the $su(2)$-$so(3)$ isomorphism as given in Appendix A delivers an $SU(2)$ transformation matrix corresponding to Λ, namely

$$H = (1+\beta^2)^{-1/2}\begin{pmatrix} 1 & -i\beta e^{-i\phi} \\ -i\beta e^{i\phi} & 1 \end{pmatrix}. \tag{2.67}$$

The particular gauge transformation

$$\Phi' = H(\phi, \beta)\Phi \tag{2.68}$$

with H given by (2.67) acts on the 2×2 linear representation

$$\Phi_u = P(\omega)\Phi,$$
$$\Phi_v = Q(\omega)\Phi,$$

to produce[2]

$$\Phi'_u = P(\omega')\Phi',$$
$$\Phi'_v = Q(\omega')\Phi'.$$

This β-*dependent* gauge transformation constitutes a prototypical matrix Darboux transformation. At the nonlinear level, it induces the auto-Bäcklund transformation (1.47) for the sine-Gordon equation (1.25).

In the present context, with adjoined temporal evolution $(2.60)_3$, the Bäcklund parameter β is allowed to depend on t in (2.67). Use of the latter expression together with (2.61) shows that

$$\begin{aligned} R' &= H_t H^{-1} + HRH^{-1} = H_t H^\dagger + HRH^\dagger \\ &= \frac{1}{2}\begin{pmatrix} ia' & -b' + ic' \\ b' + ic' & -ia' \end{pmatrix} \end{aligned} \tag{2.69}$$

where $H^\dagger = \bar{H}^{\mathsf{T}}$ and

$$\begin{aligned} a' &= a - \frac{2\beta}{1+\beta^2}[\beta(a + \phi_t) + (b\cos\phi + c\sin\phi)], \\ b' &= b + \frac{2}{1+\beta^2}[\beta(a + \phi_t)\cos\phi + \dot{\beta}\sin\phi - \beta^2\cos\phi\,(b\cos\phi + c\sin\phi)], \\ c' &= c + \frac{2}{1+\beta^2}[\beta(a + \phi_t)\sin\phi - \dot{\beta}\cos\phi - \beta^2\sin\phi\,(b\cos\phi + c\sin\phi)]. \end{aligned} \tag{2.70}$$

Particular $\{a, b, c\}$ associated with nonlinear equations compatible with the sine-Gordon equation may now be inserted in (2.70) to generate auto-Bäcklund transformations.

The Weingarten System

For the Weingarten system (2.57), with $\{l, m, n\}$ given by (2.56), the relations (2.40)–(2.42) show that

$$\{a, b, c\} = \left\{-\frac{\omega_t}{2}, \frac{\rho}{2}\omega_{ut}, \frac{\rho}{2}(\omega_{vt}\operatorname{cosec}\omega - \omega_{ut}\cot\omega)\right\}. \tag{2.71}$$

[2] This may be verified directly by calculating the matrices $P' = (H_u + HP)H^{-1}$ and $Q' = (H_v + HQ)H^{-1}$.

In order that the invariance condition

$$R' = R(\omega') \tag{2.72}$$

be satisfied, it is required that, on substitution of (2.71) in $(2.70)_1$,

$$\begin{aligned}\omega'_t &= \frac{1+\beta^2}{1-\beta^2}\omega_t \\ &\quad + \frac{2\rho\beta}{1-\beta^2}\operatorname{cosec}\omega\left[\omega_{vt}\sin\left(\frac{\omega'+\omega}{2}\right) - \omega_{ut}\sin\left(\frac{\omega'-\omega}{2}\right)\right].\end{aligned} \tag{2.73}$$

The latter relation and the classical Bäcklund relations (1.47), together with the constituent equations of the Weingarten system (2.57), show that the residual conditions $(2.70)_{2,3}$ are satisfied modulo

$$\frac{\dot\beta}{\beta} = \frac{\dot\rho}{\rho}\left(\frac{\beta^2+1}{\beta^2-1}\right), \tag{2.74}$$

whence

$$\beta(t) = k\rho(t)[1 \pm \sqrt{1 - 1/(k^2\rho^2(t))}], \tag{2.75}$$

where k is an arbitrary constant of integration, here taken to be non-zero. It is noted that the relation (2.75), which determines the Bäcklund parameter $\beta(t)$ for arbitrary time evolution of the Gaussian curvature $\mathcal{K} = -1/\rho^2(t) < 0$, is consistent with the relation (1.35), where $k = 1/L$.

It is now routine to show that the classical Bäcklund relations (1.47) and the relation (2.73) are compatible modulo the Weingarten system (2.57). Thus, the following auto-Bäcklund transformation for the Weingarten system has been established:

Theorem 2. *The Weingarten system (2.57) is invariant under the Bäcklund transformation*

$$\begin{aligned}\omega'_u &= \omega_u + \frac{2\beta}{\rho}\sin\left(\frac{\omega'+\omega}{2}\right), \\ \omega'_v &= -\omega_v + \frac{2}{\beta\rho}\sin\left(\frac{\omega'-\omega}{2}\right), \\ \omega'_t &= \frac{1+\beta^2}{1-\beta^2}\omega_t \\ &\quad + \frac{2\rho\beta}{1-\beta^2}\operatorname{cosec}\omega\left[\omega_{vt}\sin\left(\frac{\omega'+\omega}{2}\right) - \omega_{ut}\sin\left(\frac{\omega'-\omega}{2}\right)\right].\end{aligned} \tag{2.76}$$

Here, $\mathcal{K} = -1/\rho^2(t)$ *and* $\beta = \beta(t)$ *is related to* $\rho(t)$ *by (2.75).*

An Anharmonic Lattice System

For the system (2.53), on setting $\zeta = 3/(2\rho^2)$, the relations (2.40)–(2.42) yield

$$\{a, b, c\} = \left\{\frac{\omega_u}{2\rho} - \omega_t + \delta\rho\,\omega_v,\ \frac{\omega_{uu}}{2} + \delta\sin\omega,\ \frac{\omega_u^2}{4} - \frac{1}{2\rho^2} - \delta\cos\omega\right\}, \tag{2.77}$$

while both $\dot{\rho} = 0$ and $\dot{\beta} = 0$. The condition (2.72) requires, on substitution of the relations (2.77) in $(2.70)_1$, that

$$\begin{aligned}\omega'_t = \omega_t &+ \frac{\beta^2}{\rho}\omega_u + \beta\omega_{uu}\cos\left(\frac{\omega'+\omega}{2}\right) + \left(\frac{\beta}{2}\omega_u^2 + \frac{\beta^3}{\rho^2}\right)\sin\left(\frac{\omega'+\omega}{2}\right)\\ &+ \delta\left[-2\rho\,\omega_v + \frac{2}{\rho}\sin\left(\frac{\omega'-\omega}{2}\right)\right]\end{aligned} \tag{2.78}$$

together with the Bäcklund relations (1.47) wherein $\omega = \omega(u, v, t)$, $\omega' = \omega'(u, v, t)$. Indeed, the following result may be verified.

Theorem 3. *The system*

$$\begin{aligned}\omega_t &= \frac{\rho}{2}\omega_{uuu} + \frac{\rho}{4}\omega_u^3 + \delta\rho\,\omega_v,\\ \omega_{uv} &= \frac{1}{\rho^2}\sin\omega,\end{aligned} \tag{2.79}$$

is invariant under the Bäcklund transformation

$$\begin{aligned}\omega'_u &= \omega_u + \frac{2\beta}{\rho}\sin\left(\frac{\omega'+\omega}{2}\right),\\ \omega'_v &= -\omega_v + \frac{2}{\beta\rho}\sin\left(\frac{\omega'-\omega}{2}\right),\\ \omega'_t &= \omega_t + \frac{\beta^2}{\rho}\omega_u + \beta\omega_{uu}\cos\left(\frac{\omega'+\omega}{2}\right) + \left(\frac{\beta}{2}\omega_u^2 + \frac{\beta^3}{\rho^2}\right)\sin\left(\frac{\omega'+\omega}{2}\right)\\ &\quad + \delta\left[-2\rho\,\omega_v + \frac{2}{\rho}\sin\left(\frac{\omega'-\omega}{2}\right)\right].\end{aligned} \tag{2.80}$$

Here, $\dot{\rho} = \dot{\beta} = 0$.

It is observed that the 'spatial' parts of the Bäcklund transformation (2.80) coincide with the Bäcklund transformation $\mathbb{B}_\beta$ given by (1.47) for the classical sine-Gordon equation. This suggests that the permutability theorem (1.63) associated with $\mathbb{B}_\beta$ may also apply to both systems (2.57) and (2.79). Indeed, it turns out that this nonlinear superposition principle is generic to all integrable systems compatible with the classical sine-Gordon equation in the sense that they derive from compatible motions of pseudospherical surfaces. In particular, in the application of the permutability theorem for the Weingarten system, the Bäcklund parameters β_i are time-dependent and given by

$$\beta_i(t) = k_i\rho(t)\left[1 \pm \sqrt{1 - 1/\left(k_i^2\rho^2(t)\right)}\right]$$

for specified total curvature $\mathcal{K} = -1/\rho^2(t)$.

The permutability theorem (1.63) has been exploited by Konno and Sanuki [204] to generate kink and soliton solutions of the hybrid system (2.54). In Figure 2.1, the impact of a moving soliton on a stationary soliton is shown. In that case, the stationary soliton remains stationary apart from the phase shift it undergoes due to the impingement of the second pulse. In Figure 2.2, the collision process for two solitons (both moving) with amplitudes of

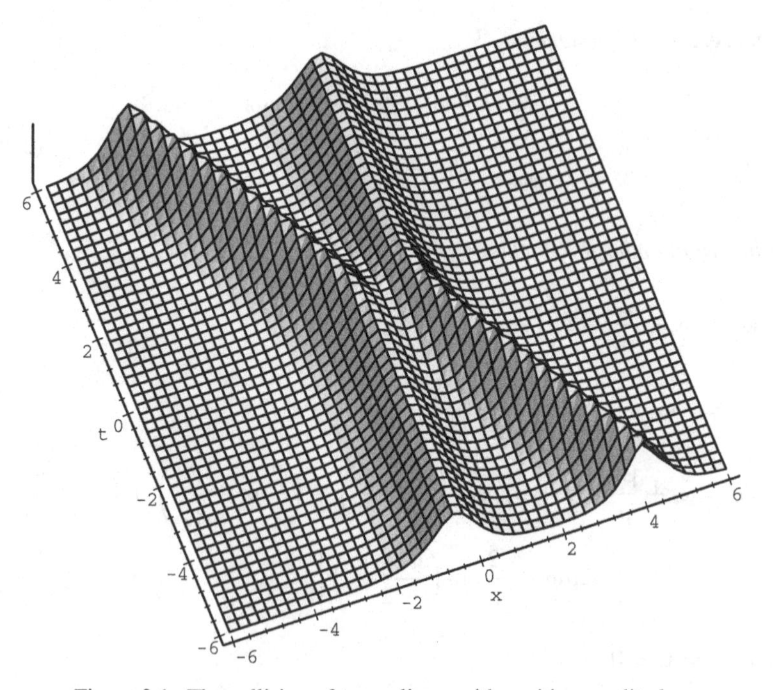

Figure 2.1. The collision of two solitons with positive amplitudes.

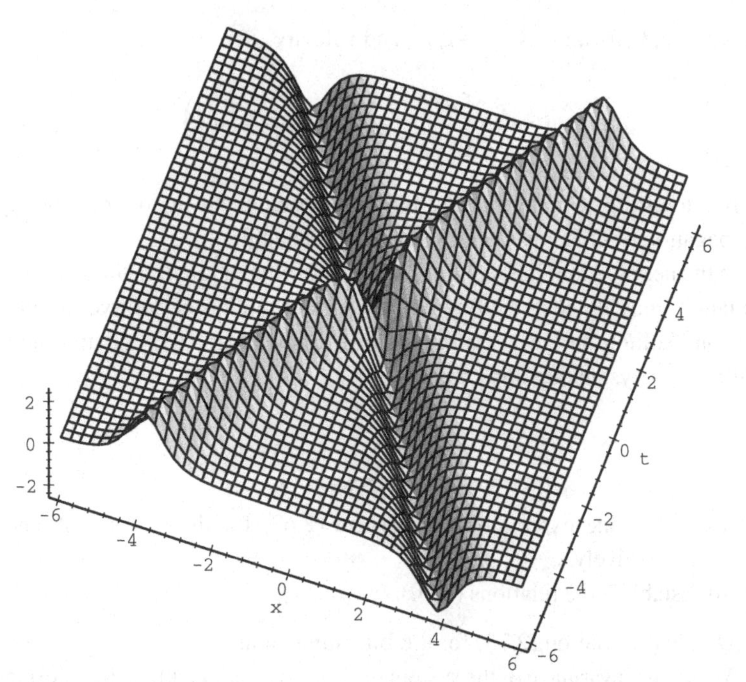

Figure 2.2. The collision of two solitons with amplitudes of opposite signs.

opposite sign is shown. Additional details on kink-antikink interaction are given in [167].

The Potential mKdV Equation

It is noted that the specialisation $\delta = 0$ in (2.80) produces the usual auto-Bäcklund transformation given by $(2.80)_{1,3}$ for the *potential* mKdV equation

$$\omega_t = \frac{\rho}{2}\omega_{uuu} + \frac{\rho}{4}\omega_u^3 \tag{2.81}$$

and thereby for the mKdV equation (2.55). Multi-soliton solutions of the mKdV equation have been constructed via the permutability theorem by Hirota and Satsuma [167]. A Bäcklund transformation and associated permutability theorem have also been constructed by Wadati [374] for a nonlinear lattice described, in the long wavelength continuum approximation, by a combined mKdV-KdV equation.

The relations (2.52) with $\delta = 0$ and $\zeta = 3/(2\rho^2)$ establish that the potential mKdV equation may be associated with the motion of a pseudospherical surface

Σ with total curvature $\mathcal{K} = -1/\rho^2$ and velocity

$$v = \left(\rho\left[\frac{\omega_u^2}{4} - \frac{3}{2\rho^2}\right], -\rho\frac{\omega_{uu}}{2}, \omega_u\right). \tag{2.82}$$

It is interesting to note that the binormal component ω_u of the velocity of propagation of Σ is governed by the mKdV equation.

In the next section, it is shown that the mKdV equation, like the sine-Gordon equation, may be generated by an appropriate motion of a curve of constant torsion. Soliton surfaces associated with both the mKdV equation and the Weingarten system are constructed.

Exercises

1. (a) Obtain the expressions (2.65) and (2.67) for the matrices Λ and H, respectively.
   (b) Establish the relations (2.70).
2. Derive the relation (2.75) for the Bäcklund parameter $\beta(t)$.
3. Verify the invariance of the system (2.79) under the Bäcklund transformation (2.80).

2.4 The mKdV Equation. Moving Curve and Soliton Surface Representations. A Solitonic Weingarten System

In this section, the mKdV equation is generated in a simple manner via a motion of an inextensible curve of zero torsion. The motion of a Dini surface associated with a single soliton solution of the potential mKdV equation (2.81) is then determined. To conclude, a solitonic triply orthogonal Weingarten system is constructed.

2.4.1 The mKdV Equation

Here, we return to the system (2.1), (2.2) consisting of the Serret-Frenet relations with adjoined time evolution of the orthonormal triad $\{\boldsymbol{t}, \boldsymbol{n}, \boldsymbol{b}\}$. It is recalled that this system has compatibility conditions given by (2.3).

If the velocity vector $v = \boldsymbol{r}_t$ of a moving curve C has the decomposition

$$v = \lambda \boldsymbol{t} + \mu \boldsymbol{n} + \nu \boldsymbol{b}, \tag{2.83}$$

then imposition of the condition $\boldsymbol{r}_{ts} = \boldsymbol{r}_{st}$ yields

$$\lambda_s \boldsymbol{t} + \lambda(\kappa \boldsymbol{n}) + \mu_s \boldsymbol{n} + \mu(\tau \boldsymbol{b} - \kappa \boldsymbol{t}) + \nu_s \boldsymbol{b} + \nu(-\tau \boldsymbol{n}) = \alpha \boldsymbol{n} + \beta \boldsymbol{b},$$

whence

$$\begin{aligned} \lambda_s - \mu\kappa &= 0, \\ \lambda\kappa + \mu_s - \nu\tau &= \alpha, \\ \mu\tau + \nu_s &= \beta. \end{aligned} \tag{2.84}$$

The temporal evolution of the curvature κ and the torsion τ of the curve C may now be expressed in terms of the components of velocity λ, μ, ν by substitution of $(2.84)_{2,3}$ into $(2.3)_{1,3}$ to obtain

$$\kappa_t = (\lambda\kappa + \mu_s - \nu\tau)_s - (\mu\tau + \nu_s)\tau, \tag{2.85}$$

$$\tau_t = \gamma_s + (\mu\tau + \nu_s)\kappa, \tag{2.86}$$

where

$$\gamma = \frac{1}{\kappa}[(\mu\tau + \nu_s)_s + \tau(\lambda\kappa + \mu_s - \nu\tau)]. \tag{2.87}$$

For an inextensible curve of zero torsion moving in such a way that $\mu = -\kappa_s$, $(2.84)_1$ shows that

$$\lambda = -\frac{\kappa^2}{2} + c_1(t) \tag{2.88}$$

where $c_1(t)$ is arbitrary. Since $\tau = 0$, the curve is planar at any instant t. If we set $c_1(t) = 0$, then $(2.84)_{2,3}$ and (2.87) yield

$$\begin{aligned} \{\alpha, \beta, \gamma\} &= \left\{-\kappa_{ss} - \frac{\kappa^3}{2}, \nu_s, \frac{\nu_{ss}}{\kappa}\right\} \\ \{\lambda, \mu, \nu\} &= \left\{-\frac{\kappa^2}{2}, -\kappa_s, \nu\right\} \end{aligned} \tag{2.89}$$

while (2.85) shows that the curvature κ evolves according to the mKdV equation

$$\boxed{\kappa_t + \kappa_{sss} + \frac{3}{2}\kappa^2\kappa_s = 0.} \tag{2.90}$$

The remaining relation (2.86) imposes the condition

$$(\nu_{ss}/\kappa)_s = -\kappa\nu_s \tag{2.91}$$

on the binormal component of the velocity.

2.4.2 Motion of a Dini Surface

To construct a pseudospherical surface moving in accordance with the potential mKdV equation (2.81), a solution of the nonlinear system

$$\begin{aligned} \omega_t &= \frac{\rho}{2}\omega_{uuu} + \frac{\rho}{4}\omega_u^3 \\ \omega_{uv} &= \frac{1}{\rho^2}\sin\omega \end{aligned} \tag{2.92}$$

is taken and the corresponding position vector $\boldsymbol{r} = \boldsymbol{r}(u, v, t)$ determined by integration of a linear representation for the mKdV equation

$$\omega'_{t'} + \frac{\rho}{16}\omega'_{u'u'u'} + \frac{3}{8}\rho\,\omega'^2\omega'_{u'} = 0, \tag{2.93}$$

namely

$$\begin{pmatrix} \boldsymbol{A} \\ \boldsymbol{B} \\ \boldsymbol{C} \end{pmatrix}_{u'} = 2\begin{pmatrix} 0 & -\omega' & 0 \\ \omega' & 0 & 1/\rho \\ 0 & -1/\rho & 0 \end{pmatrix}\begin{pmatrix} \boldsymbol{A} \\ \boldsymbol{B} \\ \boldsymbol{C} \end{pmatrix},$$

$$\begin{pmatrix} \boldsymbol{A} \\ \boldsymbol{B} \\ \boldsymbol{C} \end{pmatrix}_{t'} = \begin{pmatrix} 0 & \Gamma & -\dfrac{\omega'_{u'}}{4} \\ -\Gamma & 0 & \dfrac{1}{2\rho^2} - \dfrac{\omega'^2}{4} \\ \dfrac{\omega'_{u'}}{4} & -\left[\dfrac{1}{2\rho^2} - \dfrac{\omega'^2}{4}\right] & 0 \end{pmatrix}\begin{pmatrix} \boldsymbol{A} \\ \boldsymbol{B} \\ \boldsymbol{C} \end{pmatrix} \tag{2.94}$$

augmented by

$$\begin{pmatrix} \boldsymbol{A} \\ \boldsymbol{B} \\ \boldsymbol{C} \end{pmatrix}_{v} = \begin{pmatrix} 0 & 0 & (1/\rho)\sin\omega \\ 0 & 0 & -(1/\rho)\cos\omega \\ -(1/\rho)\sin\omega & (1/\rho)\cos\omega & 0 \end{pmatrix}\begin{pmatrix} \boldsymbol{A} \\ \boldsymbol{B} \\ \boldsymbol{C} \end{pmatrix}. \tag{2.95}$$

In (2.94),

$$\Gamma = \frac{\rho}{8}\omega'_{u'u'} + \frac{\rho}{4}\omega'^3 - \frac{\omega'}{2\rho} \tag{2.96}$$

together with $\omega' = \omega_u$, $u' = u/2$ and $t' = -t$.

Here, as an illustration, we consider the one-soliton solution of $(2.92)_2$, namely

$$\omega = 4\arctan(\exp\chi), \tag{2.97}$$

with

$$\chi = \frac{1}{\rho}\left(\beta u + \frac{v}{\beta}\right) + \xi. \tag{2.98}$$

In the present case, $\dot{\rho} = \dot{\beta} = 0$. However, if ξ is allowed to depend on t, then substitution of (2.97) into the potential mKdV equation $(2.92)_1$ gives $\dot{\xi} = \beta^3/(2\rho^2)$, so that (2.98) yields

$$\chi = \frac{1}{\rho}\left(\beta u + \frac{v}{\beta}\right) + \frac{\beta^3}{2\rho^2}t + \epsilon, \tag{2.99}$$

where ϵ is an arbitrary constant of integration.

Relation (1.85) shows that the position vector of the pseudospherical surface corresponding to the one-soliton solution of the sine-Gordon equation (1.25) is given by

$$\boldsymbol{r}^* = \begin{pmatrix} \dfrac{2\rho\beta}{1+\beta^2}\operatorname{sech}\chi\cos\left(\dfrac{u-v}{\rho}\right) \\ \dfrac{2\rho\beta}{1+\beta^2}\operatorname{sech}\chi\sin\left(\dfrac{u-v}{\rho}\right) \\ u+v-\dfrac{2\rho\beta}{1+\beta^2}\tanh\chi \end{pmatrix}. \tag{2.100}$$

This represents a moving single-soliton Dini surface. Here, χ now depends on t in the manner indicated in (2.99). It is required to determine $\boldsymbol{r} = \boldsymbol{r}(u, v, t)$ that satisfies the appropriate velocity condition, namely, on use of (2.52) with $\delta = 0$ and $\zeta = 3/(2\rho^2)$,

$$\boldsymbol{r}_t = \left(\frac{\rho}{4}\omega_u^2 - \frac{3}{2\rho}\right)\boldsymbol{A} - \frac{\rho}{2}\omega_{uu}\boldsymbol{B} + \omega_u\boldsymbol{C}. \tag{2.101}$$

This position vector $\boldsymbol{r}$ may be sought in the form [181]

$$\boldsymbol{r} = R(t)\boldsymbol{r}^* + \boldsymbol{s}(t), \tag{2.102}$$

where $R(t)$ is an appropriate rotation matrix and $\boldsymbol{s}(t)$ is a translation vector. Substitution of this relation into (2.101) yields

$$\begin{aligned}\dot{R}\boldsymbol{r}^* + \dot{\boldsymbol{s}} &= R\left[\left(\frac{\rho}{4}\omega_u^2 - \frac{3}{2\rho}\right)\boldsymbol{A}^* - \frac{\rho}{2}\omega_{uu}\boldsymbol{B}^* + \omega_u\boldsymbol{C}^* - \boldsymbol{r}_t^*\right]\\
&= R\begin{pmatrix}\dfrac{\beta}{\rho(1+\beta^2)}\operatorname{sech}\chi\sin\left(\dfrac{u-v}{\rho}\right)\\ \dfrac{-\beta}{\rho(1+\beta^2)}\operatorname{sech}\chi\cos\left(\dfrac{u-v}{\rho}\right)\\ -\dfrac{3}{2\rho}\end{pmatrix}\\
&= R\begin{pmatrix}0 & 1/(2\rho^2) & 0\\ -1/(2\rho^2) & 0 & 0\\ 0 & 0 & 0\end{pmatrix}\boldsymbol{r}^* + \begin{pmatrix}0\\0\\-3/(2\rho)\end{pmatrix}.\end{aligned}$$

Accordingly, it is required to determine $R(t)$ and $\boldsymbol{s}(t)$ such that

$$\begin{aligned}\dot{R} &= \frac{1}{2\rho^2}R\begin{pmatrix}0 & 1 & 0\\ -1 & 0 & 0\\ 0 & 0 & 0\end{pmatrix} = -\frac{1}{2\rho^2}RL_3,\\
\dot{\boldsymbol{s}} &= R\begin{pmatrix}0\\0\\-3/(2\rho)\end{pmatrix}.\end{aligned} \tag{2.103}$$

A suitable rotation matrix R and a translation vector $\boldsymbol{s}$ are seen to be given by

$$R = \begin{pmatrix}\cos(t/2\rho^2) & \sin(t/2\rho^2) & 0\\ -\sin(t/2\rho^2) & \cos(t/2\rho^2) & 0\\ 0 & 0 & 1\end{pmatrix}, \quad \boldsymbol{s} = \begin{pmatrix}0\\0\\-3t/(2\rho)\end{pmatrix}.$$

Hence, the position vector $\boldsymbol{r}$ of the moving Dini surface corresponding to the single-soliton solution of the system (2.92) is

$$\boldsymbol{r}(u, v, t) = \begin{pmatrix}\dfrac{2\rho\beta}{1+\beta^2}\operatorname{sech}\chi\cos\left(\dfrac{u-v}{\rho} - \dfrac{t}{2\rho^2}\right)\\ \dfrac{2\rho\beta}{1+\beta^2}\operatorname{sech}\chi\sin\left(\dfrac{u-v}{\rho} - \dfrac{t}{2\rho^2}\right)\\ u + v - \dfrac{3t}{2\rho} - \dfrac{2\rho\beta}{1+\beta^2}\tanh\chi\end{pmatrix}, \tag{2.104}$$

where χ is given by (2.99).

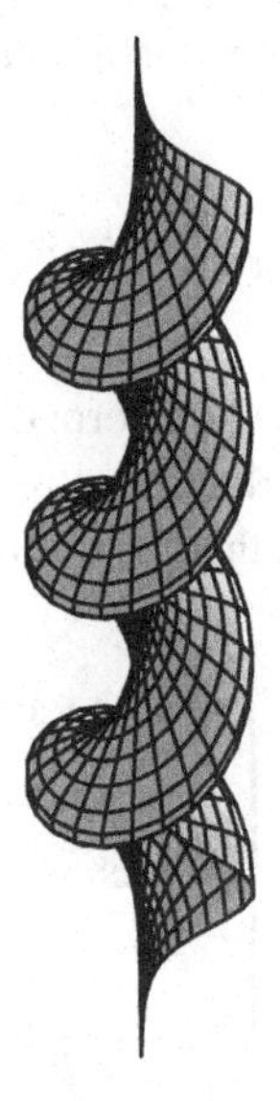

Figure 2.3. A single soliton potential mKdV surface.

It is noted that, if we set $v =$ constant in (2.104), then it provides the position vector $\boldsymbol{r}$ for the surface corresponding to the single soliton solution of the potential mKdV equation (2.81). In Figure 2.3, such a soliton surface is shown for $\rho = 1,\ \beta = 1,\ \xi = 0$ and $v = 0$.

2.4.3 A Triply Orthogonal Weingarten System

A solution of the Weingarten system of equations (2.57) corresponding to the single soliton solution of (1.25) given by (2.97), (2.98) may be constructed in an analogous manner. Here, since $\rho = \rho(t)$, the Bäcklund parameter β, as well as ξ, are now allowed to depend on t. Substitution of (2.97) into either of $(2.57)_{1,2}$ produces the relation (2.75), whence

$$\chi = k[u + v \pm (u - v)\sqrt{1 - 1/(k^2\rho(t)^2)}] + \xi, \tag{2.105}$$

and the position vector (1.85) corresponding to the single soliton solution of the sine-Gordon equation (1.25) gives, in this context,

$$\boldsymbol{r}^* = \begin{pmatrix} \dfrac{1}{k}\operatorname{sech}\chi \cos\left(\dfrac{u - v}{\rho(t)}\right) \\[2ex] \dfrac{1}{k}\operatorname{sech}\chi \sin\left(\dfrac{u - v}{\rho(t)}\right) \\[2ex] u + v - \dfrac{1}{k}\tanh\chi \end{pmatrix}, \tag{2.106}$$

where $\rho(t)$ is regarded as specified.

It is now required to determine $\boldsymbol{r} = \boldsymbol{r}(u, v, t)$ that satisfies the velocity condition (2.56), namely

$$\boldsymbol{r}_t = \frac{\rho\,\omega_t}{2}\boldsymbol{N}. \tag{2.107}$$

A position vector is again sought in the form (2.102). It turns out that, if $\xi = 0$, then we may take $\boldsymbol{r} = \boldsymbol{r}^*$. Accordingly, on reversion to curvature coordinates $x = u + v,\ y = u - v$, it is seen that the position vector

$$\boldsymbol{r}(x, y, t) = \begin{pmatrix} \dfrac{1}{k}\operatorname{sech}\chi\cos\left(\dfrac{y}{\rho}\right) \\ \dfrac{1}{k}\operatorname{sech}\chi\sin\left(\dfrac{y}{\rho}\right) \\ x - \dfrac{1}{k}\tanh\chi \end{pmatrix}, \tag{2.108}$$

where

$$\chi = k\left[x \pm y\sqrt{1 - 1/(k^2\rho^2)}\right] + \xi, \quad \xi \text{ constant} \tag{2.109}$$

and $\rho = \rho(t) \neq 0$ is arbitrary, determines a triply orthogonal system of surfaces corresponding to a single soliton solution of the Weingarten system (2.59). The coordinate surfaces $t = \text{const}$ therein are pseudospherical surfaces of Dini type. The single soliton solution may now be used as the basic component in the application of the nonlinear superposition principle (1.63). Thus, triply orthogonal coordinate systems may, in principle, now be generated corresponding to multi-soliton solutions of the Weingarten system (2.59). These N-soliton solutions represent a nonlinear superposition of single soliton solutions (2.97), with χ given by (2.109) and $k = k_i,\ i = 1, \ldots, N$.

A triply orthogonal system which consists of two families of Dini surfaces and one family of spheres with position vector as set down in [326], namely

$$\boldsymbol{r} = \begin{pmatrix} \dfrac{\cos(y - t)}{\cosh(x + y + t)} \\ \dfrac{\sin(y - t)}{\cosh(x + y + t)} \\ x - \tanh(x + y + t) \end{pmatrix} \tag{2.110}$$

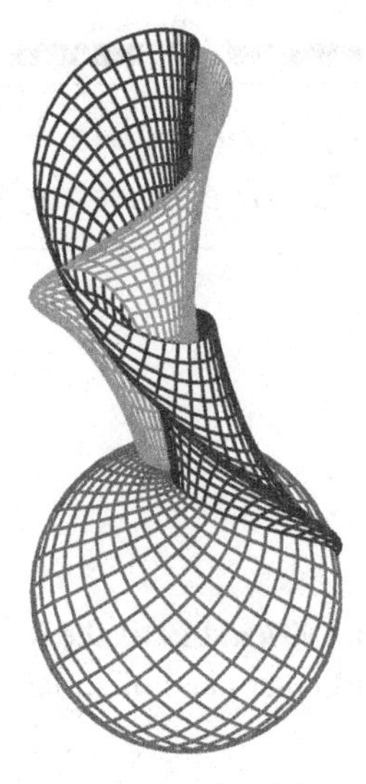

Figure 2.4. A triply orthogonal Weingarten system.

is illustrated in Figure 2.4. It is noted that this particular Weingarten system arises naturally in the context of the 'localised induction hierarchy' [283, 284] which is but another avatar of the nonlinear Schrödinger hierarchy (cf. Chapter 4).

Exercise

1. Verify that (2.94) provides a linear representation for the mKdV equation (2.93).

3

Tzitzeica Surfaces. Conjugate Nets and the Toda Lattice Scheme

In a series of papers between 1907 and 1910, the Romanian geometer Tzitzeica [369, 370] investigated a particular class of surfaces associated with the nonlinear wave equation

$$\boxed{(\ln h)_{\alpha\beta} = h - h^{-2}.} \tag{3.1}$$

Tzitzeica not only established invariance of (3.1) under a Bäcklund-type transformation, but also constructed what is essentially a linear representation incorporating a spectral parameter [370]. The rediscovery of the Tzitzeica equation (3.1) in a solitonic context had to wait until some seventy years later [111, 259].

The importance of this geometric work by Tzitzeica is not, however, confined to soliton theory. Thus, it has been shown that the Tzitzeica condition which leads to (3.1) is invariant under (equi-)affine transformations. The Tzitzeica surfaces are the analogues of spheres in affine differential geometry and, indeed, are known as affine spheres or affinsphären [39]. According to Nomizu and Sasaki [277], the origins of affine differential geometry reside in this work of Tzitzeica at the turn of the nineteenth century. Tzitzeica's early contributions in this area, along with those of Pick and contemporaries, are cited by Blaschke [39]. In 1921, Jonas [184] introduced a geometric procedure for the construction of integrals of the Tzitzeica equation. Much later in 1953, he investigated properties of another avatar of (3.1), namely the affinsphären equation [185]. The latter not only has intrinsic geometric importance, but affords a direct connection with an integrable anisentropic gasdynamics system [144–148, 336]. Further links between soliton theory and affine differential geometry are treated in [14, 17, 80, 209, 330].

Remarkably, the Tzitzeica equation is embedded in a solitonic system which had a much earlier geometric origin in the nineteenth century in the study of conjugate net systems and their iterative generation via Laplace-Darboux transformations. Thus, the integrable two-dimensional Toda lattice equation arises in the latter context and, indeed, was set down explicitly by Darboux in the nineteenth century. The Tzitzeica equation (3.1) will be seen to arise as a special reduction of the Toda lattice model. In fact, the classical study of transformations of conjugate nets has many points of contact with modern soliton theory. The monograph of Eisenhart [119] provides an invaluable compendium of the various transformations of conjugate nets. In solitonic terms, the most important is the so-called Fundamental Transformation described in 1915 by Jonas [183].

3.1 Tzitzeica Surfaces. Link to an Integrable Gasdynamics System

Here, the Tzitzeica equation and its linear representation are derived in their original geometric context. The links with the affinsphären equation and an integrable gasdynamics system are then made explicit.

3.1.1 *The Tzitzeica and Affinsphären Equations*

Hyperbolic surfaces Σ can always be parametrised in terms of real asymptotic coordinates, here denoted by α and β. In this case, the Gauss equations take the form

$$\begin{aligned} \boldsymbol{r}_{\alpha\alpha} &= \Gamma^1_{11}\boldsymbol{r}_\alpha + \Gamma^2_{11}\boldsymbol{r}_\beta, \\ \boldsymbol{r}_{\alpha\beta} &= \Gamma^1_{12}\boldsymbol{r}_\alpha + \Gamma^2_{12}\boldsymbol{r}_\beta + f\boldsymbol{N}, \\ \boldsymbol{r}_{\beta\beta} &= \Gamma^1_{22}\boldsymbol{r}_\alpha + \Gamma^2_{22}\boldsymbol{r}_\beta, \end{aligned} \tag{3.2}$$

while the Mainardi-Codazzi equations read

$$\left(\frac{f}{H}\right)_\alpha + 2\Gamma^2_{12}\frac{f}{H} = 0, \quad \left(\frac{f}{H}\right)_\beta + 2\Gamma^1_{12}\frac{f}{H} = 0, \tag{3.3}$$

and the Gaussian curvature reduces to

$$\mathcal{K} = -\frac{f^2}{H^2}. \tag{3.4}$$

Accordingly, the Mainardi-Codazzi equations may be written as

$$\Gamma^1_{12} = -\frac{1}{4}[\ln(-\mathcal{K})]_\beta, \quad \Gamma^2_{12} = -\frac{1}{4}[\ln(-\mathcal{K})]_\alpha. \tag{3.5}$$

If we denote the distance from the origin to the tangent plane to Σ at a generic point P by d, then we deduce that

$$d = \boldsymbol{N} \cdot \boldsymbol{r}, \quad d_\alpha = \boldsymbol{N}_\alpha \cdot \boldsymbol{r}, \quad d_\beta = \boldsymbol{N}_\beta \cdot \boldsymbol{r}. \tag{3.6}$$

These relations may be regarded as scaled projections of $\boldsymbol{r}$ onto the linearly independent triad $\boldsymbol{N}, \boldsymbol{N}_\alpha, \boldsymbol{N}_\beta$ and use of the Weingarten equations yields

$$\boldsymbol{r} = -\frac{d_\beta}{f}\boldsymbol{r}_\alpha - \frac{d_\alpha}{f}\boldsymbol{r}_\beta + d\boldsymbol{N}. \tag{3.7}$$

On the other hand, by virtue of (3.5), equation $(3.2)_2$ takes the form

$$\boldsymbol{r}_{\alpha\beta} = -\frac{1}{4}[\ln(-\mathcal{K})]_\beta \boldsymbol{r}_\alpha - \frac{1}{4}[\ln(-\mathcal{K})]_\alpha \boldsymbol{r}_\beta + f\boldsymbol{N} \tag{3.8}$$

which implies that

$$\boldsymbol{r}_{\alpha\beta} - \frac{f}{d}\boldsymbol{r} = -\frac{1}{4}\left[\ln\left(-\frac{\mathcal{K}}{d^4}\right)\right]_\beta \boldsymbol{r}_\alpha - \frac{1}{4}\left[\ln\left(-\frac{\mathcal{K}}{d^4}\right)\right]_\alpha \boldsymbol{r}_\beta. \tag{3.9}$$

Thus,

$$\boldsymbol{r}_{\alpha\beta} = \frac{f}{d}\boldsymbol{r} \quad \Leftrightarrow \quad -\frac{\mathcal{K}}{d^4} = \text{const} = c^2 > 0. \tag{3.10}$$

The surfaces governed by the condition

$$\mathcal{K} = -c^2 d^4 \tag{3.11}$$

are precisely those considered by Tzitzeica. Compatibility of

$$\boldsymbol{r}_{\alpha\beta} = h\boldsymbol{r}, \quad (h = f/d) \tag{3.12}$$

and the residual Gauss equations $(3.2)_{1,3}$ yields

$$\Gamma^1_{11} = \frac{h_\alpha}{h}, \quad \Gamma^2_{11} = \frac{a(\alpha)}{h}, \quad \Gamma^1_{22} = \frac{b(\beta)}{h}, \quad \Gamma^2_{22} = \frac{h_\beta}{h}, \tag{3.13}$$

where

$$(\ln h)_{\alpha\beta} = h - \frac{ab}{h^2}. \tag{3.14}$$

Since asymptotic coordinates are only defined modulo the reparametrisation $\alpha \to \tilde{\alpha}(\alpha)$, $\beta \to \tilde{\beta}(\beta)$, we may choose the functions a and b to be constants satisfying $ab = 1$ provided that $a \neq 0,\ b \neq 0$. Thus, in this 'gauge', the position vector of Tzitzeica surfaces is determined by the linear system

$$
\begin{aligned}
\boldsymbol{r}_{\alpha\alpha} &= \frac{h_\alpha}{h}\boldsymbol{r}_\alpha + \frac{\lambda}{h}\boldsymbol{r}_\beta, \\
\boldsymbol{r}_{\alpha\beta} &= h\boldsymbol{r}, \\
\boldsymbol{r}_{\beta\beta} &= \frac{h_\beta}{h}\boldsymbol{r}_\beta + \frac{\lambda^{-1}}{h}\boldsymbol{r}_\alpha,
\end{aligned}
\tag{3.15}
$$

where $\lambda = a = b^{-1}$ is a constant parameter. This system is compatible if and only if h obeys the Tzitzeica equation (3.1). The latter equation, following its rediscovery in modern soliton theory, has been the subject of extensive investigation (see e.g., [52, 140, 259, 325, 327, 336]). However, remarkably, not only the linear representation (3.15) but also an auto-Bäcklund transformation crucial to the solitonic analysis are both to be found in the original works of Tzitzeica.

Tzitzeica surfaces arise naturally in affine geometry in the form of affinsphären.[1] The latter are defined by the requirement that the affine normals $\boldsymbol{N}^a$ meet at a point which may be taken without loss of generality as the origin of the coordinate system. This is the counterpart of the definition of spheres as a surface the normals $\boldsymbol{N}$ to which all pass through a fixed point. In terms of asymptotic coordinates, it may be shown that the affine normal is parallel to the vector $\boldsymbol{r}_{\alpha\beta}$ so that affinsphären are defined by the relation (3.12) and hence coincide with Tzitzeica surfaces.

In the terminology of soliton theory, the λ-dependent linear system (3.15) is commonly called a Lax triad. Interestingly, the vector

$$
\boldsymbol{r}' = -\frac{1}{h}(\boldsymbol{r}_\alpha \times \boldsymbol{r}_\beta), \tag{3.16}
$$

which is parallel to the normal $\boldsymbol{N}$, obeys the *dual* or *adjoint* triad

$$
\begin{aligned}
\boldsymbol{r}'_{\alpha\alpha} &= \frac{h_\alpha}{h}\boldsymbol{r}'_\alpha - \frac{\lambda}{h}\boldsymbol{r}'_\beta, \\
\boldsymbol{r}'_{\alpha\beta} &= h\boldsymbol{r}', \\
\boldsymbol{r}'_{\beta\beta} &= \frac{h_\beta}{h}\boldsymbol{r}'_\beta - \frac{\lambda^{-1}}{h}\boldsymbol{r}'_\alpha.
\end{aligned}
\tag{3.17}
$$

[1] An account of affine geometry is beyond the scope of the present text. The interested reader is referred to the extensive monograph on the subject by Blaschke [39].

The above statement may be verified by differentiation and substitution for second-order derivatives as given by (3.15). Interestingly, as noted by Jonas [185], one thereby recovers the Lelieuvre formulae

$$\boldsymbol{r}'_\alpha = \boldsymbol{r} \times \boldsymbol{r}_\alpha, \quad \boldsymbol{r}'_\beta = -\boldsymbol{r} \times \boldsymbol{r}_\beta \tag{3.18}$$

associated with $\Sigma' : \boldsymbol{r}' = \boldsymbol{r}'(\alpha, \beta)$. Moreover, $\boldsymbol{r} = \boldsymbol{v}'$ is normal to Σ' and it is readily seen that

$$\frac{|\boldsymbol{r}, \boldsymbol{r}_\alpha, \boldsymbol{r}_\beta|}{h} = \frac{\boldsymbol{r} \cdot (\boldsymbol{r}_\alpha \times \boldsymbol{r}_\beta)}{h} = \text{const}, \tag{3.19}$$

whence

$$\boldsymbol{r}' \cdot \boldsymbol{r} = \text{const}. \tag{3.20}$$

The Tzitzeica surfaces represented by $\boldsymbol{r}'$ are termed *polar reciprocal* affinsphären.

The Lelieuvre formulae give rise to an important observation, namely that there exist two canonical families of conjugate lines on Tzitzeica surfaces. Thus, if we set

$$R = x, \quad u = \frac{y}{x}, \quad v = z', \tag{3.21}$$

where $\boldsymbol{r} = (x, y, z)^{\mathrm{T}}$ and $\boldsymbol{r}' = (x', y', z')^{\mathrm{T}}$, then (3.16) and the Lelieuvre formulae (3.18) deliver

$$hRv = R_\alpha v_\beta + R_\beta v_\alpha, \quad u_\alpha = \frac{1}{R^2} v_\alpha, \quad u_\beta = -\frac{1}{R^2} v_\beta, \tag{3.22}$$

whence, on use of the latter pair of relations,

$$d\alpha = \frac{R}{2v_\alpha}\left(R\,du + \frac{dv}{R}\right), \quad d\beta = \frac{R}{2v_\beta}\left(-R\,du + \frac{dv}{R}\right). \tag{3.23}$$

Insertion of these expressions into the 2nd fundamental form now produces

$$\mathrm{II} \sim d\alpha\, d\beta \sim R^2\, du^2 - \frac{dv^2}{R^2}. \tag{3.24}$$

Thus, the functions u and v define *conjugate lines* on the Tzitzeica surfaces, that is the 2nd fundamental form is purely diagonal with respect to the coordinates u and v.

Now, the relation $R_{\alpha\beta} = hR$ together with $(3.22)_1$ imply that

$$d\left[\frac{1}{v^2}(R_\alpha d\alpha - R_\beta d\beta)\right] = 0, \tag{3.25}$$

which, under the change of variables to u, v wherein

$$\partial_\alpha = v_\alpha\left(\frac{1}{R^2}\partial_u + \partial_v\right), \quad \partial_\beta = v_\beta\left(-\frac{1}{R^2}\partial_u + \partial_v\right), \tag{3.26}$$

gives

$$d\left(\frac{R^2 R_v}{v^2}du + \frac{R_u}{R^2 v^2}dv\right) = 0. \tag{3.27}$$

This relation delivers the *affinsphären equation*

$$\boxed{\left(\frac{R_u}{R^2 v^2}\right)_u = \left(\frac{R^2 R_v}{v^2}\right)_v}, \tag{3.28}$$

as originally derived by the German geometer Jonas [185]. He pointed out that the affinsphären equation (3.28) has the distinct advantage over the Tzitzeica equation that once R is known, calculation of the surface vector $\boldsymbol{r}$ involves only two quadratures. This is seen as follows.

On taking the cross-products of $\boldsymbol{r}'$ and $\boldsymbol{r}'_\alpha$, $\boldsymbol{r}'_\beta$ as given by the definition (3.16) and the Lelieuvre formulae (3.18), respectively, we obtain the Lelieuvre formulae associated with the original affinsphären

$$\boldsymbol{r}_\alpha = \boldsymbol{r}' \times \boldsymbol{r}'_\alpha, \quad \boldsymbol{r}_\beta = -\boldsymbol{r}' \times \boldsymbol{r}'_\beta, \tag{3.29}$$

where, without loss of generality, we have chosen

$$|\boldsymbol{r}, \boldsymbol{r}_\alpha, \boldsymbol{r}_\beta| = \boldsymbol{r} \cdot (\boldsymbol{r}_\alpha \times \boldsymbol{r}_\beta) = h. \tag{3.30}$$

Under the change of variables (3.26), the Lelieuvre formulae (3.29) become

$$\boldsymbol{r}_u = R^2(\boldsymbol{r}' \times \boldsymbol{r}'_v), \quad \boldsymbol{r}_v = \frac{1}{R^2}(\boldsymbol{r}' \times \boldsymbol{r}'_u), \tag{3.31}$$

the first two components of which may be brought into the form

$$\begin{aligned} \left(\frac{y'}{z'}\right)_u &= -\frac{R^2 x_v}{z'^2}, \quad \left(\frac{y'}{z'}\right)_v = -\frac{x_u}{R^2 z'^2} \\ \left(\frac{x'}{z'}\right)_u &= \frac{R^2 y_v}{z'^2}, \quad \left(\frac{x'}{z'}\right)_v = \frac{y_u}{R^2 z'^2}. \end{aligned} \tag{3.32}$$

If we now introduce the two quadratures

$$w = \int \frac{R^2 R_v}{v^2} du + \frac{R_u}{R^2 v^2} dv, \quad s = \int \frac{u R^2 R_v}{v^2} du + \frac{u R_u + R}{R^2 v^2} dv \tag{3.33}$$

and take into account that

$$\boldsymbol{r}' \cdot \boldsymbol{r} = -1, \tag{3.34}$$

then the position vectors of the affinsphäre and its dual are obtained in the form

$$\boldsymbol{r} = \begin{pmatrix} R \\ Ru \\ -v^{-1} + R(wu - s) \end{pmatrix}, \quad \boldsymbol{r}' = \begin{pmatrix} sv \\ -wv \\ v \end{pmatrix}. \tag{3.35}$$

It is interesting to note that Jonas in [185] derived the complete class of Tzitzeica surfaces of revolution. This class is associated with cnoidal solutions of the Tzitzeica equation. A typical surface of revolution is depicted in Figure 3.1.

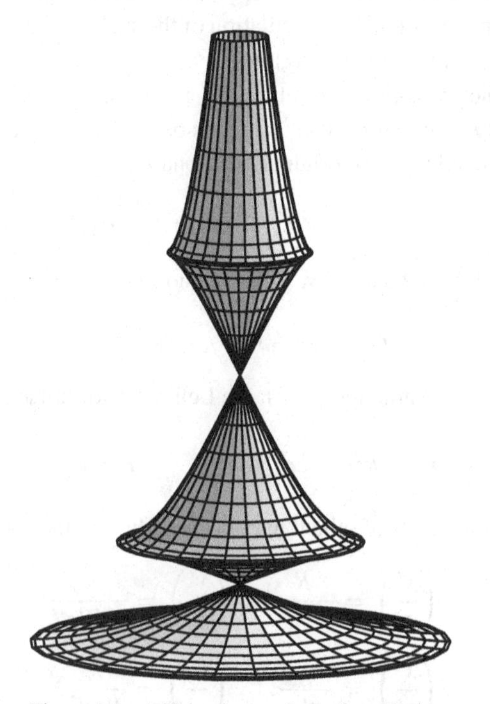

Figure 3.1. A Tzitzeica surface of revolution.

3.1.2 The Affinsphären Equation in a Gasdynamics Context

Following upon the rediscovery of the Tzitzeica equation in soliton theory, Gaffet [144–148] showed, via a Riemann invariants approach, that remarkably, for a particular class of gas laws, a 1+1-dimensional anisentropic gasdynamics system may be reduced to the Tzitzeica equation. This gasdynamics system is, accordingly, integrable and admits soliton-type solutions. In this context, the affinsphären equation has a distinct advantage over the Tzitzeica equation in that it is formulated directly in terms of the gasdynamics variables [336]. The derivation of the affinsphären equation in this anisentropic gasdynamics context is detailed herein.

The 1+1-dimensional Lagrangian equation of motion for a Prim gas [289]

$$\rho = \Pi(p)S(s), \quad \left.\frac{\partial p}{\partial \rho}\right|_s > 0 \tag{3.36}$$

takes the form

$$-S(s)\frac{\partial^2 p}{\partial \psi^2} = \frac{\partial^2}{\partial t^2}\left(\frac{1}{\Pi(p)}\right). \tag{3.37}$$

In the above, p, ρ and $s(\psi)$ designate the gas pressure, density and specific entropy, respectively. In addition, t denotes time and ψ the Lagrangian variable given by

$$d\psi = \rho(dx - q\,dt) \tag{3.38}$$

while x denotes the Eulerian space coordinate and q is the gas speed.

If we set

$$R^3 = p + \tilde{c}, \quad u = t, \quad v = \psi^{1/3} \tag{3.39}$$

and specify $\Pi(p)$ and $S(s)$ according to

$$\Pi(p) = \frac{R}{1 - \tilde{b}R}, \quad S(s) = \psi^{4/3}, \tag{3.40}$$

then the Lagrangian equation of motion (3.37) reduces directly to the affinsphären equation

$$\left(\frac{R_u}{R^2 v^2}\right)_u = \left(\frac{R^2 R_v}{v^2}\right)_v. \tag{3.41}$$

The relations (3.39), (3.40) reveal by exploiting a scaling symmetry of (3.37) that this reduction applies, more generally, to constitutive laws of the form

$$\rho = \frac{\tilde{a}(p+\tilde{c})^{1/3}\psi^{4/3}}{1-\tilde{b}(p+\tilde{c})^{1/3}} \tag{3.42}$$

where $\tilde{a}$ is a further arbitrary constant. To retrieve the gasdynamics variables corresponding to a solution $R = R(u, v)$ of the affinsphären equation (3.41), it is noted that $x = x(u, v)$ is obtained by integration of the pair of relations

$$\frac{\partial x}{\partial v} = 3\frac{v^2}{\rho} = \frac{1-\tilde{b}R}{Rv^2}, \quad \frac{\partial^2 x}{\partial u^2} = -p_\psi = -\frac{R^2 R_v}{v^2}. \tag{3.43}$$

The corresponding integrability condition $(x_v)_{uu} = (x_{uu})_v$ is satisfied by virtue of the affinsphären equation (3.41). The gasdynamics variables are then given parametrically in terms of u and v by

$$x = x(u, v), \quad t = u \tag{3.44}$$

and

$$p = R^3 - \tilde{c}, \quad \rho = \frac{3Rv^4}{1-\tilde{b}R}, \quad S(s) = v^4, \quad q = \frac{\partial x}{\partial u}. \tag{3.45}$$

Martin [242, 243] derived Monge-Ampère-type formulations of both plane steady and non-steady anisentropic gasdynamics. In the latter case, integrable Monge-Ampère equations equivalent to the Tzitzeica equation are readily obtained. To this end, it is observed that, in terms of differential forms, the affinsphären equation assumes the simple form

$$v^2 dw \wedge dv = R^2 du \wedge dR, \quad dR \wedge dv = R^2 v^2 du \wedge dw. \tag{3.46}$$

Indeed, if R and w are regarded as functions of u and v, then (3.46) is equivalent to

$$\begin{pmatrix} w \\ R \end{pmatrix}_u = \begin{pmatrix} 0 & R^2/v^2 \\ R^2v^2 & 0 \end{pmatrix} \begin{pmatrix} w \\ R \end{pmatrix}_v, \tag{3.47}$$

where w is identified as the potential of the affinsphären equation introduced in (3.33).

In the previous subsection, the position vector $\boldsymbol{r}$ of the affinsphäre was the starting point in the derivation of the affinsphären equation. Alternatively, the dual vector $\boldsymbol{r}'$ may be chosen as the basis of the derivation with $R' = x'$, $u' = y'/x'$, $v' = z$. By construction, this leads to a dual affinsphären equation which is of the same form as the affinsphären equation. This is reflected by the fact that the system of differential forms (3.46) is symmetric in R, u and v, w. Consequently, if we regard u and v as functions of w and R, we then obtain

$$\begin{pmatrix} u \\ v \end{pmatrix}_w = \begin{pmatrix} 0 & v^2/R^2 \\ v^2 R^2 & 0 \end{pmatrix} \begin{pmatrix} u \\ v \end{pmatrix}_R, \tag{3.48}$$

whence

$$\left(\frac{v_w}{v^2 R^2}\right)_w = \left(\frac{v^2 v_R}{R^2}\right)_R. \tag{3.49}$$

Thus, we have established that the matrix equation (3.47) is invariant under the hodograph transformation[2]

$$w' = u, \quad R' = v; \quad w = u', \quad R = v'. \tag{3.50}$$

This is associated with the invariance

$$(R, u, v) \leftrightarrow (v, w, R) \tag{3.51}$$

of the affinsphären equation (3.41).

On reversion to the gasdynamics variables $\tilde{p} = R^3$ and $\psi = v^3$, the system of differential forms (3.46) becomes

$$dw \wedge d\psi = du \wedge d\tilde{p}, \quad d\tilde{p} \wedge d\psi = 9(\tilde{p}\psi)^{4/3} du \wedge dw. \tag{3.52}$$

Thus, in terms of $\tilde{p}$ and ψ, one obtains

$$w_{\tilde{p}} = -u_\psi, \quad 9(\tilde{p}\psi)^{4/3}(u_{\tilde{p}} w_\psi - u_\psi w_{\tilde{p}}) = 1, \tag{3.53}$$

leading to

$$\xi_{\tilde{p}\tilde{p}}\xi_{\psi\psi} - \xi^2_{\tilde{p}\psi} + (\tilde{p}\psi)^{-4/3} = 0, \tag{3.54}$$

with $u = \frac{1}{3}\xi_{\tilde{p}}$, $w = -\frac{1}{3}\xi_\psi$. This Monge-Ampère equation represents the Martin formulation of the integrable gasdynamics system under consideration. The invariance (3.51) of the affinsphären equation corresponds to the invariance $(\tilde{p}, \psi, \xi) \to (\psi, \tilde{p}, -\xi)$ of (3.54).

[2] It is observed that invariance under a hodograph transformation has interesting application elsewhere in continuum mechanics. In the nonlinear elasticity of neo-Hookean materials, invariance of the plane strain equilibrium equations under a hodograph transformation leads to the Adkins Reciprocal Principle [4].

Alternatively, u and w may be selected as independent variables and a Legendre transformation established between (3.54) and a second Monge-Ampère equation, namely

$$\Theta_{uu}\Theta_{ww} - \Theta_{uw}^2 + (\Theta_u\Theta_w)^{4/3} = 0, \tag{3.55}$$

with $\tilde{p} = \frac{1}{27}\Theta_u$, $\psi = -\frac{1}{27}\Theta_w$. Hence, the Tzitzeica, affinsphären and Monge-Ampère–type equations (3.54), (3.55) are each solitonic equations which may be associated with the geometry of Tzitzeica surfaces.

Here, a point of contact has been established between soliton theory and a gasdynamics system. In [336], a Bäcklund transformation to be established in the next section was used to construct a class of gasdynamics solutions corresponding to the constitutive law (3.42). It is interesting to note that similar three-parameter gas laws of the type

$$\rho = \frac{\tilde{a}(p+\tilde{c})^{1/3}}{1-\tilde{b}(p+\tilde{c})^{1/3}} \tag{3.56}$$

or, equivalently,

$$p = \left(\frac{\rho_0}{\rho}\right)^3 (e+\tilde{\mu})^{-3} - \tilde{c}, \quad \tilde{\mu} = \frac{1}{\rho_0} + \frac{\tilde{b}}{\tilde{a}}, \tag{3.57}$$

where $e = \rho_0/\rho - 1$ is the stretch, were derived in homentropic gasdynamics by Loewner [237] and later independently by Cekirge and Varley [73]. In [237], Loewner systematically sought Bäcklund transformations which reduce the hodograph equations of plane, steady gasdynamics to appropriate canonical forms in subsonic, transonic and supersonic régimes. The law (3.57) was obtained as but one member of a class for which reduction is available to the classical wave equation in supersonic flow. Cekirge and Varley noted that the constitutive law (3.57) may be used to model ideally soft materials, that is, media in which the Lagrangian sound speed decreases monotonically to zero as the stretch increases without bound. It is recalled that, during expansion, a gas exhibits ideally soft behaviour. It is remarked that the class of model constitutive laws obtained originally by Loewner in [237] and subsequently by Cekirge and Varley [73], including (3.57), may be derived alternatively by a termination of Bergman series approach [72]. The latter, in turn, has connections with integrable Toda lattice systems [275]. Indeed, it has recently been shown in [338] that the model constitutive laws obtained by the Loewner or termination of Bergman series approach have a direct solitonic interpretation. It also turns out, remarkably, that an infinitesimal version of Loewner's Bäcklund transformations as originally introduced in a gasdynamics context, when suitably interpreted, leads to a novel master 2+1-dimensional soliton system [210, 211]. These recent developments have brought together the two strands of major applications of Bäcklund transformations in soliton theory and continuum mechanics which had a separate historical development and which were treated independently in the companion monograph on Bäcklund transformations by Rogers and Shadwick [311].

Exercises

1. Show that, for the Tzitzeica surfaces considered in this section,

$$E = \frac{\psi^3}{\mu^2}\left(\psi_{\alpha\alpha} - (\ln h)_\alpha \psi_\alpha + \frac{\lambda}{h}\psi_\beta\right),$$

$$F = -\frac{\psi^3}{\mu^2}(\psi_{\alpha\beta} - h\psi),$$

$$G = \frac{\psi^3}{\mu^2}\left(\psi_{\beta\beta} - (\ln h)_\beta \psi_\beta + \frac{\lambda^{-1}}{h}\psi_\alpha\right),$$

where $\psi = 1/d$.

2. (a) Show that the direction of the usual normal $\boldsymbol{N}$ coincides with the direction of the Laplace-Beltrami operator acting on the position vector $\boldsymbol{r}$, viz

$$\boldsymbol{N} \sim \Delta \boldsymbol{r} = \frac{1}{H}\left[\left(\frac{G\boldsymbol{r}_\alpha - F\boldsymbol{r}_\beta}{H}\right)_\alpha + \left(\frac{E\boldsymbol{r}_\beta - F\boldsymbol{r}_\alpha}{H}\right)_\beta\right],$$

with $H^2 = EG - F^2$.

(b) Show that the direction of the affine normal is given by

$$\boldsymbol{N}^a \sim \Delta^a \boldsymbol{r},$$

where Δ^a is the Laplace-Beltrami operator associated with the second fundamental form, that is

$$\Delta^a = \frac{1}{\tilde{h}}\left[\frac{\partial}{\partial \alpha}\left(\frac{g}{\tilde{h}}\frac{\partial}{\partial \alpha} - \frac{f}{\tilde{h}}\frac{\partial}{\partial \beta}\right) + \frac{\partial}{\partial \beta}\left(\frac{e}{\tilde{h}}\frac{\partial}{\partial \beta} - \frac{f}{\tilde{h}}\frac{\partial}{\partial \alpha}\right)\right],$$

with $\tilde{h}^2 = eg - f^2$. This underlines the analogy between the normal $\boldsymbol{N}$ and the affine normal $\boldsymbol{N}^a$ (see [39]).

(c) Use the above expression for the affine normal to verify that

$$\boldsymbol{N}^a \,||\, \boldsymbol{r}_{\alpha\beta}$$

in terms of asymptotic coordinates.

3. Use the linear representation for the Tzitzeica equation to show that [185]

$$h = 2\frac{RR_v}{v}\sqrt[3]{\frac{v^2}{P^2 - Q^2}},$$

where

$$P = \frac{R_{uu}}{R^2} - 2\frac{R_u^2}{R^3} + R^2 R_{vv}, \quad Q = 2\left(R_{uv} - \frac{R_u R_v}{R}\right).$$

4. (a) Establish that Tzitzeica surfaces of revolution are associated with solutions of the 'stationary' Tzitzeica equation

$$(\ln h)'' = h - h^{-2}, \quad h = h(\alpha + \beta).$$

This may be solved in terms of elliptic functions, since it admits the first integral

$$h'^2 = 2h^3 + c_0 h^2 + 1.$$

(b) Show that, in cylindrical coordinates

$$x = \rho \cos\varphi, \quad y = \rho \sin\varphi, \quad z,$$

the Gaussian curvature $\mathcal{K}$ and the distance d from the origin to the tangent plane of a surface of revolution $\boldsymbol{r} = \boldsymbol{r}(\rho, z)$ read

$$\mathcal{K} = \frac{z'z''}{\rho(1 + z'^2)^2}, \quad d = \frac{z - \rho z'}{\sqrt{1 + z'^2}}$$

and hence conclude that Tzitzeica surfaces of revolution are governed by

$$z'z'' + c^2\rho(z - \rho z')^4 = 0,$$

where $z = z(\rho)$.

5. Prove that the integrable gasdynamics system associated with the affinsphären equation may be described by the Monge-Ampère equation

$$\Lambda_{qq}\Lambda_{\bar{x}\bar{x}} - \Lambda_{q\bar{x}}^2 = \Lambda^{-4},$$

where q is the gas speed, $\bar{x} = x - qt$ and $\Lambda = \alpha\psi^{\delta}$, with $\delta = -1/3$, $\alpha\delta = \pm\sqrt{3}$.

3.2 Construction of Tzitzeica Surfaces. An Induced Bäcklund Transformation

Tzitzeica not only set down a linear representation but also presented a Bäcklund transformation for the nonlinear equation (3.1) that now bears his name. However, Tzitzeica suppressed the details of how that Bäcklund transformation was obtained. Here, the Bäcklund transformation is generated in the context of a simple, purely geometric construction of chains of Tzitzeica surfaces. Thus, we investigate as to whether two Tzitzeica surfaces Σ and Σ' represented by the pairs $(\boldsymbol{r}, h)$ and $(\boldsymbol{r}', h')$ may be related in such a way that the line segment which connects corresponding generic points P and P' on the surfaces is tangential to the surface Σ, that is

$$\boldsymbol{r}' - \boldsymbol{r} \perp \boldsymbol{N}. \tag{3.58}$$

In analytic terms, this requirement translates into the condition

$$\boldsymbol{r}' = \boldsymbol{r} + \frac{a}{h}\boldsymbol{r}_\alpha + \frac{b}{h}\boldsymbol{r}_\beta, \tag{3.59}$$

where $a(\alpha, \beta)$ and $b(\alpha, \beta)$ are to be so chosen that Σ' is an affinsphäre and parametrised in the same manner as Σ. In other words, the surface vector $\boldsymbol{r}'$ has to satisfy Gauss equations of the form

$$\begin{aligned} \boldsymbol{r}'_{\alpha\alpha} &= \frac{h'_\alpha}{h'}\boldsymbol{r}'_\alpha + \frac{\lambda}{h'}\boldsymbol{r}'_\beta, \\ \boldsymbol{r}'_{\alpha\beta} &= h'\boldsymbol{r}', \\ \boldsymbol{r}'_{\beta\beta} &= \frac{h'_\beta}{h'}\boldsymbol{r}'_\beta + \frac{\lambda^{-1}}{h'}\boldsymbol{r}'_\alpha, \end{aligned} \tag{3.60}$$

so that h' is a solution of the companion Tzitzeica equation

$$(\ln h')_{\alpha\beta} = h' - h'^{-2}. \tag{3.61}$$

This implies, in turn, that if either of the surfaces may be chosen arbitrarily, the relation (3.59) between Σ and Σ' will then induce an auto-Bäcklund transformation for the Tzitzeica equation.

Insertion of the position vector $\boldsymbol{r}'$ into each of the Gauss equations (3.60) and evaluation modulo the original Gauss equations (3.15) produces a relation of the form

$$E_0 \boldsymbol{r} + E_1 \boldsymbol{r}_\alpha + E_2 \boldsymbol{r}_\beta = \boldsymbol{0}. \tag{3.62}$$

Since $\{\boldsymbol{r}_\alpha, \boldsymbol{r}_\beta, \boldsymbol{N}\}$ constitutes a linearly independent set, by virtue of the general relation (3.7), so does $\{\boldsymbol{r}_\alpha, \boldsymbol{r}_\beta, \boldsymbol{r}\}$ whence $E_i = 0$ in (3.62). Indeed, the hyperbolic equation $(3.60)_2$ delivers

$$\begin{aligned} a_{\alpha\beta} &= \frac{h_\beta}{h} a_\alpha + h' a - \frac{a}{h^2} - \lambda^{-1} \left(\frac{b}{h}\right)_\alpha, \\ b_{\alpha\beta} &= \frac{h_\alpha}{h} b_\beta + h' b - \frac{b}{h^2} - \lambda \left(\frac{a}{h}\right)_\beta, \\ h' &= h + a_\alpha + b_\beta, \end{aligned} \tag{3.63}$$

while the $\boldsymbol{r}_\alpha$-component of $(3.60)_1$ and the $\boldsymbol{r}_\beta$-component of $(3.60)_3$ yield

$$\begin{aligned} a_{\alpha\alpha} &= \frac{a_\alpha}{b_\beta} \left[\frac{h_\alpha}{h} b_\beta + h' b - \frac{b}{h^2} - \lambda \left(\frac{a}{h}\right)_\beta \right], \\ b_{\beta\beta} &= \frac{b_\beta}{a_\alpha} \left[\frac{h_\beta}{h} a_\alpha + h' a - \frac{a}{h^2} - \lambda^{-1} \left(\frac{b}{h}\right)_\alpha \right]. \end{aligned} \tag{3.64}$$

Comparison of $(3.63)_{1,2}$ and (3.64) now shows that

$$b_\beta a_{\alpha\alpha} = a_\alpha b_{\alpha\beta}, \quad a_\alpha b_{\beta\beta} = b_\beta a_{\alpha\beta} \tag{3.65}$$

whence

$$b_\beta = c a_\alpha, \quad c = \text{const.} \tag{3.66}$$

On introduction of the potential p via

$$a = p_\beta, \quad b = c p_\alpha, \tag{3.67}$$

the relations $(3.63)_{1,2}$ may be integrated once to give

$$
\begin{aligned}
p_{\alpha\alpha} &= \frac{1+c}{2} p_\alpha^2 + \frac{h_\alpha}{h} p_\alpha - \frac{\lambda c^{-1}}{h} p_\beta + A(\alpha), \\
p_{\beta\beta} &= \frac{1+c}{2} p_\beta^2 + \frac{h_\beta}{h} p_\beta - \frac{\lambda^{-1} c}{h} p_\alpha + B(\beta).
\end{aligned}
\tag{3.68}
$$

These equations can, in turn, be linearised by the change of variables

$$
\phi = \exp\left(-\frac{1+c}{2} p\right), \quad c = -\frac{\lambda}{\mu}, \tag{3.69}
$$

leading to

$$
\begin{aligned}
\phi_{\alpha\alpha} &= \frac{h_\alpha}{h}\phi_\alpha + \frac{\mu}{h}\phi_\beta + \frac{\lambda-\mu}{2\mu} A(\alpha), \\
\phi_{\beta\beta} &= \frac{h_\beta}{h}\phi_\beta + \frac{\mu^{-1}}{h}\phi_\alpha + \frac{\lambda-\mu}{2\mu} B(\beta).
\end{aligned}
\tag{3.70}
$$

Moreover, the compatibility condition $(\phi_{\alpha\alpha})_{\beta\beta} = (\phi_{\beta\beta})_{\alpha\alpha}$ reveals that $A(\alpha)$ and $B(\beta)$ have to vanish identically for a generic solution h of the Tzitzeica equation. It is then readily verified that the above system admits the first integral

$$
\phi_{\alpha\beta} = (h + c_0)\phi, \quad c_0 = \text{const} \tag{3.71}
$$

and $\boldsymbol{r}'$ as given by (3.59), (3.67), (3.69) satisfies the Gauss equations (3.60) provided that $c_0 = 0$. Thus, we have established the following result:

Theorem 4 (The Tzitzeica Transformation). *The Gauss equations (3.15) and the Tzitzeica equation (3.1) are invariant under*

$$
\begin{aligned}
\boldsymbol{r} \to \boldsymbol{r}' &= \boldsymbol{r} - \frac{2}{(\lambda-\mu)h}\left(\lambda \frac{\phi_\alpha}{\phi}\boldsymbol{r}_\beta - \mu \frac{\phi_\beta}{\phi}\boldsymbol{r}_\alpha\right), \\
h \to h' &= h - 2(\ln\phi)_{\alpha\beta},
\end{aligned}
\tag{3.72}
$$

where ϕ is a particular solution of (3.15) with parameter μ.

The Tzitzeica transformation is a particular variant of the classical Moutard transformation [264] which was originally introduced in connection with a search for the sequential reduction of linear hyperbolic equations to a canonical equation whose solution was known. The basic form of the Moutard result, which together with its variants will be subsequently seen to have extensive application in soliton theory, is as follows:

Theorem 5 (The Moutard Transformation). *The hyperbolic equation*

$$\boldsymbol{r}_{\alpha\beta} = h\boldsymbol{r} \tag{3.73}$$

is form-invariant under the transformation

$$\begin{aligned} \boldsymbol{r} &\to \boldsymbol{r}' = \boldsymbol{r} - 2\frac{\boldsymbol{m}}{\phi}, \\ h &\to h' = h - 2(\ln\phi)_{\alpha\beta}, \end{aligned} \tag{3.74}$$

where the bilinear potential $\boldsymbol{m}$ *is defined by*

$$\boldsymbol{m}_\alpha = \phi_\alpha \boldsymbol{r}, \quad \boldsymbol{m}_\beta = \phi \boldsymbol{r}_\beta \tag{3.75}$$

and ϕ *is a particular solution of the scalar version of (3.73).*

The connection with Tzitzeica's transformation is as follows. If the Moutard equation (3.73) is supplemented by the Gauss equations

$$\begin{aligned} \boldsymbol{r}_{\alpha\alpha} &= \frac{h_\alpha}{h}\boldsymbol{r}_\alpha + \frac{\lambda}{h}\boldsymbol{r}_\beta, \\ \boldsymbol{r}_{\beta\beta} &= \frac{h_\beta}{h}\boldsymbol{r}_\beta + \frac{\lambda^{-1}}{h}\boldsymbol{r}_\alpha \end{aligned} \tag{3.76}$$

and ϕ is a solution thereof with parameter μ, then the potential $\boldsymbol{m}$ may be calculated explicitly. It becomes

$$(\lambda + \mu)\boldsymbol{m} = \mu\phi\boldsymbol{r} + \lambda\frac{\phi_\alpha}{h}\boldsymbol{r}_\beta - \mu\frac{\phi_\beta}{h}\boldsymbol{r}_\alpha + \text{const.} \tag{3.77}$$

Thus, with the constant of integration set to zero, we obtain

$$\boldsymbol{r}' = \frac{\lambda - \mu}{\lambda + \mu}\left[\boldsymbol{r} - \frac{2}{(\lambda - \mu)h}\left(\lambda\frac{\phi_\alpha}{\phi}\boldsymbol{r}_\beta - \mu\frac{\phi_\beta}{\phi}\boldsymbol{r}_\alpha\right)\right], \tag{3.78}$$

which is nothing but the Tzitzeica transform up to an irrelevant constant factor.

It is emphasized that the 'standard' form (1.174) of the classical Moutard transformation is readily obtained by introducing the skew-symmetric potential

$$\boldsymbol{S} = \phi\boldsymbol{r} - 2\boldsymbol{m} \tag{3.79}$$

so that

$$\boldsymbol{r}' = \frac{\boldsymbol{S}}{\phi} \tag{3.80}$$

with defining relations

$$\boldsymbol{S}_\alpha = \phi \boldsymbol{r}_\alpha - \phi_\alpha \boldsymbol{r}, \quad \boldsymbol{S}_\beta = \phi_\beta \boldsymbol{r} - \phi \boldsymbol{r}_\beta. \tag{3.81}$$

However, the alternative form $(3.74)_1$ shows that there exists a close connection between the Moutard transformation and the so-called Fundamental Transformation to be discussed in detail in Section 5.4.

It is evident that the Tzitzeica transformation can be iterated in an algorithmic manner. Likewise, the Moutard transformation may be so iterated. In fact, the link between the Tzitzeica and Moutard transformations may be exploited to construct closed formulae for the iterated Tzitzeica transformation. In this way, one may construct multi-parameter Tzitzeica surfaces and corresponding solutions of the Tzitzeica equation. In particular, the multi-soliton solutions of the Tzitzeica equation derived by Michailov [259] give rise to solitonic surfaces of affinsphären type. Furthermore, permutability theorems associated with the double Tzitzeica and triple Moutard transformations play an important role in connection with the construction of discrete integrable systems [273, 274, 325, 327, 330].

Here, we focus on the simplest application of the Tzitzeica transformation, namely its action on the trivial solution

$$h = 1. \tag{3.82}$$

Without loss of generality, we may set $\lambda = 1$ so that the corresponding Gauss equations have particular solution

$$\begin{aligned} x &= \cos\left[\tfrac{1}{2}\sqrt{3}(\alpha - \beta)\right] \exp\left[-\tfrac{1}{2}(\alpha + \beta)\right], \\ y &= \sin\left[\tfrac{1}{2}\sqrt{3}(\alpha - \beta)\right] \exp\left[-\tfrac{1}{2}(\alpha + \beta)\right], \\ z &= \exp(\alpha + \beta). \end{aligned} \tag{3.83}$$

This represents a surface of revolution and is displayed in Figure 3.2. In view of its shape, we have coined the term *Jonas' hexenhut*.[3] The general solution is obtained by applying the linear transformation

$$\boldsymbol{r} \to D\boldsymbol{r}, \tag{3.84}$$

where D is a non-singular constant matrix.

[3] The hexenhut arises naturally in hydrodynamics as a free surface bounding axi-symmetric jets impinging normally on a plate and of 'whirlpools' [134, 239].

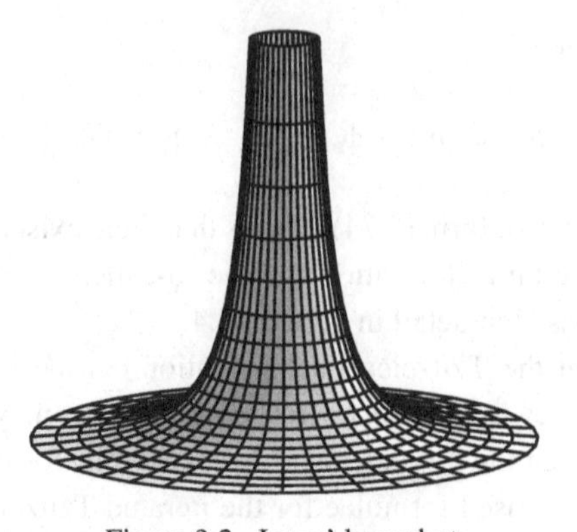

Figure 3.2. Jonas' hexenhut.

The one-soliton solution of the Tzitzeica equation is obtained by choosing the particular solution

$$\phi = (\exp i\delta)\cosh\gamma \tag{3.85}$$

with

$$\gamma = \tfrac{1}{2}\sqrt{3}\big(\nu^{1/3}\alpha + \nu^{-1/3}\beta\big), \quad \delta = \tfrac{1}{2}\big(\nu^{1/3}\alpha - \nu^{-1/3}\beta\big), \quad \mu = i\nu. \tag{3.86}$$

Even though ϕ is complex, the Tzitzeica transform $(3.72)_2$, that is

$$h' = 1 - \frac{3}{2\cosh^2\gamma}, \tag{3.87}$$

proves to be real. Apart from an additive constant, this constitutes the typical gibbous one-soliton solution. Since h is real, a new family of surfaces Σ' is generated by inserting the position vector with components (3.83) into the transformation formula $(3.72)_1$ and taking the real part. We then obtain, up to irrelevant constant factors,

$$\begin{aligned}
x' &= [c_1\cos\kappa - c_2\sin\kappa + (c_3\cos\kappa - c_4\sin\kappa)\tanh\gamma]\exp\big[-\tfrac{1}{2}(\alpha+\beta)\big],\\
y' &= [c_1\sin\kappa + c_2\cos\kappa + (c_3\sin\kappa + c_4\cos\kappa)\tanh\gamma]\exp\big[-\tfrac{1}{2}(\alpha+\beta)\big],\\
z' &= (c_5 - c_6\tanh\gamma)\exp(\alpha+\beta), \qquad \kappa = \tfrac{1}{2}\sqrt{3}(\alpha-\beta),
\end{aligned} \tag{3.88}$$

with constants

$$\begin{aligned}
c_1 &= 2 + 2\nu^2 - \nu^{2/3} - \nu^{4/3}, & c_2 &= \sqrt{3}\big(\nu^{2/3} - \nu^{4/3}\big),\\
c_3 &= \sqrt{3}\big(\nu^{1/3} + \nu^{5/3}\big), & c_4 &= 3\big(\nu^{1/3} - \nu^{5/3}\big),\\
c_5 &= 1 + \nu^2 + \nu^{2/3} + \nu^{4/3}, & c_6 &= \sqrt{3}\big(\nu^{1/3} + \nu^{5/3}\big).
\end{aligned} \tag{3.89}$$

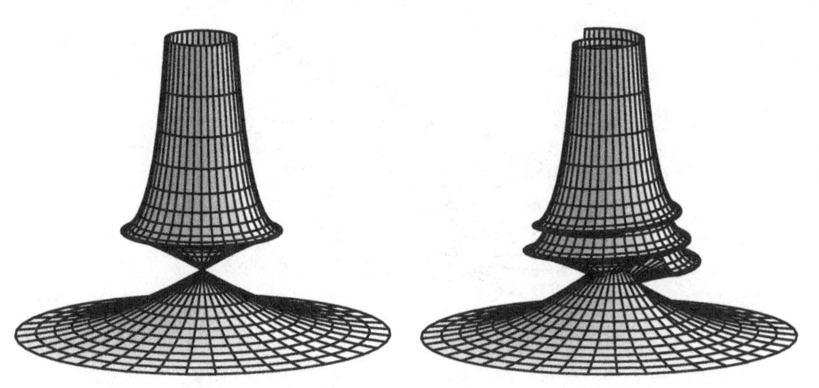

Figure 3.3. One-soliton Tzitzeica surfaces.

The surface corresponding to $\nu = 1$ is a surface of revolution and represents the stationary one-soliton solution of the Tzitzeica equation. This surface we have termed *Jonas' Kelch*. For $\nu \neq 1$, the surfaces encapsulate moving one-soliton solutions. The stationary one-soliton Tzitzeica surface and a typical moving one-soliton Tzitzeica surface are depicted in Figure 3.3. Here, as well as in the previous case, the parametric lines are $\alpha + \beta = \text{const}$ and $\alpha - \beta = \text{const}$. It is noted that the edges and vertices of the one-soliton surfaces are described by the zeros of

$$h' = -\tfrac{1}{2}[1 - \sqrt{3}\tanh\gamma][1 + \sqrt{3}\tanh\gamma]. \tag{3.90}$$

Moreover, the linear transformation (3.84) may be regarded as a deformation of Tzitzeica surfaces. A typical stationary one-soliton Tzitzeica surface with 'shear' is shown in Figure 3.4.

Two-soliton solutions of the Tzitzeica equation may be obtained by means of a second application of the Tzitzeica transformation. In particular, if one chooses the Bäcklund parameters appropriately, one obtains the stationary breather

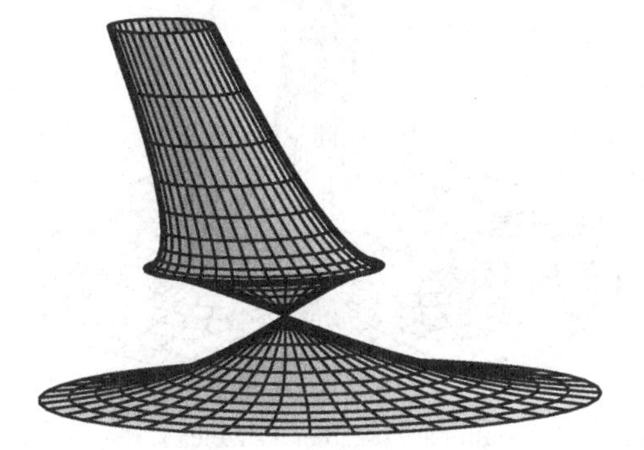

Figure 3.4. A one-soliton Tzitzeica surface with 'shear'.

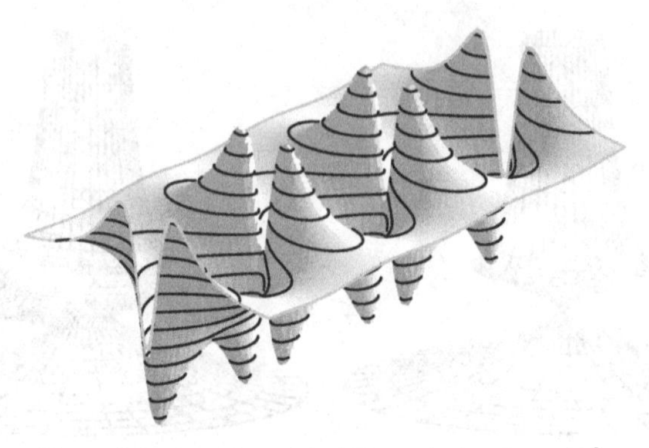

Figure 3.5. A stationary breather solution for $\nu = \frac{3}{4}$.

solution

$$h'' = 1 - \frac{3\sin 2\nu(\sin 2\nu + pq\cosh\gamma\cos\delta)}{(p\sin\nu\cosh\gamma + q\cos\nu\cos\delta)^2}, \tag{3.91}$$

where

$$\begin{aligned} \gamma &= \sqrt{3}(\alpha+\beta)\cos\nu, \quad p^2 = 4\sin^2\nu - 1, \\ \delta &= \sqrt{3}(\alpha-\beta)\sin\nu, \quad q^2 = 4\cos^2\nu - 1. \end{aligned} \tag{3.92}$$

This solution is localised in $\alpha+\beta$ and periodic in $\alpha-\beta$ as shown in Figure 3.5.

The Tzitzeica transformation permits the construction of explicit formulae for the position vector of affinsphären corresponding to solutions of the underlying Tzitzeica equation. A typical stationary breather Tzitzeica surface is displayed in Figure 3.6.

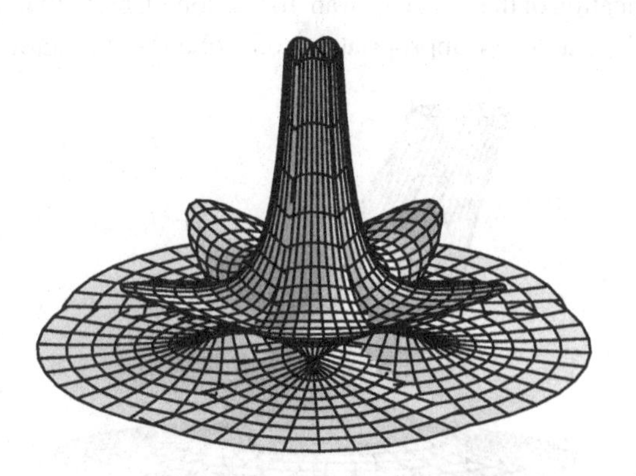

Figure 3.6. A stationary breather Tzitzeica surface for $\nu = \frac{5}{6}$.

Exercise

1. (a) Verify that

$$\phi' = \phi^{-1}$$

is a particular solution of the Gauss equations (3.60) with parameter $\mu' = -\mu$.

(b) It has been shown that the Tzitzeica transformation (3.72) is uniquely determined by the condition that $\delta \boldsymbol{r} = \boldsymbol{r}' - \boldsymbol{r}$ be tangential to the surface Σ. Derive the inverse Tzitzeica transformation

$$\boldsymbol{r} = \boldsymbol{r}' - \frac{2}{(\lambda - \mu')h'} \left(\lambda \frac{\phi'_\alpha}{\phi'} \boldsymbol{r}'_\beta - \mu' \frac{\phi'_\beta}{\phi'} \boldsymbol{r}'_\alpha \right)$$

to establish that $\delta \boldsymbol{r}$ is also tangential to the second surface Σ'.

3.3 Laplace-Darboux Transformations. The Two-Dimensional Toda Lattice. Conjugate Nets

Laplace-Darboux transformations owe their origin to work by Laplace on reduction and invariance properties associated with the canonical hyperbolic equation

$$\Delta r = 0, \tag{3.93}$$

where Δ is the operator

$$\Delta = \partial_u \partial_v + a\partial_u + b\partial_v + c \tag{3.94}$$

and a, b, c are real, scalar functions of u and v. In geometric terms, it was subsequently established by Darboux [93] that these transformations have an interesting interpretation in the theory of conjugate nets. A Bäcklund transformation for the generation of conjugate nets is to be found in the classical literature. Extensive accounts of the transformation theory of conjugate nets are contained in the treatises of both Eisenhart [119] and Lane [227].

Here, our interest resides in remarkable connections between the transformation theory of conjugate nets and certain solitonic systems. Thus, the integrable two-dimensional Toda lattice model arises naturally in connection with sequences of Laplace-Darboux transformations applied to (3.93) or, equivalently, to associated conjugate nets. This lattice system contains as particular reductions both the hyperbolic sinh-Gordon equation and the Tzitzeica equation descriptive of the soliton surfaces as discussed in the preceding section.

In addition, it is known that the classical Darboux system which determines conjugate coordinate systems in $\mathbb{R}^3$ is solitonic. In fact, it incorporates as important reductions not only the SIT system but also a $2+1$-dimensional integrable version of the sine-Gordon equation [214, 340]. An integrable matrix analogue of the Darboux system has been set down by Zakharov and Manakov [392].

In what follows, Laplace-Darboux transformations are introduced along with the key associated notion of Laplace-Darboux invariants. Iterated application of such transformations is shown to lead naturally to the two-dimensional Toda lattice model. The concept of periodicity in Laplace-Darboux sequences is introduced. Salient aspects of the underlying Lie-algebraic structure of Toda lattice systems pertinent to particular symmetry reductions are then outlined. To conclude, the geometric interpretation of Laplace-Darboux transformations with regard to the construction of sequences of conjugate nets is described.

3.3.1 Laplace-Darboux Transformations

The operator Δ as given by (3.94) admits two key decompositions which are important in subsequent developments, namely

$$\Delta = (\partial_u + b)(\partial_v + a) - h \tag{3.95}$$

and

$$\Delta = (\partial_v + a)(\partial_u + b) - k, \tag{3.96}$$

where

$$h = a_u + ba - c, \quad k = b_v + ab - c. \tag{3.97}$$

Thus, the presence of non-zero h, k prevents the reduction of Δ to a factorisable form which would allow immediate integration of (3.93).

The quantities h, k are invariant under the action of gauge transformations of (3.93), namely mappings

$$r \rightarrow r' = gr, \tag{3.98}$$

where g is an arbitrary, invertible function of u and v. This corresponds, at the operator level, to the result that h, k are invariant under the class of gauge transformations

$$\mathcal{G} : \Delta \to \Delta' = g\Delta g^{-1}. \tag{3.99}$$

The gauge transformations $\mathcal{G}$ arise from changes (3.98) in the dependent variable r which preserve the form of Δ. Thus, the invariants h, k label an equivalence class $[h, k]$ of operators linked by gauge transformations [19].

The introduction of Laplace-Darboux transformations is motivated by a search for sequential reduction of (3.93) to an associated equation for which one of the Laplace invariants h, k is zero. The two kinds of classical *Laplace-Darboux transformations*, here denoted by σ_1 and σ_{-1}, originate in the representation of $\Delta r = 0$ as a first-order system in two distinct ways corresponding to the two decompositions (3.95), (3.96), viz. [164]

$$\left.\begin{aligned} (\partial_u + b)r_1 - h(\Delta)r &= 0 \\ r_1 - (\partial_v + a)r &= 0 \end{aligned}\right\} \sigma_1 \tag{3.100}$$

and

$$\left.\begin{aligned} (\partial_v + a)r_{-1} - k(\Delta)r &= 0 \\ r_{-1} - (\partial_u + b)r &= 0 \end{aligned}\right\} \sigma_{-1}. \tag{3.101}$$

It is then readily shown that, under the Laplace-Darboux transformations σ_1 and σ_{-1} given by (3.100) and (3.101), $\Delta \to \Delta_1$ and $\Delta \to \Delta_{-1}$, respectively, where

$$\begin{aligned} \Delta_1 &= \partial_u\partial_v + (a - [\ln h(\Delta)]_v)\partial_u + b\partial_v + (c - a_u + b_v - b[\ln h(\Delta)]_v) \\ \Delta_{-1} &= \partial_u\partial_v + a\partial_u + (b - [\ln k(\Delta)]_u)\partial_v + (c - b_v + a_u - a[\ln k(\Delta)]_u) \end{aligned} \tag{3.102}$$

while the Laplace-Darboux invariants of the operators Δ_1 and Δ_{-1} are given by

$$\begin{aligned} h(\Delta_1) &= 2h(\Delta) - k(\Delta) - [\ln h(\Delta)]_{uv}, \quad k(\Delta_1) = h(\Delta), \\ h(\Delta_{-1}) &= k(\Delta), \quad k(\Delta_{-1}) = 2k(\Delta) - h(\Delta) - [\ln k(\Delta)]_{uv}. \end{aligned} \tag{3.103}$$

3.3.2 *Iteration of Laplace-Darboux Transformations. The Two-Dimensional Toda Lattice*

Let $[h_n, k_n]$ denote the equivalence classes obtained by n applications of the map σ_1. The relations (3.103) show that

$$h_{n+1} = 2h_n - k_n - (\ln h_n)_{uv}, \quad k_{n+1} = h_n, \tag{3.104}$$

whence we obtain the recurrence relations

$$\boxed{(\ln h_n)_{uv} = -h_{n+1} + 2h_n - h_{n-1}, \quad n \in \mathbb{Z}.} \tag{3.105}$$

These determine the integrable two-dimensional *Toda lattice* system.

The Toda lattice system admits various types of reductions due to its underlying Lie-algebraic structure. Thus, if we denote the operators which result from the repeated action of σ_1 and σ_{-1} by

$$\Delta_n = \sigma_1^n(\Delta), \quad \Delta_{-n} = \sigma_{-1}^n(\Delta) \tag{3.106}$$

then the operator Δ is said to have period p if

$$\Delta_p = \Delta. \tag{3.107}$$

In terms of the Laplace-Darboux invariants, periodicity is represented by

$$h_{n+p} = h_n. \tag{3.108}$$

Accordingly, the *periodic* Toda lattice reads

$$\begin{pmatrix} \ln h_1 \\ \vdots \\ \ln h_p \end{pmatrix}_{uv} = \begin{pmatrix} 2 & -1 & & -1 \\ -1 & 2 & \ddots & \\ & \ddots & \ddots & -1 \\ -1 & & -1 & 2 \end{pmatrix} \begin{pmatrix} h_1 \\ \vdots \\ h_p \end{pmatrix}, \tag{3.109}$$

where the dots in the above matrix indicate entries -1 and 2 with all other entries zero. This implies, in turn, that

$$\sum_{k=1}^{p} (\ln h_k)_{uv} = 0 \tag{3.110}$$

and hence

$$\prod_{k=1}^{p} h_k = f(u)g(v). \tag{3.111}$$

The right-hand side of (3.111) may be normalised to ± 1 if p is even and 1 if p is odd on introduction of an appropriate change of variables

$$(u, v, h_k) \to (u'(u), v'(v), F(u)G(v)h_k). \tag{3.112}$$

It is evident that the Toda lattice of period p constitutes a system of $p-1$ equations for $p-1$ unknowns.

The constant matrix in the periodic Toda lattice (3.109) is a particular example of a *Cartan matrix* of rank $p-1$ associated with the infinite-dimensional affine Kac-Moody algebra $A_{p-1}^{(1)}$ [186]. In the case of $A_1^{(1)}$ when the period is 2, the Toda lattice reduces to the sinh-Gordon equation

$$\varphi_{uv} = 4\sinh\varphi, \tag{3.113}$$

if one chooses the normalisation

$$h_2 = h_1^{-1} = e^{-\varphi}. \tag{3.114}$$

Period 3 delivers the Fordy-Gibbons system [140]

$$\begin{aligned}(\ln h_1)_{uv} &= -h_2 + 2h_1 - \frac{1}{h_1 h_2},\\ (\ln h_2)_{uv} &= -\frac{1}{h_1 h_2} + 2h_2 - h_1\end{aligned} \tag{3.115}$$

with the normalisation

$$h_1 h_2 h_3 = 1. \tag{3.116}$$

Moreover, the associated Kac-Moody algebra $A_2^{(1)}$ contains the twisted Kac-Moody subalgebra $A_2^{(2)}$ which is reflected by the admissible constraint $h_1 = h_2 = h$ on the Fordy-Gibbons system (3.115). Thus, in this case, the latter reduces to the Tzitzeica equation

$$(\ln h)_{uv} = h - h^{-2}. \tag{3.117}$$

In conclusion, it is noted that the Toda lattice (3.105) admits reductions associated with a variety of finite-dimensional simple Lie algebras as classified by Killing and Cartan [179] or with infinite-dimensional affine Kac-Moody algebras. The results have been described by Athorne in [19].

3.3.3 *The Two-Dimensional Toda Lattice: Its Linear Representation and Bäcklund Transformation*

The transformation formula (3.102) shows that the condition $b = 0$ is invariant under the Laplace-Darboux transformation σ_1. Thus, in what follows, we focus on the Toda lattice generated by the Laplace-Darboux transformation

$$\phi_{n+1} = (\partial_v + a_n)\phi_n \equiv \mathcal{L}\phi_n, \tag{3.118}$$

acting on the hyperbolic equation

$$\phi_{nuv} + a_n \phi_{nu} + c_n \phi_n = 0. \tag{3.119}$$

The Laplace-Darboux relations may then be brought into the form

$$\phi_{nu} = h_{n-1}\phi_{n-1}, \quad \phi_{nv} = \phi_{n+1} - a_n \phi_n \tag{3.120}$$

and are compatible if and only if

$$a_{nu} = h_n - h_{n-1}, \quad a_n - a_{n-1} = -(\ln h_{n-1})_v, \tag{3.121}$$

leading to the condition

$$(\ln h_{n-1})_{uv} = -h_n + 2h_{n-1} - h_{n-2}. \tag{3.122}$$

Consequently, the linear system (3.120) may be regarded as a Lax pair for the Toda lattice (3.122).

To derive a Bäcklund transformation for the Toda lattice, it is observed that since the Toda lattice encapsulates the Laplace-Darboux invariants h_n obtained from the iterative action of the Laplace-Darboux transformation $\mathcal{L}$ on the seed equation

$$\phi_{uv} + a\phi_u + c\phi = 0, \tag{3.123}$$

it follows that any form-invariance of the latter generates another hierarchy of hyperbolic equations of the form (3.119) with associated Laplace-Darboux invariants h_n'. Here, we seek invariances of the form $\phi \to \phi'$, where

$$\phi' = (\partial_v + A)\phi \tag{3.124}$$

and $A = A(u, v)$ is as yet unspecified. If the expression (3.124) is inserted into the primed version of (3.123), then evaluation modulo (3.123) produces the system

$$\begin{aligned} c' - c + A_u &= 0, \\ (A - a)(a' - a) + (A - a)_v &= 0, \\ (A_u - c)_v + a'(A_u - c) + c'A - c(A - a) &= 0. \end{aligned} \tag{3.125}$$

There are now two cases to distinguish corresponding to $A = a$ or $A \neq a$. If $A = a$, then we retrieve the Laplace-Darboux transformation

$$\begin{aligned} \phi' &= \mathcal{L}\phi = (\partial_v + a)\phi, \\ a' &= a - [\ln(a_u - c)]_v, \quad c' = c - a_u. \end{aligned} \tag{3.126}$$

In the case $A \neq a$, the first two relations in (3.125) may be regarded as the definitions of c' and a', while the remaining relation is equivalent to the conservation law

$$\left(\frac{A_u - c}{A - a}\right)_v = A_u. \tag{3.127}$$

Introduction of a potential ψ according to

$$A = -(\ln \psi)_v, \quad \frac{A_u - c}{A - a} = -(\ln \psi)_u \tag{3.128}$$

produces the linear equation

$$\psi_{uv} + a\psi_u + c\psi = 0. \tag{3.129}$$

Thus, we obtain the *Darboux-type* transformation

$$\begin{gathered} \phi' = \mathcal{B}\phi = \left(\partial_v - \frac{\psi_v}{\psi}\right)\phi, \\ a' = a - \left[\ln\left(\frac{\psi_v + a\psi}{\psi}\right)\right]_v, \quad c' = c + (\ln \psi)_{uv}, \end{gathered} \tag{3.130}$$

where ψ is another solution of the seed equation (3.123). Consequently, another Laplace-Darboux hierarchy of eigenfunctions is generated in the form

$$\phi'_n = \mathcal{L}^n \mathcal{B}\phi, \tag{3.131}$$

with associated coefficients a'_n, c'_n and Laplace-Darboux invariants h'_n.

Remarkably, the Laplace-Darboux transformation $\mathcal{L}$ and the Darboux transformation $\mathcal{B}$ commute, that is

$$\mathcal{L}\mathcal{B} = \mathcal{B}\mathcal{L}, \tag{3.132}$$

which implies that

$$\phi'_n = \mathcal{B}\phi_n. \tag{3.133}$$

We therefore conclude that the Darboux transformation $\mathcal{B}$ applied to the Laplace-Darboux hierarchy (3.119) generates a new hierarchy of hyperbolic equations whose members are again Laplace-Darboux transforms of each other. In terms of the Lax pair (3.120) for the Toda lattice, this result reads as follows:

Theorem 6. *The Lax pair (3.120) and the two-dimensional Toda lattice (3.122) are invariant under*

$$\begin{aligned} \phi_n &\to \phi_n' = \phi_{n+1} - \frac{\psi_{n+1}}{\psi_n}\phi_n, \\ a_n &\to a_n' = a_{n+1} - \frac{\psi_{n+2}}{\psi_{n+1}} + \frac{\psi_{n+1}}{\psi_n}, \\ h_n &\to h_n' = \frac{\psi_{n+2}\,\psi_n}{\psi_{n+1}^2} h_n, \end{aligned} \tag{3.134}$$

where ψ_n is another solution of (3.120).

In the periodic case, it is consistent to assume that

$$\phi_{n+p} = \lambda\phi_n, \tag{3.135}$$

where λ is an arbitrary constant. The Lax pair (3.120) then adopts the form

$$\begin{aligned} \begin{pmatrix} \phi_1 \\ \vdots \\ \phi_p \end{pmatrix}_u &= \begin{pmatrix} 0 & & & \lambda^{-1}h_p \\ h_1 & 0 & & \\ & \ddots & \ddots & \\ & & h_{p-1} & 0 \end{pmatrix} \begin{pmatrix} \phi_1 \\ \vdots \\ \phi_p \end{pmatrix}, \\ \begin{pmatrix} \phi_1 \\ \vdots \\ \phi_p \end{pmatrix}_v &= \begin{pmatrix} -a_1 & 1 & & \\ & -a_2 & \ddots & \\ & & \ddots & 1 \\ \lambda & & & -a_p \end{pmatrix} \begin{pmatrix} \phi_1 \\ \vdots \\ \phi_p \end{pmatrix} \end{aligned} \tag{3.136}$$

and compatibility produces the periodic Toda lattice (3.109). In addition, the transformation formulae (3.134) show that

$$\phi_{n+p}' = \lambda\phi_n', \quad h_{n+p}' = h_n', \quad \prod_{k=1}^{p} h_k' = \prod_{k=1}^{p} h_k \tag{3.137}$$

if ψ_n is a solution of (3.136) with arbitrary parameter μ, that is

$$\psi_{n+p} = \mu\psi_n. \tag{3.138}$$

We are led to the important conclusion that the Darboux transformation $\mathcal{B}$ preserves periodicity and leaves invariant any specified normalisation of the form (3.111).

3.3.4 Conjugate Nets

The requirement that the parametric lines on a surface Σ form a conjugate system leads to $f = 0$ whence the Gauss equation $(1.4)_2$ becomes

$$\boldsymbol{r}_{uv} = \Gamma^1_{12}\boldsymbol{r}_u + \Gamma^2_{12}\boldsymbol{r}_v. \tag{3.139}$$

It is established in Eisenhart [119] that if $r_i(u, v)$, $i = 1, 2, 3$ be three linearly independent real solutions of an equation of the type (3.93) with $c = 0$, then

$$\boldsymbol{r} = (r_1(u, v), r_2(u, v), r_3(u, v)) \tag{3.140}$$

determines a surface Σ upon which the parametric curves form a conjugate system. The Gauss equation (3.139) shows that

$$a = -\Gamma^1_{12}, \quad b = -\Gamma^2_{12}. \tag{3.141}$$

It is observed in passing that in a projective space $\mathbb{P}^3$, conjugate systems of coordinate lines are associated with any hyperbolic equation of the type (3.93), (3.94) [136].

Here, we adopt a variant of the Laplace-Darboux transformation as given in Eisenhart [119]. Thus, if $\Sigma : \boldsymbol{r} = \boldsymbol{r}(u, v)$ is a surface on which the parametric lines constitute a conjugate net $\mathcal{N}$, then $\Sigma_1 : \boldsymbol{r}_1 = \boldsymbol{r}_1(u, v)$ and $\Sigma_{-1} : \boldsymbol{r}_{-1} = \boldsymbol{r}_{-1}(u, v)$ likewise sustain u, v as a conjugate coordinate system where the position vectors $\boldsymbol{r}_1$ and $\boldsymbol{r}_{-1}$ are given by the Laplace-Darboux transformations

$$\boldsymbol{r}_1 = \boldsymbol{r} + \frac{1}{a}\boldsymbol{r}_v, \quad \boldsymbol{r}_{-1} = \boldsymbol{r} + \frac{1}{b}\boldsymbol{r}_u. \tag{3.142}$$

It is noted that these Laplace-Darboux transformations may be decomposed according to

$$\boldsymbol{r}_1 = \frac{1}{a}\sigma_1\boldsymbol{r}, \quad \boldsymbol{r}_{-1} = \frac{1}{b}\sigma_{-1}\boldsymbol{r}. \tag{3.143}$$

To interpret geometrically the Laplace-Darboux transformations considered here, it is observed that

$$\boldsymbol{r}_{1u} = \left[\left(\frac{1}{a}\right)_u - \frac{b}{a}\right]\boldsymbol{r}_v, \quad \boldsymbol{r}_{-1v} = \left[\left(\frac{1}{b}\right)_v - \frac{a}{b}\right]\boldsymbol{r}_u. \tag{3.144}$$

Hence, the tangents to the curves $u =$ const of the conjugate net $\mathcal{N}$ are tangents to the curves $v =$ const of a conjugate net $\mathcal{N}_1$ on Σ_1. Likewise, the tangents to the curves $v =$ const of the conjugate net $\mathcal{N}$ on Σ are tangents to the curves

$u =$ const of a conjugate net $\mathcal{N}_{-1}$ on Σ_{-1}. The Laplace-Darboux transformations (3.142) may be applied iteratively to produce a suite of surfaces on which the parametric lines constitute conjugate nets.

This concludes our introduction to the subject of Laplace-Darboux transformations. It is noted, however, that matrix versions of Laplace-Darboux transformations may also be introduced [207]. These were applied sequentially in [337] to reduce a novel class of parameter-dependent Ernst-type equations to the canonical member, namely the Ernst equation of general relativity. The geometry of these matrix Laplace-Darboux transformations has been discussed in [332] in connection with generalized Weingarten surfaces and Bäcklund transformations for Painlevé equations. The local and global geometry of Painlevé equations has recently been the subject of extensive research [41–44, 333].

Exercises

1. Show that

$$h(g^{-1}\Delta g) = h(\Delta), \quad k(g^{-1}\Delta g) = k(\Delta).$$

2. (a) Verify that, under the Laplace-Darboux transformations σ_1 and σ_{-1} defined by (3.100) and (3.101), $\Delta \to \Delta_1$ and $\Delta \to \Delta_{-1}$, respectively, where Δ_1 and Δ_{-1} are given by (3.102).
   (b) Show that the Laplace-Darboux invariants of the operators Δ_1 and Δ_{-1} are given by (3.103).
   (c) Prove that

$$(\Delta_{-1})_1 = k(\Delta)^{-1}\Delta k(\Delta)^{-1}, \quad (\Delta_1)_{-1} = h(\Delta)^{-1}\Delta h\, h(\Delta)^{-1}.$$

   Deduce that σ_1 and σ_{-1} are inverse maps on the set of equivalence classes $[h, k]$.

3. (a) Verify that conjugation of the Laplace-Darboux transformation $\mathcal{L}$ and the Darboux transformation $\mathcal{B}$ is independent of the order in which the transformations are applied.
   (b) Establish the relations (3.134).

4

Hasimoto Surfaces and the Nonlinear Schrödinger Equation. Geometry and Associated Soliton Equations

The nonlinear Schrödinger (NLS) equation

$$\boxed{iq_t + q_{xx} + \nu|q|^2 q = 0} \tag{4.1}$$

models a wide range of physical phenomena. It seems to have been first set down explicitly by Kelley [194], Baspalov and Talanov [24], and Talanov [360] in independent studies of self-focussing of optical beams in nonlinear media. However, as remarked in Chapter 2 in connection with the integrable motion of curves, Da Rios [97] – in 1906 in an investigation of the motion of an isolated vortex filament in an unbounded liquid – derived a pair of coupled nonlinear evolution equations which may be combined to produce the NLS equation.

The derivation of the NLS equation in nonlinear optics was soon followed by its occurrence in the study of the modulation of monochromatic waves [18, 161, 191, 361] and the propagation of Langmuir waves in plasma [142, 178, 348]. The NLS equation also arises in relation to the Ginzburg-Landau equation in superconductivity [99] while its occurrence in low temperature physics has been documented by Tsuzuki [368]. It likewise appears in the study of the propagation of nonlinear wave packets in weakly inhomogeneous plasma [76, 263]. The NLS equation has been derived in the analysis of deep-water gravity waves by Zakharov [390] and subsequently in that context by both Hasimoto and Ono [163] and Davey [96] using multiple scale techniques. Later, Yuen and Lake [387] rederived the same equation using Whitham's averaged variational principle. The role of the NLS equation as a canonical form has been described in work on slowly varying solitary waves by Grimshaw [154]. It has been recently derived in connection with a capillarity model [13].

The above catalogue of physical applications of the NLS equation emphasises its importance in physics. Here, it is shown that, remarkably, it may be derived in a purely geometric manner. The manner of derivation is, however, related to that by Hasimoto [162] in the physical context of the three-dimensional motion of a vortex in an inviscid fluid. It may be established that particular such vortex motions involving no change of form correspond to travelling wave solutions of the NLS equation.

The soliton surfaces associated with the NLS equation are here termed Hasimoto surfaces. Appropriate compatible motions of Hasimoto surfaces produce the unpumped Maxwell-Bloch system which, in turn, is associated with the integrable stimulated Raman scattering (SRS) and self-induced transparency (SIT) equations [349, 350]. The integrable Pohlmeyer-Lund-Regge system [240, 285] derived in 1976 in a study of the dynamics of relativistic vortices may also, in its turn, be shown to be linked to the SIT system.

Here, the Da Rios equations and their composition, the NLS equation, are derived in a purely geometric manner via a binormal motion of an inextensible curve. Hasimoto surfaces are constructed, and salient geometric properties are obtained. Connections between the NLS equation and other soliton systems, including the Heisenberg spin equation, are catalogued. A geometric formulation with its origin in a kinematic study of hydrodynamics by Marris and Passman [247] in 1969 is then used to derive an auto-Bäcklund transformation for the NLS equation. Breather and 'smoke ring' solutions of the NLS equation are generated thereby. This formalism has recently been used by Rogers [300] to show that, remarkably, a canonical geometrically constrained hydrodynamics system which was investigated in detail by Gilbarg [151], Prim [288], Howard [175], Wasserman [379] and Marris [246] encapsulates the NLS equation. Equivalence with the NLS equation subject to a geometric constraint has subsequently been demonstrated in [313]. Building upon this work, Schief [335] has derived new exact solutions of the hydrodynamics or an equivalent magnetohydrostatics system wherein the constant pressure and magnetic surfaces constitute nested toroids or, more generally, helicoids.

4.1 Binormal Motion and the Nonlinear Schrödinger Equation. The Heisenberg Spin Equation

The work of Hasimoto in 1972 was concerned with an approximation to the self-induced motion of a thin isolated vortex filament travelling without stretching in an incompressible fluid.[1] Therein, if the position vector of the vortex filament is $\boldsymbol{r} = \boldsymbol{r}(s, t)$, then the curve velocity relation

$$v = \boldsymbol{r}_t = \kappa \boldsymbol{b} \tag{4.2}$$

[1] A lucid survey of geometric aspects of knotted vortex filament theory with mention of the NLS equation connection is contained in Keener and Tyson [193].

holds. Thus, in the notation of Section 2.4, the $\{\boldsymbol{t}, \boldsymbol{n}, \boldsymbol{b}\}$ components of $\boldsymbol{r}_t$ are given by

$$\{\lambda, \mu, \nu\} = \{0, 0, \kappa\}, \tag{4.3}$$

whence the relations (2.84) yield

$$\alpha = -\kappa\tau, \tag{4.4}$$

$$\beta = \kappa_s, \tag{4.5}$$

while the compatibility conditions (2.3) give

$$\gamma = \frac{\kappa_{ss} - \kappa\tau^2}{\kappa} \tag{4.6}$$

together with the time evolution for the curvature κ and the torsion τ, namely

$$\kappa_t = -2\kappa_s\tau - \kappa\tau_s \tag{4.7}$$

and

$$\tau_t = \left(-\tau^2 + \frac{\kappa_{ss}}{\kappa} + \frac{\kappa^2}{2}\right)_s, \tag{4.8}$$

respectively. The coupled nonlinear equations (4.7), (4.8) constitute the original Da Rios system as derived in [97] in 1906. Here, it is recalled that $\{\alpha, \beta, \gamma\}$ denote the components in the time evolution (2.2) of the unit triad $\{\boldsymbol{t}, \boldsymbol{n}, \boldsymbol{b}\}$, so chosen as to be compatible with the Serret-Frenet relations (2.1).

If we now introduce the Hasimoto transformation

$$q = \kappa e^{i\sigma}, \tag{4.9}$$

where

$$\sigma = \int_{s_0}^{s} \tau(s^*, t)ds^*, \tag{4.10}$$

then, on use of (4.7), (4.8), it is seen that

$$\begin{aligned} q_t &= \left[\kappa_t + i\kappa\left(-\tau^2 + \frac{\kappa_{ss}}{\kappa} + \frac{\kappa^2}{2} - T(t)\right)\right]e^{i\sigma}, \\ q|q|^2 &= \kappa^3 e^{i\sigma}, \\ q_{ss} &= (\kappa_{ss} + 2i\kappa_s\tau + i\kappa\tau_s - \kappa\tau^2)e^{i\sigma}, \end{aligned} \tag{4.11}$$

where

$$T(t) = \left. \left(-\tau^2 + \frac{\kappa_{ss}}{\kappa} + \frac{\kappa^2}{2} \right) \right|_{s_0} . \tag{4.12}$$

The relations (4.11) yield

$$q_t = i \left[q_{ss} + \frac{1}{2} q |q|^2 - T(t) q \right], \tag{4.13}$$

whence, on setting

$$q^* = q \exp\left(i \int_0^t T(t^*) dt^* \right), \tag{4.14}$$

the NLS equation

$$i q_t^* + q_{ss}^* + \frac{1}{2} |q^*| q^* = 0 \tag{4.15}$$

results. If $T(t) = 0$, then the unstarred version of the NLS equation (4.15) admits the linear representation

$$\begin{pmatrix} \boldsymbol{t} \\ \boldsymbol{n} \\ \boldsymbol{b} \end{pmatrix}_s = \begin{pmatrix} 0 & \kappa & 0 \\ -\kappa & 0 & \tau \\ 0 & -\tau & 0 \end{pmatrix} \begin{pmatrix} \boldsymbol{t} \\ \boldsymbol{n} \\ \boldsymbol{b} \end{pmatrix},$$
$$\begin{pmatrix} \boldsymbol{t} \\ \boldsymbol{n} \\ \boldsymbol{b} \end{pmatrix}_t = \begin{pmatrix} 0 & -\kappa\tau & \kappa_s \\ \kappa\tau & 0 & \dfrac{\kappa_{ss}}{\kappa} - \tau^2 \\ -\kappa_s & -\dfrac{\kappa_{ss}}{\kappa} + \tau^2 & 0 \end{pmatrix} \begin{pmatrix} \boldsymbol{t} \\ \boldsymbol{n} \\ \boldsymbol{b} \end{pmatrix}. \tag{4.16}$$

The surfaces swept out by the binormal motions (4.2) are termed *Hasimoto surfaces* or more commonly *NLS surfaces*.

4.1.1 A Single Soliton NLS Surface

Here, we discuss the original Hasimoto solution of the NLS equation, corresponding to a solitary wave propagating with a constant velocity c along a vortex filament which is straight as $s \to \infty$, that is,

$$\kappa \to 0 \quad \text{as} \quad s \to \infty. \tag{4.17}$$

A solution of the evolution equations (4.7), (4.8) with $\kappa = \kappa(\xi)$ and $\tau = \tau(\xi)$, where $\xi = s - ct$ is sought. On substitution, we obtain

$$c\kappa' = 2\kappa'\tau + \kappa\tau' \tag{4.18}$$

and

$$-c\tau' = \left[-\tau^2 + \frac{\kappa''}{\kappa} + \frac{\kappa^2}{2}\right]'. \tag{4.19}$$

Integration of (4.18) and use of the boundary condition (4.17) together with the assumption that τ is bounded as $s \to \infty$ yields

$$(c - 2\tau)\kappa^2 = 0,$$

whence, discarding the trivial case $\kappa = 0$, it is seen that

$$\tau = \frac{c}{2} = \tau_0, \tag{4.20}$$

where τ_0 is constant.

Accordingly, the torsion is constant along the vortex filament and the velocity of propagation of the solitary wave along the filament is twice the torsion. If $\kappa''/\kappa > 0$,[2] then the relation (4.19) yields

$$\frac{\kappa''}{\kappa} + \frac{\kappa^2}{2} = \epsilon^2, \tag{4.21}$$

where ϵ is an arbitrary constant. This nonlinear ordinary differential equation,[3] augmented by the boundary condition (4.17), admits the solution

$$\kappa = 2\epsilon \operatorname{sech}(\epsilon\xi). \tag{4.22}$$

Hence,

$$q = 2\epsilon \operatorname{sech}[\epsilon(s - 2\tau_0 t)] \exp[i\tau_0(s - s_0)] \tag{4.23}$$

is a solution of (4.13) with $T(t) = -\tau_0^2 + \epsilon^2$. The corresponding single soliton solution of the NLS equation (4.15) is given by

$$q^* = 2\epsilon \operatorname{sech}[\epsilon(s - 2\tau_0 t)] \exp i\left[\tau_0 (s - s_0) + \left(\epsilon^2 - \tau_0^2\right)t\right]. \tag{4.24}$$

[2] This is subsequently shown to be the case if the NLS soliton surface is hyperbolic, i.e., $\mathcal{K} < 0$.
[3] The general solution of (4.21) may be expressed in terms of elliptic functions.

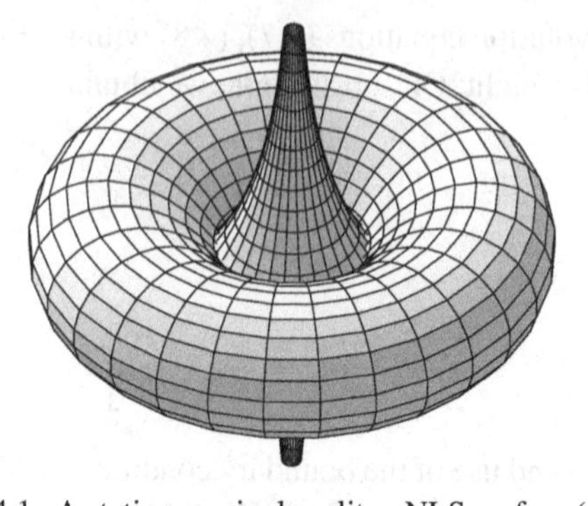

Figure 4.1. A stationary single soliton NLS surface ($\tau = 0$).

To construct the soliton surface associated with the single soliton solution (4.23) of the NLS equation (4.13), it is required to determine its generic position vector $\boldsymbol{r} = \boldsymbol{r}(s, t)$ via its linear representation (4.16) with τ and κ given by (4.20) and (4.22), respectively. In Chapter 6, a general procedure will be given for the construction of the position vector $\boldsymbol{r}$ of soliton surfaces. Here, it is merely recorded that the position vector of the soliton surface associated with the single soliton solution of the NLS equation is given by

$$\boldsymbol{r}(s,t) = 2\begin{pmatrix} \dfrac{s}{2} - \dfrac{\epsilon}{\epsilon^2+\tau_0^2}\tanh(\epsilon\xi) \\ -\dfrac{\epsilon}{\epsilon^2+\tau_0^2}\operatorname{sech}(\epsilon\xi)\cos\left[\tau_0 s + \left(\epsilon^2 - \tau_0^2\right)t\right] \\ -\dfrac{\epsilon}{\epsilon^2+\tau_0^2}\operatorname{sech}(\epsilon\xi)\sin\left[\tau_0 s + \left(\epsilon^2 - \tau_0^2\right)t\right] \end{pmatrix}. \tag{4.25}$$

In Figures 4.1 and 4.2, surfaces associated with particular stationary and moving single solitons of the NLS equation are displayed. A scaling has been made therein to obtain the solution corresponding to the version of the NLS equation

$$i\tilde{q}_t + \tilde{q}_{ss} + 2|\tilde{q}|^2\tilde{q} = 0,$$

as adopted in the original work on soliton surfaces by Sym [356].

Surfaces associated with breather-type solutions of the NLS equation periodic in time have been recently generated via a Bäcklund transformation in [314]. A gallery of NLS breather surfaces is presented in Figure 4.3.

4.1.2 Geometric Properties

In what follows, we establish certain geometric properties of Hasimoto surfaces $\Sigma : \boldsymbol{r} = \boldsymbol{r}(s, t)$ swept out by the vortex filament as time t evolves. Properties of the coordinate curves $s = \text{const}$ and $t = \text{const}$ on Σ are also recorded.

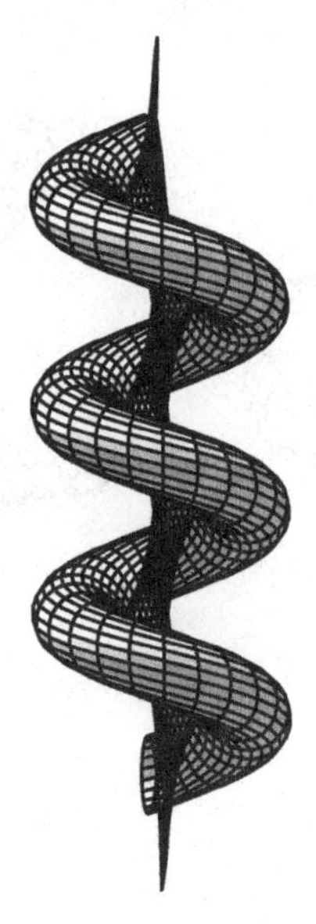

Figure 4.2. A moving single soliton NLS surface ($\tau = \tau_0 \neq 0$).

Here, in view of the geometric relation (4.2),

$$\begin{aligned} \mathrm{I} = d\boldsymbol{r} \cdot d\boldsymbol{r} &= (\boldsymbol{r}_s ds + \boldsymbol{r}_t dt) \cdot (\boldsymbol{r}_s ds + \boldsymbol{r}_t dt) \\ &= (\boldsymbol{t} ds + \kappa \boldsymbol{b} dt) \cdot (\boldsymbol{t} ds + \kappa \boldsymbol{b} dt) \\ &= ds^2 + \kappa^2 dt^2, \end{aligned} \tag{4.26}$$

so that

$$E = 1, \quad F = 0, \quad G = \kappa^2. \tag{4.27}$$

Moreover,

$$\boldsymbol{N} = \frac{\boldsymbol{r}_s \times \boldsymbol{r}_t}{|\boldsymbol{r}_s \times \boldsymbol{r}_t|} = \frac{\boldsymbol{t} \times \kappa \boldsymbol{b}}{\kappa} = -\boldsymbol{n},$$

whence

$$\begin{aligned} \mathrm{II} = -d\boldsymbol{r} \cdot d\boldsymbol{N} &= (\boldsymbol{t} ds + \kappa \boldsymbol{b} dt) \cdot (\boldsymbol{n}_s ds + \boldsymbol{n}_t dt) \\ &= -\kappa ds^2 + 2\kappa\tau ds dt + (\kappa_{ss} - \kappa\tau^2) dt^2, \end{aligned} \tag{4.28}$$

so that

$$e = -\kappa, \quad f = \kappa\tau, \quad g = \kappa_{ss} - \kappa\tau^2.$$

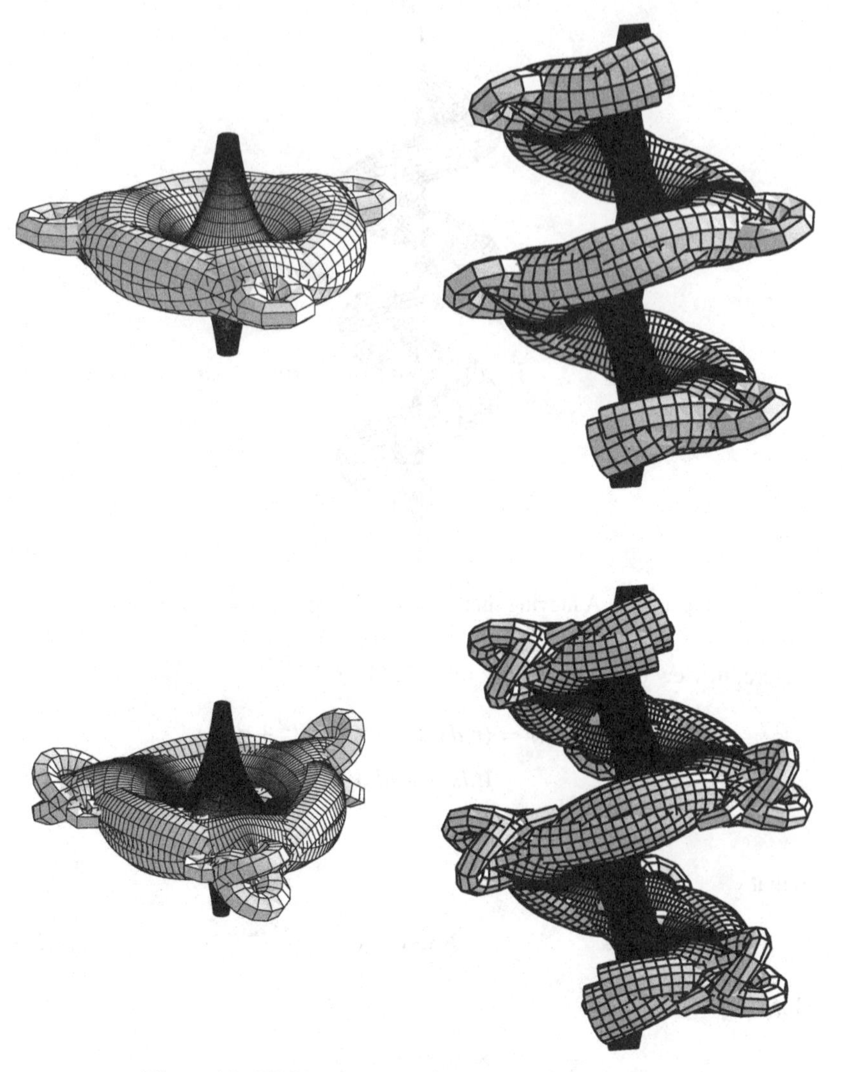

Figure 4.3. NLS stationary and moving breather surfaces.

The Total and Mean Curvature

The total curvature $\mathcal{K}$ of the Hasimoto surface is given by

$$\mathcal{K} = (eg - f^2)/(EG - F^2) = -\kappa_{ss}/\kappa. \tag{4.29}$$

Thus, the NLS soliton surface is elliptic or hyperbolic according to whether $\kappa_{ss}/\kappa < 0$ or $\kappa_{ss}/\kappa > 0$, respectively. The mean curvature $\mathcal{J} = 2\mathcal{M}$ of the surface $\Sigma : \boldsymbol{r} = \boldsymbol{r}(u, v)$ is given by[4]

$$\mathcal{J} = -\mathrm{div}_\Sigma \boldsymbol{N} = \boldsymbol{\nabla} \cdot \boldsymbol{n}, \tag{4.30}$$

where $\boldsymbol{\nabla}$ is the surface gradient

$$\boldsymbol{\nabla} = \frac{1}{H^2}\left[\boldsymbol{r}_u\left(G\frac{\partial}{\partial u} - F\frac{\partial}{\partial v}\right) + \boldsymbol{r}_v\left(E\frac{\partial}{\partial v} - F\frac{\partial}{\partial u}\right)\right]. \tag{4.31}$$

Hence, the mean curvature $\mathcal{J}$ of the Hasimoto surface is given by

$$\mathcal{J} = \frac{1}{\kappa^2}[\boldsymbol{r}_s \cdot (\kappa^2 \boldsymbol{n}_s) + \boldsymbol{r}_t \cdot \boldsymbol{n}_t],$$

that is

$$\mathcal{J} = \frac{1}{\kappa}\left(\frac{\kappa_{ss}}{\kappa} - \kappa^2 - \tau^2\right). \tag{4.32}$$

A Geodesic Property

The geodesic curvature of the parametric curves $v = \text{const}$ on the NLS surface is given by, on use of [380],

$$\kappa_g|_{v=\text{const}} = -\mathrm{div}_\Sigma\left(\frac{\boldsymbol{r}_2}{\sqrt{G}}\right) = -\frac{E_v}{2E\sqrt{G}}, \tag{4.33}$$

so that, here, from the relations (4.27),

$$\kappa_g|_{t=\text{const}} = -\mathrm{div}_\Sigma \boldsymbol{b} = 0. \tag{4.34}$$

Thus, each coordinate line $v = t = \text{const}$ is a geodesic on the NLS surface.

The geodesic curvature of the parametric curves $u = \text{const}$ is given by

$$\kappa_g|_{u=\text{const}} = \mathrm{div}_\Sigma\left(\frac{\boldsymbol{r}_1}{\sqrt{E}}\right) = \frac{G_u}{2G\sqrt{E}}, \tag{4.35}$$

so that, for the present NLS surfaces,

$$\kappa_g|_{s=\text{const}} = \mathrm{div}_\Sigma \boldsymbol{t} = \frac{\kappa_s}{\kappa}. \tag{4.36}$$

[4] It should be noted that in Eisenhart [118] and Weatherburn [380, 381] $\mathcal{J}$ is termed the mean curvature while in Blaschke [39] it is $\mathcal{M}$.

A Parallel Property

A set of curves with unit tangent $\boldsymbol{T}$ on a surface Σ is a family of parallels if and only if [381]

$$\mathrm{div}_\Sigma \boldsymbol{T} = 0.$$

Accordingly, by virtue of (4.34), the coordinate lines $u = s = \text{const}$ are parallel curves on the NLS surfaces.

4.1.3 The Heisenberg Spin Equation

It is readily shown that the NLS equation may be set in correspondence with a 1+1-dimensional version of the *Landau-Lifschitz* equation

$$\frac{\partial \boldsymbol{S}}{\partial t} = \boldsymbol{S} \times \boldsymbol{\nabla}^2 \boldsymbol{S}, \tag{4.37}$$

namely, the *Heisenberg spin equation* [222, 359]

$$\boxed{\frac{\partial \boldsymbol{S}}{\partial t} = \boldsymbol{S} \times \frac{\partial^2 \boldsymbol{S}}{\partial s^2},} \tag{4.38}$$

where the spin field $\boldsymbol{S} = (S_1, S_2, S_3)$ is a unit vector so that

$$\boldsymbol{S}^2 = 1. \tag{4.39}$$

If the classical spin vector $\boldsymbol{S}$ is identified with the unit tangent $\boldsymbol{t}$, then the equation of motion (4.38) yields,

$$\boldsymbol{t}_t = \boldsymbol{t} \times (\kappa \boldsymbol{n})_s = -\kappa\tau \boldsymbol{n} + \kappa_s \boldsymbol{b}. \tag{4.40}$$

The time evolution of $\boldsymbol{n}$ and $\boldsymbol{b}$ may now be calculated to obtain

$$\boldsymbol{e}_{it} = \boldsymbol{\omega} \times \boldsymbol{e}_i, \tag{4.41}$$

where $\boldsymbol{\omega} = \omega_1 \boldsymbol{e}_1 + \omega_2 \boldsymbol{e}_2 + \omega_3 \boldsymbol{e}_3$ with

$$\omega_1 = \frac{\kappa_{ss}}{\kappa} - \tau^2, \quad \omega_2 = -\kappa_s, \quad \omega_3 = -\kappa\tau. \tag{4.42}$$

In the above, the notation

$$\boldsymbol{e}_1 = \boldsymbol{t}, \quad \boldsymbol{e}_2 = \boldsymbol{n}, \quad \boldsymbol{e}_3 = \boldsymbol{b} \tag{4.43}$$

has been adopted.

The system (4.41) together with the Serret-Frenet equations, namely[5]

$$\boldsymbol{e}_{is} = \boldsymbol{D} \times \boldsymbol{e}_i, \quad \boldsymbol{D} = \tau \boldsymbol{e}_1 + \kappa \boldsymbol{e}_3 \tag{4.44}$$

coincides with the NLS linear representation (4.16).

It is remarked that the Heisenberg spin equation (4.38) possesses the conservation law

$$\frac{\partial \epsilon}{\partial t} + \frac{\partial j}{\partial s} = 0, \tag{4.45}$$

where $\epsilon(s, t)$ is the energy density and $j(s, t)$ is the current density given, respectively, by

$$\epsilon(s, t) = \frac{1}{2} \left| \frac{\partial \boldsymbol{S}}{\partial s} \right|^2, \tag{4.46}$$

$$j(s, t) = \boldsymbol{S} \cdot \left(\frac{\partial \boldsymbol{S}}{\partial s} \times \frac{\partial^2 \boldsymbol{S}}{\partial s^2} \right). \tag{4.47}$$

The physical quantities ϵ and j are given in terms of the curvature κ and torsion τ by

$$\epsilon = \frac{1}{2}\kappa^2, \quad j = \kappa^2 \tau. \tag{4.48}$$

Exercise

1. Verify that the orthonormal triad $\{\boldsymbol{t}, \boldsymbol{n}, \boldsymbol{b}\}$ associated with the position vector $\boldsymbol{r}$ given by (4.25) satisfies the linear system (4.16).

4.2 The Pohlmeyer-Lund-Regge Model. SIT and SRS Connections. Compatibility with the NLS Equation

In 1976, in a study of the dynamics of relativistic vortices, Lund and Regge [240] were led to a nonlinear coupled solitonic system which, in a particular reduction, produces the classical sine-Gordon equation. This system was obtained independently in the same year by Pohlmeyer [285] in an investigation of the nonlinear sigma model of field theory. It is now commonly known as the Pohlmeyer-Lund-Regge model. However, the study of the mathematically related sharpline self-induced transparency (SIT) system has earlier origins in the

[5] $\boldsymbol{D}$ is known as the Darboux vector.

work of McCall and Hahn [255] in 1967. Multi-soliton solutions of this system were subsequently discussed by Caudrey et al. [69, 70]. It has been shown by Steudel [349] that the stimulated Raman scattering (SRS) equations, with neglect of the population at the Raman level, are connected to the sharpline SIT equations. In [350], Steudel had earlier applied the Wahlquist-Estabrook procedure to construct a linear representation for the SIT system. Solitons associated with stimulated Raman scattering in caesium were thereby constructed. Related work on the solution of initial value problems in stimulated Raman scattering and two-photon propagation has been conducted by Kaup [192].

Here, the interrelations between the Pohlmeyer-Lund-Regge, SIT and SRS systems are made explicit. Importantly, it is also shown that the unpumped Maxwell-Bloch system which generates the SIT equations is compatible with the NLS equation in the same way that the sine-Gordon and mKdV equations have been shown to be compatible via appropriate motion of pseudospherical surfaces. Thus, the unpumped Maxwell-Bloch system may be associated with suitable motions of Hasimoto surfaces.

4.2.1 The Pohlmeyer-Lund-Regge Model

A parametrisation of generic surfaces $\Sigma : \boldsymbol{r} = \boldsymbol{r}(\xi, \eta)$ with 1st and 2nd fundamental forms

$$\begin{aligned} \mathrm{I} &= \cos^2\theta\, d\xi^2 + \sin^2\theta\, d\eta^2 \\ \mathrm{II} &= e\, d\xi^2 + 2f\, d\xi d\eta + g\, d\eta^2 \end{aligned} \tag{4.49}$$

is considered. The associated Gauss equations (1.4) are

$$\begin{aligned} \boldsymbol{r}_{\xi\xi} &= -\theta_\xi \tan\theta\, \boldsymbol{r}_\xi + \theta_\eta \cot\theta\, \boldsymbol{r}_\eta + e\boldsymbol{N}, \\ \boldsymbol{r}_{\xi\eta} &= -\theta_\eta \tan\theta\, \boldsymbol{r}_\xi + \theta_\xi \cot\theta\, \boldsymbol{r}_\eta + f\boldsymbol{N}, \\ \boldsymbol{r}_{\eta\eta} &= -\theta_\xi \tan\theta\, \boldsymbol{r}_\xi + \theta_\eta \cot\theta\, \boldsymbol{r}_\eta + g\boldsymbol{N}, \end{aligned} \tag{4.50}$$

while the Weingarten equations (1.5) become

$$\begin{aligned} \boldsymbol{N}_\xi &= -\frac{e}{\cos^2\theta}\boldsymbol{r}_\xi - \frac{f}{\sin^2\theta}\boldsymbol{r}_\eta, \\ \boldsymbol{N}_\eta &= \frac{f}{\cos^2\theta}\boldsymbol{r}_\xi + \frac{g}{\sin^2\theta}\boldsymbol{r}_\eta. \end{aligned} \tag{4.51}$$

The Mainardi-Codazzi equations (1.10) reduce to

$$\begin{aligned}\frac{\partial}{\partial \xi}(f \tan\theta) &= \frac{\partial}{\partial \eta}\left(\frac{e+g}{2}\tan\theta\right) + \sin^2\theta \frac{\partial}{\partial \eta}\left(\frac{e-g}{2\cos\theta\sin\theta}\right)\\ \frac{\partial}{\partial \eta}(f \cot\theta) &= \frac{\partial}{\partial \xi}\left(\frac{e+g}{2}\tan\theta\right) - \cos^2\theta \frac{\partial}{\partial \xi}\left(\frac{e-g}{2\cos\theta\sin\theta}\right)\end{aligned} \tag{4.52}$$

while the total curvature $\mathcal{K}$ is given by

$$\mathcal{K} = \frac{eg - f^2}{\cos^2\theta \sin^2\theta} = \frac{\theta_{\eta\eta} - \theta_{\xi\xi}}{\cos\theta\sin\theta},$$

whence

$$eg - f^2 = \cos\theta\sin\theta\,(\theta_{\eta\eta} - \theta_{\xi\xi}). \tag{4.53}$$

Moreover, the relations

$$\begin{aligned}\boldsymbol{r}_{\xi\xi} - \boldsymbol{r}_{\eta\eta} &= (e-g)\boldsymbol{N}\\ \boldsymbol{r}_\xi \times \boldsymbol{r}_\eta = |\boldsymbol{r}_\xi \times \boldsymbol{r}_\eta|\boldsymbol{N} &= \pm\sin\theta\cos\theta\boldsymbol{N}\end{aligned} \tag{4.54}$$

obtain.

If we now impose the restriction

$$e - g = \epsilon \sin 2\theta, \quad \epsilon \text{ constant}, \tag{4.55}$$

then the Mainardi-Codazzi equations (4.52) reduce to the pair of conservation laws

$$\begin{aligned}\frac{\partial}{\partial \xi}(f \tan\theta) &= \frac{\partial}{\partial \eta}\left(\frac{e+g}{2}\tan\theta\right),\\ \frac{\partial}{\partial \eta}(f \cot\theta) &= \frac{\partial}{\partial \xi}\left(\frac{e+g}{2}\cot\theta\right).\end{aligned} \tag{4.56}$$

Equation $(4.56)_1$ allows the introduction of a potential $\phi(\xi, \eta)$ such that

$$\phi_\xi = \frac{e+g}{2}\tan\theta, \quad \phi_\eta = f\tan\theta, \tag{4.57}$$

whence, on substitution into (4.53) and $(4.56)_2$ we obtain

$$\boxed{\begin{aligned}\theta_{\xi\xi} - \theta_{\eta\eta} - \epsilon^2\cos\theta\sin\theta + \frac{\cos\theta}{\sin^3\theta}\left(\phi_\xi^2 - \phi_\eta^2\right) &= 0,\\ \frac{\partial}{\partial \xi}(\phi_\xi \cot^2\theta) &= \frac{\partial}{\partial \eta}(\phi_\eta \cot^2\theta).\end{aligned}} \tag{4.58}$$

This is the *Pohlmeyer-Lund-Regge* system as originally set down in 1976. By virtue of (4.54), it is seen that the constraint (4.55) implies the concise equation

$$\boldsymbol{r}_{\xi\xi} - \boldsymbol{r}_{\eta\eta} = \pm 2\epsilon \boldsymbol{r}_{\xi} \times \boldsymbol{r}_{\eta} \tag{4.59}$$

for the position vector $\boldsymbol{r}$ to the surfaces Σ.

4.2.2 The SIT Connection

The sharpline SIT equations arise out of the unpumped Maxwell-Bloch system [349]

$$\boxed{\begin{aligned} E_x &= P, \\ P_t &= EN, \\ N_t &= -\frac{1}{2}(\bar{E}P + E\bar{P}), \\ N^2 + P\bar{P} &= 1, \end{aligned}} \tag{4.60}$$

where E, P designate the slowly varying amplitudes of the electric field and polarisation, respectively, N is the atomic inversion and x, t are appropriately scaled space and retarded time variables. Here, $\bar{E}$, $\bar{P}$ denote the complex conjugates of E and P, respectively.

The relation $(4.60)_4$ allows us to set

$$N = \cos\chi \tag{4.61}$$

together with

$$P = e^{i\nu}\sin\chi, \tag{4.62}$$

whence the residual equations in (4.60) yield

$$\begin{aligned} \chi_{tx} &= \sin\chi + \nu_t \nu_x \tan\chi, \\ \nu_{tx} &= -\nu_x \chi_t \cot\chi - \nu_t \chi_x (\cos\chi \sin\chi)^{-1}. \end{aligned} \tag{4.63}$$

This is the standard sharpline SIT system.

The SIT equation $(4.63)_2$ possesses the conservation law

$$(\nu_x \cos\chi)_t = (\nu_t / \cos\chi)_x, \tag{4.64}$$

so that a potential $\mu(x, t)$ may be introduced according to

$$\mu_x = \nu_x \cos\chi, \quad \mu_t = \nu_t / \cos\chi. \tag{4.65}$$

In terms of χ and μ, the SIT system (4.63) becomes

$$\begin{aligned} \chi_{tx} &= \sin\chi + \mu_t \mu_x \tan\chi, \\ \mu_{tx} &= -\mu_t \chi_x \cot\chi - \mu_x \chi_t (\cos\chi \sin\chi)^{-1}. \end{aligned} \tag{4.66}$$

This represents an SIT system (4.63) with $\nu \to \mu$ and the roles of x and t interchanged. Thus, the quantities

$$\begin{aligned} \tilde{N} &= \cos\chi, \quad \tilde{P} = e^{i\mu} \sin\chi, \\ \tilde{E} &= (\cos\chi)^{-1} (e^{i\mu} \sin\chi)_x \end{aligned} \tag{4.67}$$

satisfy the associated unpumped Maxwell-Bloch system

$$\begin{aligned} \tilde{E}_t &= \tilde{P}, \\ \tilde{P}_x &= \tilde{E}\tilde{N}, \\ \tilde{N}_x &= -\frac{1}{2}(\bar{\tilde{E}}\tilde{P} + \tilde{E}\bar{\tilde{P}}), \\ \tilde{N}^2 &+ \tilde{P}\bar{\tilde{P}} = 1. \end{aligned} \tag{4.68}$$

The conservation law (4.64) also allows the introduction of a potential ζ such that

$$\begin{aligned} 2\zeta_x &= \nu_x(\cos\chi + 1), \\ 2\zeta_t &= \nu_t \left(\frac{1}{\cos\chi} + 1 \right), \end{aligned} \tag{4.69}$$

whence, on use of (4.65), it is seen that $\zeta = (\mu + \nu)/2$. Relations (4.69) yield

$$\nu_x = \frac{2\zeta_x}{\cos\chi + 1}, \quad \nu_t = \frac{2\zeta_t \cos\chi}{\cos\chi + 1} \tag{4.70}$$

and substitution into $(4.63)_1$ together with appeal to the compatibility condition $\nu_{xt} = \nu_{tx}$ gives

$$\chi_{tx} = \sin\chi + \frac{4\zeta_x \zeta_t \sin\chi}{(\cos\chi + 1)^2}, \quad \zeta_{tx} = -\frac{1}{\sin\chi}(\zeta_x \chi_t + \zeta_t \chi_x). \tag{4.71}$$

Thus, on setting $\Sigma = \chi/2$, the system

$$\begin{aligned} \Sigma_{tx} - \zeta_t \zeta_x \frac{\sin \Sigma}{\cos^3 \Sigma} &= \sin \Sigma \cos \Sigma, \\ \zeta_{tx} + \frac{\zeta_x \Sigma_t + \zeta_t \Sigma_x}{\sin \Sigma \cos \Sigma} &= 0 \end{aligned} \tag{4.72}$$

results. If we now introduce

$$\Sigma = \theta + \frac{\pi}{2}, \quad \zeta = \phi, \tag{4.73}$$

and

$$x = -\frac{\epsilon^2}{2}(\xi + \eta), \quad t = \frac{1}{2}(\xi - \eta), \tag{4.74}$$

then (4.72) coincides with the Pohlmeyer-Lund-Regge system (4.58). To summarise, the latter system is connected to the sharpline SIT system (4.63) via the transformation

$$\begin{aligned} 2\,d\phi &= v_x(\cos\chi + 1)\,dx + v_t\left(\frac{1}{\cos\chi} + 1\right)dt, \\ 2\theta &= \chi - \pi \end{aligned} \tag{4.75}$$

together with the change of independent variables (4.74).

4.2.3 The SRS Connection

The SRS system adopts the form [349]

$$\frac{\partial A_1}{\partial X} = -SA_2, \quad \frac{\partial A_2}{\partial X} = \bar{S}A_1, \quad \frac{\partial S}{\partial T} = A_1\bar{A}_2, \tag{4.76}$$

where A_1, A_2 are the electric field amplitudes of the pump and Stokes waves, respectively, while S is the Raman amplitude. The total intensity

$$I = |A_1|^2 + |A_2|^2 \tag{4.77}$$

depends on T alone. If we set

$$\tilde{E} = 2S, \quad \tilde{P} = 2\frac{A_1\bar{A}_2}{I}, \quad \tilde{N} = \frac{|A_1|^2 - |A_2|^2}{I} \tag{4.78}$$

together with

$$x = X, \quad t = \int_0^T I(T')dT' \tag{4.79}$$

then the unpumped Maxwell-Bloch system (4.68) is recovered.

4.2.4 Compatibility of the Maxwell-Bloch System with the NLS Equation

To establish a connection between the unpumped Maxwell-Bloch system and the NLS equation, it proves convenient to return to the 3×3 linear representation (4.16). It is observed that its compatibility conditions deliver time evolution equations for κ and τ, respectively, given by (4.7), (4.8), which are invariant under the transformation

$$\begin{aligned} &\kappa \to \kappa^* = \kappa, && \tau \to \tau^* = \tau + \lambda \\ &s \to s^* = s + 2\lambda t, && t \to t^* = t. \end{aligned} \tag{4.80}$$

Under these changes of variable, the geometric constraint (4.2) becomes

$$\boldsymbol{r}_{t^*} = \kappa^* \boldsymbol{b} - 2\lambda \boldsymbol{t}, \tag{4.81}$$

so that introduction of the parameter λ produces a sliding motion contribution.

At the linear level, the invariance under (4.80) injects a 'spectral' parameter λ into the representation (4.16), leading to

$$\begin{aligned} \begin{pmatrix} \boldsymbol{t} \\ \boldsymbol{n} \\ \boldsymbol{b} \end{pmatrix}_{s^*} &= \begin{pmatrix} 0 & \kappa^* & 0 \\ -\kappa^* & 0 & \tau^* - \lambda \\ 0 & -\tau^* + \lambda & 0 \end{pmatrix} \begin{pmatrix} \boldsymbol{t} \\ \boldsymbol{n} \\ \boldsymbol{b} \end{pmatrix}, \\ \begin{pmatrix} \boldsymbol{t} \\ \boldsymbol{n} \\ \boldsymbol{b} \end{pmatrix}_{t^*} &= \begin{pmatrix} 0 & -\kappa^*(\tau^* + \lambda) & \kappa^*_{s^*} \\ \kappa^*(\tau^* + \lambda) & 0 & \Xi \\ -\kappa^*_{s^*} & -\Xi & 0 \end{pmatrix} \begin{pmatrix} \boldsymbol{t} \\ \boldsymbol{n} \\ \boldsymbol{b} \end{pmatrix} \end{aligned} \tag{4.82}$$

where

$$\Xi = \frac{\kappa^*_{s^*s^*}}{\kappa^*} - \tau^{*^2} + \lambda^2. \tag{4.83}$$

Under the correspondence $L_k \leftrightarrow e_k$, this 3×3 linear representation for the

NLS equation goes over to the 2×2 linear representation

$$\begin{aligned}
\begin{pmatrix} \Phi_1 \\ \Phi_2 \end{pmatrix}_{s^*} &= \left[-(\tau^* - \lambda)\frac{\sigma_1}{2i} - \kappa^* \frac{\sigma_3}{2i} \right] \begin{pmatrix} \Phi_1 \\ \Phi_2 \end{pmatrix} \\
&= \frac{1}{2i} \begin{pmatrix} -\kappa^* & -(\tau^* - \lambda) \\ -(\tau^* - \lambda) & \kappa^* \end{pmatrix} \begin{pmatrix} \Phi_1 \\ \Phi_2 \end{pmatrix}, \\
\begin{pmatrix} \Phi_1 \\ \Phi_2 \end{pmatrix}_{t^*} &= \left[-\Xi \frac{\sigma_1}{2i} + \kappa^*_{s^*} \frac{\sigma_2}{2i} + \kappa^*(\tau^* + \lambda)\frac{\sigma_3}{2i} \right] \begin{pmatrix} \Phi_1 \\ \Phi_2 \end{pmatrix} \\
&= \frac{1}{2i} \begin{pmatrix} \kappa^*(\tau^* + \lambda) & -\Xi - i\kappa^*_{s^*} \\ -\Xi + i\kappa^*_{s^*} & -\kappa^*(\tau^* + \lambda) \end{pmatrix} \begin{pmatrix} \Phi_1 \\ \Phi_2 \end{pmatrix}.
\end{aligned} \tag{4.84}$$

A rotation corresponding to $L_1 \to e_3$, $L_2 \to e_1$, $L_3 \to e_2$ applied to (4.82) produces a system gauge equivalent to (4.84), namely

$$\begin{aligned}
\begin{pmatrix} \tilde{\Phi}_1 \\ \tilde{\Phi}_2 \end{pmatrix}_{s^*} &= \frac{1}{2} \begin{pmatrix} i(\tau^* - \lambda) & \kappa^* \\ -\kappa^* & -i(\tau^* - \lambda) \end{pmatrix} \begin{pmatrix} \tilde{\Phi}_1 \\ \tilde{\Phi}_2 \end{pmatrix}, \\
\begin{pmatrix} \tilde{\Phi}_1 \\ \tilde{\Phi}_2 \end{pmatrix}_{t^*} &= \frac{1}{2i} \begin{pmatrix} -\Xi & \kappa^*_{s^*} - i\kappa^*(\tau^* + \lambda) \\ \kappa^*_{s^*} + i\kappa^*(\tau^* + \lambda) & \Xi \end{pmatrix} \begin{pmatrix} \tilde{\Phi}_1 \\ \tilde{\Phi}_2 \end{pmatrix}.
\end{aligned} \tag{4.85}$$

On setting

$$\begin{aligned}
\tilde{\Phi}_1 &= v_2 e^{\frac{1}{2}i \int \tau^* ds^*}, \\
\tilde{\Phi}_2 &= v_1 e^{-\frac{1}{2}i \int \tau^* ds^*}
\end{aligned} \tag{4.86}$$

in (4.85), the standard AKNS linear representation for the NLS equation results, namely

$$\begin{aligned}
\begin{pmatrix} v_1 \\ v_2 \end{pmatrix}_{s^*} &= \frac{1}{2} \left[\begin{pmatrix} 0 & -q \\ \bar{q} & 0 \end{pmatrix} + i\lambda \begin{pmatrix} 1 & 0 \\ 0 & -1 \end{pmatrix} \right] \begin{pmatrix} v_1 \\ v_2 \end{pmatrix}, \\
\begin{pmatrix} v_1 \\ v_2 \end{pmatrix}_{t^*} &= \frac{1}{2} \begin{pmatrix} A & B \\ -\bar{B} & -A \end{pmatrix} \begin{pmatrix} v_1 \\ v_2 \end{pmatrix},
\end{aligned} \tag{4.87}$$

where

$$q = \kappa^* e^{i \int \tau^* ds^*} \tag{4.88}$$

and

$$A = \frac{1}{2} i |q|^2 - i\lambda^2, \quad B = -i q_{s^*} + q\lambda. \tag{4.89}$$

If we now adjoin an evolution

$$\begin{pmatrix} v_1 \\ v_2 \end{pmatrix}_x = \frac{1}{2i\lambda} \begin{pmatrix} N & -P \\ -\bar{P} & -N \end{pmatrix} \begin{pmatrix} v_1 \\ v_2 \end{pmatrix} \tag{4.90}$$

to the AKNS representation (4.87) for the NLS equation then compatibility of $(4.87)_1$, and (4.90) produces the unpumped Maxwell-Bloch system

$$\begin{aligned} E_x &= P, \\ P_{s^*} &= EN, \\ N_{s^*} &= -\frac{1}{2}(\bar{E}P + E\bar{P}), \end{aligned} \tag{4.91}$$

where

$$E = -q = -\kappa^* e^{i\int \tau^* ds^*}. \tag{4.92}$$

The residual relation (cf. 4.60)

$$N^2 + P\bar{P} = 1$$

is a consequence of (4.91) and appropriate scaling. Thus, we obtain the important result that the unpumped Maxwell Bloch system may be derived via compatibility with the NLS equation. It is recalled that both the mKdV equation and the Weingarten system have been earlier obtained via compatibility with the classical sine-Gordon equation.

4.3 Geometry of the NLS Equation. The Auto-Bäcklund Transformation

In Section 4.1, a geometric derivation of the NLS equation was presented based on a particular evolution of a curve in space. Here, this canonical equation is derived in a geometric setting adopted in a kinematic analysis of certain hydrodynamic motions by Marris and Passman [247] and subsequently applied in magnetohydrodynamics by Rogers and Kingston [305]. A derivation of the auto-Bäcklund transformation for the NLS equation is obtained at the level of the Hasimoto surface. Smoke ring-type solutions are thereby generated as in [312].

The starting point is the system governing the directional derivatives of the orthonormal triad $\{\boldsymbol{t}, \boldsymbol{n}, \boldsymbol{b}\}$, namely (Marris and Passman [247])

$$\begin{aligned}
\frac{\delta}{\delta s}\begin{pmatrix}\boldsymbol{t}\\ \boldsymbol{n}\\ \boldsymbol{b}\end{pmatrix} &= \begin{pmatrix}0 & \kappa & 0\\ -\kappa & 0 & \tau\\ 0 & -\tau & 0\end{pmatrix}\begin{pmatrix}\boldsymbol{t}\\ \boldsymbol{n}\\ \boldsymbol{b}\end{pmatrix},\\
\frac{\delta}{\delta n}\begin{pmatrix}\boldsymbol{t}\\ \boldsymbol{n}\\ \boldsymbol{b}\end{pmatrix} &= \begin{pmatrix}0 & \theta_{ns} & (\Omega_b+\tau)\\ -\theta_{ns} & 0 & -\operatorname{div}\boldsymbol{b}\\ -(\Omega_b+\tau) & \operatorname{div}\boldsymbol{b} & 0\end{pmatrix}\begin{pmatrix}\boldsymbol{t}\\ \boldsymbol{n}\\ \boldsymbol{b}\end{pmatrix},\\
\frac{\delta}{\delta b}\begin{pmatrix}\boldsymbol{t}\\ \boldsymbol{n}\\ \boldsymbol{b}\end{pmatrix} &= \begin{pmatrix}0 & -(\Omega_n+\tau) & \theta_{bs}\\ \Omega_n+\tau & 0 & \kappa+\operatorname{div}\boldsymbol{n}\\ -\theta_{bs} & -(\kappa+\operatorname{div}\boldsymbol{n}) & 0\end{pmatrix}\begin{pmatrix}\boldsymbol{t}\\ \boldsymbol{n}\\ \boldsymbol{b}\end{pmatrix}.
\end{aligned} \tag{4.93}$$

The notation of [247] is adopted throughout, so that, in particular, $\delta/\delta s$, $\delta/\delta n$ and $\delta/\delta b$ denote directional derivatives in the tangential, principal normal and binormal directions, respectively. Thus, $(4.93)_1$ represents the usual Serret-Frenet relations while $(4.93)_2$ and $(4.93)_3$ provide the directional derivatives of the orthonormal triad $\{\boldsymbol{t}, \boldsymbol{n}, \boldsymbol{b}\}$ in the $\boldsymbol{n}$- and $\boldsymbol{b}$-directions, respectively. Accordingly,

$$\operatorname{grad} = \boldsymbol{t}\frac{\delta}{\delta s} + \boldsymbol{n}\frac{\delta}{\delta n} + \boldsymbol{b}\frac{\delta}{\delta b} \tag{4.94}$$

while θ_{bs} and θ_{ns} are the quantities originally introduced by Bjørgum [38] via

$$\theta_{ns} = \boldsymbol{n}\cdot\frac{\delta\boldsymbol{t}}{\delta n}, \quad \theta_{bs} = \boldsymbol{b}\cdot\frac{\delta\boldsymbol{t}}{\delta b}, \tag{4.95}$$

and

$$\begin{aligned}
\operatorname{div}\boldsymbol{t} &= \left(\boldsymbol{t}\frac{\delta}{\delta s} + \boldsymbol{n}\frac{\delta}{\delta n} + \boldsymbol{b}\frac{\delta}{\delta b}\right)\cdot\boldsymbol{t}\\
&= \boldsymbol{t}\cdot(\kappa\boldsymbol{n}) + \boldsymbol{n}\cdot\frac{\delta\boldsymbol{t}}{\delta n} + \boldsymbol{b}\cdot\frac{\delta\boldsymbol{t}}{\delta b} = \theta_{ns} + \theta_{bs},
\end{aligned} \tag{4.96}$$

$$\begin{aligned}
\operatorname{div}\boldsymbol{n} &= \left(\boldsymbol{t}\frac{\delta}{\delta s} + \boldsymbol{n}\frac{\delta}{\delta n} + \boldsymbol{b}\frac{\delta}{\delta b}\right)\cdot\boldsymbol{n}\\
&= \boldsymbol{t}\cdot(-\kappa\boldsymbol{t} + \tau\boldsymbol{b}) + \boldsymbol{n}\cdot\frac{\delta\boldsymbol{n}}{\delta n} + \boldsymbol{b}\cdot\frac{\delta\boldsymbol{n}}{\delta b} = -\kappa + \boldsymbol{b}\cdot\frac{\delta\boldsymbol{n}}{\delta b}
\end{aligned} \tag{4.97}$$

$$\begin{aligned}
\operatorname{div}\boldsymbol{b} &= \left(\boldsymbol{t}\frac{\delta}{\delta s} + \boldsymbol{n}\frac{\delta}{\delta n} + \boldsymbol{b}\frac{\delta}{\delta b}\right)\cdot\boldsymbol{b}\\
&= \boldsymbol{t}\cdot(-\tau\boldsymbol{n}) + \boldsymbol{n}\cdot\frac{\delta\boldsymbol{b}}{\delta n} = -\boldsymbol{b}\cdot\frac{\delta\boldsymbol{n}}{\delta n}.
\end{aligned} \tag{4.98}$$

Moreover,

$$\begin{aligned}
\operatorname{curl} \boldsymbol{t} &= \left(\boldsymbol{t} \times \frac{\delta}{\delta s} + \boldsymbol{n} \times \frac{\delta}{\delta n} + \boldsymbol{b} \times \frac{\delta}{\delta b}\right)\boldsymbol{t} \\
&= \boldsymbol{t} \times \frac{\delta \boldsymbol{t}}{\delta s} + \boldsymbol{n} \times \frac{\delta \boldsymbol{t}}{\delta n} + \boldsymbol{b} \times \frac{\delta \boldsymbol{t}}{\delta b} \\
&= \boldsymbol{t} \times (\kappa \boldsymbol{n}) + \boldsymbol{n} \times \left[\theta_{ns}\boldsymbol{n} + \left(\boldsymbol{b} \cdot \frac{\delta \boldsymbol{t}}{\delta n}\right)\boldsymbol{b}\right] \\
&\qquad + \boldsymbol{b} \times \left[\left(\boldsymbol{n} \cdot \frac{\delta \boldsymbol{t}}{\delta b}\right)\boldsymbol{n} + \left(\boldsymbol{b} \cdot \frac{\delta \boldsymbol{t}}{\delta b}\right)\boldsymbol{b}\right] \\
&= \Omega_s \boldsymbol{t} + \kappa \boldsymbol{b},
\end{aligned} \tag{4.99}$$

where

$$\Omega_s = \boldsymbol{t} \cdot \operatorname{curl} \boldsymbol{t} = \boldsymbol{b} \cdot \frac{\delta \boldsymbol{t}}{\delta n} - \boldsymbol{n} \cdot \frac{\delta \boldsymbol{t}}{\delta b} \tag{4.100}$$

is called the abnormality of the $\boldsymbol{t}$-field. The relation (4.99) is of considerable importance in the sequel. It was originally obtained in 1927 by Masotti [249] and rediscovered independently by Emde [120] and Bjørgum [38].

Similarly,

$$\begin{aligned}
\operatorname{curl} \boldsymbol{n} &= \left(\boldsymbol{t} \times \frac{\delta}{\delta s} + \boldsymbol{n} \times \frac{\delta}{\delta n} + \boldsymbol{b} \times \frac{\delta}{\delta b}\right)\boldsymbol{n} \\
&= \boldsymbol{t} \times (-\kappa \boldsymbol{t} + \tau \boldsymbol{b}) + \boldsymbol{n} \times \left[\left(\boldsymbol{t} \cdot \frac{\delta \boldsymbol{n}}{\delta n}\right)\boldsymbol{t} + \left(\boldsymbol{b} \cdot \frac{\delta \boldsymbol{n}}{\delta n}\right)\boldsymbol{b}\right] \\
&\qquad + \boldsymbol{b} \times \left(\boldsymbol{t} \cdot \frac{\delta \boldsymbol{n}}{\delta n}\right)\boldsymbol{t} \\
&= -(\operatorname{div} \boldsymbol{b})\boldsymbol{t} + \Omega_n \boldsymbol{n} + \theta_{ns} \boldsymbol{b},
\end{aligned} \tag{4.101}$$

where

$$\Omega_n = \boldsymbol{n} \cdot \operatorname{curl} \boldsymbol{n} = \boldsymbol{t} \cdot \frac{\delta \boldsymbol{n}}{\delta b} - \tau \tag{4.102}$$

is termed the abnormality of the $\boldsymbol{n}$-field. Finally,

$$\begin{aligned}
\operatorname{curl} \boldsymbol{b} &= \left(\boldsymbol{t} \times \frac{\delta}{\delta s} + \boldsymbol{n} \times \frac{\delta}{\delta n} + \boldsymbol{b} \times \frac{\delta}{\delta b}\right)\boldsymbol{b} \\
&= \boldsymbol{t} \times (-\tau \boldsymbol{n}) + \boldsymbol{n} \times \left(\boldsymbol{t} \cdot \frac{\delta \boldsymbol{b}}{\delta n}\right)\boldsymbol{t}
\end{aligned} \tag{4.103}$$

$$\begin{aligned}
&\qquad + \boldsymbol{b} \times \left[\left(\boldsymbol{t} \cdot \frac{\delta \boldsymbol{b}}{\delta b}\right)\boldsymbol{t} + \left(\boldsymbol{n} \cdot \frac{\delta \boldsymbol{b}}{\delta b}\right)\boldsymbol{n}\right] \\
&= (\kappa + \operatorname{div} \boldsymbol{n})\boldsymbol{t} - \theta_{bs} \boldsymbol{n} + \Omega_b \boldsymbol{b},
\end{aligned} \tag{4.104}$$

where

$$\Omega_b = -\tau - \boldsymbol{t} \cdot \frac{\delta \boldsymbol{b}}{\delta n} \tag{4.105}$$

is the abnormality of the $\boldsymbol{b}$-field.

Now, the identity curl grad $\phi = \boldsymbol{0}$ yields

$$\frac{\delta\phi}{\delta s}\operatorname{curl}\boldsymbol{t} + \operatorname{grad}\left(\frac{\delta\phi}{\delta s}\right) \times \boldsymbol{t} + \frac{\delta\phi}{\delta n}\operatorname{curl}\boldsymbol{n} + \operatorname{grad}\left(\frac{\delta\phi}{\delta n}\right) \times \boldsymbol{n}$$
$$+ \frac{\delta\phi}{\delta b}\operatorname{curl}\boldsymbol{b} + \operatorname{grad}\left(\frac{\delta\phi}{\delta b}\right) \times \boldsymbol{b} = \boldsymbol{0},$$

whence

$$\left(\frac{\delta^2\phi}{\delta n\delta b} - \frac{\delta^2\phi}{\delta b\delta n}\right)\boldsymbol{t} + \left(\frac{\delta^2\phi}{\delta b\delta s} - \frac{\delta^2\phi}{\delta s\delta b}\right)\boldsymbol{n} + \left(\frac{\delta^2\phi}{\delta s\delta n} - \frac{\delta^2\phi}{\delta n\delta s}\right)\boldsymbol{b}$$
$$+ \frac{\delta\phi}{\delta s}\operatorname{curl}\boldsymbol{t} + \frac{\delta\phi}{\delta n}\operatorname{curl}\boldsymbol{n} + \frac{\delta\phi}{\delta b}\operatorname{curl}\boldsymbol{b} = \boldsymbol{0}. \tag{4.106}$$

In the above, the convention $\dfrac{\delta^2\phi}{\delta n\delta b} = \dfrac{\delta}{\delta n}\left(\dfrac{\delta\phi}{\delta b}\right)$ has been adopted. On use of the relations (4.99), (4.101) and (4.103) in (4.106), it follows that

$$\begin{aligned}
\frac{\delta^2\phi}{\delta n\delta b} - \frac{\delta^2\phi}{\delta b\delta n} &= -\frac{\delta\phi}{\delta s}\Omega_s + \frac{\delta\phi}{\delta n}\operatorname{div}\boldsymbol{b} - \frac{\delta\phi}{\delta b}(\kappa + \operatorname{div}\boldsymbol{n}),\\
\frac{\delta^2\phi}{\delta b\delta s} - \frac{\delta^2\phi}{\delta s\delta b} &= -\frac{\delta\phi}{\delta n}\Omega_n + \frac{\delta\phi}{\delta b}\theta_{bs},\\
\frac{\delta^2\phi}{\delta s\delta n} - \frac{\delta^2\phi}{\delta n\delta s} &= -\frac{\delta\phi}{\delta s}\kappa - \frac{\delta\phi}{\delta n}\theta_{ns} - \frac{\delta\phi}{\delta b}\Omega_b.
\end{aligned} \tag{4.107}$$

Thus, the second-order mixed intrinsic derivatives, in general, do not commute and so the quantities s, n and b represent anholonomic coordinates [373]. The compatibility of the linear system (4.93) now imposes, on use of the relations (4.107), a set of nine conditions on the eight geometric quantities κ, τ, Ω_s, Ω_n, div $\boldsymbol{n}$, div $\boldsymbol{b}$, θ_{ns} and θ_{bs}, namely [248]:

$$\frac{\delta}{\delta b}\theta_{ns} + \frac{\delta}{\delta n}(\tau + \Omega_n) = (\kappa + \operatorname{div}\boldsymbol{n})(\Omega_s - 2\Omega_n - 2\tau)$$
$$+(\theta_{bs} - \theta_{ns})\operatorname{div}\boldsymbol{b} + \Omega_s\kappa, \tag{4.108}$$

$$\frac{\delta}{\delta b}(\tau + \Omega_n - \Omega_s) + \frac{\delta}{\delta n}\theta_{bs} = (\kappa + \operatorname{div}\boldsymbol{n})(\theta_{ns} - \theta_{bs})$$
$$+(\Omega_s - 2\Omega_n - 2\tau)\operatorname{div}\boldsymbol{b}, \tag{4.109}$$

$$\frac{\delta}{\delta b}(\operatorname{div}\boldsymbol{b}) + \frac{\delta}{\delta n}(\kappa + \operatorname{div}\boldsymbol{n}) = (\tau + \Omega_n)(\tau + \Omega_n - \Omega_s) - \theta_{ns}\theta_{bs} - \tau\Omega_s - (\operatorname{div}\boldsymbol{b})^2 - (\kappa + \operatorname{div}\boldsymbol{n})^2, \tag{4.110}$$

$$\frac{\delta}{\delta s}(\tau + \Omega_n) + \frac{\delta\kappa}{\delta b} = -\Omega_n\theta_{ns} - (2\tau + \Omega_n)\theta_{bs}, \tag{4.111}$$

$$\frac{\delta}{\delta s}\theta_{bs} = -\theta_{bs}^2 + \kappa(\kappa + \operatorname{div}\boldsymbol{n}) - \Omega_n(\tau + \Omega_n - \Omega_s) + \tau(\tau + \Omega_n), \tag{4.112}$$

$$\frac{\delta}{\delta s}(\kappa + \operatorname{div}\boldsymbol{n}) - \frac{\delta\tau}{\delta b} = -\Omega_n \operatorname{div}\boldsymbol{b} - \theta_{bs}(2\kappa + \operatorname{div}\boldsymbol{n}), \tag{4.113}$$

$$\frac{\delta\kappa}{\delta n} - \frac{\delta}{\delta s}\theta_{ns} = \kappa^2 + \theta_{ns}^2 + (\tau + \Omega_n)(3\tau + \Omega_n) - \Omega_s(2\tau + \Omega_n), \tag{4.114}$$

$$\frac{\delta}{\delta s}(\tau + \Omega_n - \Omega_s) = -\theta_{ns}(\Omega_n - \Omega_s) + \kappa \operatorname{div}\boldsymbol{b} + \theta_{bs}(-2\tau - \Omega_n + \Omega_s), \tag{4.115}$$

$$\frac{\delta\tau}{\delta n} + \frac{\delta}{\delta s}(\operatorname{div}\boldsymbol{b}) = -\kappa(\Omega_n - \Omega_s) - \theta_{ns}\operatorname{div}\boldsymbol{b} + (\kappa + \operatorname{div}\boldsymbol{n})(-2\tau - \Omega_n + \Omega_s). \tag{4.116}$$

An equivalent system to the above was obtained independently in a viscometric study by Yin and Pipkin [386].

In conclusion, it is observed that the important relation

$$\Omega_s - \tau = \frac{1}{2}(\Omega_s + \Omega_n + \Omega_b) \tag{4.117}$$

is obtained by combination of the relations (4.100), (4.102) and (4.105). This result is recorded in the treatise of Weatherburn [381] published in 1930. Therein, Ω_s, Ω_n and Ω_b are termed the total moments of the $\boldsymbol{t}$, $\boldsymbol{n}$ and $\boldsymbol{b}$ congruences, respectively. It is noted there that the relation (4.117) incorporates various important theorems as corollaries, including Dupin's theorem which states that the curves of intersection of the surfaces of a triply orthogonal system are lines of curvature on these surfaces.

4.3.1 The Nonlinear Schrödinger Equation

In what follows, the vanishing abnormality condition

$$\Omega_n = 0 \tag{4.118}$$

is imposed. Hydrodynamic and magnetohydrodynamic motions with such a geometric constraint have been investigated, in turn, by Marris and Passman [247] and Rogers and Kingston [305]. Here, however, our purpose is to derive the celebrated nonlinear Schrödinger equation via such a restriction as in [312].

The condition (4.118) represents the necessary and sufficient condition for the existence of a one-parameter family of surfaces containing the s-lines and b-lines. It is equivalent to the requirement

$$\boldsymbol{n} = \psi \operatorname{grad} U, \tag{4.119}$$

where the 'foliation' has constituent members $U(\boldsymbol{r}) = \text{const}$ and ψ is the distance function.

Now, the geodesic curvature of a family of curves with unit tangent $\boldsymbol{T}$ on a surface Σ is given by [380]

$$\kappa_g = \boldsymbol{N} \cdot \operatorname{curl}_\Sigma \boldsymbol{T}. \tag{4.120}$$

In the case of the above foliation, since the s-lines and b-lines lie on the constituent surfaces Σ, it follows that $\boldsymbol{n}$ is perpendicular to Σ. Accordingly, $\boldsymbol{n}$ is parallel to $\boldsymbol{N}$ and the Masotti relation (4.99) shows that

$$\boldsymbol{N} \cdot \operatorname{curl}_\Sigma \boldsymbol{t} = \boldsymbol{N} \cdot \operatorname{curl} \boldsymbol{t} = 0, \tag{4.121}$$

whence the geodesic curvature of the s-lines is zero. Thus, the s-lines are geodesics on the members of the foliation. Furthermore, the orthogonal trajectories of a family of geodesics constitute a family of parallels [381]. Hence, the b-lines are necessarily parallel curves on members of the foliation.

The geodesic curvature of the b-lines is given by

$$\kappa_{bg} = \boldsymbol{N} \cdot \operatorname{curl} \boldsymbol{b} = -\boldsymbol{n} \cdot \operatorname{curl} \boldsymbol{b} = \theta_{bs} \tag{4.122}$$

so that the b-lines are also geodesics iff $\theta_{bs} = 0$. When $\theta_{bs} = 0$, the members Σ of the foliation sustain orthogonal geodesics, namely the s-lines and b-lines, and are, as will be seen, necessarily developables.

The total curvature $\mathcal{K}$ of the surfaces Σ containing the s-lines and b-lines is given by [380]

$$\mathcal{K} = |\boldsymbol{N}, \mathrm{curl}_\Sigma \boldsymbol{t}, \mathrm{curl}_\Sigma \boldsymbol{b}|, \tag{4.123}$$

where

$$\mathrm{curl}_\Sigma \boldsymbol{t} = \left(\boldsymbol{t}\frac{\delta}{\delta s} + \boldsymbol{b}\frac{\delta}{\delta b} \right) \times \boldsymbol{t} = (\Omega_n + \tau)\boldsymbol{t} + \kappa \boldsymbol{b} \tag{4.124}$$

and

$$\mathrm{curl}_\Sigma \boldsymbol{b} = \left(\boldsymbol{t}\frac{\delta}{\delta s} + \boldsymbol{b}\frac{\delta}{\delta b} \right) \times \boldsymbol{b} = (\kappa + \mathrm{div}\, \boldsymbol{n})\boldsymbol{t} - \theta_{bs}\boldsymbol{n} - \tau \boldsymbol{b}, \tag{4.125}$$

so that

$$\mathcal{K} = -\kappa(\kappa + \mathrm{div}\, \boldsymbol{n}) - \tau^2. \tag{4.126}$$

The mean curvature $\mathcal{J} = 2\mathcal{M}$ of the surfaces Σ of the foliation is given by [380]

$$\mathcal{J} = -\mathrm{div}_\Sigma \boldsymbol{N} = -\mathrm{div}\, \boldsymbol{N} = \mathrm{div}\, \boldsymbol{n}. \tag{4.127}$$

We now turn to the compatibility conditions (4.111)–(4.113) for the linear equations $(4.93)_1$ and $(4.93)_3$. In the present case with $\Omega_n = 0$, these reduce to the nonlinear system

$$\frac{\delta\tau}{\delta s} + \frac{\delta\kappa}{\delta b} = -2\tau\theta_{bs}, \tag{4.128}$$

$$\frac{\delta}{\delta s}(\kappa + \mathrm{div}\, \boldsymbol{n}) - \frac{\delta\tau}{\delta b} = -\theta_{bs}(2\kappa + \mathrm{div}\, \boldsymbol{n}), \tag{4.129}$$

$$\frac{\delta}{\delta s}\theta_{bs} = -\theta_{bs}^2 + \kappa(\kappa + \mathrm{div}\, \boldsymbol{n}) + \tau^2. \tag{4.130}$$

These incorporate the Gauss-Mainardi-Codazzi equations for the constituent members Σ of the foliation. In particular, (4.130) embodies the Theorema Egregium and here delivers an expression for the Gaussian curvature entirely in terms of the geodesic curvature θ_{bs} of the b-lines, viz

$$\mathcal{K} = -\frac{\delta\theta_{bs}}{\delta s} - \theta_{bs}^2. \tag{4.131}$$

If $\theta_{bs} = 0$ then $\mathcal{K} = 0$ so that, as noted previously, the surfaces Σ of the foliation become developables. The relation (4.127) together with (4.130) show that the mean curvature $\mathcal{J}$ of members of the foliation is given by

$$\mathcal{J} = \left[\frac{\delta}{\delta s}\theta_{bs} + \theta_{bs}^2 - \tau^2 - \kappa^2 \right] \Big/ \kappa. \tag{4.132}$$

If the s-lines and b-lines are now taken as parametric curves on the member surfaces Σ of the foliation, then the surface metric adopts the geodesic form [247]

$$\mathrm{I}_\Sigma = ds^2 + g(s,b)db^2, \tag{4.133}$$

while the two-parameter surface gradient for Σ is given by

$$\mathrm{grad}_\Sigma = \boldsymbol{t}\frac{\delta}{\delta s} + \boldsymbol{b}\frac{\delta}{\delta b} = \boldsymbol{t}\frac{\partial}{\partial s} + \frac{\boldsymbol{b}}{g^{1/2}}\frac{\partial}{\partial b}. \tag{4.134}$$

The Gauss equations $(4.93)_{1,3}$ yield

$$\frac{\partial}{\partial s}\begin{pmatrix} \boldsymbol{t} \\ \boldsymbol{n} \\ \boldsymbol{b} \end{pmatrix} = \begin{pmatrix} 0 & \kappa & 0 \\ -\kappa & 0 & \tau \\ 0 & -\tau & 0 \end{pmatrix}\begin{pmatrix} \boldsymbol{t} \\ \boldsymbol{n} \\ \boldsymbol{b} \end{pmatrix}, \tag{4.135}$$

$$g^{-1/2}\frac{\partial}{\partial b}\begin{pmatrix} \boldsymbol{t} \\ \boldsymbol{n} \\ \boldsymbol{b} \end{pmatrix} = \begin{pmatrix} 0 & -\tau & \theta_{bs} \\ \tau & 0 & \kappa + \operatorname{div}\boldsymbol{n} \\ -\theta_{bs} & -(\kappa + \operatorname{div}\boldsymbol{n}) & 0 \end{pmatrix}\begin{pmatrix} \boldsymbol{t} \\ \boldsymbol{n} \\ \boldsymbol{b} \end{pmatrix} \tag{4.136}$$

and the expression (4.134) encapsulates the relations

$$\frac{\partial \boldsymbol{r}}{\partial s} = \boldsymbol{t}, \quad \frac{\partial \boldsymbol{r}}{\partial b} = g^{1/2}\boldsymbol{b} \tag{4.137}$$

for the position vector $\boldsymbol{r}$ to the surface Σ. Accordingly, the linear system (4.135), (4.136) implies that

$$\frac{1}{g^{1/2}}\frac{\partial^2 \boldsymbol{r}}{\partial b\,\partial s} = -\tau\boldsymbol{n} + \theta_{bs}\boldsymbol{b} \tag{4.138}$$

and

$$\frac{\partial^2 \boldsymbol{r}}{\partial s\,\partial b} = -g^{1/2}\tau\boldsymbol{n} + \frac{\partial g^{1/2}}{\partial s}\boldsymbol{b} \tag{4.139}$$

so that we retrieve the important relation

$$\theta_{bs} = \frac{\partial}{\partial s}\left(\ln g^{1/2}\right). \tag{4.140}$$

The Gauss-Mainardi-Codazzi equations (4.128)–(4.130) now become

$$\frac{\partial}{\partial s}(g\tau) + g^{1/2}\frac{\partial \kappa}{\partial b} = 0 \tag{4.141}$$

$$\frac{\partial \tau}{\partial b} = \frac{\partial}{\partial s}\left[g^{1/2}(\kappa + \operatorname{div} n)\right] + \kappa\frac{\partial g^{1/2}}{\partial s} \tag{4.142}$$

$$[\kappa(\kappa + \operatorname{div}\boldsymbol{n}) + \tau^2]g^{1/2} = \frac{\partial^2 g^{1/2}}{\partial s^2}. \tag{4.143}$$

On elimination of $\kappa + \operatorname{div}\boldsymbol{n}$ between (4.142) and (4.143), it is seen that

$$\frac{\partial \tau}{\partial b} = \frac{\partial}{\partial s}\left[\frac{1}{\kappa}\left(\frac{\partial^2 g^{1/2}}{\partial s^2} - \tau^2 g^{1/2}\right)\right] + \kappa\frac{\partial g^{1/2}}{\partial s}. \tag{4.144}$$

If we now set

$$g^{1/2} = \lambda\kappa, \tag{4.145}$$

where λ varies only in the direction normal to the member surfaces of the foliation, then the pair of equations (4.141) and (4.144), on scaling $\lambda b \to b$, produce the classical Da Rios system

$$\boxed{\begin{aligned} \frac{\partial \kappa}{\partial b} &= -2\frac{\partial \kappa}{\partial s}\tau - \kappa\frac{\partial \tau}{\partial s} \\ \frac{\partial \tau}{\partial s} &= \frac{\partial}{\partial s}\left(\frac{1}{\kappa}\frac{\partial^2 \kappa}{\partial s^2} + \frac{\kappa^2}{2} - \tau^2\right). \end{aligned}} \tag{4.146}$$

On introduction of the Hasimoto transformation (cf. (4.9))

$$q = \kappa e^{i\sigma}, \quad \sigma = \int \tau\, ds, \tag{4.147}$$

the pair of equations (4.146) yields the standard NLS equation

$$i\frac{\partial q}{\partial b} + \frac{\partial^2 q}{\partial s^2} + \frac{1}{2}|q|^2 q = 0, \tag{4.148}$$

where the boundary flux term of the type (4.12) has been absorbed into q.

4.3.2 *The Auto-Bäcklund Transformation*

In the present framework, a surface $\Sigma : \boldsymbol{r} = \boldsymbol{r}(s, b)$ is associated with the NLS equation if and only if the relations

$$\boldsymbol{r}_s^2 = 1, \quad \boldsymbol{r}_s \cdot \boldsymbol{r}_b = 0, \quad \boldsymbol{r}_{ss}^2 = \boldsymbol{r}_b^2 \tag{4.149}$$

hold. This may be seen as follows. If $\{\boldsymbol{t}, \boldsymbol{n}, \boldsymbol{b}\}$ is the orthonormal triad corresponding to the s-parametric lines, then the Serret-Frenet equation

$$\boldsymbol{r}_{ss} = \boldsymbol{t}_s = \kappa \boldsymbol{n} \tag{4.150}$$

and $(4.149)_3$ yield

$$\boldsymbol{r}_b^2 = \kappa^2. \tag{4.151}$$

Thus, the 1st fundamental form of Σ reads

$$\mathrm{I} = ds^2 + \kappa^2 db^2. \tag{4.152}$$

On the other hand, condition $(4.149)_2$ and its differential consequence

$$\boldsymbol{r}_{ss} \cdot \boldsymbol{r}_b = 0 \tag{4.153}$$

show that $\boldsymbol{r}_b \cdot \boldsymbol{t} = \boldsymbol{r}_b \cdot \boldsymbol{n} = 0$ which implies, without loss of generality, that

$$\boldsymbol{r}_b = \kappa \boldsymbol{b}. \tag{4.154}$$

The metric (4.152) and the ‘velocity’ condition (4.154) define the class of Hasimoto surfaces. It is therefore concluded that any invariance of the constraints (4.149) induces an invariance of the NLS equation.

Here, an auto-Bäcklund transformation for the surface Σ is sought in the form [312]

$$\boldsymbol{r}' = \boldsymbol{r} + \alpha \boldsymbol{t} + \beta \boldsymbol{n} + \gamma \boldsymbol{b}, \quad |\boldsymbol{r}' - \boldsymbol{r}| = \text{const.} \tag{4.155}$$

On use of the parametrisation

$$\boldsymbol{r}' = \boldsymbol{r} + L(\cos\theta\, \boldsymbol{t} + \sin\theta \cos\varphi\, \boldsymbol{n} + \sin\theta \sin\varphi\, \boldsymbol{b}), \tag{4.156}$$

invariance of the defining relations (4.149) results in the compatible system

$$
\begin{aligned}
\theta_s &= -\kappa \cos\varphi + \frac{1+\cos\chi}{L}\sin\theta, \\
\varphi_s &= \kappa \cot\theta \sin\varphi - \tau + \frac{\sin\chi}{L}, \\
\theta_b &= -\kappa_s \sin\varphi + \kappa\tau \cos\varphi - 2\frac{(1+\cos\chi)\sin\chi}{L^2}\sin\theta \\
&\quad + \kappa\frac{\cos\varphi\sin\chi - (1+\cos\chi)\sin\varphi\cos\theta}{L}, \\
\varphi_b &= -\frac{\kappa_{ss}}{\kappa} + \tau^2 - (\kappa_s\cos\varphi + \kappa\tau\sin\varphi)\cot\theta + 2\frac{(1+\cos\chi)\cos\chi}{L^2} \\
&\quad - \kappa\frac{\cos\theta\sin\varphi\sin\chi + (1+\cos\chi)\cos\varphi}{L\sin\theta},
\end{aligned}
\tag{4.157}
$$

where χ is an arbitrary constant. It is noted that the calculations leading to the relations (4.157) are considerably reduced if one takes into account the invariance of the redundant condition (4.153).

It emerges that the system (4.157) may be linearised via the decomposition

$$
\xi = \frac{\phi^1}{\phi^2} = \cot\frac{\theta}{2}\, e^{i\varphi},
\tag{4.158}
$$

where $\Phi = (\phi^1 \ \ \phi^2)^{\mathsf{T}}$ is a solution of

$$
\begin{aligned}
\Phi_s &= \frac{1}{2}\begin{pmatrix} -i(\tau-\lambda) & \kappa \\ -\kappa & i(\tau-\lambda) \end{pmatrix}\Phi, \\
\Phi_b &= \frac{1}{2}\begin{pmatrix} -i(\kappa_{ss}/\kappa - \tau^2 + \lambda^2) & i\kappa_s - (\kappa\tau + \kappa\lambda) \\ i\kappa_s + (\kappa\tau + \kappa\lambda) & i(\kappa_{ss}/\kappa - \tau^2 + \lambda^2) \end{pmatrix}\Phi
\end{aligned}
\tag{4.159}
$$

and

$$
\lambda = \frac{\sin\chi + i(1+\cos\chi)}{L}.
\tag{4.160}
$$

In terms of Φ, the Bäcklund transformation (4.156) takes the form

$$
\boldsymbol{r}' = \boldsymbol{r} + 2\frac{\Im(\lambda)}{|\lambda|^2}(n_1\boldsymbol{t} + n_2\boldsymbol{n} + n_3\boldsymbol{b}),
\tag{4.161}
$$

where

$$n_1 = \frac{|\xi|^2 - 1}{|\xi|^2 + 1}, \quad n_2 = \frac{2\Re(\xi)}{|\xi|^2 + 1}, \quad n_3 = \frac{2\Im(\xi)}{|\xi|^2 + 1}. \tag{4.162}$$

Moreover, the relations

$$\kappa'^2 = \boldsymbol{r}'^2_b, \quad \tau' = -\frac{\boldsymbol{r}'_{ss} \cdot \boldsymbol{r}'_{sb}}{\boldsymbol{r}'^2_{ss}} \tag{4.163}$$

deliver the new solution $\{\kappa', \tau'\}$ via

$$\begin{aligned} \kappa'^2 &= \left(\kappa - 4\Im(\lambda)\frac{\xi}{|\xi|^2+1}\right)\left(\kappa - 4\Im(\lambda)\frac{\bar{\xi}}{|\xi|^2+1}\right), \\ \tau' &= \tau + \frac{i}{2}\left[\ln\left(\frac{\kappa - 4\Im(\lambda)\dfrac{\bar{\xi}}{|\xi|^2+1}}{\kappa - 4\Im(\lambda)\dfrac{\xi}{|\xi|^2+1}}\right)\right]_s, \end{aligned} \tag{4.164}$$

whence

$$q' = \kappa' e^{i\int \tau' ds} = q - 4\Im(\lambda)\frac{\tilde{\xi}}{|\tilde{\xi}|^2+1}, \qquad \tilde{\xi} = \xi e^{i\int \tau\, ds}. \tag{4.165}$$

The latter relation suggests the introduction of the gauge-transformation

$$\Psi = \begin{pmatrix} e^{\frac{1}{2}i\int \tau\, ds} & 0 \\ 0 & e^{-\frac{1}{2}i\int \tau\, ds} \end{pmatrix}\Phi, \qquad \tilde{\xi} = \frac{\psi^1}{\psi^2}, \tag{4.166}$$

whereupon the linear representation (4.159) delivers the standard Lax pair

$$\begin{aligned} \Psi_s &= \frac{1}{2}\begin{pmatrix} i\lambda & q \\ -\bar{q} & -i\lambda \end{pmatrix}\Psi, \\ \Psi_b &= \frac{1}{2}\begin{pmatrix} i\left[\frac{1}{2}|q|^2 - \lambda^2\right] & iq_s - \lambda q \\ i\bar{q}_s + \lambda\bar{q} & -i\left[\frac{1}{2}|q|^2 - \lambda^2\right] \end{pmatrix}\Psi \end{aligned} \tag{4.167}$$

for the NLS equation. The relation (4.165) constitutes its Bäcklund transformation as set down in [225].

Darboux transformations have the attractive property that they may be iterated in a purely algebraic manner. Indeed, if one reinterprets the Bäcklund transformation for the Hasimoto surfaces as a Darboux transformation acting on the triad $\{\boldsymbol{t}, \boldsymbol{n}, \boldsymbol{b}\}$, one can construct an infinite suite of surfaces once a seed surface is known. For instance, a straight line corresponding to the solution $\kappa = \tau = 0$ may be regarded as the simplest degenerate Hasimoto surface. Iterative action of the Bäcklund transformation then produces a multi-parameter class of Hasimoto surfaces, and in particular those associated with the multi-soliton solutions of the NLS equation [48].

Here, we focus on the case

$$\kappa = \text{const}, \quad \tau = 0, \quad q = \kappa e^{\frac{1}{2} i \kappa^2 t} \tag{4.168}$$

which corresponds to a cylinder as seed surface. The solution of the Serret-Frenet equations (4.135), (4.136) is then given by

$$\begin{pmatrix} \boldsymbol{t} \\ \boldsymbol{n} \\ \boldsymbol{b} \end{pmatrix} = \begin{pmatrix} \boldsymbol{t}_0 \cos(\kappa s) + \boldsymbol{n}_0 \sin(\kappa s) \\ \boldsymbol{n}_0 \cos(\kappa s) - \boldsymbol{t}_0 \sin(\kappa s) \\ \boldsymbol{b}_0 \end{pmatrix} \tag{4.169}$$

so that the relations

$$\boldsymbol{r}_s = \boldsymbol{t}, \quad \boldsymbol{r}_b = \kappa \boldsymbol{b} \tag{4.170}$$

yield

$$\boldsymbol{r} = \frac{\boldsymbol{t}_0}{\kappa} \sin(\kappa s) - \frac{\boldsymbol{n}_0}{\kappa} \cos(\kappa s) + \kappa b \boldsymbol{b}_0, \tag{4.171}$$

where $\boldsymbol{t}_0$, $\boldsymbol{n}_0$ and $\boldsymbol{b}_0$ are vector-valued orthonormal constants of integration.

The general solution of the Lax pair (4.159) associated with the seed solution (4.168) is of the form

$$\Phi = \Phi(z), \quad z = s - \lambda b. \tag{4.172}$$

This implies that ξ is governed by the Riccati equation

$$\dot{\xi} = i\lambda\xi + \frac{1}{2}\kappa(1 + \xi^2), \tag{4.173}$$

with solution

$$\xi = -i\frac{\lambda}{\kappa} + \beta \tan(\mu z + z_0), \tag{4.174}$$

where

$$\beta = \sqrt{1 + \frac{\lambda^2}{\kappa^2}}, \quad \mu = \frac{1}{2}\kappa\beta \tag{4.175}$$

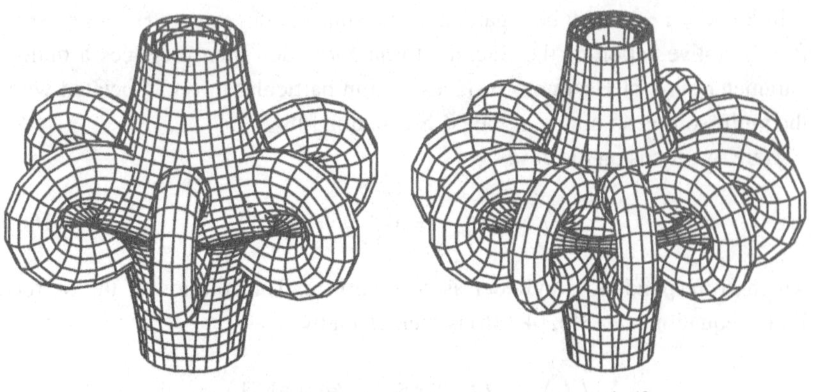

Figure 4.4. NLS 'smoke ring' surfaces for $\frac{m}{n} = \frac{6}{7}, \frac{10}{11}$.

with z_0 an arbitrary constant. Thus, the position vector of the new surfaces Σ' is given by

$$\begin{aligned} \boldsymbol{r}' = \boldsymbol{r} + L\{&[n_1 \cos(\kappa s) - n_2 \sin(\kappa s)]\boldsymbol{t}_0 \\ &+ [n_1 \sin(\kappa s) + n_2 \cos(\kappa s)]\boldsymbol{n}_0 + n_3 \boldsymbol{b}_0]\}. \end{aligned} \tag{4.176}$$

The above class of Hasimoto surfaces has been discussed by Cieśliński et al. [85] in connection with the motion of a vortex filament in the localised-induction approximation. Of particular interest is the subclass defined by

$$\lambda = i\kappa\sqrt{1 - \frac{m^2}{n^2}}, \qquad 0 < m < n, \quad m, n \in \mathbb{N} \tag{4.177}$$

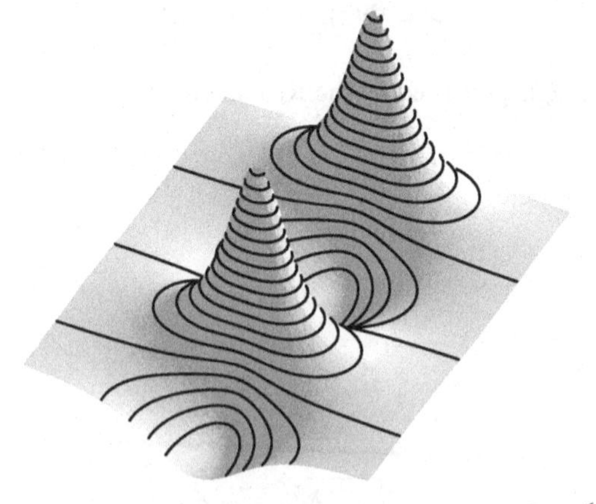

Figure 4.5. An NLS spatial breather solution for $\frac{m}{n} = \frac{6}{7}$.

which guarantees periodicity of the position vector $\boldsymbol{r}'$ in the spatial variable s so that the surfaces admit a discrete rotational symmetry. In Figure 4.4, two such surfaces are displayed for $\kappa = 2$. It is noted, in conclusion, that $\kappa'^2 = |q'|^2$ is not only periodic in s but also approaches κ^2 exponentially as $b \to \pm\infty$. Thus, the subclass (4.177) corresponds to 'spatial' breather solutions of the NLS equation (see Figure 4.5).

Exercises

1. Consider the hydrodynamic system [260]

$$\operatorname{div} \boldsymbol{q} = 0$$
$$\rho(\boldsymbol{q} \cdot \nabla)\boldsymbol{q} + \nabla p = \boldsymbol{0},$$

where $\boldsymbol{q}$ is the steady fluid velocity and p, ρ denote pressure and constant density, respectively. If $\boldsymbol{q} = \mathrm{q}\boldsymbol{t}$, where $\boldsymbol{t}$ is the unit tangent to the streamlines, then derive the following intrinsic compatibility conditions for the pressure distribution:

$$2\left(\frac{\delta}{\delta n} \ln q\right) \operatorname{div} \boldsymbol{t} = -\frac{\delta \kappa}{\delta s} + \kappa \theta_{bs} - \frac{\delta}{\delta n} \operatorname{div} \boldsymbol{t} + 2\kappa \operatorname{div} \boldsymbol{t}$$
$$2\left(\frac{\delta}{\delta b} \ln q\right) \operatorname{div} \boldsymbol{t} = \kappa \Omega_n - \frac{\delta}{\delta b} \operatorname{div} \boldsymbol{t}$$
$$2\kappa\left(\frac{\delta}{\delta b} \ln q\right) = -\operatorname{div}(\kappa \boldsymbol{b}) - \Omega_s \operatorname{div} \boldsymbol{t} = \frac{\delta \Omega_s}{\delta s}.$$

2. Use the compatibility conditions of Exercise 1 to show that if $\operatorname{div} \boldsymbol{t} = 0$, then on individual constant pressure surfaces, the unit tangent vector $\boldsymbol{t}$ obeys the Heisenberg spin equation

$$\frac{\partial \boldsymbol{t}}{\partial b} = \boldsymbol{t} \times \frac{\partial^2 \boldsymbol{t}}{\partial s^2},$$

where s denotes arc length of the geodesic streamlines and b appropriately parametrises their orthogonal trajectories [300, 313].

5

Isothermic Surfaces. The Calapso and Zoomeron Equations

This chapter deals with another class of surfaces that has a solitonic connection, namely isothermic surfaces. Their original study seems to be due to Lamé carried out in the period 1837–1852. Subsequent important contributions were made by Bour [54] in 1862, by Darboux [95] in 1899, Calapso [62] in 1903 and Bianchi [36] in 1905. There is an early account of the classical theory of isothermic surfaces in an encyclopaedia contribution by Voss [372], while later more extensive discussions are provided in the 'Leçons sur la théorie générale des surfaces' of Darboux [93] and in the 'Lezioni di geometria differenziale' of Bianchi [37]. An historical review of the subject has been given by Klimczewski et al. [200].

Here, our treatment will be directed to those properties of classical isothermic surfaces and their recent generalisations which bear upon the subject of modern soliton theory.

5.1 The Gauss-Mainardi-Codazzi Equations for Isothermic Surfaces. The Calapso Equation. Dual Isothermic Surfaces

In what follows, we adopt a *conformal* parametrisation of a surface $\Sigma : \boldsymbol{r} = \boldsymbol{r}(x, y)$ so that the 1st fundamental form reads

$$\mathrm{I} = E(dx^2 + dy^2). \tag{5.1}$$

If, in addition, the 2nd fundamental form is purely diagonal, that is

$$\mathrm{II} = e\,dx^2 + g\,dy^2, \tag{5.2}$$

then the coordinate lines are said to give an *isothermic* parametrisation of Σ. The class of surfaces upon which such a coordinate system can be established is said to be isothermic and x, y then constitute curvature coordinates. The principal curvatures κ_1 and κ_2 are given by

$$\kappa_1 = \frac{e}{E}, \quad \kappa_2 = \frac{g}{E}. \tag{5.3}$$

Equivalently, isothermic surfaces may be defined by the requirement that the lines of curvature be conformal.

Constant mean curvature (CMC) surfaces and, more generally, Bonnet surfaces are isothermic, as are quadrics and surfaces of revolution. In the latter connection, any surface of revolution

$$\boldsymbol{r} = \begin{pmatrix} r\cos\eta \\ r\sin\eta \\ \phi(r) \end{pmatrix} \tag{5.4}$$

parametrised in terms of lines of curvature $r = \text{const}$ (parallels) and $\eta = \text{const}$ (meridians) is seen to be isothermic since its metric

$$\mathrm{I} = [1 + \phi'(r)^2]dr^2 + r^2 d\eta^2 \tag{5.5}$$

can be reduced to the conformally flat form

$$\mathrm{I} = r^2(dx^2 + dy^2) \tag{5.6}$$

by setting

$$dx = \frac{1}{r}\sqrt{1 + \phi'(r)^2}\, dr, \quad dy = d\eta. \tag{5.7}$$

It is noted that the latter transformation merely reparametrises the meridians.

In general, if one sets $E = G = e^{2\theta}$, then the fundamental forms for isothermic surfaces become

$$\mathrm{I} = e^{2\theta}(dx^2 + dy^2), \quad \mathrm{II} = e^{2\theta}(\kappa_1 dx^2 + \kappa_2 dy^2) \tag{5.8}$$

with associated Gauss-Weingarten equations

$$\begin{gathered} \begin{pmatrix} \boldsymbol{r}_x \\ \boldsymbol{r}_y \\ \boldsymbol{N} \end{pmatrix}_x = \begin{pmatrix} \theta_x & -\theta_y & \kappa_1 e^{2\theta} \\ \theta_y & \theta_x & 0 \\ -\kappa_1 & 0 & 0 \end{pmatrix} \begin{pmatrix} \boldsymbol{r}_x \\ \boldsymbol{r}_y \\ \boldsymbol{N} \end{pmatrix}, \\ \begin{pmatrix} \boldsymbol{r}_x \\ \boldsymbol{r}_y \\ \boldsymbol{N} \end{pmatrix}_y = \begin{pmatrix} \theta_y & \theta_x & 0 \\ -\theta_x & \theta_y & \kappa_2 e^{2\theta} \\ 0 & -\kappa_2 & 0 \end{pmatrix} \begin{pmatrix} \boldsymbol{r}_x \\ \boldsymbol{r}_y \\ \boldsymbol{N} \end{pmatrix}. \end{gathered} \tag{5.9}$$

The compatibility conditions for the latter produce the Gauss-Mainardi-Codazzi equations

$$\boxed{\begin{aligned} \theta_{xx} + \theta_{yy} + \kappa_1\kappa_2 e^{2\theta} &= 0,\\ \kappa_{1y} + (\kappa_1 - \kappa_2)\theta_y &= 0,\\ \kappa_{2x} + (\kappa_2 - \kappa_1)\theta_x &= 0 \end{aligned}} \tag{5.10}$$

associated with isothermic surfaces. The nonlinear system (5.10) has been extensively studied in the classical literature [37, 95]. Remarkably, it will be seen to be yet another solitonic system with concomitant Bäcklund transformation, associated permutability theorem and admittance of a linear representation with a spectral parameter. That it is indeed an integrable system was recognised in 1995 by Cieśliński et al. [86]

Calapso [62] in 1903 showed that the isothermic system (5.10) allows a reduction to a single fourth-order nonlinear equation. Thus, if we set

$$h_1 = -\kappa_1 e^{\theta}, \quad h_2 = -\kappa_2 e^{\theta} \tag{5.11}$$

then the isothermic system (5.10) reads

$$\begin{aligned} \theta_{xx} + \theta_{yy} + h_1 h_2 = 0,\\ h_{1y} = \theta_y h_2, \quad h_{2x} = \theta_x h_1. \end{aligned} \tag{5.12}$$

The latter relations show that

$$h_{1xy} = \theta_x\theta_y h_1 + \theta_{xy} h_2, \quad h_{2xy} = \theta_x\theta_y h_2 + \theta_{xy} h_1, \tag{5.13}$$

whence if we set

$$z = \frac{1}{\sqrt{2}}(h_1 + h_2), \tag{5.14}$$

then

$$\frac{z_{xy}}{z} = \frac{(e^{\theta})_{xy}}{e^{\theta}} = \theta_{xy} + \theta_x\theta_y. \tag{5.15}$$

Hence, on use of the Gauss equation $(5.10)_1$,

$$\begin{aligned} \Delta\left(\frac{z_{xy}}{z}\right) &= (\Delta\theta)_{xy} + \theta_x(\Delta\theta)_y + \theta_y(\Delta\theta)_x + 2\theta_{xy}\Delta\theta\\ &= -(h_1h_2)_{xy} - \theta_x(h_1h_2)_y - \theta_y(h_1h_2)_x - 2\theta_{xy}(h_1h_2) \end{aligned} \tag{5.16}$$

while

$$(z^2)_{xy} = (h_1 h_2)_{xy} + \theta_x (h_1 h_2)_y + \theta_y (h_1 h_2)_x + 2\theta_{xy}(h_1 h_2), \tag{5.17}$$

where $\Delta = \partial_x^2 + \partial_y^2$. On addition of the relations (5.16) and (5.17), the *Calapso* equation

$$\boxed{\Delta\left(\frac{z_{xy}}{z}\right) + (z^2)_{xy} = 0} \tag{5.18}$$

results.[1] It is noted that the transition $y \to iy$ takes this classical equation to the *zoomeron* equation

$$\Box\left(\frac{z_{xy}}{z}\right) + (z^2)_{xy} = 0, \quad \Box = \partial_x^2 - \partial_y^2 \tag{5.19}$$

to be later set down in a solitonic context in 1976 as a specialisation of the matrix boomeron equation by Calogero and Degasperis [63, 64].

To conclude this section, we record the simple involutory symmetry

$$(\theta, h_1, h_2) \to (-\theta, -h_1, h_2) \tag{5.20}$$

of the system (5.12) which constitutes a classical invariance of the isothermic system (5.10) under the transformation [119]

$$\theta^* = -\theta, \quad \kappa_1^* = -e^{2\theta}\kappa_1, \quad \kappa_2^* = e^{2\theta}\kappa_2. \tag{5.21}$$

Thus, there exists a dual isothermic surface $\Sigma^* : \boldsymbol{r}^* = \boldsymbol{r}^*(x, y)$ corresponding to an isothermic surface $\Sigma : \boldsymbol{r} = \boldsymbol{r}(x, y)$. It has

$$\boldsymbol{r}_x^* = e^{-2\theta}\boldsymbol{r}_x, \quad \boldsymbol{r}_y^* = -e^{-2\theta}\boldsymbol{r}_y, \quad \boldsymbol{N}^* = -\boldsymbol{N}, \tag{5.22}$$

and associated fundamental forms

$$\begin{aligned} \mathrm{I}^* &= e^{-2\theta}(dx^2 + dy^2), \\ \mathrm{II}^* &= -\kappa_1 dx^2 + \kappa_2 dy^2. \end{aligned} \tag{5.23}$$

It is noted that the integrability condition $\boldsymbol{r}_{xy}^* = \boldsymbol{r}_{yx}^*$ for relations $(5.22)_{1,2}$ is indeed satisfied by virtue of the Gauss-Weingarten equations (5.9). The dual surface Σ^* is sometimes called the Christoffel transform of the surface Σ. The

[1] Calapso recorded that an equation equivalent to (5.18) was set down is a thesis of Rothe in 1897.

involution (5.20) implies that the solution of the Calapso equation associated with Σ^* is given by

$$z^* = \frac{1}{\sqrt{2}}(h_2 - h_1) \tag{5.24}$$

and hence may be regarded as dual to z.

Exercises

1. *The Minimal Surface of Enneper (1864).* Show that the Enneper surface

$$\boldsymbol{r} = (3x + 3xy^2 - x^3, 3y + 3x^2y - y^3, 3x^2 - 3y^2)$$

is isothermic and that its lines of curvature $x = \text{const}$ and $y = \text{const}$ are planar.

2. Show that the lines of curvature on a surface of constant mean curvature $\pm 1/a$ form an isothermic system and that the parameters x, y can be chosen such that its metric is given by

$$\mathrm{I} = a^2 e^{\pm\omega}(dx^2 + dy^2),$$

where ω is a solution of the sinh-Gordon equation

$$\omega_{xx} + \omega_{yy} + \sinh\omega = 0.$$

3. Show that an isothermic surface is transformed under inversion

$$\boldsymbol{r} \to \frac{\boldsymbol{r}}{|\boldsymbol{r}|^2}$$

into another isothermic surface.

5.2 The Geometry of Isothermic Surfaces in $\mathbb{R}^{n+2}$

The notion of isothermic surfaces may be readily extended to higher dimensions. In fact, the analysis of isothermic surfaces in Euclidean spaces of arbitrary dimension is formally identical with that of classical isothermic surfaces. In the following, we shall be concerned with the properties of two-dimensional surfaces $\Sigma : \boldsymbol{r} = \boldsymbol{r}(x, y)$ embedded in Euclidean spaces $\mathbb{R}^{n+2}$, that is mappings

$$\boldsymbol{r} : \mathbb{R}^2 \to \mathbb{R}^{n+2}. \tag{5.25}$$

If we choose a frame $(\boldsymbol{X}, \boldsymbol{Y}, \underline{\boldsymbol{N}})$ consisting of the normalised tangent vectors

$$\boldsymbol{X} = \frac{\boldsymbol{r}_x}{|\boldsymbol{r}_x|}, \quad \boldsymbol{Y} = \frac{\boldsymbol{r}_y}{|\boldsymbol{r}_y|} \tag{5.26}$$

and n linearly independent unit vectors

$$\underline{\boldsymbol{N}} = (\boldsymbol{N}^1, \ldots, \boldsymbol{N}^n) \tag{5.27}$$

comprising the normal bundle, that is $\boldsymbol{N}^i \cdot \boldsymbol{X} = \boldsymbol{N}^i \cdot \boldsymbol{Y} = 0$ or

$$\underline{\boldsymbol{N}} \cdot \boldsymbol{X} = \underline{\boldsymbol{N}} \cdot \boldsymbol{Y} = 0, \tag{5.28}$$

then one may associate with a surface $\Sigma \subset \mathbb{R}^{n+2}$ a 1st fundamental form I and 2nd fundamental forms $\underline{\mathrm{II}} = (\mathrm{II}^1, \ldots, \mathrm{II}^n)$ defined by

$$\mathrm{I} = d\boldsymbol{r} \cdot d\boldsymbol{r}, \quad \underline{\mathrm{II}} = -d\underline{\boldsymbol{N}} \cdot d\boldsymbol{r}. \tag{5.29}$$

A generic surface $\Sigma \subset \mathbb{R}^{n+2}$ is completely encoded in its fundamental forms I and $\underline{\mathrm{II}}$ [116].

5.2.1 *Conjugate and Orthogonal Coordinates*

We now assume that the surface Σ may be parametrised in terms of conjugate coordinates [119]. This is always possible for any surface in $\mathbb{R}^3$ or $\mathbb{R}^4$. In the latter case, the conjugate coordinate system is determined uniquely. However, for surfaces in $\mathbb{R}^{n+2}$, $n > 2$, this assumption imposes a constraint.

Conjugate coordinates are defined by the requirement that the 2nd fundamental forms $\underline{\mathrm{II}}$ be purely diagonal, that is

$$\underline{\boldsymbol{N}}_x \cdot \boldsymbol{r}_y = \underline{\boldsymbol{N}}_y \cdot \boldsymbol{r}_x = 0. \tag{5.30}$$

Consequently,

$$\underline{\boldsymbol{N}} \cdot \boldsymbol{r}_{xy} = 0, \tag{5.31}$$

whence the position vector $\boldsymbol{r}$ satisfies the usual point equation (cf. Section 3.3) for surfaces parametrised in terms of conjugate coordinates, namely

$$\boldsymbol{r}_{xy} = A\boldsymbol{r}_x + B\boldsymbol{r}_y. \tag{5.32}$$

Conversely, if there exist functions A and B such that the position vector $\boldsymbol{r}$ of a surface Σ obeys (5.32), then the coordinates on Σ are conjugate.

If, in addition, the coordinate lines on Σ are orthogonal, that is

$$\boldsymbol{r}_x \cdot \boldsymbol{r}_y = 0, \tag{5.33}$$

then the fundamental forms I, $\underline{\mathrm{II}}$ are all purely diagonal and therefore admit the parametrisation

$$\begin{aligned} \mathrm{I} &= H_1^2 dx^2 + H_2^2 dy^2 \\ \underline{\mathrm{II}} &= H_1^2 \underline{\kappa}_1 dx^2 + H_2^2 \underline{\kappa}_2 dy^2. \end{aligned} \tag{5.34}$$

Here, the quantities $\underline{\kappa}_1$ and $\underline{\kappa}_2$ constitute vector analogues of the principal curvatures in $\mathbb{R}^3$. It is now readily shown [116] that one may choose an orthonormal frame $(\boldsymbol{X}, \boldsymbol{Y}, \underline{\boldsymbol{N}})$ such that the position vector $\boldsymbol{r}$ is governed by the Gauss-Weingarten equations

$$\begin{aligned} \begin{pmatrix} \boldsymbol{X} \\ \boldsymbol{Y} \\ \underline{\boldsymbol{N}} \end{pmatrix}_x &= \begin{pmatrix} 0 & -p & -\underline{h}_1^{\mathsf{T}} \\ p & 0 & 0 \\ \underline{h}_1 & 0 & 0 \end{pmatrix} \begin{pmatrix} \boldsymbol{X} \\ \boldsymbol{Y} \\ \underline{\boldsymbol{N}} \end{pmatrix} \\ \begin{pmatrix} \boldsymbol{X} \\ \boldsymbol{Y} \\ \underline{\boldsymbol{N}} \end{pmatrix}_y &= \begin{pmatrix} 0 & q & 0 \\ -q & 0 & -\underline{h}_2^{\mathsf{T}} \\ 0 & \underline{h}_2 & 0 \end{pmatrix} \begin{pmatrix} \boldsymbol{X} \\ \boldsymbol{Y} \\ \underline{\boldsymbol{N}} \end{pmatrix}. \end{aligned} \tag{5.35}$$

Here, p, q and $\underline{h}_1$, $\underline{h}_2$ are given by the relations

$$H_{1y} = pH_2, \quad H_{2x} = qH_1 \tag{5.36}$$

and

$$\underline{h}_1 = -H_1\underline{\kappa}_1, \quad \underline{h}_2 = -H_2\underline{\kappa}_2, \tag{5.37}$$

respectively, and for any two vectors $\underline{f} = (f^1, \ldots, f^n)$, $\underline{g} = (g^1, \ldots, g^n)$, it is understood that

$$\underline{f}^{\mathsf{T}}\underline{g} = \sum_{i=1}^{n} f^i g^i. \tag{5.38}$$

The compatibility condition for (5.35) produces the Gauss-Mainardi-Codazzi equations

$$p_y + q_x + \underline{h}_1^{\mathsf{T}}\underline{h}_2 = 0, \quad \underline{h}_{1y} = p\underline{h}_2, \quad \underline{h}_{2x} = q\underline{h}_1. \tag{5.39}$$

It is noted that both $\underline{h}_1$, $\underline{h}_2$ and H_1, H_2 obey the same *linear* system. Conversely, any solution of the system (5.36) and (5.39) determines uniquely (up to a rotation) the orthonormal frame $(\boldsymbol{X}, \boldsymbol{Y}, \underline{\boldsymbol{N}})$ via the Gauss-Weingarten equations (5.35). Accordingly, the corresponding surface Σ is defined up to its position in space by the relations (5.26), viz

$$\boldsymbol{r}_x = H_1 \boldsymbol{X}, \quad \boldsymbol{r}_y = H_2 \boldsymbol{Y}. \tag{5.40}$$

5.2.2 *Isothermic Surfaces*

In analogy with the classical case, a surface $\Sigma \subset \mathbb{R}^{n+2}$ is called isothermic if it admits a coordinate system which is both conjugate and conformal. Thus, the 1st fundamental form is proportional to the flat metric:

$$\mathrm{I} = e^{2\theta}(dx^2 + dy^2), \quad H_1 = H_2 = e^{\theta}. \tag{5.41}$$

This implies that

$$p = \theta_y, \quad q = \theta_x \tag{5.42}$$

so that the Gauss-Mainardi-Codazzi equations become

$$\Delta\theta + \underline{h}_1^{\mathrm{T}} \underline{h}_2 = 0, \quad \underline{h}_{1y} = \theta_y \underline{h}_2, \quad \underline{h}_{2x} = \theta_x \underline{h}_1 \tag{5.43}$$

with the Laplacian $\Delta = \partial_x^2 + \partial_y^2$. The system (5.43) is here termed the *vector isothermic system*. It represents a generalisation to $\mathbb{R}^{n+2}$ of the classical isothermic system (5.12).

As in the classical case, any isothermic surface Σ admits a dual isothermic surface. Thus, consider the family of surfaces Σ^* on which there exist conjugate lines which are 'parallel' to those on a given isothermic surface Σ. By definition (cf. Subsection 5.4.1), for any such Σ^* there exist functions π_1 and π_2 for which the corresponding position vector $\boldsymbol{r}^*$ is obtained via

$$\boldsymbol{r}_x^* = \pi_1 \boldsymbol{X}, \quad \boldsymbol{r}_y^* = \pi_2 \boldsymbol{Y}. \tag{5.44}$$

Cross-differentiation of the latter and use of the Gauss-Weingarten equations (5.35) yield

$$\pi_{1y} = \theta_y \pi_2, \quad \pi_{2x} = \theta_x \pi_1, \tag{5.45}$$

which, once again, is of the form (5.36). If we now require Σ^* to be isothermic, then $\boldsymbol{r}_x^* \cdot \boldsymbol{r}_x^* = \boldsymbol{r}_y^* \cdot \boldsymbol{r}_y^*$ and hence

$$\pi_1 = \pm\pi_2. \tag{5.46}$$

The plus sign in the above corresponds to $\pi_1 = \pi_2 = e^\theta$ without loss of generality and in this case the original surface Σ is retrieved since

$$\boldsymbol{r}_x = e^\theta \boldsymbol{X}, \quad \boldsymbol{r}_y = e^\theta \boldsymbol{Y}. \tag{5.47}$$

However, if the minus sign is chosen, then

$$\pi_1 = -\pi_2 = e^{-\theta} \tag{5.48}$$

and

$$\boldsymbol{r}^*_x = e^{-2\theta}\boldsymbol{r}_x, \quad \boldsymbol{r}^*_y = -e^{-2\theta}\boldsymbol{r}_y. \tag{5.49}$$

It is natural, following the terminology in $\mathbb{R}^3$, to refer to the isothermic surface Σ^* as the Christoffel transform of Σ. The corresponding fundamental forms read

$$\begin{aligned} \mathrm{I}^* &= e^{-2\theta}(dx^2 + dy^2) \\ \underline{\mathrm{II}}^* &= -\underline{\kappa}_1 dx^2 + \underline{\kappa}_2 dy^2. \end{aligned} \tag{5.50}$$

Comparison with the fundamental forms (5.34), (5.41) delivers relations

$$\theta^* = -\theta, \quad \underline{h}^*_1 = -\underline{h}_1, \quad \underline{h}^*_2 = \underline{h}_2, \tag{5.51}$$

which provide the vectorial analogue of the involution (5.20).

5.2.3 Specialisations and Generalisations

The preceding analysis leading to the Christoffel transform of an isothermic surface gives rise to canonical reductions of the vector isothermic system. We first consider the classical case $n = 1$. Since π_i and h_i satisfy the same linear equations, we deduce that any linear combination of the solutions of (5.46), that is

$$h_1 = c_1 e^\theta + c_2 e^{-\theta}, \quad h_2 = c_1 e^\theta - c_2 e^{-\theta}, \tag{5.52}$$

identically satisfies the last two equations of the isothermic system (5.43). The remaining equation reduces to the elliptic sinh-Gordon equation

$$\Delta\theta + c_1^2 e^{2\theta} - c_2^2 e^{-2\theta} = 0 \tag{5.53}$$

and the mean curvature $\mathcal{H}$ of the surface becomes

$$\mathcal{H} = \kappa_1 + \kappa_2 = -e^{-\theta}(h_1 + h_2) = -2c_1. \tag{5.54}$$

Accordingly, the specialisation (5.52) gives rise to constant mean curvature surfaces (cf. Section 1.5). Moreover, if $c_1 = 0$, then the mean curvature vanishes and Σ is minimal. It is known that the surfaces Σ^* dual to minimal surfaces are spheres. Conversely, if $c_2 = 0$, Σ is a sphere and Σ^* is minimal. In both cases, (5.53) constitutes the elliptic Liouville equation.

In the vector case, it is admissible to assume that (5.52) holds for one component of $\underline{h}_1$ and $\underline{h}_2$. Thus, in the simplest case, we may set

$$\underline{h}_i = \left(h_i^1, \ldots, h_i^{n-1}, e^\theta\right) = (\hat{\underline{h}}_i, e^\theta), \tag{5.55}$$

leading to

$$\Delta\theta + \hat{\underline{h}}_1^\mathsf{T}\hat{\underline{h}}_2 + e^{2\theta} = 0, \quad \hat{\underline{h}}_{1y} = \theta_y \hat{\underline{h}}_2, \quad \hat{\underline{h}}_{2x} = \theta_x \hat{\underline{h}}_1. \tag{5.56}$$

The Gauss-Weingarten equations (5.35) then deliver

$$\boldsymbol{N}_x^n = e^\theta \boldsymbol{X}, \quad \boldsymbol{N}_y^n = e^\theta \boldsymbol{Y}, \tag{5.57}$$

which, by virtue of (5.47), implies that

$$\boldsymbol{r} = \boldsymbol{N}^n \tag{5.58}$$

without loss of generality. Hence, the isothermic surface Σ is embedded in an $n+1$-dimensional sphere, that is

$$\boldsymbol{r} : \mathbb{R}^2 \to \mathbb{S}^{n+1} \subset \mathbb{R}^{n+2}. \tag{5.59}$$

In other words, the vector isothermic system also incorporates isothermic surfaces in Riemannian spaces of constant positive curvature [116].

In conclusion, we note that the transition $y \to iy$ takes the vector isothermic system (5.43) to

$$\Box\theta + \underline{h}_1^\mathsf{T}\underline{h}_2 = 0, \quad \underline{h}_{1y} = \theta_y \underline{h}_2, \quad \underline{h}_{2x} = \theta_x \underline{h}_1, \tag{5.60}$$

where $\Box$ designates the d'Alembert operator $\Box = \partial_x^2 - \partial_y^2$. In geometric terms, this system is descriptive of isothermic surfaces in Minkowski space $\mathbb{M}^{n+2}$ with a space-like normal bundle. The isothermic systems in $\mathbb{M}^3$ and $\mathbb{M}^4$ are related to the solitonic zoomeron and boomeron equations set down by Calogero and Degasperis [63, 64].

Exercise

1. If $\underline{N}$ denotes an orthonormal basis of the normal bundle, then the Gauss-Weingarten equations associated with surfaces which admit conjugate and orthogonal coordinates adopt the form

$$\begin{pmatrix} X \\ Y \\ \underline{N} \end{pmatrix}_x = \begin{pmatrix} 0 & -p & -\underline{h}_1^{\mathrm{T}} \\ p & 0 & 0 \\ \underline{h}_1 & 0 & D_1 \end{pmatrix} \begin{pmatrix} X \\ Y \\ \underline{N} \end{pmatrix}, \quad D_1^{\mathrm{T}} = -D_1$$

$$\begin{pmatrix} X \\ Y \\ \underline{N} \end{pmatrix}_y = \begin{pmatrix} 0 & q & 0 \\ -q & 0 & -\underline{h}_2^{\mathrm{T}} \\ 0 & \underline{h}_2 & D_2 \end{pmatrix} \begin{pmatrix} X \\ Y \\ \underline{N} \end{pmatrix}, \quad D_2^{\mathrm{T}} = -D_2.$$

Derive the corresponding Gauss-Mainardi-Codazzi equations and show that there exists an orthogonal matrix $O = (O^i_j)$ such that

$$D_1 \to 0, \quad D_2 \to 0$$

under the transformation

$$N^i \to O^i_j N^j.$$

5.3 The Vector Calapso System. Its Scalar Lax Pair

In Section 5.1, Calapso's fourth-order equation associated with isothermic surfaces in $\mathbb{R}^3$ was derived. Here, we introduce a vector version of the Calapso equation and construct an associated compact scalar (non-local) Lax pair.

5.3.1 The Vector Calapso System

To derive an analogue of Calapso's equation for isothermic surfaces in $\mathbb{R}^{n+2}$, the vector-valued function

$$\underline{z} = \frac{1}{\sqrt{2}}(\underline{h}_1 + \underline{h}_2), \tag{5.61}$$

is introduced. The procedure which led to the classical Calapso equation may now be modified as follows. Differentiation of $(5.43)_2$ and $(5.43)_3$ with respect to x and y, respectively, yields

$$\underline{z}_{xy} = \rho \underline{z}, \quad \rho = e^{-\theta}(e^{\theta})_{xy} \tag{5.62}$$

and evaluation of $\Delta\rho$ by means of $(5.43)_1$ produces

$$\Delta\rho + (\underline{z}^2)_{xy} = 0 \tag{5.63}$$

with the natural abbreviation $\underline{z}^2 = \underline{z}^\mathsf{T}\underline{z}$. We refer to the system

$$\boxed{\underline{z}_{xy} = \rho\underline{z}, \quad \Delta\rho + (\underline{z}^2)_{xy} = 0} \tag{5.64}$$

as the *vector Calapso system* [59, 60, 331] since in the classical case $n = 1$, the function ρ may be eliminated to obtain the Calapso equation (5.18). It is noted that[2]

$$\underline{z}^* = \frac{1}{\sqrt{2}}(\underline{h}_1 - \underline{h}_2), \quad \rho^* = e^\theta(e^{-\theta})_{xy} \tag{5.65}$$

constitute another solution of the vector Calapso system by virtue of the invariance (5.51) of the vector isothermic system and the discrete symmetry $\underline{z} \to -\underline{z}$.

From a soliton-theoretical point of view, isothermic surfaces in $\mathbb{R}^4$ are of particular interest. Indeed, if we introduce the complex-valued function u according to

$$\underline{z} = (\Re(u), \Im(u)), \tag{5.66}$$

then the vector Calapso system may be regarded as the stationary reduction $(u_t = 0)$ of the integrable (Benney-Roskes-)Davey-Stewartson II equation [29, 98]

$$iu_t = u_{xy} - \rho u, \quad \Delta\rho + (|u|^2)_{xy} = 0. \tag{5.67}$$

Thus, the stationary Davey-Stewartson II equation describes isothermic surfaces in $\mathbb{R}^4$. This interesting connection was made by Ferapontov [128] in the case of the classical Calapso equation which corresponds to the *real* reduction of the Davey-Stewartson II equation. Moreover, the transition $(y, \rho) \to i(y, -\rho)$ associated with isothermic surfaces in Minkowski space $\mathbb{M}^4$ yields

$$\underline{u}_{xy} = \rho\underline{u}, \quad \Box\rho + (|u|^2)_{xy} = 0. \tag{5.68}$$

This constitutes the stationary reduction of the integrable Davey-Stewartson III equation [50, 320, 342]

$$iu_t = u_{xy} - \rho u, \quad \Box\rho + (|u|^2)_{xy} = 0. \tag{5.69}$$

[2] Strictly speaking, $\underline{z}^*$ should denote the quantity $(\underline{h}_1^* + \underline{h}_2^*)/\sqrt{2}$ which is $(5.65)_1$ except for a minus sign.

Remarkably, as observed in [100], the vector Calapso system (5.68) is equivalent to a particular case of the matrix *boomeron* equation introduced independently in a solitonic context by Calogero and Degasperis [63, 64].

5.3.2 A Scalar Lax Pair

The vector Calapso system may be regarded as a Moutard equation $(5.64)_1$ for $\underline{z}$ coupled via the potential ρ with the Poisson equation $(5.64)_2$. This interpretation is exploited here to construct a Lax pair for the vector Calapso equation. Thus, we introduce the 'scaled' position vector $\boldsymbol{\psi}$ of the Christoffel transform Σ^* according to

$$\boldsymbol{\psi} = e^{\theta}\boldsymbol{r}^*, \tag{5.70}$$

whence, use of (5.47) and (5.49) yields

$$\boldsymbol{X} = \boldsymbol{\psi}_x - \theta_x\boldsymbol{\psi}, \quad \boldsymbol{Y} = -\boldsymbol{\psi}_y + \theta_y\boldsymbol{\psi}. \tag{5.71}$$

These relations may be used to eliminate the tangent vectors $\boldsymbol{X}$ and $\boldsymbol{Y}$ from the Gauss-Weingarten equations (5.35). It is noted that the expressions for both $\boldsymbol{X}_y$ and $\boldsymbol{Y}_x$ deliver a vector equation of the form $(5.64)_1$, namely

$$\boldsymbol{\psi}_{xy} = \rho\boldsymbol{\psi}, \quad \rho = e^{-\theta}(e^{\theta})_{xy}. \tag{5.72}$$

Thus, both $\underline{z}$ and $\boldsymbol{\psi}$ satisfy a Moutard equation with the same potential ρ. The Moutard equation admits a bilinear potential (cf. Chapter 1) which, in the present context, involves the set of normals $\underline{\boldsymbol{N}}$. Specifically, if we set

$$\underline{\boldsymbol{S}} = \underline{z}^*\boldsymbol{\psi} - \sqrt{2}\underline{\boldsymbol{N}} \tag{5.73}$$

then the Gauss-Weingarten equations imply that

$$\underline{\boldsymbol{S}}_x = \underline{z}_x\boldsymbol{\psi} - \underline{z}\boldsymbol{\psi}_x, \quad \underline{\boldsymbol{S}}_y = \underline{z}\boldsymbol{\psi}_y - \underline{z}_y\boldsymbol{\psi}. \tag{5.74}$$

The remaining Gauss-Weingarten equations are readily shown to produce

$$\begin{aligned} \Box\boldsymbol{\psi} + f\boldsymbol{\psi} &= \underline{z}^{\mathsf{T}}\underline{\boldsymbol{S}}(\underline{z}, \boldsymbol{\psi}) \\ \Delta\boldsymbol{\psi} - 2\nabla\theta\cdot\nabla\boldsymbol{\psi} + g\boldsymbol{\psi} &= \underline{z}^{*\mathsf{T}}\underline{\boldsymbol{S}}(\underline{z}, \boldsymbol{\psi}), \end{aligned} \tag{5.75}$$

wherein the functional dependence of $\underline{\boldsymbol{S}}$ on $\underline{z}$ and $\boldsymbol{\psi}$ is indicated and the functions f, g are given by

$$f = \underline{z}^{\mathsf{T}}\underline{z}^* - e^{-\theta}\Box e^{\theta}, \quad g = \underline{z}^{*2} + e^{\theta}\Delta e^{-\theta}. \tag{5.76}$$

By construction, the system (5.75) is nothing but a reformulation of the Gauss-Weingarten equations (5.35).

In general terms, it is observed that the bilinear potential $\underline{\mathbf{S}}$ is well-defined if and only if $\underline{z}$ is a solution of the Moutard equation $(5.64)_1$, while the compatibility condition $\Box(\boldsymbol{\psi}_{xy}) = (\Box\boldsymbol{\psi})_{xy}$ produces

$$f_x = 2\rho_y + (\underline{z}^2)_x, \quad f_y = -2\rho_x - (\underline{z}^2)_y \tag{5.77}$$

together with $(5.64)_2$. The latter embodies the integrability condition for the pair of equations (5.77). Thus, the vector Calapso system (5.64) is retrieved and the function f constitutes an associated potential which, in fact, appears in Calapso's original work [62].

Since the general solution of (5.77) depends on an arbitrary additive constant, the invariance

$$f \to f + k \tag{5.78}$$

may now be exploited to inject an arbitrary 'spectral' parameter into the linear system under consideration to obtain the following result:

Theorem 7 (A Lax pair for the vector Calapso system). *The scalar (non-local) Lax pair*

$$\psi_{xy} = \rho\psi, \quad \Box\psi + (f+k)\psi = \underline{z}^{\mathrm{T}}\underline{S}(\underline{z}, \psi), \tag{5.79}$$

where k is an arbitrary constant and

$$\underline{S}_x = \underline{z}_x\psi - \underline{z}\psi_x, \quad \underline{S}_y = \underline{z}\psi_y - \underline{z}_y\psi, \tag{5.80}$$

is compatible if and only if $(\underline{z}, \rho)$ is a solution of the vector Calapso system (5.64) and f is the corresponding potential obeying (5.77). The linear system (5.79) admits the first integral

$$2\psi\,\Delta\psi - 2(\nabla\psi)^2 + \underline{z}^2\psi^2 - \underline{S}^2 = I = \text{const.} \tag{5.81}$$

If ψ_0 is an eigenfunction, that is a solution of (5.79) with parameter k_0, subject to $I = 0$, then

$$\underline{z}^* = \frac{\underline{S}(\underline{z}, \psi_0)}{\psi_0}, \quad \rho^* = \psi_0\left(\psi_0^{-1}\right)_{xy} \tag{5.82}$$

constitutes another solution of the vector Calapso system and

$$\theta = \ln\psi_0, \quad \underline{h}_1 = \frac{1}{\sqrt{2}}(\underline{z} + \underline{z}^*), \quad \underline{h}_2 = \frac{1}{\sqrt{2}}(\underline{z} - \underline{z}^*) \tag{5.83}$$

satisfies the vector isothermic system (5.43).

Proof. We have already shown that the Lax pair (5.79) is compatible if and only if the vector Calapso system is satisfied. Alternatively, one may introduce the $(n+4)$-dimensional vector $\phi = (\psi, \psi_x, \psi_y, \Delta\psi, \underline{S})^T$ and rewrite (5.79), (5.80) as a matrix Frobenius system of the form

$$\phi_x = F\phi, \quad \phi_y = G\phi. \tag{5.84}$$

The compatibility condition

$$F_y - G_x + [F, G] = 0 \tag{5.85}$$

then delivers the vector Calapso system together with the Frobenius system for f.

The first integral (5.81) may be verified directly and plays an important role in the construction of a Bäcklund transformation for the vector Calapso system. It also provides the link between the vector isothermic and Calapso systems. Thus, let $\psi_0 = e^{\theta}$ be an eigenfunction and $\underline{S}_0 = \underline{S}(\underline{z}, \psi_0)$ the corresponding bilinear potential. Then, (5.81) may be brought into the form

$$2\Delta\theta + \underline{z}^2 - \underline{S}_0^2 e^{-2\theta} = I e^{-2\theta} \tag{5.86}$$

which reduces to $(5.43)_1$ in the case $I = 0$ if $\underline{z}^*$ and $\underline{h}_1, \underline{h}_2$ are defined as in (5.82) and (5.83), respectively. It is readily shown that $\underline{h}_1$ and $\underline{h}_2$ indeed satisfy the remaining equations $(5.43)_{2,3}$ which, in turn, implies that (5.82) constitutes another solution of the vector Calapso system (cf. (5.65)). Moreover, if I is positive, then (5.86) may be linked to isothermic surfaces in $\mathbb{R}^{n+3}$ for which the dual surfaces are embedded in the sphere $\mathbb{S}^{n+2}$ (cf. Subsection 5.2.3). □

5.3.3 Reductions

We conclude this section with the important observation that since $z = -e^{-\theta}\mathcal{H}/\sqrt{2}$, the class of solutions

$$z = 0, \quad \Delta\rho = 0 \tag{5.87}$$

of the Calapso equation corresponds to minimal surfaces (cf. Exercise 1). The class of solutions of the Calapso equation associated with constant mean curvature surfaces is obtained as follows. Since z obeys the first equation of the Lax pair (5.79), it is consistent to demand that z indeed be an eigenfunction of the Calapso equation. This constraint on the Calapso equation allows us to express the potential f explicitly in terms of z by means of $(5.79)_2$ according to

$$f = S - k - \frac{\Box z}{z}, \tag{5.88}$$

where $S = S(z, z)$ is now a constant. Insertion of (5.88) into the relations (5.77) produces two third-order equations which may be integrated to obtain the first integral (5.81), viz

$$\Delta(\ln z) + \frac{1}{2}z^2 - \frac{S^2 + I}{2z^2} = 0. \tag{5.89}$$

This is nothing but the elliptic sinh-Gordon equation (5.53) with $\theta = \ln z$.

Exercise

1. Show that if z satisfies the Calapso equation and $(5.79)_2$, then the solution of (5.77) is given by (5.88) where $\theta = \ln z$ obeys the sinh-Gordon equation

$$\Delta\theta + \frac{1}{2}e^{2\theta} - ce^{-2\theta} = 0$$

and c is a constant of integration.

5.4 The Fundamental Transformation

The main body of classical results on transformations of conjugate nets is to be found in the monograph of Eisenhart [119]. Therein, Laplace-Darboux, radial, Levy and Fundamental Transformations are all treated extensively. The relevance of certain of these classical geometric transformations to modern soliton theory was signalled in [212]. Their discrete analogues have been subsequently developed in that context by Konopelchenko and Schief [213] and Doliwa et al. [113]. In the following, we shall be concerned with the Fundamental Transformation as recorded in 1915 by Jonas [183].

5.4.1 Parallel Nets. The Combescure Transformation

Here, we return to the conjugate net (point) equation written in the form

$$\boldsymbol{r}_{xy} = (\ln H_1)_y \boldsymbol{r}_x + (\ln H_2)_x \boldsymbol{r}_y. \tag{5.90}$$

It has been seen in Section 3.3 that the vector-valued function $\boldsymbol{r}$ may be identified with the position vector of a surface Σ on which the coordinate lines $x = \text{const}$ and $y = \text{const}$ form a conjugate net $\mathcal{N}$. A net $\mathcal{N}'$ on a surface Σ' is said to be parallel to $\mathcal{N}$ if the tangent vectors to the coordinate lines on Σ and Σ' are parallel at corresponding points, that is there exist functions h and l

such that

$$\mathcal{C}: \quad \boldsymbol{r}'_x = h\boldsymbol{r}_x, \quad \boldsymbol{r}'_y = l\boldsymbol{r}_y. \tag{5.91}$$

Since $\boldsymbol{r}'$ satisfies the point equation

$$\boldsymbol{r}'_{xy} = (\ln H'_1)_y \boldsymbol{r}'_x + (\ln H'_2)_x \boldsymbol{r}'_y \tag{5.92}$$

with the coefficients

$$H'_1 = hH_1, \quad H'_2 = lH_2, \tag{5.93}$$

it is concluded that the net $\mathcal{N}'$ is also conjugate. The transition from $\mathcal{N}$ to $\mathcal{N}'$ is commonly referred to as *Combescure transformation*. The existence of parallel nets is guaranteed if the compatibility condition for (5.91) is satisfied, that is

$$h_y = (l - h)(\ln H_1)_y, \quad l_x = (h - l)(\ln H_2)_x. \tag{5.94}$$

The latter linear system is adjoint to the point equation (5.90).

5.4.2 The Radial Transformation

Further conjugate nets may be constructed on use of a scalar solution ϕ of the point equation (5.90). Indeed, the *radial transform*

$$\mathcal{R}: \quad \boldsymbol{r}^* = \frac{\boldsymbol{r}}{\phi} \tag{5.95}$$

satisfies the conjugate net equation

$$\boldsymbol{r}^*_{xy} = (\ln H^*_1)_y \boldsymbol{r}^*_x + (\ln H^*_2)_x \boldsymbol{r}^*_y \tag{5.96}$$

with coefficients

$$H^*_1 = \frac{H_1}{\phi}, \quad H^*_2 = \frac{H_2}{\phi}. \tag{5.97}$$

In the terminology of soliton theory, the radial transformation constitutes a special gauge transformation.

A particular solution of the adjoint system (5.94) associated with the radial transform $\boldsymbol{r}^*$ is readily shown to be

$$h^* = \phi' - h\phi, \quad l^* = \phi' - l\phi, \tag{5.98}$$

where the quantity ϕ' represents a scalar analogue of $\boldsymbol{r}'$ defined by

$$\phi'_x = h\phi_x, \quad \phi'_y = l\phi_y. \tag{5.99}$$

It is also noted that

$$\phi^* = \frac{1}{\phi} \tag{5.100}$$

is a scalar solution of the point equation (5.96), since any constant constitutes a trivial scalar solution of the seed equation (5.90).

5.4.3 *The Fundamental Transformation*

If we now map the conjugate net $\mathcal{N}^*$ to a conjugate net $\mathcal{N}^{*\prime}$ via the Combescure transformation associated with the pair (h^*, l^*), then the position vector of the surface $\Sigma^{*\prime}$ satisfies the relations

$$\begin{aligned} \boldsymbol{r}^{*\prime}_x &= h^*\boldsymbol{r}^* = (\phi' - h\phi)\left(\frac{\boldsymbol{r}}{\phi}\right)_x \\ \boldsymbol{r}^{*\prime}_y &= l^*\boldsymbol{r}^* = (\phi' - l\phi)\left(\frac{\boldsymbol{r}}{\phi}\right)_y. \end{aligned} \tag{5.101}$$

These equations may be integrated explicitly to yield, without loss of generality,

$$\boldsymbol{r}^{*\prime} = \frac{\phi'}{\phi}\boldsymbol{r} - \boldsymbol{r}'. \tag{5.102}$$

Accordingly, a scalar solution of the associated point equation

$$\boldsymbol{r}^{*\prime}_{xy} = (\ln H_1^{*\prime})_y \boldsymbol{r}^{*\prime}_x + (\ln H_2^{*\prime})_x \boldsymbol{r}^{*\prime}_y \tag{5.103}$$

with

$$H_1^{*\prime} = (\phi' - h\phi)\frac{H_1}{\phi}, \quad H_2^{*\prime} = (\phi' - l\phi)\frac{H_2}{\phi} \tag{5.104}$$

is given by

$$\phi^{*\prime} = \frac{\phi'}{\phi}. \tag{5.105}$$

Thus, a second application of the radial transformation gives rise to the conjugate net $\tilde{\mathcal{N}} = \mathcal{N}^{*\prime *}$ with corresponding position vector

$$\tilde{\boldsymbol{r}} = \boldsymbol{r}^{*\prime *} = \frac{\boldsymbol{r}^{*\prime}}{\phi^{*\prime}}. \tag{5.106}$$

The latter is known as the *Fundamental Transform* of $\boldsymbol{r}$.

Theorem 8 (The Fundamental Transformation). *The conjugate net equation (5.90) is invariant under the Fundamental Transformation*

$$\mathcal{F}: \quad \begin{cases} \boldsymbol{r} \to \tilde{\boldsymbol{r}} = \boldsymbol{r} - \dfrac{\phi}{\phi'}\boldsymbol{r}' \\ H_1 \to \tilde{H}_1 = \left(1 - h\dfrac{\phi}{\phi'}\right) H_1 \\ H_2 \to \tilde{H}_2 = \left(1 - l\dfrac{\phi}{\phi'}\right) H_2. \end{cases} \tag{5.107}$$

$\mathcal{F}$ may be decomposed into a Combescure and two radial transformations according to

$$\mathcal{F} = \mathcal{R} \circ \mathcal{C} \circ \mathcal{R}. \tag{5.108}$$

A permutability theorem associated with the Fundamental Transformation and its application to isothermic surfaces is discussed in the following sections.

Exercises

1. Show that the *Levy transforms* with respect to ϕ, viz

$$\boldsymbol{r}_1 = \boldsymbol{r} - \frac{\phi}{\phi_y}\boldsymbol{r}_y$$

and

$$\boldsymbol{r}_2 = \boldsymbol{r} - \frac{\phi}{\phi_x}\boldsymbol{r}_x$$

take the conjugate net $\mathcal{N}$ with point equation (5.90) to the associated conjugate nets with point equations

$$\boldsymbol{r}_{1xy} = [\ln\{\alpha(\ln\beta)_x\}]_y \boldsymbol{r}_{1x} + (\ln\beta)_x \boldsymbol{r}_{1y}$$

and

$$\boldsymbol{r}_{2xy} = (\ln\alpha)_y \boldsymbol{r}_{2x} + [\ln\{\beta(\ln\alpha)_y\}]_x \boldsymbol{r}_{2y},$$

respectively, where

$$\alpha = H_1\frac{\phi}{\phi_x}, \quad \beta = H_2\frac{\phi}{\phi_y}.$$

2. Show that the tangents to the curves $x = \text{const}$ and $y = \text{const}$ at corresponding points on the conjugate net $\mathcal{N}$ and its radial transform $\mathcal{N}^*$ with respect to ϕ meet in points whose position vectors are, respectively,

$$\boldsymbol{r} - \frac{(\phi - 1)}{(\phi - 1)_y}\boldsymbol{r}_y$$

and

$$\boldsymbol{r} - \frac{(\phi - 1)}{(\phi - 1)_x}\boldsymbol{r}_x$$

corresponding to the Levy transforms of $\mathcal{N}$ with respect to $\phi - 1$.

5.5 A Bäcklund Transformation for Isothermic Surfaces

This section is concerned with a generalisation of the classical Bäcklund transformation for isothermic surfaces in $\mathbb{R}^3$ as set down originally by Darboux [95] and subsequently discussed in detail by Bianchi [36]. Treatments of Darboux's Bäcklund transformation are also to be found in [59–61, 165]. Its formulation in terms of a matrix Darboux transformation (cf. Chapter 7) is due to Cieśliński [83]. The Bäcklund transformation for isothermic surfaces in spaces of arbitrary dimension is directly analogous to Darboux's classical transformation if one replaces appropriate scalar functions by vector-valued functions [331]. Here, a soliton-theoretical derivation of the Bäcklund transformation is presented based on the classical Fundamental Transformation.

5.5.1 The Fundamental Transformation for Conjugate Coordinates

In the previous section, it has been shown that the point equation (5.90) is form-invariant under the Fundamental Transformation. Although the latter was derived in the classical context of $\mathbb{R}^3$, it is evident that Theorem 8 remains valid in spaces of arbitrary dimension. In this connection, we consider two equivalent forms of the point equation, namely the original second-order formulation

$$\boldsymbol{r}_{xy} = (\ln H_1)_y \boldsymbol{r}_x + (\ln H_2)_x \boldsymbol{r}_y \tag{5.109}$$

and the first-order form

$$\boldsymbol{X}_y = q\boldsymbol{Y}, \quad \boldsymbol{Y}_x = p\boldsymbol{X}, \tag{5.110}$$

which is obtained by introduction of the quantities

$$\boldsymbol{X} = \frac{\boldsymbol{r}_x}{H_1}, \quad \boldsymbol{Y} = \frac{\boldsymbol{r}_y}{H_2}, \quad p = \frac{H_{1y}}{H_2}, \quad q = \frac{H_{2x}}{H_1}. \tag{5.111}$$

It is noted that (5.110) represents part of the Gauss-Weingarten equations (5.35) while the relations (5.111) incorporate (5.36) and (5.40). However, an interpretation of $\boldsymbol{X}$, $\boldsymbol{Y}$ and H_i as unit vectors and metric coefficients respectively is not to hand at this stage.

It is recalled that a conjugate coordinate system on a surface Σ' is termed a parallel conjugate net if, at corresponding points, the coordinate tangent vectors of Σ and Σ' are parallel, that is there exist functions X' and Y' such that

$$\boldsymbol{r}'_x = X'\boldsymbol{X}, \quad \boldsymbol{r}'_y = Y'\boldsymbol{Y}. \tag{5.112}$$

Any solution of the linear system constituting the compatibility conditions for (5.112), namely

$$X'_y = pY', \quad Y'_x = qX', \tag{5.113}$$

accordingly gives rise to parallel conjugate nets. Here, the quantities X' and Y' are related to those used in the previous section by

$$X' = H_1 h, \quad Y' = H_2 l. \tag{5.114}$$

It has been seen that the linear system (5.113) may be regarded as adjoint to (5.110). In soliton-theoretic terminology, we refer to solutions of the systems (5.109) and (5.110) as eigenfunctions while the solutions of (5.113) are termed adjoint eigenfunctions. The parallel net $\boldsymbol{r}'$ is therefore defined via squared eigenfunctions. We observe that H_1 and H_2 are adjoint eigenfunctions by virtue of $(5.111)_{3,4}$.

In the above notation, we obtain [331]:

Theorem 9 (The Fundamental Transformation). *Let $\boldsymbol{r}$ be a conjugate net with tangent vectors $\boldsymbol{X}$ and $\boldsymbol{Y}$ and r, X, Y be scalar solutions of (5.109), (5.110) with*

$$r_x = H_1 X, \quad r_y = H_2 Y. \tag{5.115}$$

Let $\boldsymbol{r}'$ be a parallel net with associated solutions X', Y' of (5.113). Then, a second conjugate net $\tilde{\boldsymbol{r}}$ is given by the Fundamental Transform

$$\begin{gathered}\tilde{\boldsymbol{r}} = \boldsymbol{r} - r\frac{\boldsymbol{r}'}{r'}, \quad \tilde{\boldsymbol{X}} = \boldsymbol{X} - X\frac{\boldsymbol{r}'}{r'}, \quad \tilde{\boldsymbol{Y}} = \boldsymbol{Y} - Y\frac{\boldsymbol{r}'}{r'} \\ \tilde{H}_1 = H_1 - X'\frac{r}{r'}, \quad \tilde{H}_2 = H_2 - Y'\frac{r}{r'} \\ \tilde{p} = p - \frac{X'Y}{r'}, \quad \tilde{q} = q - \frac{Y'X}{r'},\end{gathered} \tag{5.116}$$

where r' is a scalar bilinear potential obeying the scalar analogue of (5.112), viz

$$r'_x = X'X, \quad r'_y = Y'Y. \tag{5.117}$$

In the next subsection, it will be necessary to know how the Fundamental Transformation acts on parallel nets and their corresponding adjoint eigenfunctions. Considerations of symmetry lead us to expect that parallel nets and adjoint eigenfunctions transform in the same manner as the original net $\boldsymbol{r}$ and the adjoint eigenfunctions H_i respectively. Indeed, this is the case, and the analogues of the transformation laws $(5.116)_{1,4,5}$ are given below.

Corollary 1. *With the assumptions of Theorem 5.9, let $\mathbf{r}''$ be a parallel net obeying*

$$\boldsymbol{r}''_x = X''\mathbf{X}, \quad \boldsymbol{r}''_y = Y''\mathbf{Y} \tag{5.118}$$

for some adjoint eigenfunctions X'', Y''. If r'' is a corresponding scalar solution of

$$r''_x = X''X, \quad r''_y = Y''Y, \tag{5.119}$$

then the Fundamental Transforms of $\mathbf{r}'$ and X'', Y'' are given by

$$\tilde{\boldsymbol{r}}'' = \boldsymbol{r}'' - r''\frac{\boldsymbol{r}'}{r'}, \quad \tilde{X}'' = X'' - X'\frac{r''}{r'}, \quad \tilde{Y}'' = Y'' - Y'\frac{r''}{r'}. \tag{5.120}$$

Proof. It is readily shown that $\tilde{\boldsymbol{r}}''$ and $\tilde{X}'', \tilde{Y}''$, as given by (5.120), satisfy the relations

$$\tilde{\boldsymbol{r}}''_x = \tilde{X}''\tilde{\mathbf{X}}, \quad \tilde{\boldsymbol{r}}''_y = \tilde{Y}''\tilde{\mathbf{Y}} \tag{5.121}$$

and hence $\tilde{\boldsymbol{r}}''$ is parallel to $\tilde{\boldsymbol{r}}$. □

5.5.2 The Ribaucour Transformation

If the coordinate system is orthogonal, then the Gauss-Weingarten equations (5.35) hold. In particular, the normals $\underline{\boldsymbol{N}}$ satisfy the system

$$\underline{\boldsymbol{N}}_x = \underline{h}_1\boldsymbol{X}, \quad \underline{\boldsymbol{N}}_y = \underline{h}_2\boldsymbol{Y}, \tag{5.122}$$

where $\underline{h}_1$ and $\underline{h}_2$ are solutions of $(5.39)_{2,3}$, that is

$$\underline{h}_{1y} = p\underline{h}_2, \quad \underline{h}_{2x} = q\underline{h}_1. \tag{5.123}$$

Comparison with (5.112), (5.113) shows that the normals $\boldsymbol{N}^i, i = 1, \ldots, n$ define parallel nets with corresponding adjoint eigenfunctions h_1^i and h_2^i. Thus, Corollary 1 implies that the Fundamental Transforms of $\underline{\boldsymbol{N}}$ and $\underline{h}_1$, $\underline{h}_2$ read

$$\tilde{\underline{\boldsymbol{N}}} = \underline{\boldsymbol{N}} - \underline{N}\frac{\boldsymbol{r}'}{r'}, \quad \tilde{\underline{h}}_1 = \underline{h}_1 - X'\frac{\underline{N}}{r'}, \quad \tilde{\underline{h}}_2 = \underline{h}_2 - Y'\frac{\underline{N}}{r'} \tag{5.124}$$

with

$$\underline{N}_x = \underline{h}_1 X, \quad \underline{N}_y = \underline{h}_2 Y. \tag{5.125}$$

It remains to examine under what circumstances the vectors $\tilde{\boldsymbol{X}}$, $\tilde{\boldsymbol{Y}}$ and $\tilde{\underline{\boldsymbol{N}}}$ as given by $(5.116)_{2,3}$, $(5.124)_1$ are orthonormal.

Evaluation of the orthonormality conditions

$$\tilde{\boldsymbol{X}}^2 = 1, \quad \tilde{\boldsymbol{X}} \cdot \tilde{\boldsymbol{Y}} = 0, \quad \tilde{\boldsymbol{Y}}^2 = 1 \tag{5.126}$$

leads to the constraints

$$X = 2\frac{r'}{\boldsymbol{r}'^2}\boldsymbol{X} \cdot \boldsymbol{r}', \quad Y = 2\frac{r'}{\boldsymbol{r}'^2}\boldsymbol{Y} \cdot \boldsymbol{r}' \tag{5.127}$$

which, in turn, may be formulated as

$$\left(\frac{\boldsymbol{r}'^2}{r'}\right)_x = 0, \quad \left(\frac{\boldsymbol{r}'^2}{r'}\right)_y = 0. \tag{5.128}$$

Without loss of generality, we may take

$$r' = \frac{1}{2}\boldsymbol{r}'^2 \tag{5.129}$$

so that

$$X = \boldsymbol{X} \cdot \boldsymbol{r}', \quad Y = \boldsymbol{Y} \cdot \boldsymbol{r}'. \tag{5.130}$$

Similarly, the relations

$$\tilde{\boldsymbol{N}}^i \cdot \tilde{\boldsymbol{N}}^j = \delta^{ij}, \quad \tilde{\underline{\boldsymbol{N}}} \cdot \tilde{\boldsymbol{X}} = 0, \quad \tilde{\underline{\boldsymbol{N}}} \cdot \tilde{\boldsymbol{Y}} = 0 \tag{5.131}$$

produce

$$\underline{N} = \underline{\boldsymbol{N}} \cdot \boldsymbol{r}'. \tag{5.132}$$

Finally, (5.130) yields

$$X_x = X' - pY - \underline{h}_1^{\mathsf{T}}\underline{N}, \quad Y_y = Y' - qX - \underline{h}_2^{\mathsf{T}}\underline{N} \tag{5.133}$$

by virtue of the Gauss-Weingarten equations (5.35). Hence, we have the following generalisation of the classical *Ribaucour transformation* for curvature nets [119].

Theorem 10 (The Ribaucour transformation). *The Fundamental Transformation leaves form-invariant the Gauss-Weingarten equations (5.35) together with the Gauss-Mainardi-Codazzi equations (5.39) if the parallel net* $\mathbf{r}'$ *and its scalar companion* r' *are chosen to be*

$$\mathbf{r}' = X\mathbf{X} + Y\mathbf{Y} + \underline{N}^{\mathsf{T}}\underline{\mathbf{N}}, \quad r' = \frac{1}{2}\mathbf{r}'^2 = \frac{1}{2}(X^2 + Y^2 + \underline{N}^2) \qquad (5.134)$$

with corresponding adjoint eigenfunctions

$$X' = X_x + pY + \underline{h}_1^{\mathsf{T}}\underline{N}, \quad Y' = Y_y + qX + \underline{h}_2^{\mathsf{T}}\underline{N}. \qquad (5.135)$$

The Ribaucour transforms of $\underline{\mathbf{N}}$ *and* $\underline{h}_i$ *are given by*

$$\tilde{\underline{\mathbf{N}}} = \underline{\mathbf{N}} - \underline{N}\frac{\mathbf{r}'}{r'}, \quad \tilde{\underline{h}}_1 = \underline{h}_1 - X'\frac{\underline{N}}{r'}, \quad \tilde{\underline{h}}_2 = \underline{h}_2 - Y'\frac{\underline{N}}{r'}, \qquad (5.136)$$

where $\underline{N}$ *is defined by (5.125).*

Proof. A short calculation reveals that X' and Y' are indeed solutions of the adjoint system (5.113). The form of the parallel net $\mathbf{r}'$ is dictated by the relations (5.130) and (5.132), since $(\mathbf{X}, \mathbf{Y}, \underline{\mathbf{N}})$ form an orthonormal frame. In fact, it is readily verified that $\mathbf{r}'$ and r' obey the systems (5.112) and (5.117), respectively. □

It is interesting to note that, in the classical case $n=1$, the relation $N(\tilde{\mathbf{r}} - \mathbf{r}) = r(\tilde{\mathbf{N}} - \mathbf{N})$ holds. Consequently, the normals to the surfaces Σ and $\tilde{\Sigma}$ at corresponding points not only intersect but also intersect at the same distance to the surfaces. Σ and $\tilde{\Sigma}$ are therefore the sheets of the envelope of a two-parameter family of spheres, namely the so-called Ribaucour sphere congruence [95]. In fact, it was just this property that originally led Ribaucour to the construction of the Fundamental Transformation for lines of curvature.

5.5.3 A Bäcklund Transformation for Isothermic Surfaces

The Ribaucour transformation maps within the class of isothermic surfaces if the condition $H_1 = H_2$ is preserved. In the present form of the Ribaucour transformation, this implies that $X' = Y'$. However, the Gauss-Weingarten equations

(5.35) and Gauss-Mainardi-Codazzi equations (5.39) are invariant under the involution

$$(\mathbf{X}, H_1, p, q, \underline{h}_1) \rightarrow -(\mathbf{X}, H_1, p, q, \underline{h}_1) \tag{5.137}$$

with the remaining quantities being unchanged. Thus, composition with the Ribaucour transformation gives rise to the constraint $\tilde{H}_1 = -\tilde{H}_2$, viz.

$$r' = \frac{1}{2}(X' + Y')re^{-\theta}. \tag{5.138}$$

Accordingly,

$$X'_x - \theta_x Y' = \frac{X' - Y'}{r} X e^{\theta}, \quad Y'_y - \theta_y X' = \frac{Y' - X'}{r} Y e^{\theta}, \tag{5.139}$$

which may be written as

$$\left(\frac{X' - Y'}{r} e^{\theta}\right)_x = 0, \quad \left(\frac{X' - Y'}{r} e^{\theta}\right)_y = 0, \tag{5.140}$$

whence

$$X' - Y' = 2mre^{-\theta}, \quad m = \text{const.} \tag{5.141}$$

If we now introduce a function r^* defined by

$$X' + Y' = 2mr^* e^{\theta}, \tag{5.142}$$

then the constraint (5.138) becomes

$$r' = mrr^*, \tag{5.143}$$

while the relations (5.139) read

$$r^*_x = e^{-\theta} X, \; r^*_y = -e^{-\theta} Y. \tag{5.144}$$

We therefore conclude that the function r^* is nothing but a scalar version of the Christoffel transform $\mathbf{r}^*$ as defined by (5.49).

We can now state the analogue of Darboux's classical Bäcklund transformation for isothermic surfaces [331].

Theorem 11 (A Bäcklund transformation for isothermic surfaces). *Let* $\mathbf{r}$ *be the position vector of an isothermic surface* Σ. *Then, a second isothermic* *surface $\tilde{\Sigma}$ is given by*

$$\tilde{\boldsymbol{r}} = \boldsymbol{r} - \frac{1}{mr^*}(X\boldsymbol{X} + Y\boldsymbol{Y} + \underline{N}^{\mathsf{T}}\underline{\boldsymbol{N}}), \tag{5.145}$$

where X, Y, $\underline{N}$ and r, r^* are solutions of the compatible linear system

$$\begin{pmatrix} X \\ Y \\ \underline{N} \\ r \\ r^* \end{pmatrix}_x = \begin{pmatrix} 0 & -\theta_y & -\underline{h}_1^{\mathsf{T}} & me^{-\theta} & me^{\theta} \\ \theta_y & 0 & 0 & 0 & 0 \\ \underline{h}_1 & 0 & 0 & 0 & 0 \\ e^{\theta} & 0 & 0 & 0 & 0 \\ e^{-\theta} & 0 & 0 & 0 & 0 \end{pmatrix} \begin{pmatrix} X \\ Y \\ \underline{N} \\ r \\ r^* \end{pmatrix}$$
$$\begin{pmatrix} X \\ Y \\ \underline{N} \\ r \\ r^* \end{pmatrix}_y = \begin{pmatrix} 0 & \theta_x & 0 & 0 & 0 \\ -\theta_x & 0 & -\underline{h}_2^{\mathsf{T}} & -me^{-\theta} & me^{\theta} \\ 0 & \underline{h}_2 & 0 & 0 & 0 \\ 0 & e^{\theta} & 0 & 0 & 0 \\ 0 & -e^{-\theta} & 0 & 0 & 0 \end{pmatrix} \begin{pmatrix} X \\ Y \\ \underline{N} \\ r \\ r^* \end{pmatrix} \tag{5.146}$$

satisfying the admissible constraint

$$X^2 + Y^2 + \underline{N}^2 = 2mrr^*. \tag{5.147}$$

The second solution of the vector isothermic system (5.43) reads

$$e^{\tilde{\theta}} = \frac{r}{r^*}e^{-\theta}$$
$$\tilde{\underline{h}}_1 = -\underline{h}_1 + \left(\frac{e^{\theta}}{r} + \frac{e^{-\theta}}{r^*}\right)\underline{N}, \quad \tilde{\underline{h}}_2 = \underline{h}_2 - \left(\frac{e^{\theta}}{r} - \frac{e^{-\theta}}{r^*}\right)\underline{N}. \tag{5.148}$$

Proof. The linear system (5.146) is obtained by comparison of the expressions (5.135) and (5.141), (5.142) for the adjoint eigenfunctions X' and Y'. The system is indeed compatible modulo the vector isothermic system. Remarkably, for $m = 0$, (5.146) constitutes a scalar version of the Gauss-Weingarten equations together with the first-order relations for the position vector and its Christoffel transform. The constraint (5.147) is the result of equating the two expressions $(5.134)_2$ and (5.143) for r'. It represents a particular case of the first integral

$$X^2 + Y^2 + \underline{N}^2 - 2mrr^* = \text{const.} \tag{5.149}$$

The transformation laws (5.148) are derived from (5.116) and (5.136) modulo the involution (5.137). □

Exercises

1. Show that the quantities $\tilde{\boldsymbol{r}}''$ and $\tilde{X}''$, $\tilde{Y}''$ as given in Corollary 1 satisfy the relations (5.121).
2. Verify that $\boldsymbol{r}'$ and r' as defined in Theorem 10 satisfy the systems (5.112) and (5.117), respectively.
3. Show that

$$X^2 + Y^2 + \underline{N}^2 - 2mrr^*$$

is constant modulo the linear system (5.146).

5.6 Permutability Theorems and Their Geometric Implications

In the previous section, we have seen that the Fundamental Transformation may be specialised to accommodate additional constraints imposed on conjugate nets. Thus, the requirements of orthogonality and conformality have been shown to induce natural and admissible conditions on the Fundamental Transformation. It will now be shown that these algebraic conditions impose hidden geometric constraints which are unveiled in the context of permutability theorems associated with iterated Bäcklund transformations. For notational convenience, the prime on parallel nets and adjoint eigenfunctions is replaced by an overbar. Superscripts label the quantities which generate the Bäcklund transformations and subscripts are attached to the corresponding Bäcklund transforms.

5.6.1 A Permutability Theorem for Conjugate Nets. Planarity

Let $\boldsymbol{r}$ be a conjugate net and $\boldsymbol{X}$, $\boldsymbol{Y}$ be the corresponding tangent vectors satisfying

$$\boldsymbol{r}_x = H_1\boldsymbol{X}, \quad \boldsymbol{r}_y = H_2\boldsymbol{Y}, \quad \boldsymbol{X}_y = q\boldsymbol{Y}, \quad \boldsymbol{Y}_x = p\boldsymbol{X}, \tag{5.150}$$

where p and q are related to H_1 and H_2 in the usual manner. If r^1, X^1, Y^1 are eigenfunctions obeying

$$r^1_x = H_1X^1, \quad r^1_y = H_2Y^1, \quad X^1_y = qY^1, \quad Y^1_x = pX^1, \tag{5.151}$$

and $\bar{\boldsymbol{r}}^1$ is a parallel net associated with some adjoint eigenfunctions $\bar{X}^1$, $\bar{Y}^1$, that is

$$\bar{\boldsymbol{r}}^1_x = \bar{X}^1\boldsymbol{X}, \quad \bar{\boldsymbol{r}}^1_y = \bar{Y}^1\boldsymbol{Y}, \tag{5.152}$$

then the Fundamental Transform $\Sigma_1 = \mathbb{B}_1(\Sigma)$ of Σ is given by

$$\mathbb{B}_1 : \quad \boldsymbol{r}_1 = \boldsymbol{r} - r^1\frac{\bar{\boldsymbol{r}}^1}{\bar{r}^{11}}, \tag{5.153}$$

where the scalar function $\bar{r}^{11}$ is defined by

$$\bar{r}^{11}_x = \bar{X}^1 X^1, \quad \bar{r}^{11}_y = \bar{Y}^1 Y^1. \tag{5.154}$$

Similarly, a second set of quantities $\{r^2, X^2, Y^2, \bar{X}^2, \bar{Y}^2, \bar{\boldsymbol{r}}^2, \bar{r}^{22}\}$ defines a third surface $\Sigma_2 = \mathbb{B}_2(\Sigma)$ represented by

$$\mathbb{B}_2 : \quad \boldsymbol{r}_2 = \boldsymbol{r} - r^2 \frac{\bar{\boldsymbol{r}}^2}{\bar{r}^{22}}. \tag{5.155}$$

Application of another Fundamental Transformation to the surface Σ_1 now requires the knowledge of (adjoint) eigenfunctions and a parallel net associated with Σ_1. These may be readily constructed if one takes into account the transformation properties of the Fundamental Transformation. Firstly, since the eigenfunctions r^2 and X^2, Y^2 are scalar analogues of the position vector $\boldsymbol{r}$ and the tangent vectors $\boldsymbol{X}$, $\boldsymbol{Y}$, respectively, their Fundamental Transforms are given by the scalar counterparts of $(5.116)_{1,2,3}$, that is

$$r^2_1 = r^2 - r^1 \frac{\bar{r}^{12}}{\bar{r}^{11}}, \quad X^2_1 = X^2 - X^1 \frac{\bar{r}^{12}}{\bar{r}^{11}}, \quad Y^2_1 = Y^2 - Y^1 \frac{\bar{r}^{12}}{\bar{r}^{11}}, \tag{5.156}$$

where the bilinear potential $\bar{r}^{12}$ satisfies the usual first-order relations

$$\bar{r}^{12}_x = \bar{X}^1 X^2, \quad \bar{r}^{12}_y = \bar{Y}^1 Y^2. \tag{5.157}$$

Secondly, application of Corollary 1 to the parallel net $\bar{\boldsymbol{r}}^2$ provides us with a net parallel to Σ_1 and corresponding eigenfunctions. These read

$$\bar{\boldsymbol{r}}^2_1 = \bar{\boldsymbol{r}}^2 - \bar{r}^{21} \frac{\bar{\boldsymbol{r}}^1}{\bar{r}^{11}}, \quad \bar{X}^2_1 = \bar{X}^2 - \bar{X}^1 \frac{\bar{r}^{21}}{\bar{r}^{11}}, \quad \bar{Y}^2_1 = \bar{Y}^2 - \bar{Y}^1 \frac{\bar{r}^{21}}{\bar{r}^{11}} \tag{5.158}$$

with

$$\bar{r}^{21}_x = \bar{X}^2 X^1, \quad \bar{r}^{21}_y = \bar{Y}^2 Y^1 \tag{5.159}$$

and the scalar counterpart $\bar{r}^{22}$ transforms as

$$\bar{r}^{22}_1 = \bar{r}^{22} - \bar{r}^{21} \frac{\bar{r}^{12}}{\bar{r}^{11}}. \tag{5.160}$$

By construction, the set of quantities $\{r^2_1, X^2_1, Y^2_1, \bar{X}^2_1, \bar{Y}^2_1, \bar{\boldsymbol{r}}^2_1, \bar{r}^{22}_1\}$ is related to the surface Σ_1 in the same way as $\{r^i, X^i, Y^i, \bar{X}^i, \bar{Y}^i, \bar{\boldsymbol{r}}^i, \bar{r}^{ii}\}$, $i = 1, 2$ are related to the surface Σ. Thus, another application of the Fundamental Transformation generates a fourth surface $\Sigma_{12} = \mathbb{B}_2^1(\Sigma_1)$ given by

$$\mathbb{B}_2^1 : \quad \boldsymbol{r}_{12} = \boldsymbol{r}_1 - r_1^2 \frac{\bar{\boldsymbol{r}}_1^2}{\bar{r}_1^{22}}, \tag{5.161}$$

where the indices on $\mathbb{B}_2^1$ indicate that the Bäcklund transformation is generated by the $\mathbb{B}_1$-transforms of quantities bearing an index 2. Similarly, application of another Fundamental Transformation generated by the $\mathbb{B}_2$-transforms of quantities carrying an index 1 produces a fifth surface $\Sigma_{21} = \mathbb{B}_1^2(\Sigma_2)$ represented by

$$\mathbb{B}_1^2 : \quad \boldsymbol{r}_{21} = \boldsymbol{r}_2 - r_2^1 \frac{\bar{\boldsymbol{r}}_2^1}{\bar{r}_2^{11}}. \tag{5.162}$$

Remarkably, the surfaces Σ_{12} and Σ_{21} are identical. This key geometric property underpins the classical permutability theorem for the Fundamental Transformation [119, 183].

Theorem 12 (A permutability theorem for the Fundamental Transformation). *The Fundamental Transformations $\mathbb{B}_1$, $\mathbb{B}_2$ and $\mathbb{B}_2^1$, $\mathbb{B}_1^2$ are such that the corresponding Bianchi diagram closes, that is*

$$\mathbb{B}_2^1 \circ \mathbb{B}_1 = \mathbb{B}_1^2 \circ \mathbb{B}_2. \tag{5.163}$$

The position vector of the surface $\Sigma_{12} = \Sigma_{21}$ is given by

$$\boldsymbol{r}_{12} = \boldsymbol{r}_{21} = \frac{\begin{vmatrix} \boldsymbol{r} & r^1 & r^2 \\ \bar{\boldsymbol{r}}^1 & \bar{r}^{11} & \bar{r}^{12} \\ \bar{\boldsymbol{r}}^2 & \bar{r}^{21} & \bar{r}^{22} \end{vmatrix}}{\begin{vmatrix} \bar{r}^{11} & \bar{r}^{12} \\ \bar{r}^{21} & \bar{r}^{22} \end{vmatrix}}. \tag{5.164}$$

Proof. Insertion of $\boldsymbol{r}_1$, r_1^2 and $\bar{\boldsymbol{r}}_1^2$, $\bar{r}_1^{22}$ as given by (5.153), (5.156)$_1$ and (5.158)$_1$, (5.160), respectively, into (5.161) produces the expression (5.164) for the position vector $\boldsymbol{r}_{12}$. The symmetry in the indices 1 and 2 implies that $\boldsymbol{r}_{12} = \boldsymbol{r}_{21}$ and hence the Bianchi diagram closes. □

The above theorem implies that

$$\boldsymbol{r}_{12} - \boldsymbol{r} = \alpha \bar{\boldsymbol{r}}^1 + \beta \bar{\boldsymbol{r}}^2, \tag{5.165}$$

where the functions α and β are obtained from (5.164). On solving (5.153) and

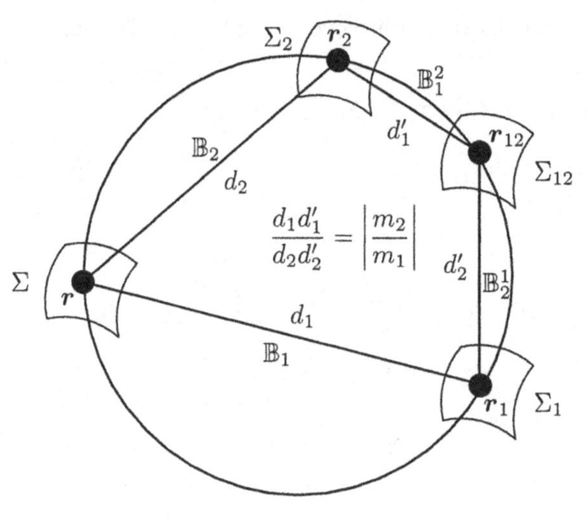

Figure 5.1. Planarity, cyclicity and constant cross-ratio of the Bianchi quadrilateral.

(5.155) for $\bar{\boldsymbol{r}}^1$ and $\bar{\boldsymbol{r}}^2$, respectively, we obtain

$$\boldsymbol{r}_{12} - \boldsymbol{r} = a(\boldsymbol{r}_1 - \boldsymbol{r}) + b(\boldsymbol{r}_2 - \boldsymbol{r}) \tag{5.166}$$

with $a = -\alpha\bar{r}^{11}/r^1$ and $b = -\beta\bar{r}^{22}/r^2$ and hence the vectors $\boldsymbol{r}_{12} - \boldsymbol{r}$ and $\boldsymbol{r}_1 - \boldsymbol{r}$, $\boldsymbol{r}_2 - \boldsymbol{r}$ are co-planar. Thus, we have the following 'physical realisation' of the Bianchi diagram associated with the Fundamental Transformation.

Corollary 2 (Planarity of the Bianchi quadrilateral). *The position vector* $\boldsymbol{r}$ *and its Fundamental Transforms* $\boldsymbol{r}_1, \boldsymbol{r}_2, \boldsymbol{r}_{12}$ *satisfy a linear equation of the form (5.166). For any fixed choice of the conjugate parameters* (x, y)*, the vertices of the Bianchi quadrilateral* $\boldsymbol{r}(x, y), \boldsymbol{r}_1(x, y), \boldsymbol{r}_2(x, y), \boldsymbol{r}_{12}(x, y)$ *lie on a plane (cf. Figure 5.1).*

5.6.2 A Permutability Theorem for Orthogonal Conjugate Nets. Cyclicity

We have shown that the requirement of orthogonality leads to the admissible constraints (5.134), (5.135) on the Fundamental Transformation. In order for the permutability theorem to be valid in this restricted case, these constraints must also be preserved by the Fundamental Transformation. Thus, in the present notation, the expressions

$$\begin{aligned} \bar{\boldsymbol{r}}^2 &= X^2\boldsymbol{X} + Y^2\boldsymbol{Y} + \underline{N}^{2\mathrm{T}}\underline{N}, \quad \bar{r}^{22} = \frac{1}{2}(\bar{\boldsymbol{r}}^2)^2 \\ \bar{X}^2 &= X^2_x + pY^2 + \underline{h}_1^{\mathrm{T}}\underline{N}^2, \qquad \bar{Y}^2 = Y^2_y + qX^2 + \underline{h}_2^{\mathrm{T}}\underline{N}^2 \end{aligned} \tag{5.167}$$

must be invariant under the Ribaucour transformation $\mathbb{B}_1$, that is

$$\begin{aligned} \bar{\boldsymbol{r}}_1^2 &= X_1^2 \boldsymbol{X}_1 + Y_1^2 \boldsymbol{Y}_1 + \underline{N}_1^{2^{\mathrm{T}}} \underline{\boldsymbol{N}}_1, \quad \bar{r}_1^{22} = \frac{1}{2}\left(\bar{\boldsymbol{r}}_1^2\right)^2 \\ \bar{X}_1^2 &= X_{1x}^2 + p_1 Y_1^2 + \underline{h}_{1,1}^{\mathrm{T}} \underline{N}_1^2, \qquad \bar{Y}_1^2 = Y_{1y}^2 + q_1 X_1^2 + \underline{h}_{2,1}^{\mathrm{T}} \underline{N}_1^2, \end{aligned} \tag{5.168}$$

where

$$\begin{aligned} \boldsymbol{X}_1 &= \boldsymbol{X} - X^1 \frac{\bar{\boldsymbol{r}}^1}{\bar{r}^{11}}, \quad \boldsymbol{Y}_1 = \boldsymbol{Y} - Y^1 \frac{\bar{\boldsymbol{r}}^1}{\bar{r}^{11}}, \quad \underline{\boldsymbol{N}}_1 = \underline{\boldsymbol{N}} - \underline{N}^1 \frac{\bar{\boldsymbol{r}}^1}{\bar{r}^{11}} \\ p_1 &= p - \frac{\bar{X}^1 Y^1}{\bar{r}^{11}}, \quad q_1 = q - \frac{\bar{Y}^1 X^1}{\bar{r}^{11}} \\ \underline{h}_{1,1} &= \underline{h}_1 - \bar{X}^1 \frac{\underline{N}^1}{\bar{r}^{11}}, \quad \underline{h}_{2,1} = \underline{h}_2 - \bar{Y}^1 \frac{\underline{N}^1}{\bar{r}^{11}}. \end{aligned} \tag{5.169}$$

It is readily verified that evaluation of the consistency conditions (5.168) leads to one additional constraint which proves only relevant in the course of iteration of the Fundamental Transformation. For completeness, it is noted that

$$\underline{N}_1^2 = \underline{N}^2 - \underline{N}^1 \frac{\bar{r}^{12}}{\bar{r}^{11}}. \tag{5.170}$$

Theorem 13 (A permutability theorem for the Ribaucour transformation). *The Bianchi diagram associated with the Ribaucour transformation is closed modulo the particular first integral*

$$\bar{r}^{12} + \bar{r}^{21} = \bar{\boldsymbol{r}}^1 \cdot \bar{\boldsymbol{r}}^2 = X^1 X^2 + Y^1 Y^2 + \underline{N}^{1^{\mathrm{T}}} \underline{N}^2. \tag{5.171}$$

Proof. Differentiation reveals that

$$\bar{\boldsymbol{r}}^1 \cdot \bar{\boldsymbol{r}}^2 - \bar{r}^{12} - \bar{r}^{21} = \text{const} \tag{5.172}$$

and hence the constraint (5.171) is indeed admissible. □

The above theorem gives rise to a geometric observation originally made in the classical case by Bianchi [37]. Thus, the constraint (5.171) implies that

$$\bar{\boldsymbol{r}}_2^1 \cdot \bar{\boldsymbol{r}}^2 + \bar{\boldsymbol{r}}_1^2 \cdot \bar{\boldsymbol{r}}^1 = 2(\bar{\boldsymbol{r}}^1 \cdot \bar{\boldsymbol{r}}^2 - \bar{r}^{12} - \bar{r}^{21}) = 0, \tag{5.173}$$

where $\bar{\boldsymbol{r}}_1^2$ and $\bar{\boldsymbol{r}}_2^1$ are given by $(5.158)_1$ and $(5.158)_1|_{1\leftrightarrow 2}$, respectively, while the generic Fundamental Transformation admits the relation

$$\begin{aligned}
&\bar{\boldsymbol{r}}^1_2 \cdot C\bar{\boldsymbol{r}}^2 + \bar{\boldsymbol{r}}^2_1 \cdot C\bar{\boldsymbol{r}}^1 \\
&\quad = \bar{\boldsymbol{r}}^1 \cdot C\bar{\boldsymbol{r}}^2 + \bar{\boldsymbol{r}}^2 \cdot C\bar{\boldsymbol{r}}^1 - \frac{\bar{r}^{21}}{\bar{r}^{11}}\bar{\boldsymbol{r}}^1 \cdot C\bar{\boldsymbol{r}}^1 - \frac{\bar{r}^{12}}{\bar{r}^{22}}\bar{\boldsymbol{r}}^2 \cdot C\bar{\boldsymbol{r}}^2 = 0
\end{aligned} \tag{5.174}$$

for any constant skew-symmetric matrix C. The above algebraic relations now give rise to the result [331]:

Theorem 14 (Cyclicity of the Bianchi quadrilateral). *The vertices of the Bianchi quadrilateral associated with the Ribaucour transformation lie on a circle (cf. Figure 5.1).*

Proof. Here, it proves convenient to introduce the notion of the *cross-ratio* of four points on a plane which we identify with the complex plane, that is

$$P(\boldsymbol{r}) = a + ib \in \mathbb{C} \quad \leftrightarrow \quad \boldsymbol{r} = \begin{pmatrix} a \\ b \end{pmatrix} \in \mathbb{R}^2. \tag{5.175}$$

The cross-ratio of four points $\boldsymbol{r}, \boldsymbol{r}_1, \boldsymbol{r}_2, \boldsymbol{r}_{12}$ is defined by the complex quantity

$$\frac{P(\boldsymbol{r}_1 - \boldsymbol{r})P(\boldsymbol{r}_{12} - \boldsymbol{r}_2)}{P(\boldsymbol{r}_2 - \boldsymbol{r})P(\boldsymbol{r}_{12} - \boldsymbol{r}_1)}. \tag{5.176}$$

The cross-ratio is known to be real if and only if the four points lie on a circle (or on a straight line which may be regarded as a circle of infinite radius). Thus, in the present context, the vectors $\boldsymbol{r}_1 - \boldsymbol{r}$, $\boldsymbol{r}_2 - \boldsymbol{r}$, $\boldsymbol{r}_{12} - \boldsymbol{r}_1$ and $\boldsymbol{r}_{12} - \boldsymbol{r}_2$ lie on a plane which we identify with the complex plane. Since the quantities $\bar{\boldsymbol{r}}^1, \bar{\boldsymbol{r}}^2, \bar{\boldsymbol{r}}^1_2$ and $\bar{\boldsymbol{r}}^2_1$ are parallel to the edges of the Bianchi quadrilateral, we deduce that the Bianchi quadrilateral is inscribed in a circle if the cross-ratio

$$\frac{P(\bar{\boldsymbol{r}}^1)P(\bar{\boldsymbol{r}}^1_2)}{P(\bar{\boldsymbol{r}}^2)P(\bar{\boldsymbol{r}}^2_1)} \tag{5.177}$$

is real. Without loss of generality, we may therefore assume that the vectors $\bar{\boldsymbol{r}}^1, \bar{\boldsymbol{r}}^2, \bar{\boldsymbol{r}}^1_2$ and $\bar{\boldsymbol{r}}^2_1$ are two-dimensional and

$$C = \begin{pmatrix} 0 & -1 \\ 1 & 0 \end{pmatrix}. \tag{5.178}$$

An elementary calculation solely based on the identities (5.173) and (5.174) now shows that (5.177) is indeed real. □

5.6.3 A Permutability Theorem for Isothermic Surfaces. Constant Cross-Ratio

In the case of isothermic surfaces, it is necessary to show that the constraints (5.141)–(5.143) are preserved by the composition of the Ribaucour transformation and the involution (5.137). To this end, we note that the Christoffel transform $\boldsymbol{r}^*$ is but a particular parallel net and hence transforms in the same manner as any other parallel net. Accordingly, the Ribaucour transform $\mathbb{B}_1$ of the scalar quantity r^{*2} is given by

$$r_1^{*2} = r^{*2} - r^{*1}\frac{\bar{r}^{12}}{\bar{r}^{11}}. \tag{5.179}$$

Now, invariance of (5.143) requires that

$$\bar{r}_1^{22} = m_2 r_1^2 r_1^{*2}, \tag{5.180}$$

which reduces to

$$\frac{\bar{r}^{12}}{m_1} + \frac{\bar{r}^{21}}{m_2} = r^1 r^{*2} + r^2 r^{*1}. \tag{5.181}$$

Comparison with the constraint (5.171) therefore delivers

$$\begin{aligned}\bar{r}^{12} &= \frac{m_1}{m_1 - m_2}[\bar{\boldsymbol{r}}^1 \cdot \bar{\boldsymbol{r}}^2 - m_2(r^1 r^{*2} + r^{*1} r^2)]\\ \bar{r}^{21} &= \frac{m_2}{m_2 - m_1}[\bar{\boldsymbol{r}}^1 \cdot \bar{\boldsymbol{r}}^2 - m_1(r^1 r^{*2} + r^{*1} r^2)]\end{aligned} \tag{5.182}$$

and these are, in fact, solutions of the defining relations (5.157) and (5.159), respectively. It remains to point out that preservation of the relations (5.141) and (5.142) does not impose any further constraints on the Ribaucour transformation. Thus, we are now in a position to formulate the analogue of Bianchi's permutability theorem [36] for classical isothermic surfaces [331].[2]

Theorem 15 (A permutability theorem for isothermic surfaces). *For a given seed isothermic surface Σ corresponding to a vector-valued solution $(\boldsymbol{X}, \boldsymbol{Y}, \underline{\boldsymbol{N}}, \boldsymbol{r}, \boldsymbol{r}^*)$ of the linear system (5.146) with parameter $m = 0$, let $(X^1, Y^1, \underline{N}^1, r^1, r^{*1})$ be a solution of (5.146) with parameter $m = m_1$ subject to the constraint (5.147).*

[2] For convenience, we have dropped the index 2 on the relevant quantities.

Then, the linear system (5.146) is invariant under

$$\mathbb{B}_1 : \begin{cases} (X, Y, \underline{N}, r, r^*, \theta, \underline{h}_1, \underline{h}_2) \\ \downarrow \\ (X_1, Y_1, \underline{N}_1, r_1, r_1^*, \theta_1, \underline{h}_{1,1}, \underline{h}_{2,1}), \end{cases} \tag{5.183}$$

where

$$\begin{gathered} X_1 = -X + \frac{X^1}{r^1 r^{*1}} M^1, \quad Y_1 = Y - \frac{Y^1}{r^1 r^{*1}} M^1, \quad \underline{N}_1 = \underline{N} - \frac{\underline{N}^1}{r^1 r^{*1}} M^1 \\ r_1 = r - \frac{1}{r^{*1}} M^1, \quad r_1^* = r^* - \frac{1}{r^1} M^1, \quad e^{\theta_1} = \frac{r^1}{r^{*1}} e^{-\theta} \\ \underline{h}_{1,1} = -\underline{h}_1 + \left(\frac{e^\theta}{r^1} + \frac{e^{-\theta}}{r^{*1}} \right) \underline{N}^1, \quad \underline{h}_{2,1} = \underline{h}_2 - \left(\frac{e^\theta}{r^1} - \frac{e^{-\theta}}{r^{*1}} \right) \underline{N}^1 \end{gathered} \tag{5.184}$$

and

$$M^1 = \frac{X^1 X + Y^1 Y + \underline{N}^{1\mathrm{T}} \underline{N} - m(r^1 r^* + r^{*1} r)}{m_1 - m}. \tag{5.185}$$

The isothermic surface Σ_1, its Christoffel transform Σ_1^* and the orthonormal frame $(\boldsymbol{X}_1, \boldsymbol{Y}_1, \underline{\boldsymbol{N}}_1)$ are obtained from (5.184) and (5.185) by means of the substitution $(X, Y, \underline{N}, r, r^*, m) \to (\boldsymbol{X}, \boldsymbol{Y}, \underline{\boldsymbol{N}}, \boldsymbol{r}, \boldsymbol{r}^*, 0)$. The Bianchi diagram associated with two Bäcklund transformations $\mathbb{B}_1$ and $\mathbb{B}_2$ with parameters m_1 and m_2 is closed, that is

$$\mathbb{B}_2 \circ \mathbb{B}_1 = \mathbb{B}_1 \circ \mathbb{B}_2. \tag{5.186}$$

We remark that, at each step of the iteration procedure, the Bäcklund transformations $\mathbb{B}_i$ carry only one index since we now regard the transformations as being distinguished by the Bäcklund parameters m_i rather than the corresponding eigenfunctions.

In the previous subsection, we have shown that the cross-ratio of the Bianchi quadrilateral is real. In this case, the cross-ratio admits the geometric interpretation

$$\left| \frac{P(\boldsymbol{r}_1 - \boldsymbol{r}) P(\boldsymbol{r}_{12} - \boldsymbol{r}_2)}{P(\boldsymbol{r}_2 - \boldsymbol{r}) P(\boldsymbol{r}_{12} - \boldsymbol{r}_1)} \right| = \frac{d_1 d_1'}{d_2 d_2'}, \tag{5.187}$$

where

$$\begin{aligned} d_1 &= |\boldsymbol{r}_1 - \boldsymbol{r}|, \quad d_1' = |\boldsymbol{r}_{12} - \boldsymbol{r}_2| \\ d_2 &= |\boldsymbol{r}_2 - \boldsymbol{r}|, \quad d_2' = |\boldsymbol{r}_{12} - \boldsymbol{r}_1|. \end{aligned} \tag{5.188}$$

The sign of the cross-ratio determines whether or not the edges of the quadrilaterals intersect. A non-intersecting (embedded) Bianchi quadrilateral is depicted in Figure 5.1 corresponding to a negative cross-ratio. The lengths of the edges of the Bianchi quadrilateral are readily expressed in terms of θ and its Bäcklund transforms. For instance, the transformation law (5.145) for the position vector associated with the Bäcklund transformation $\mathbb{B}_1$ delivers

$$(d_1)^2 = (\boldsymbol{r}_1 - \boldsymbol{r})^2 = \frac{2}{m_1}\frac{r^1}{r^{*1}} = \frac{2}{m_1}e^{\theta_1+\theta} \tag{5.189}$$

by virtue of $(5.184)_6$. Similarly, we find that

$$(d_2)^2 = \frac{2}{m_2}e^{\theta_2+\theta}, \quad (d_1')^2 = \frac{2}{m_1}e^{\theta_{12}+\theta_2}, \quad (d_2')^2 = \frac{2}{m_2}e^{\theta_{21}+\theta_1}. \tag{5.190}$$

Moreover, closure of the Bianchi diagram implies that $\theta_{12} = \theta_{21}$ and hence

$$\frac{d_1 d_1'}{d_2 d_2'} = \left|\frac{m_2}{m_1}\right|. \tag{5.191}$$

The sign of the cross-ratio may be calculated directly using Theorem 15. In this way, we obtain a theorem which generalises a classical result due to Demoulin [101], viz:

Corollary 3 (Constant cross-ratio of the Bianchi quadrilateral). *The cross-ratio of the Bianchi quadrilateral is independent of the coordinates x and y and is given by*

$$\frac{P(\boldsymbol{r}_1 - \boldsymbol{r})P(\boldsymbol{r}_{12} - \boldsymbol{r}_2)}{P(\boldsymbol{r}_2 - \boldsymbol{r})P(\boldsymbol{r}_{12} - \boldsymbol{r}_1)} = \frac{m_2}{m_1} \tag{5.192}$$

(cf. Figure 5.1).

Exercises

1. Derive the position vector (5.164) for the second generation isothermic surface Σ_{12}.
2. Verify that the relations (5.168) are satisfied modulo the constraint (5.171). Establish the validity of the relation (5.172).
3. Prove that if $\boldsymbol{a}, \boldsymbol{b}, \boldsymbol{c}, \boldsymbol{d}$ are four two-dimensional vectors satisfying

$$\boldsymbol{a}\cdot\boldsymbol{d} + \boldsymbol{b}\cdot\boldsymbol{c} = 0, \quad \boldsymbol{a}\cdot C\boldsymbol{d} + \boldsymbol{b}\cdot C\boldsymbol{c} = 0, \quad C = \begin{pmatrix} 0 & -1 \\ 1 & 0 \end{pmatrix}$$

then the cross-ratio

$$\frac{P(\boldsymbol{a})P(\boldsymbol{b})}{P(\boldsymbol{c})P(\boldsymbol{d})}$$

is real.

4. Show that the relations (5.141) and (5.142) are preserved by the Bäcklund transformation for isothermic surfaces iff the constraint (5.181) holds. Verify that (5.182) are indeed solutions of the defining relations (5.157) and (5.159).
5. Use Theorem 15 to establish the cross-ratio relation (5.192).

5.7 An Explicit Permutability Theorem for the Vector Calapso System

It has been established that the Ribaucour transformation for isothermic surfaces gives rise to a permutability theorem with concomitant geometric properties. This permutability theorem was formulated at the linear (surface) level. However, an explicit permutability theorem at the nonlinear level, that is for the isothermic system itself, does not seem to appear in the classical literature. Here, it is shown that such a superposition principle is readily obtained for the vector Calapso system if one formulates the Ribaucour transformation in terms of the classical Moutard transformation. The permutability theorem for the vector Calapso system so derived is of remarkable simplicity.

5.7.1 The Ribaucour-Moutard Connection

Here, we return to the classical Moutard transformation as discussed in Chapter 1:

Theorem 16 (The Moutard transformation). *The Moutard equation*

$$\psi_{xy} = \rho\psi \tag{5.193}$$

is invariant under

$$\psi \to \psi_1 = \frac{S(\psi, \psi^1)}{\psi^1}, \quad \rho \to \rho_1 = \rho - 2(\ln \psi^1)_{xy}, \tag{5.194}$$

where ψ^1 is another solution of (5.193) and the bilinear potential $S(\psi, \psi^1)$ is defined by

$$S_x = \psi_x\psi^1 - \psi\psi^1_x, \quad S_y = \psi\psi^1_y - \psi_y\psi^1. \tag{5.195}$$

If we now define the quantities (cf. Section 5.3)

$$\psi = e^{\theta} r^*, \quad \underline{S} = \underline{z}^* \psi - \sqrt{2}\underline{N} \tag{5.196}$$

then the linear system (5.146) implies that the relations (5.71) hold for the scalar quantities X and Y. Evaluation of X_y, Y_x and $X_x + Y_y$ produces the Lax pair (5.79) for the vector Calapso system with $k = -2m$ and $X_x - Y_y$ may be used to express r in terms of ψ and $\underline{S}$. The Lax pair (5.79) may be shown to be equivalent to the linear representation (5.146) and the first integrals (5.81) and (5.149) coincide. Furthermore, the transformation laws $(5.184)_{5,6}$ for r^* and θ give rise to a transformation law for ψ which is, up to an irrelevant constant factor, exactly of the form $(5.194)_1$ with a particular bilinear potential $S(\psi, \psi^1)$. The action of the Bäcklund transformation on the vector Calapso equation is suggested by the observation that, since $\underline{z}$ is a (vector-valued) solution of the Moutard equation (5.193), it should transform in the same manner as the eigenfunction ψ. Indeed, this proves to be the case and the Bäcklund transformation adopts the following form [331]:

Theorem 17 (A Bäcklund transformation for the vector Calapso system). *The vector Calapso system (5.64) is invariant under the Moutard-type transformation*

$$\underline{z} \to \underline{z}_1 = \frac{\underline{S}(\underline{z}, \psi^1)}{\psi^1}, \quad \rho \to \rho_1 = \rho - 2(\ln \psi^1)_{xy}, \tag{5.197}$$

where ψ^1 is a solution of the Lax pair (5.79) subject to the first integral $I = 0$.

Proof. Since $(5.197)_{1,2}$ is a vector version of the classical Moutard transformation (5.194), it is evident that $(5.64)_1$ is preserved. Moreover, the first integral $I = 0$ implies that

$$\underline{z}_1^2 = \frac{\underline{S}^2}{(\psi^1)^2} = \underline{z}^2 + 2\Delta(\ln \psi^1) \tag{5.198}$$

and hence

$$\begin{aligned}\left(\underline{z}_1^2\right)_{xy} + \Delta\rho_1 &= [\underline{z}^2 + 2\Delta(\ln \psi^1)]_{xy} + \Delta\rho - 2\Delta(\ln \psi^1)_{xy} \\ &= (\underline{z}^2)_{xy} + \Delta\rho\,.\end{aligned} \tag{5.199}$$

□

It is remarked that the above Bäcklund transformation has been foreshadowed in Theorem 7 since at the level of the vector Calapso system, the dual solution

$(\underline{z}^*, \rho^*)$ of the vector Calapso system as given by (5.82) and the Bäcklund transform $(\underline{z}_1, \rho_1)$ are generated in the same manner. However, if we make the identification $\psi_0 = e^\theta$, where θ is the solution of the vector isothermic system, then the particular choice $\psi^1 = \psi_0$ implies that the dual solution and the Bäcklund transform are identical. This is in agreement with the fact that the Christoffel transform Σ^* of an isothermic surface Σ may be regarded as a degenerate Ribaucour transform of Σ [61]. This observation is elaborated on in Section 5.8 in connection with Dupin cyclides.

5.7.2 A Permutability Theorem

The derivation of a permutability theorem for the vector Calapso system is now based on a direct consequence of Theorem 17, viz.:

Corollary 4. *The Bäcklund transforms of the solution of the vector Calapso system written in the form*

$$\underline{z}_{xy} = a_y \underline{z}, \quad a_y = \epsilon b_x, \quad a_x + \epsilon b_y + \underline{z}^2 = 0, \tag{5.200}$$

where $\epsilon^2 = 1$ *and* $\rho = a_y = \epsilon b_x$, $f = -a_x + \epsilon b_y$, *read*

$$\mathbb{B}_1 : \ \underline{z}_1 = \frac{\underline{S}(\underline{z}, \psi^1)}{\psi^1}, \quad \begin{array}{l} a_1 = a - 2(\ln \psi^1)_x \\ b_1 = -b + 2\epsilon(\ln \psi^1)_y \end{array} \tag{5.201}$$

with $\epsilon_1 = -\epsilon$.

The constant $\epsilon = \pm 1$ has been introduced to cast the permutability theorem into a compact vectorial form. The first step in this procedure is to eliminate the eigenfunction ψ^1 and the corresponding bilinear potential $\underline{S}(\underline{z}, \psi^1)$ from the Lax pair and the defining relations for $\underline{S}$ by means of the transformation formulae (5.201). Thus, if we solve the latter for ψ^1_x, ψ^1_y and $\underline{S}$, then the relation $(5.80)_1$ and the Lax pair (5.79) become

$$\begin{aligned} (\underline{z}_1 - \underline{z})_x &= \frac{1}{2}(a_1 - a)(\underline{z}_1 + \underline{z}), \quad (b_1 - b)_x = \frac{1}{2}(a_1 - a)(b_1 + b) \\ (a_1 + a)_x &= \frac{1}{4}(a_1 - a)^2 - \frac{1}{4}(b_1 + b)^2 - \frac{1}{2}(\underline{z}_1 + \underline{z})^2 + k_1. \end{aligned} \tag{5.202}$$

A Bäcklund transformation $\mathbb{B}_2$ associated with a parameter k_2 gives rise to similar relations and the action of $\mathbb{B}_1$, $\mathbb{B}_2$ on the Bäcklund transforms generated by $\mathbb{B}_2$, $\mathbb{B}_1$, respectively, results in another six relations. For instance, the three additional copies of $(5.202)_1$ are given by

$$\begin{aligned} \mathbb{B}_2: \quad (\underline{z}_2 - \underline{z})_x &= \frac{1}{2}(a_2 - a)(\underline{z}_2 + \underline{z}) \\ \mathbb{B}_1: \quad (\underline{z}_{21} - \underline{z}_2)_x &= \frac{1}{2}(a_{21} - a_2)(\underline{z}_{21} + \underline{z}_2) \\ \mathbb{B}_2: \quad (\underline{z}_{12} - \underline{z}_1)_x &= \frac{1}{2}(a_{12} - a_1)(\underline{z}_{12} + \underline{z}_1). \end{aligned} \tag{5.203}$$

The relations $(5.202)_1$ and (5.203) are the analogues of those employed in Section 1.3 in connection with the classical permutability theorem for pseudospherical surfaces. Thus, if we take into account that the Bianchi diagram commutes, then the operation $(5.202)_1 - (5.203)_1 - (5.203)_2 + (5.203)_3$ produces the algebraic relation

$$\begin{aligned} &(a_1 - a)(\underline{z}_1 + \underline{z}) - (a_{12} - a_2)(\underline{z}_{12} + \underline{z}_2) \\ &- (a_2 - a)(\underline{z}_2 + \underline{z}) + (a_{12} - a_1)(\underline{z}_{12} + \underline{z}_1) = 0. \end{aligned} \tag{5.204}$$

The remaining relations $(5.202)_{2,3}$ and their counterparts may be manipulated analogously to derive another two purely algebraic relations. This algebraic system of three equations may now be solved for a_{12}, b_{12} and $\underline{z}_{12}$ to obtain

$$\begin{gathered} a_{12} - a = \hat{\rho}(a_1 - a_2), \quad b_{12} - b = \hat{\rho}(b_1 - b_2), \quad \underline{z}_{12} - \underline{z} = \hat{\rho}(\underline{z}_1 - \underline{z}_2) \\ \hat{\rho} = 4\frac{k_2 - k_1}{(a_1 - a_2)^2 + (b_1 - b_2)^2 + 2(\underline{z}_1 - \underline{z}_2)^2}. \end{gathered} \tag{5.205}$$

Remarkably, this superposition principle is symmetric in a, b and $\sqrt{2}\underline{z}$ and therefore admits the following formulation [331]:

Theorem 18 (A permutability theorem for the vector Calapso system). *If the solutions $\underline{u}_1$ and $\underline{u}_2$ of the vector Calapso system (5.200) are related to the seed solution $\underline{u} = (a, b, \sqrt{2}\underline{z})$ by the Bäcklund transformations $\mathbb{B}_1$ and $\mathbb{B}_2$, then $\underline{u}_{12}$ as given by*

$$\boxed{\underline{u}_{12} - \underline{u} = \frac{4(k_2 - k_1)}{(\underline{u}_1 - \underline{u}_2)^2}(\underline{u}_1 - \underline{u}_2)} \tag{5.206}$$

is also a solution of the vector Calapso system and constitutes the Bäcklund transform of both $\underline{u}_1$ and $\underline{u}_2$, that is, $\underline{u}_{12} = \mathbb{B}_2(\underline{u}_1) = \mathbb{B}_1(\underline{u}_2)$.

The above superposition principle represents the analogue of the classical permutability theorem (1.63) for the sine-Gordon equation. It is interesting to note that, in the case $n = 2$ and $b \to ib$ (cf. Subsection 5.3.1), the permutability theorem in the form (5.205) constitutes a simplified version of that set down for the boomeron equation in [64]. The permutability theorem therein contains scalar and vector components and was derived in connection with a matrix version of the classical Darboux transformation applied to a matrix Schrödinger equation.

Exercises

1. Use Theorem 17 to prove Corollary 4.
2. Derive the Bäcklund equations (5.202) and manipulate the relations $(5.202)_1$ and (5.203) and their counterparts for a and b in the manner indicated above to obtain the permutability theorem (5.205).

5.8 Particular Isothermic Surfaces. One-Soliton Surfaces and Cyclides

Here, we construct explicit solutions of the classical isothermic system and Calapso equation by means of the Darboux-Ribaucour and Moutard transformations, respectively. Application of these transformations to trivial seed solutions is shown to lead to 'one-soliton' isothermic surfaces and Dupin cyclides. The corresponding localised solutions of the zoomeron equation are related to the important dromion solutions of the Davey-Stewartson III equation via a simple Lie point symmetry.

5.8.1 One-Soliton Isothermic Surfaces

It is evident that planes constitute isothermic surfaces in $\mathbb{R}^3$. In fact, the parametrisation

$$\boldsymbol{r} = \begin{pmatrix} x \\ y \\ 0 \end{pmatrix}, \quad \boldsymbol{X} = \begin{pmatrix} 1 \\ 0 \\ 0 \end{pmatrix}, \quad \boldsymbol{Y} = \begin{pmatrix} 0 \\ 1 \\ 0 \end{pmatrix}, \quad \boldsymbol{N} = \begin{pmatrix} 0 \\ 0 \\ 1 \end{pmatrix} \tag{5.207}$$

is readily shown to satisfy the Gauss-Weingarten equations (5.35), (5.42) with corresponding solution

$$\theta = 0, \quad h_1 = 0, \quad h_2 = 0 \tag{5.208}$$

of the isothermic system. In this case, the linear system (5.146) reduces to

$$\begin{aligned} r_x = X, \quad r_x^* = X, \quad & X_x = m(r^* + r), \quad X = X(x) \\ r_y = Y, \quad r_y^* = -Y, \quad & Y_y = m(r^* - r), \quad Y = Y(y) \end{aligned} \tag{5.209}$$

together with $N = N_0$ so that

$$X_{xx} = \omega^2 X, \quad Y_{yy} = -\omega^2 Y \qquad (\omega^2 = 2m) \tag{5.210}$$

and hence, if $\omega \neq 0$,

$$\begin{aligned} X &= X_0 \sinh \omega x, \quad & r &= \frac{X_0}{\omega} \cosh \omega x - \frac{Y_0}{\omega} \cos \omega y \\ Y &= Y_0 \sin \omega y, \quad & r^* &= \frac{X_0}{\omega} \cosh \omega x + \frac{Y_0}{\omega} \cos \omega y \end{aligned} \tag{5.211}$$

without loss of generality. Accordingly, the constraint (5.147) yields

$$X_0^2 = Y_0^2 + N_0^2 \tag{5.212}$$

and the position vector of the new isothermic surface $\tilde{\Sigma}$ reads [86]

$$\tilde{\boldsymbol{r}} = \begin{pmatrix} x \\ y \\ 0 \end{pmatrix} - \frac{2/\omega}{X_0 \cosh \omega x + Y_0 \cos \omega y} \begin{pmatrix} X_0 \sinh \omega x \\ Y_0 \sin \omega y \\ N_0 \end{pmatrix}. \tag{5.213}$$

The above surfaces are here termed one-soliton isothermic surfaces. The corresponding solution of the classical Calapso equation is readily derived by means of the transformation (5.148) and is

$$\tilde{z} = \frac{\sqrt{2}\, \omega N_0}{X_0 \cosh \omega x + Y_0 \cos \omega y}. \tag{5.214}$$

Prototypical members of this class of isothermic surfaces are depicted in Figure 5.2 for $Y_0/N_0 = 0$, $1/4$, $1/2$, $3/4$. In the case $Y_0 = 0$, $X_0 = N_0$, a cylindrical isothermic surface is obtained as in [86]. It is interesting to note that, in this special case, the base curve

$$\tilde{\boldsymbol{r}} = \begin{pmatrix} x - \dfrac{2}{\omega} \tanh \omega x \\ -\dfrac{2}{\omega} \dfrac{1}{\cosh \omega x} \end{pmatrix} \tag{5.215}$$

adopts the form of a loop soliton (cf. Chapter 6).

5.8.2 A Class of Solutions Generated by the Moutard Transformation

The simplest seed solution of the classical Calapso equation (5.18) is given by

$$z = \text{const} \quad \Rightarrow \quad \rho = 0, \quad f = \text{const} \tag{5.216}$$

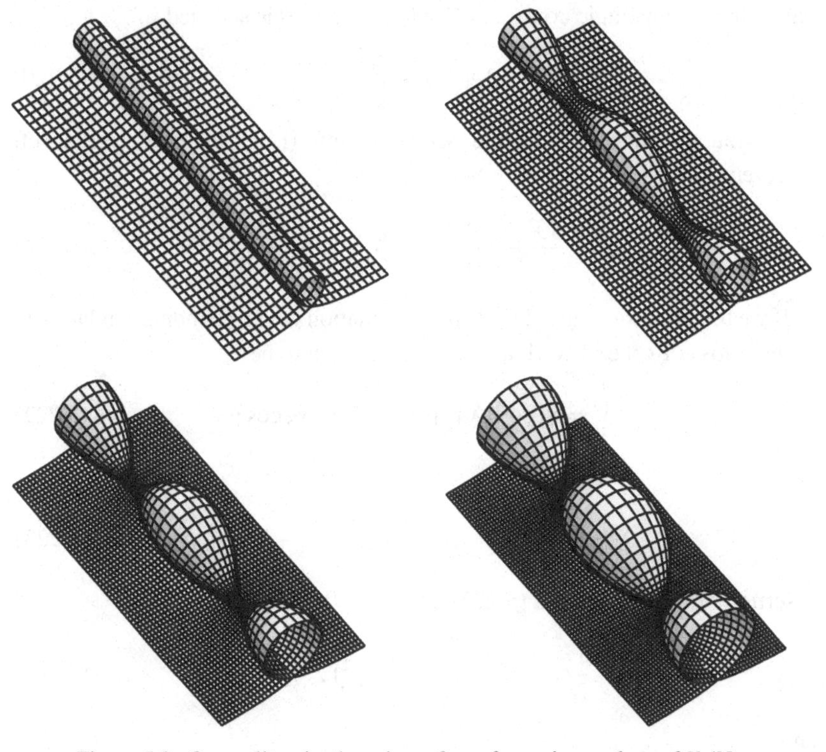

Figure 5.2. One-soliton isothermic surfaces for various values of Y_0/N_0.

with corresponding solution of the Lax pair (5.79), (5.80) of the form

$$\psi = \mathcal{X}(x) + \mathcal{Y}(y), \quad S = z(\mathcal{Y}(y) - \mathcal{X}(x)) + s, \tag{5.217}$$

where s is an arbitrary constant of integration. Insertion of these expressions into $(5.79)_2$ produces

$$\begin{aligned} \mathcal{X}_{xx} + (z^2 + l)\mathcal{X} &= \alpha + \frac{1}{2}sz \\ \mathcal{Y}_{yy} + (z^2 - l)\mathcal{Y} &= \alpha - \frac{1}{2}sz \end{aligned} \tag{5.218}$$

where $l = f + k$ and α is a constant of separation. The first integrals

$$\begin{aligned} \mathcal{X}_x^2 + (z^2 + l)\mathcal{X}^2 &= (2\alpha + sz)\mathcal{X} + \beta \\ \mathcal{Y}_y^2 + (z^2 - l)\mathcal{Y}^2 &= (2\alpha - sz)\mathcal{Y} + \gamma \end{aligned} \tag{5.219}$$

show that the quadratic constraint (5.81) with $I = 0$ is satisfied iff

$$2\beta + 2\gamma + s^2 = 0. \tag{5.220}$$

The Moutard transform of the seed solution (z, ρ) therefore reads (cf. Theorem 17)

$$\tilde{z} = \frac{z(\mathcal{Y} - \mathcal{X}) + s}{\mathcal{X} + \mathcal{Y}}, \quad \tilde{\rho} = 2\frac{\mathcal{X}_x \mathcal{Y}_y}{(\mathcal{X} + \mathcal{Y})^2}. \tag{5.221}$$

If we focus on solutions of the Calapso equation which are non-singular then, without loss of generality, $\mathcal{X}$ and $\mathcal{Y}$ may be taken to be

$$\mathcal{X} = \gamma_1 \cosh \lambda x + \gamma_2, \quad \mathcal{Y} = \gamma_3 \cos \mu y \tag{5.222}$$

with

$$\lambda^2 = -z^2 - l, \quad \mu^2 = z^2 - l, \quad -l > z^2. \tag{5.223}$$

Insertion of the expressions (5.222) into (5.219) produces

$$\alpha = \frac{1}{2} s z, \quad \beta = \lambda^2 \left(\gamma_2^2 - \gamma_1^2\right), \quad \gamma = \mu^2 \gamma_3^2 \tag{5.224}$$

and

$$\lambda^2 \gamma_2 = -s z. \tag{5.225}$$

The relations (5.223), (5.224) may be regarded as definitions of z, l and α, β, γ, respectively. The remaining relation (5.225) determines s if $z \neq 0$. Thus, there exist two subcases.

The Case $z = 0$

Here, $\gamma_2 = 0$ and the relations (5.223) reveal that $\lambda = \mu$ while the constraint (5.220) reads

$$s^2 = 2\lambda^2 \left(\gamma_1^2 - \gamma_3^2\right). \tag{5.226}$$

The latter implies that $\gamma_1 \neq 0$ and hence the new solution of the Calapso equation is given by

$$\tilde{z} = \frac{\lambda\sqrt{2(1 - \gamma_3^2)}}{\cosh \lambda x + \gamma_3 \cos \lambda y} \tag{5.227}$$

Figure 5.3. A 'one-soliton' solution of the Calapso equation.

without loss of generality. From a geometric point of view, λ is irrelevant since scalings of the form $(x, y, \tilde{z}) \to (cx, cy, \tilde{z}/c)$ merely correspond to scalings of the ambient space $\mathbb{R}^3$. Thus, $\tilde{z}$ essentially embodies a one-parameter class of solutions of the Calapso equation. If $\gamma_3 = 0$ then $\tilde{z}$ is independent of y and exponentially decaying as $x \to \pm\infty$. For non-vanishing γ_3, the solution is periodic in y and exponentially decaying in the x-direction. This is illustrated in Figure 5.3.

An algebraically localised solution (lump) is obtained in the limit

$$\lambda = \epsilon, \quad \gamma_3 = \epsilon^2 - 1, \quad \epsilon \to 0, \tag{5.228}$$

namely (vide Figure 5.4)

$$\tilde{z} = \frac{4}{2 + x^2 + y^2}. \tag{5.229}$$

It is subsequently shown that this solution of the Calapso equation may be associated via the Christoffel transformation with the classical Enneper minimal surface. In this connection, it is noted that the Enneper surface is not dual to

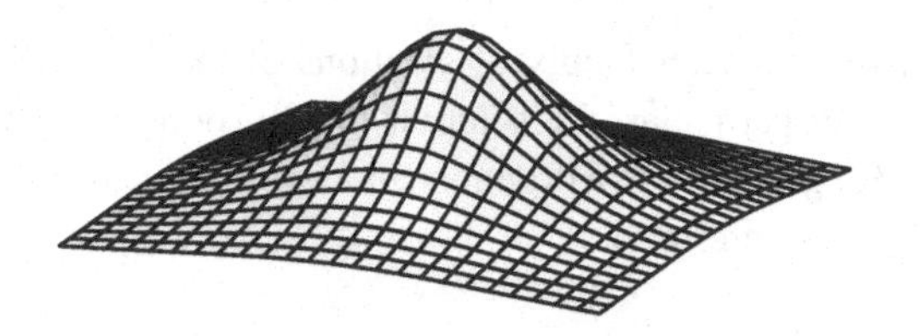

Figure 5.4. A 'lump' solution of the Calapso equation.

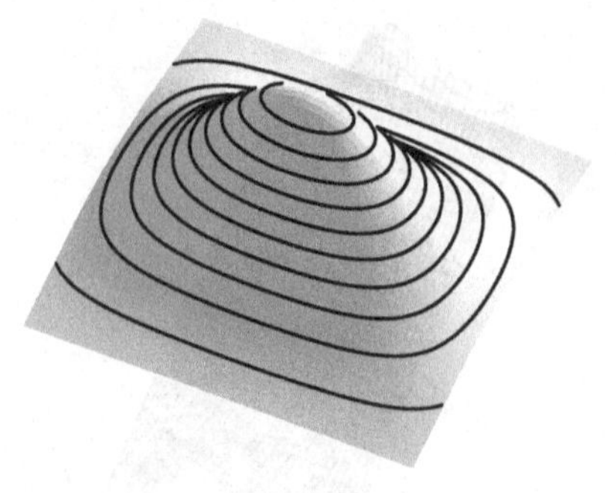

Figure 5.5. A dromion-type solution of the zoomeron equation.

any of the one-soliton isothermic surfaces introduced hitherto.[3] However, at the level of the Calapso equation, the two classes are identical, as can be seen by comparison of (5.214) and (5.227). This is a consequence of the fact that the mapping between solutions of the Calapso equation and the isothermic system is not one-to-one.

The substitution $y \to iy$ takes the solution (5.227) of the Calapso equation to the solution

$$\tilde{z} = \frac{\lambda\sqrt{2(1-\gamma_3^2)}}{\cosh\lambda x + \gamma_3 \cosh\lambda y} \tag{5.230}$$

of the zoomeron equation

$$\Box\left(\frac{z_{xy}}{z}\right) + (z^2)_{xy} = 0. \tag{5.231}$$

The quantity $\tilde{z}$ is now exponentially localised and non-singular for non-negative γ_3 as depicted in Figure 5.5. Indeed, a connection with coherent structures as described in [49] is readily established. Thus, if $z(x, y)$ is a solution of the zoomeron equation (5.231), then

$$u = e^{i(\nu x+\mu y+\mu\nu t)} z(x + \mu t, y + \nu t), \quad \rho = \frac{|u|_{xy}}{|u|} \tag{5.232}$$

constitutes a two-parameter family of solutions of the Davey-Stewartson III equation (5.69). In particular, the solution (5.230) boosted by the Lie-point symmetry (5.232) gives rise to a dromion solution of the Davey-Stewartson III equation [282] (cf. Exercise 1).

[3] It is noted parenthetically that the Christoffel transform of a minimal surface is a sphere.

The Case $z \neq 0$

The relations (5.224), (5.225) here imply that

$$s = -\lambda^2 \frac{\gamma_2}{z}, \quad z^2 = \frac{1}{2}(\mu^2 - \lambda^2), \tag{5.233}$$

whence the constraint (5.220) yields

$$\mu^2\lambda^2\gamma_2^2 + (\mu^2 - \lambda^2)\left(\mu^2\gamma_3^2 - \lambda^2\gamma_1^2\right) = 0. \tag{5.234}$$

The new solution of the Calapso equation therefore reads

$$\tilde{z} = -z + \frac{2z\gamma_3 \cos \mu y + s}{\gamma_1 \cosh \lambda x + \gamma_3 \cos \mu y + \gamma_2}, \tag{5.235}$$

where s, z are given by (5.233) and the constants γ_i are constrained by the relation (5.234). If we set aside scalings of the ambient space $\mathbb{R}^3$, then it is seen that (5.235) depends on two ratios of λ, μ, γ_i and hence constitutes a two-parameter family of solutions of the Calapso equation. Once again, the solutions are periodic in y and approach $-z$ exponentially as $x \to \pm\infty$. A prototypical solution is depicted in Figure 5.6.

In conclusion, we remark that the integrability conditions for the system

$$2\Delta(\ln \tau) = z^2 + \delta, \quad 2\Box(\ln \tau) = f, \quad 2(\ln \tau)_{xy} = -\rho, \tag{5.236}$$

where δ is an arbitrary constant, are satisfied modulo the classical Calapso system and the relations (5.77) for $n = 1$. Accordingly, the existence of a 'τ function' satisfying the above relations is guaranteed. If we now define

Figure 5.6. A solution of the Calapso equation corresponding to $z \neq 0$.

a function σ by

$$\sigma = z\tau \tag{5.237}$$

then the Calapso system $(5.64)_{n=1}$ may be brought into the bilinear form[4]

$$D_x D_y \sigma \cdot \tau = 0, \quad \left(D_x^2 + D_y^2\right)\tau \cdot \tau = \sigma^2 + \delta\tau^2, \tag{5.238}$$

where D_x and D_y are bilinear operators introduced by Hirota [166] and defined by

$$D_x^n D_y^m a \cdot b = (\partial_x - \partial_{x'})^n (\partial_y - \partial_{y'})^m a(x, y) b(x', y')|_{x'=x, y'=y}. \tag{5.239}$$

The ansatz

$$\tau = \mathcal{X}(x) + \mathcal{Y}(y), \quad \sigma = \hat{\mathcal{X}}(x) + \hat{\mathcal{Y}}(y) \tag{5.240}$$

is then readily shown to lead precisely to the class (5.221).

It will be shown in the sequel that the class of solutions (5.235) (and (5.227) as a degenerate case) may be associated with classical Dupin cyclides.

5.8.3 Dupin Cyclides

The class of surfaces known as the Dupin cyclides were introduced by the mathematician and naval architect Dupin [114]. Dupin cyclides are characterised by the requirement that all lines of curvature thereon be circles. They were extensively investigated in the nineteenth century by such luminaries as Maxwell [253] in 1868 and Cayley [71] in 1873. There are many types of cyclides. One important type, namely the ring cyclide, may be thought of simply as a deformed torus. There are also self-intersecting forms known as the horned cyclide and spindle cyclide. The Dupin cyclides include as special cases all the surfaces conventionally used in computer-aided design, namely planes, circular cylinders, cones, spheres, natural quadrics and tori. Indeed, in recent years, there has been very considerable interest in the use of Dupin cyclides in computer-aided engineering design [9–12, 46, 103, 244, 245, 256, 278, 287, 347].

Here, we shall be interested in the occurrence of Dupin cyclides as generated via Bäcklund transformations in a solitonic context [100]. It is remarked parenthetically that Dupin hypersurfaces have been shown to arise in connection with integrable Hamiltonian systems of hydrodynamic type by Ferapontov [127].

[4] An account of the Hirota bilinear operator formalism is outside the scope of the present work. The interested reader may repair to the monograph on the subject by Matsuno [250]. The basic properties of bilinear operators are set down in an appendix in the companion volume on Bäcklund transformations by Rogers and Shadwick [311].

The mapping between solutions of the Calapso equation and the isothermic system is not one-to-one. However, given a solution of the Calapso system, a privileged solution of the isothermic system is obtained by identifying the Bäcklund transform $\tilde{z}$ with the Christoffel transform z^* (vide Theorem 7). A solution of the isothermic system corresponding to $z = \text{const}$ is accordingly given by

$$e^{\theta} = \psi, \quad h_1 = \frac{1}{\sqrt{2}}(z + \tilde{z}), \quad h_2 = \frac{1}{\sqrt{2}}(z - \tilde{z}) \tag{5.241}$$

and that associated with $\tilde{z}$ by[5]

$$e^{-\tilde{\theta}} = \psi, \quad \tilde{h}_1 = h_1, \quad \tilde{h}_2 = -h_2. \tag{5.242}$$

Insertion of the expressions $(5.217)_1$ and $(5.221)_1$ therefore produces the new solution

$$e^{\tilde{\theta}} = \frac{1}{\mathcal{X} + \mathcal{Y}}, \quad \tilde{h}_1 = \frac{s + 2z\mathcal{Y}}{\sqrt{2}(\mathcal{X} + \mathcal{Y})}, \quad \tilde{h}_2 = \frac{s - 2z\mathcal{X}}{\sqrt{2}(\mathcal{X} + \mathcal{Y})} \tag{5.243}$$

of the isothermic system (5.12).

The curvatures and torsions of the lines of curvature on isothermic surfaces are given by, on use of [380],

$$\begin{aligned}
\kappa^{(x)2} &= \frac{|\boldsymbol{r}_x \times \boldsymbol{r}_{xx}|^2}{\boldsymbol{r}_x^6} = e^{-2\theta}(\theta_y^2 + h_1^2) \\
\kappa^{(y)2} &= \frac{|\boldsymbol{r}_y \times \boldsymbol{r}_{yy}|^2}{\boldsymbol{r}_y^6} = e^{-2\theta}(\theta_x^2 + h_2^2) \\
\tau^{(x)} &= \frac{[\boldsymbol{r}_x\, \boldsymbol{r}_{xx}\, \boldsymbol{r}_{xxx}]}{|\boldsymbol{r}_x \times \boldsymbol{r}_{xx}|^2} = e^{-\theta}\left(\frac{\theta_y h_{1x} - \theta_{xy} h_1}{\theta_y^2 + h_1^2}\right) \\
\tau^{(y)} &= \frac{[\boldsymbol{r}_y\, \boldsymbol{r}_{yy}\, \boldsymbol{r}_{yyy}]}{|\boldsymbol{r}_y \times \boldsymbol{r}_{yy}|^2} = e^{-\theta}\left(\frac{\theta_x h_{2y} - \theta_{xy} h_2}{\theta_x^2 + h_2^2}\right)
\end{aligned} \tag{5.244}$$

by virtue of the Gauss-Weingarten equations (5.9). In the present case, it is readily verified that

$$\tilde{\kappa}_x^{(x)} = 0, \quad \tilde{\kappa}_y^{(y)} = 0, \quad \tilde{\tau}^{(x)} = 0, \quad \tilde{\tau}^{(y)} = 0 \tag{5.245}$$

so that the lines of curvature are circles. In fact, all surfaces on which the lines of curvature form circles are incorporated in the class (5.243). These surfaces have

[5] Note that we use the convention $z^* = -\frac{1}{\sqrt{2}}(h_1^* + h_2^*)$.

come to be known as Dupin cyclides. They admit the compact parametrisation [114]

$$\tilde{\boldsymbol{r}} = \frac{1}{a - c\cos\alpha\cos\beta}\begin{pmatrix} \mu(c - a\cos\alpha\cos\beta) + b^2\cos\alpha \\ b\sin\alpha(a - \mu\cos\beta) \\ b\sin\beta(c\cos\alpha - \mu) \end{pmatrix} \tag{5.246}$$

$$a^2 = b^2 + c^2$$

with curvature coordinates α and β. The latter are not conformal since

$$\frac{\tilde{\boldsymbol{r}}_\alpha^2}{\tilde{\boldsymbol{r}}_\beta^2} = \left(\frac{a - \mu\cos\beta}{c\cos\alpha - \mu}\right)^2 \neq 1 \tag{5.247}$$

in general. However, the transformation

$$\begin{aligned} \alpha &= 2\arctan\left[\sqrt{\frac{c-\mu}{c+\mu}}\tanh\left(\frac{\sqrt{c^2-\mu^2}}{2}x\right)\right] \\ \beta &= 2\arctan\left[\sqrt{\frac{a-\mu}{a+\mu}}\tan\left(\frac{\sqrt{a^2-\mu^2}}{2}y\right)\right] \end{aligned} \tag{5.248}$$

yields

$$\tilde{\boldsymbol{r}}_x^2 = \tilde{\boldsymbol{r}}_y^2 \tag{5.249}$$

and hence a parametrisation in terms of the usual conformal curvature coordinates x and y is obtained. By construction, the position vector $\tilde{\boldsymbol{r}}$ as given by (5.246) and (5.248) obeys the Gauss-Weingarten equations (5.9) corresponding to the solution $(\tilde{\theta}, \tilde{h}_1, \tilde{h}_2)$ on appropriate identification of certain constants. The class of solutions of the Calapso equation associated with Dupin cyclides is therefore given by (5.227), (5.235), corresponding to the seeds $z = 0$ and $z = \text{const} \neq 0$, respectively.

The significance of the parameters a, c and μ is readily established. Thus, the curves $\beta = \pi$ and $\beta = 0$ constitute two circles on the plane $Z = 0$. These are given by

$$(X \pm c)^2 + Y^2 = (a \pm \mu)^2. \tag{5.250}$$

The two circles intersect if $c^2 > \mu^2$. A corresponding cyclide is displayed in Figure 5.7. In the case $c^2 = \mu^2$, touching circles are obtained. A typical cyclide is depicted in Figure 5.8. If $c^2 < \mu^2$, then the smaller circle is enclosed by the larger one and the associated cyclide may be regarded as a deformed torus (Figure 5.9). Indeed, the Dupin cyclide degenerates to a torus as $c \to 0$. In this case, the two circles are concentric (Figure 5.10).

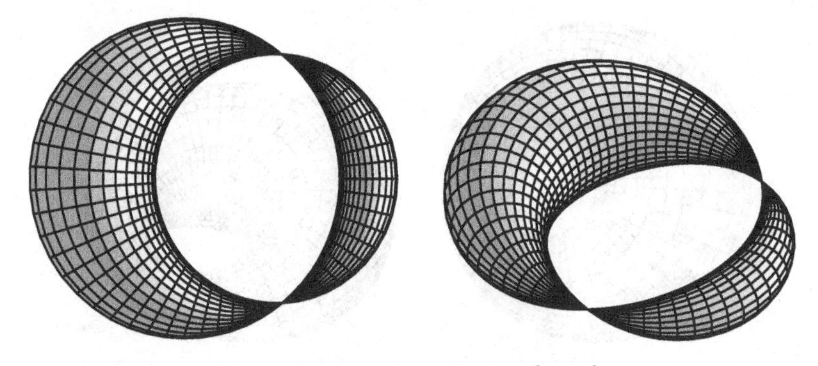

Figure 5.7. A Dupin cyclide for $c^2 > \mu^2$.

A particular parametrisation of a sphere is obtained in the limit

$$\alpha = \epsilon x, \quad \beta = \epsilon y, \quad a = 2 + \epsilon^2, \quad c = 2, \quad \mu = 0, \quad \epsilon \to 0, \tag{5.251}$$

namely

$$\tilde{\boldsymbol{r}} = \frac{4}{1 + x^2 + y^2}\begin{pmatrix} 1 \\ x \\ y \end{pmatrix}. \tag{5.252}$$

Accordingly, the Christoffel transform $\tilde{\Sigma}^*$, which is obtained by integration of the expressions

$$\tilde{\boldsymbol{r}}^*_x = \frac{\tilde{\boldsymbol{r}}_x}{\tilde{\boldsymbol{r}}^2_x}, \quad \tilde{\boldsymbol{r}}^*_y = -\frac{\tilde{\boldsymbol{r}}_y}{\tilde{\boldsymbol{r}}^2_y}, \tag{5.253}$$

constitutes a minimal surface. Indeed, these relations yield

$$\tilde{\boldsymbol{r}}^* = \frac{1}{12}\begin{pmatrix} 3y^2 - 3x^2 \\ 3x + 3xy^2 - x^3 \\ -3y - 3yx^2 + y^3 \end{pmatrix} \tag{5.254}$$

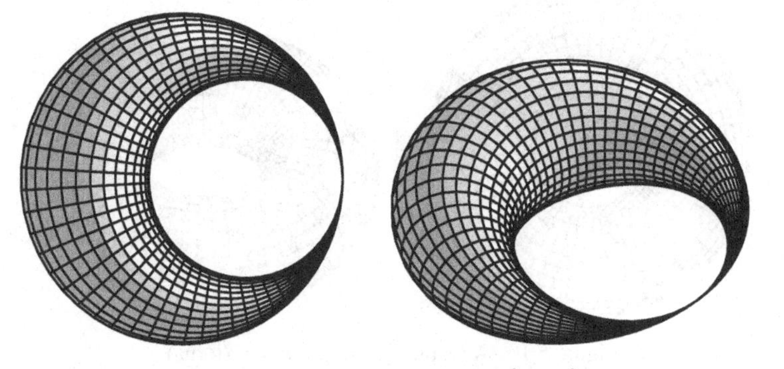

Figure 5.8. A Dupin cyclide for $c^2 = \mu^2$.

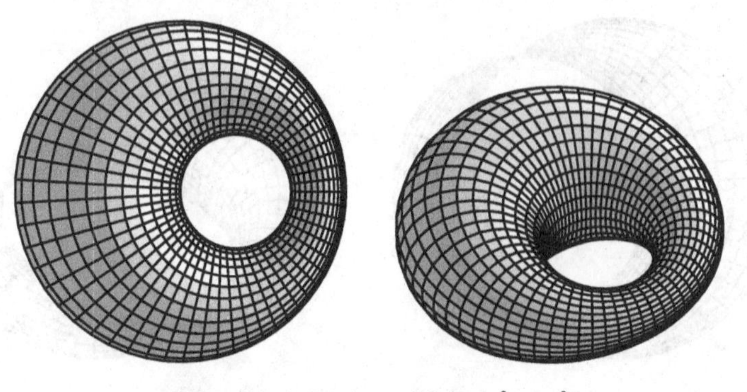

Figure 5.9. A Dupin cyclide for $c^2 < \mu^2$.

so that $\tilde{\boldsymbol{r}}^*$ parametrises the classical Enneper surface (cf. Section 5.1, Exercise 1). The latter is displayed in Figure 5.11. Moreover, since we have identified the Bäcklund transform $\tilde{\Sigma}$ with the Christoffel transform Σ^*, the Enneper surface constitutes a particular seed surface Σ corresponding to the zero-solution of the Calapso equation. In fact, the solution of the isothermic system associated with the sphere (5.252) is

$$e^{\tilde{\theta}} = \frac{4}{1+x^2+y^2}, \quad \tilde{h}_1 = \tilde{h}_2 = \frac{2}{1+x^2+y^2} \tag{5.255}$$

so that the solution of the Calapso equation reads

$$\tilde{z} = \frac{2\sqrt{2}}{1+x^2+y^2}. \tag{5.256}$$

Up to the invariance $(x, y, \tilde{z}) \rightarrow (x/\sqrt{2}, y/\sqrt{2}, \sqrt{2}\,\tilde{z})$ of the Calapso equation, this is nothing but the particular Bäcklund transform (5.229) generated from

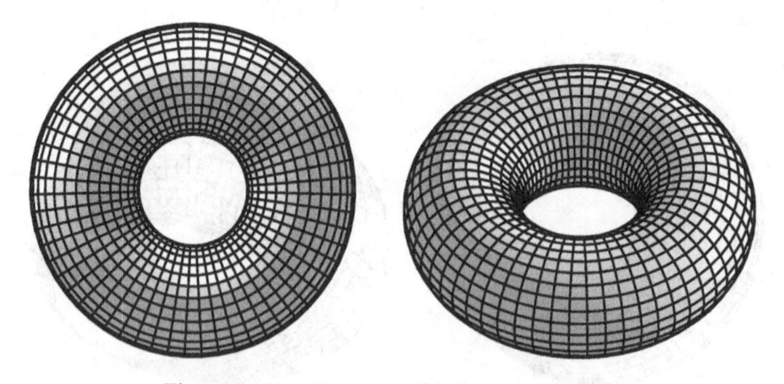

Figure 5.10. A Dupin cyclide for $c = 0$ (torus).

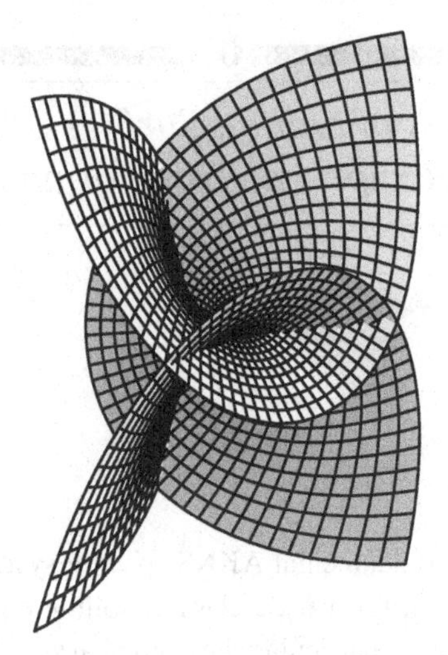

Figure 5.11. The Enneper surface.

the seed solution $z = 0$.

Exercises

1. Show that if $u = u(x, y, t)$, $\rho = \rho(x, y, t)$ is a solution of the Davey-Stewartson III equation (5.69), then

$$\tilde{u} = e^{i(\nu x + \mu y + \mu \nu t)} u(x + \mu t, y + \nu t, t), \quad \tilde{\rho} = \rho(x + \mu t, y + \nu t, t)$$

constitutes a two-parameter family of solutions of the Davey-Stewartson III equation.

2. Derive the bilinear form (5.238) of the Calapso system and find all solutions of the form (5.240).
3. Verify that the lines of curvature associated with the class (5.243) are indeed circles.
4. Derive the representation (5.250) and verify that the two circles intersect iff $c^2 > \mu^2$. Show that the points of intersection are given by

$$X = \frac{a}{c}\mu, \quad Y = \pm\frac{b}{c}\sqrt{c^2 - \mu^2}.$$

6

General Aspects of Soliton Surfaces. Role of Gauge and Reciprocal Transformations

It was in 1973 that the fundamental AKNS spectral system was set down [1]. This linear representation for a wide class of soliton equations yields up via compatibility conditions, in particular, the canonical sine-Gordon, mKdV, KdV and NLS equations. Hard upon this work, in 1976, came that of Lund and Regge [240] and Pohlmeyer [285], which established a connection between the geometry of privileged classes of surfaces and soliton theory. Thus, what is now known as the Pohlmeyer-Lund-Regge solitonic system was generated via a Gauss-Mainardi-Codazzi system, with the corresponding Gauss-Weingarten equations viewed as a 3×3 linear representation. Moreover, Lund and Regge adopted a spinor formulation to make direct connection with 2×2 representations as embodied in the AKNS system.

The next major development in the geometry of soliton theory came in 1982 with the pioneering work of Sym [353] when he introduced the notion of soliton surfaces. Thus, the soliton surfaces associated with the sine-Gordon and NLS equations are, in turn, the pseudospherical and Hasimoto surfaces. In general, the one-parameter class of soliton surfaces $\Sigma : \boldsymbol{r} = \boldsymbol{r}(u, v)$ associated with a particular solution of a solitonic equation is obtained by insertion of that solution into the relevant Gauss-Weingarten equations and integration thereof to determine the position vector $\boldsymbol{r}$. However, this direct integration approach may be circumvented by an ingenious method described in [356]. This Sym-Tafel procedure depends crucially on the presence of a 'spectral parameter' in the linear representation for the soliton equation in question. It will be described here for a broad subclass of soliton equations generated by the AKNS spectral system.

The position vectors of soliton surfaces may be interpreted as solutions of 'eigenfunction equations', which are themselves integrable [206]. This result is established for the NLS hierarchy here generated via the action of a recursion operator. An associated solitonic hierarchy is constructed thereby which has base member the celebrated Heisenberg spin equation.

The soliton surfaces associated with the NLS hierarchy have the geometric property that they are swept out by the motion of what are termed loop solitons [376]. This is revealed by suitable parametrisation of the soliton surfaces in terms of coordinates introduced via reciprocal transformations. It is recalled that reciprocal transformations originated in the study of important invariance properties in gasdynamics (vide [311]). Here, by contrast, we describe the role of such transformations in soliton theory and, in particular, in the link between the Dym, mKdV and KdV hierarchies. A permutability theorem is set down for the potential KdV equation and connection made with the ϵ-algorithm that arises in numerical analysis. The geometry underlying the generation of the mKdV hierarchy by the planar motion of curves is then described.

To conclude, the purely binormal motion of inextensible curves of constant curvature or torsion is shown to lead to extended versions of the Dym and sine-Gordon equations, respectively. The soliton surfaces generated via the motion of these curves admit dual soliton surfaces and Bäcklund transformations with the constant length property.

6.1 The AKNS 2 × 2 Spectral System

6.1.1 The Position Vector of Pseudospherical Surfaces

Here, we return to the fundamental forms (1.24) associated with pseudospherical surfaces given in terms of asymptotic coordinates. A spectral parameter λ is injected therein via the invariance (2.31). For simplicity, it is assumed that $\rho = 1$ so that the Gaussian curvature is $\mathcal{K} = -1$. The associated AKNS representation as generated by the Gauss-Weingarten equations takes the form (2.32), namely

$$\Phi_{,\mu} = g_\mu \Phi, \quad \mu = 1, 2 \tag{6.1}$$

with

$$\begin{aligned} g_1 &= \frac{i}{2}(\omega_{,1}\,\sigma_2 + \lambda\sigma_3), \\ g_2 &= \frac{i}{2\lambda}(\sin\omega\,\sigma_1 - \cos\omega\,\sigma_3). \end{aligned} \tag{6.2}$$

Here, the notation $_{,\mu} = \partial/\partial x^\mu$, $\mu = 1, 2$ has been adopted where $x^1 = u$, $x^2 = v$ and the tilde has been dropped for notational convenience. The σ_i designate the Pauli matrices as given in (2.25). The correspondence $e_k = \dfrac{\sigma_k}{2i} \leftrightarrow L_k$, as introduced in Section 2.2, may now be used to obtain an alternative 3×3 linear representation in terms of the matrices L_i. Thus, the matrices g_μ admits the

Pauli decomposition

$$g_\mu = \boldsymbol{g}_\mu \cdot \boldsymbol{e}, \quad \boldsymbol{e} = \begin{pmatrix} e_1 \\ e_2 \\ e_3 \end{pmatrix}, \tag{6.3}$$

where the vectors

$$\boldsymbol{g}_\mu = \begin{pmatrix} g_{\mu 1} \\ g_{\mu 2} \\ g_{\mu 3} \end{pmatrix}, \tag{6.4}$$

are determined via (6.2) by the decomposition (6.3) involving the matrix-valued scalar product

$$\boldsymbol{g}_\mu \cdot \boldsymbol{e} = g_{\mu 1} e_1 + g_{\mu 2} e_2 + g_{\mu 3} e_3. \tag{6.5}$$

The identification $e_k \leftrightarrow L_k$ now produces the $so(3)$ analogue of (6.1), namely

$$\hat{\Phi}_{,\mu} = (\boldsymbol{g}_\mu \cdot \boldsymbol{L})\hat{\Phi}, \tag{6.6}$$

where

$$\boldsymbol{L} = \begin{pmatrix} L_1 \\ L_2 \\ L_3 \end{pmatrix}. \tag{6.7}$$

The connection between the $so(3)$ linear representation and the unit triad formalism associated with the Gauss-Weingarten equations has been described in Section 2.2. The key observation here is that only knowledge of the matrices g_μ is necessary for the calculation of the fundamental forms I, II. Thus, from (6.2),

$$g_{1,\lambda} = \frac{i}{2}\sigma_3, \quad g_{2,\lambda} = \frac{i}{2\lambda^2}(\sin\omega\,\sigma_1 - \cos\omega\,\sigma_3) \tag{6.8}$$

with ${}_{,\lambda} = \partial/\partial\lambda$, whence it follows that

$$-2\mathrm{Tr}(g_{\mu,\lambda}\, g_{\nu,\lambda}) dx^\mu dx^\nu|_{\lambda=1} = du^2 + 2\cos\omega\, du dv + dv^2. \tag{6.9}$$

In the above, Einstein's summation convention has again been adopted and 'Tr' designates the trace of a matrix.

Comparison of (1.24) and (6.9) now shows that the 1st fundamental form associated with pseudospherical surfaces in asymptotic coordinates may be written as

$$\mathrm{I} = -2\mathrm{Tr}(g_{\mu,\lambda}\, g_{\nu,\lambda})dx^{\mu}dx^{\nu}|_{\lambda=1} \tag{6.10}$$

or, equivalently, as

$$\mathrm{I} = (\boldsymbol{g}_{\mu,\lambda}\cdot \boldsymbol{g}_{\nu,\lambda})dx^{\mu}dx^{\nu}|_{\lambda=1} \tag{6.11}$$

in terms of the vectors $\boldsymbol{g}_{\mu}$.

On the other hand, if $\boldsymbol{r} = (X(u, v),\ Y(u, v),\ Z(u, v))^{\mathsf{T}}$ is the generic position vector of a surface Σ then, in terms of the matrix

$$r = Xe_1 + Ye_2 + Ze_3 = \boldsymbol{r}\cdot\boldsymbol{e}, \tag{6.12}$$

the 1st fundamental form $(1.2)_1$ may be written as

$$\mathrm{I} = -2\mathrm{Tr}(r_{,\mu}r_{,\nu})dx^{\mu}dx^{\nu}|_{\lambda=1} \tag{6.13}$$

where, again, $_{,\mu}$ denotes $\partial/\partial x^{\mu}$, $\mu = 1, 2$. It is noted that

$$-2\,\mathrm{Tr}(e_i e_k) = \delta_{ik}. \tag{6.14}$$

Since the trace of matrices is invariant under a similarity transformation, comparison of (6.10) and (6.13) suggests the ansatz

$$r_{,\mu} = G^{-1}g_{\mu,\lambda}G, \tag{6.15}$$

where the gauge matrix G remains to be determined. On use of the integrability condition for the AKNS system (6.1), namely the 'Gauss-Mainardi-Codazzi' equations

$$g_{1,2} - g_{2,1} + [g_1,\ g_2] = 0, \tag{6.16}$$

the compatibility condition $r_{,1,2} = r_{,2,1}$ applied to (6.15) produces the commutator relation

$$[G_{,1}G^{-1} - g_1,\ g_{2,\lambda}] = [G_{,2}G^{-1} - g_2,\ g_{1,\lambda}]. \tag{6.17}$$

The latter is identically satisfied by the choice

$$G = \Phi, \tag{6.18}$$

where Φ is the 2×2 *eigenfunction matrix* in the AKNS representation (6.1).

On insertion of (6.18) into (6.15), it remains to integrate the equations

$$r_{,\mu} = \Phi^{-1} g_{\mu,\lambda} \Phi, \quad \mu = 1, 2, \tag{6.19}$$

which are now guaranteed to be compatible. But, modulo (6.16) so that $\Phi_{,\lambda,\mu} = \Phi_{,\mu,\lambda}$, it is seen that

$$\begin{aligned} (\Phi^{-1}\Phi_{,\lambda})_{,\mu} &= -\Phi^{-1}\Phi_{,\mu}\Phi^{-1}\Phi_{,\lambda} + \Phi^{-1}\Phi_{,\lambda,\mu} \\ &= -\Phi^{-1} g_\mu \Phi_{,\lambda} + \Phi^{-1}(g_{\mu,\lambda}\Phi + g_\mu \Phi_{,\lambda}) = \Phi^{-1} g_{\mu,\lambda}\Phi, \end{aligned} \tag{6.20}$$

and the relation (6.19) yields

$$r_{,\mu} = (\Phi^{-1}\Phi_{,\lambda})_{,\mu}. \tag{6.21}$$

Integration of (6.21) produces the key Sym-Tafel relation [356][1]

$$\boxed{r = \Phi^{-1}\Phi_{,\lambda}} \tag{6.22}$$

up to a translation in space. The generic position vector $\boldsymbol{r}$ of the soliton surface is now retrieved via the decomposition (6.12), that is

$$\boldsymbol{r} = -2\,\mathrm{Tr}(r\boldsymbol{e}) = i\,\mathrm{Tr}(\Phi^{-1}\Phi_{,\lambda}\boldsymbol{\sigma}). \tag{6.23}$$

It has been established that the position vector $\boldsymbol{r}$ given by (6.23) leads to the 1st fundamental form corresponding to pseudospherical surfaces in asymptotic coordinates. It remains to show that the 2nd fundamental form associated with this $\boldsymbol{r}$ is also that for such pseudospherical surfaces. To this end, it is convenient to express the 2nd fundamental form

$$\mathrm{II} = h_{\mu\nu}\, dx^\mu\, dx^\nu = (\boldsymbol{r}_{,\mu,\nu} \cdot \boldsymbol{N}) dx^\mu dx^\nu \tag{6.24}$$

in terms of the 'position matrix' r. In this connection, it can be shown that

$$h_{\mu\nu} = -\frac{1}{\det^{1/2}[r_{,1}, r_{,2}]}\,\mathrm{Tr}([r_{,1}, r_{,2}] r_{,\mu,\nu}). \tag{6.25}$$

and use of $r_{,\mu}$ as given by (6.19) together with the relation

$$(\Phi^{-1} g_{\mu,\lambda}\Phi)_{,\nu} = \Phi^{-1}(g_{\mu,\nu,\lambda} + [g_{\mu,\lambda}, g_\nu])\Phi \tag{6.26}$$

[1] Soliton immersions and generalisations of the Sym-Tafel formula to surfaces on Lie groups and Lie algebras have been the subject of recent research [84, 110, 138, 139, 208].

produces

$$\begin{aligned} \mathrm{II} &= h_{\mu\nu} dx^{\mu} dx^{\nu}, \\ h_{\mu\nu} &= -\frac{1}{\det^{1/2}[g_{1,\lambda}, g_{2,\lambda}]} \operatorname{Tr}([g_{1\lambda}, g_{2,\lambda}](g_{\mu,\nu,\lambda} + [g_{\mu,\lambda}, g_{\nu}])). \end{aligned} \tag{6.27}$$

On insertion of the expressions (6.2) for g_{μ} into (6.27), we obtain

$$h_{\mu\nu} dx^{\mu} dx^{\nu}|_{\lambda=1} = 2 \sin \omega \, du dv, \tag{6.28}$$

whence the 2nd fundamental form $(1.24)_2$ for pseudospherical surfaces with $\mathcal{K} = -1/\rho^2 = -1$ is retrieved. Accordingly, it has been established that the position vector $\boldsymbol{r}$ as given by (6.23) does indeed deliver pseudospherical surfaces expressed in asymptotic coordinates. It is important to note that the position vector $\boldsymbol{r}$ may now be calculated via a derivative of the eigenfunction Φ with respect to the parameter λ rather than by direct integration of the Gauss-Weingarten equations with respect to the asymptotic coordinates u and v. This result and its extensions prove to be of great utility in the construction of soliton surfaces.

6.1.2 The su(2) Linear Representation and Its Associated Soliton Surfaces. The AKNS Case $r = -\bar{q}$

In the preceding, the derivation of the expressions (6.10) and (6.27) for the fundamental forms I and II, as well as (6.23) for the position vector $\boldsymbol{r}$ for the associated pseudospherical surface, was motivated by the AKNS representation (6.1), (6.2) for the classical sine-Gordon equation. However, it is evident that the derivation of the formulae (6.10) and (6.27) for I and II is valid for any matrices $g_{\mu} \in su(2)$ subject only to the compatibility condition (6.16) which enshrines the underlying nonlinear equations. It is noted that no knowledge of the eigenfunction matrix Φ is required in the calculation of I and II. The nonlinear condition (6.16) guarantees the existence of Φ satisfying the AKNS linear matrix system (6.1). This eigenfunction matrix Φ may then be used to calculate, via the expression (6.23), the position vector $\boldsymbol{r}$ of the surface associated with I and II. This more general result is encapsulated in the following:

Theorem 19. *Let the matrices* $g_{\mu} \in su(2)$, $\mu = 1, 2$ *depend parametrically on a real parameter* λ *and obey the nonlinear equations*

$$g_{1,2} - g_{2,1} + [g_1, g_2] = 0. \tag{6.29}$$

Then there exists a matrix-valued function $\Phi \in SU(2)$ *satisfying the* linear *equations*

$$\Phi_{,\mu} = g_\mu \Phi. \tag{6.30}$$

Moreover, the vector-valued function $\boldsymbol{r}$ *given implicitly by*

$$r = \boldsymbol{r}\cdot\boldsymbol{e} = \Phi^{-1}\Phi_{,\lambda} \tag{6.31}$$

defines a surface in $\mathbb{R}^3$ *whose fundamental forms read*

$$\boxed{\begin{aligned} \mathrm{I} &= -2\,\mathrm{Tr}(g_{\mu,\lambda}\, g_{\nu,\lambda})\, dx^\mu\, dx^\nu, \\ \mathrm{II} &= -\frac{1}{\det^{1/2}[g_{1,\lambda}, g_{2,\lambda}]}\,\mathrm{Tr}([g_{1,\lambda}, g_{2,\lambda}](g_{\mu,\nu,\lambda} + [g_{\mu,\lambda}, g_\nu]))\, dx^\mu\, dx^\nu. \end{aligned}} \tag{6.32}$$

By construction, the Mainardi-Codazzi and Gauss equations are satisfied modulo the nonlinear compatibility condition (6.29). It has been pointed out by Sym [356] that the Euclidean space $\mathbb{R}^3$ may be identified with the vector space spanned by the Lie algebra $su(2) \cong so(3)$ and the 1st fundamental form $(6.32)_1$ is induced by the corresponding Killing-Cartan metric. This observation may be exploited to generalise Theorem 19 to $m \times m$ linear problems associated with semi-simple Lie algebras. In particular, if the underlying structure is the Lie algebra $sl(2) \cong so(2, 1)$ then the corresponding surfaces are embedded in a 2+1-dimensional Minkowski space. Indeed, soliton surfaces associated with the KdV equation and Ernst equation of general relativity arise in a Minkowski space context [353, 358]. Here, in the main, we confine ourselves to the classical geometry of a three-dimensional Euclidean space $\mathbb{R}^3$.

It is natural to enquire as to what class of equations is represented by the nonlinear matrix condition (6.29). In this connection, it is observed that the linear $su(2)$ representations of the sine-Gordon equation and the compatible mKdV equation, namely (2.32), together with the linear $su(2)$ representation (4.87) for the NLS equation, share the property that their 'spatial' parts are of the form

$$\Phi_x = \frac{1}{2}\left[\begin{pmatrix} 0 & q \\ r & 0 \end{pmatrix} + i\lambda \begin{pmatrix} 1 & 0 \\ 0 & -1 \end{pmatrix}\right]\Phi, \tag{6.33}$$

where $r = -\bar{q}$. This so-called *scattering problem* has been used by Ablowitz et al. [2] to derive a broad class of solitonic equations, which are amenable to the

Inverse Scattering Transform. The latter constitutes an extension of the classical Fourier transform method which allows the solution of certain initial value problems for 1+1-dimensional soliton equations [3]. The canonical members of the integrable system are the sine-Gordon ($r = -q$), KdV ($r = -1$), mKdV ($r = \pm q$) and NLS ($r = \pm\bar{q}$) equations. The method of derivation of this celebrated AKNS system involves supplementing the scattering problem (6.33) by an appropriate time evolution in Φ to obtain nonlinear equations for r and q via compatibility. Here, we restrict our attention to the case $r = -\bar{q}$ associated with the $su(2)$ Lie algebra.

Let the time evolution

$$\Phi_t = \frac{1}{2}\begin{pmatrix} iA(\lambda) & B(\lambda) \\ -\bar{B}(\lambda) & -iA(\lambda) \end{pmatrix} \Phi \tag{6.34}$$

be adjoined to (6.33) where A, B are real and complex-valued polynomials, respectively, of degree N in the real spectral parameter λ. Compatibility of (6.33) and (6.34) produces the nonlinear equations

$$\begin{aligned} q_t - B_x - iqA + i\lambda B &= 0, \\ 2A_x + i(\bar{q}B - q\bar{B}) &= 0. \end{aligned} \tag{6.35}$$

Expansion of A and B according to

$$A = \sum_{k=0}^{N} A_k \lambda^k, \quad B = \sum_{k=0}^{N} B_k \lambda^k \tag{6.36}$$

immediately yields

$$B_{N-1} = qA_N, \quad B_N = 0, \quad A_N = A_N(t), \tag{6.37}$$

together with the recursion relations

$$\begin{aligned} &q_t - B_{0x} - iqA_0 = 0, \\ &2A_{kx} + i(\bar{q}B_k - q\bar{B}_k) = 0, \qquad k = 0, \dots, N-1 \\ &B_{k+1x} - iB_k + iqA_{k+1} = 0, \qquad k = 0, \dots, N-2. \end{aligned} \tag{6.38}$$

Now, on integration of $(6.38)_2$, we obtain

$$2A_k = -i\partial_x^{-1}(\bar{q}B_k - q\bar{B}_k), \qquad k = 0, \dots, N-1 \tag{6.39}$$

and insertion into $(6.38)_3$ produces the vector recurrence relations

$$\begin{pmatrix} B_k \\ \bar{B}_k \end{pmatrix} = L \begin{pmatrix} B_{k+1} \\ \bar{B}_{k+1} \end{pmatrix}, \qquad k = 0, \dots, N-2, \tag{6.40}$$

where

$$L = i\begin{pmatrix} -\partial_x - \frac{1}{2}q\partial_x^{-1}\bar{q} & \frac{1}{2}q\partial_x^{-1}q \\ -\frac{1}{2}\bar{q}\partial_x^{-1}\bar{q} & \partial_x + \frac{1}{2}\bar{q}\partial_x^{-1}q \end{pmatrix}. \tag{6.41}$$

The matrix-valued operator L is termed a *recursion operator* for reasons to become apparent later. The recurrence relations (6.40) subject to the condition $(6.37)_1$ determine the coefficients B_k uniquely according to

$$\begin{pmatrix} B_k \\ \bar{B}_k \end{pmatrix} = A_N L^{N-k-1}\begin{pmatrix} q \\ \bar{q} \end{pmatrix}, \qquad k = 0, \dots, N-2. \tag{6.42}$$

The remaining equation $(6.38)_1$, together with its complex conjugate, produces a vector evolution equation for q and $\bar{q}$ only, viz.

$$\begin{pmatrix} q \\ -\bar{q} \end{pmatrix}_t = iA_N L^N\begin{pmatrix} q \\ \bar{q} \end{pmatrix}. \tag{6.43}$$

The system (6.43) generates an *NLS hierarchy*. Repeated action of the recursion operator L on the time-independent part of the NLS equation produces the nonlinear integrable equations of order N determined by (6.43). It is remarkable that at each step of this procedure (6.43) produces only differential equations even though L constitutes an integro-differential operator. For $N = 2$, $A_N = -1$, the system (6.43) generates the NLS equation

$$iq_t + q_{xx} + \frac{1}{2}|q|^2 q = 0. \tag{6.44}$$

The members of the NLS hierarchy (6.43) are mutually compatible. Thus, if we regard the eigenfunction Φ as a function of an infinite number of 'times' $t_1 = x, t_2, t_3, \dots$, that is

$$\Phi = \Phi(t_1, t_2, t_3, \dots), \tag{6.45}$$

then the 'temporal' evolutions

$$\Phi_{t_n} = \frac{1}{2}\begin{pmatrix} iA^{(n)} & B^{(n)} \\ -\bar{B}^{(n)} & -iA^{(n)} \end{pmatrix}\Phi, \qquad n = 2, 3, \dots \tag{6.46}$$

with

$$A^{(n)} = \sum_{k=0}^{n} A_k^{(n)}\lambda^k, \quad B^{(n)} = \sum_{k=0}^{n} B_k^{(n)}\lambda^k \tag{6.47}$$

may be adjoined to the 'spatial part' of the linear representation for the canonical NLS equation, namely

$$\Phi_{t_1} = \frac{1}{2}\begin{pmatrix} i\lambda & q \\ -\bar{q} & -i\lambda \end{pmatrix}\Phi. \tag{6.48}$$

Accordingly, the compatibility conditions

$$\Phi_{t_n t_1} = \Phi_{t_1 t_n} \tag{6.49}$$

produce

$$B^{(n)}_{n-1} = qA^{(n)}_n, \quad B^{(n)}_n = 0, \quad A^{(n)}_n = A^{(n)}_n(t) \tag{6.50}$$

together with the recurrence relations

$$\begin{aligned} q_{t_n} - B^{(n)}_{0,x} - iqA^{(n)}_0 &= 0, \\ 2A^{(n)}_{kx} + i(\bar{q}B^{(n)}_k - q\bar{B}^{(n)}_k) &= 0, \qquad k = 0, \ldots, n-1 \\ B^{(n)}_{k+1x} - iB^{(n)}_k + iqA^{(n)}_{k+1} &= 0, \qquad k = 0, \ldots, n-2. \end{aligned} \tag{6.51}$$

By construction, the relations (6.50), (6.51) coincide with (6.37), (6.38) for $N = n = 2, 3, \ldots$ and hence generate the higher order members of the NLS hierarchy. It is noted that (6.48) together with the time evolution (6.46) with $n = 2$, $A^{(2)} = -1$ and the identification $t_2 = t$ produces the usual AKNS linear representation for the NLS equation (6.44).

If we now set

$$\mathrm{K}_n[q, \bar{q}] = (-1)^{n+1} iL^n \begin{pmatrix} q \\ \bar{q} \end{pmatrix}, \quad A^{(n)}_n = (-1)^{n+1} \tag{6.52}$$

then it may be demonstrated that, modulo the NLS hierarchy

$$\begin{pmatrix} q \\ -\bar{q} \end{pmatrix}_{t_n} = \mathrm{K}_n[q, \bar{q}], \tag{6.53}$$

the compatibility conditions

$$\partial_{t_m}\mathrm{K}_n = \partial_{t_n}\mathrm{K}_m \tag{6.54}$$

are satisfied. This reflects the fact that the remaining compatibility conditions

$$\Phi_{t_n t_m} = \Phi_{t_m t_n} \tag{6.55}$$

are also satisfied modulo the NLS hierarchy (6.53). In the terminology of soliton theory [271], the *flows* K_n are said to *commute* and the K_n are said to represent *symmetries* of the canonical NLS equation (6.44). The simplest case $n = 1$ produces as base member the translation symmetry

$$q_{t_1} = q_x. \tag{6.56}$$

The mKdV Hierarchy

The NLS hierarchy (6.43) may be specialised to real q for odd N. This generates the mKdV hierarchy to be considered in more detail in Section 6.3. In particular, it is readily shown that the specialisation $N = 3$ with $A_N = 1$ in (6.43) produces the mKdV equation

$$q_t + q_{xxx} + \frac{3}{2}q^2 q_x = 0. \tag{6.57}$$

The NLS Fundamental Forms

The AKNS linear representation with spatial part (6.48) and time evolution (6.46) is in the $su(2)$ form as required by Theorem 19. Thus, the NLS and mKdV hierarchies are naturally associated with surfaces in $\mathbb{R}^3$. In particular, the time evolution (6.46) for the case $n = 2$ with $A_2 = -1$ corresponding to the NLS equation (6.44) reads

$$\Phi_{t_2} = \frac{1}{2}\begin{pmatrix} -i\lambda^2 + \frac{1}{2}i|q|^2 & -\lambda q + iq_x \\ \lambda\bar{q} + i\bar{q}_x & i\lambda^2 - \frac{1}{2}i|q|^2 \end{pmatrix}\Phi, \tag{6.58}$$

with $t_2 = t$. Insertion of the $su(2)$ matrices g_μ as given by (6.48) and (6.58) into the expressions (6.32) for the fundamental forms produces

$$\begin{aligned} \mathrm{I} &= dx^2 - 4\lambda\, dxdt + (\kappa^2 + 4\lambda^2)\, dt^2, \\ \mathrm{II} &= -\kappa\, dx^2 + 2\kappa(\tau + \lambda)\, dxdt + [\kappa_{xx} - \kappa(\tau + \lambda)^2]\, dt^2, \end{aligned} \tag{6.59}$$

where the usual decomposition

$$q = \kappa \exp\left(i\partial_x^{-1}\tau\right) \tag{6.60}$$

has been used. The total curvature of the NLS surface accordingly adopts the form

$$\mathcal{K} = -\frac{\kappa_{xx}}{\kappa}, \tag{6.61}$$

which is in agreement with (4.29).

The fundamental forms I and II as given by (6.59) coincide with those obtained in a geometric manner in Section 4.1, namely (4.26) and (4.28) boosted by the invariance (4.80) of the NLS equation. Consequently, the surfaces swept out by the curve moving with velocity $\boldsymbol{r}_t = \kappa \boldsymbol{b}$ may be identified with the NLS surfaces as generated by Theorem 19.

In this section, it has been established that nonlinear equations with $su(2)$ linear representations as underlying structure may be naturally associated with surfaces in three-dimensional Euclidean space. Moreover, Theorem 19 may be used to reproduce pseudospherical and Hasimoto surfaces associated, in turn, with the sine-Gordon and NLS equations. In these cases, the coordinate lines on the surfaces have been previously shown to be asymptotic lines or geodesic and parallel curves. The following section is concerned with a detailed investigation of systems of coordinates on soliton surfaces connected with gauge and reciprocal transformations. In particular, it will be shown that certain soliton surfaces may be regarded as being swept out by moving loop soliton curves.

Exercises

1. (a) Establish the commutator relation (6.17).

   (b) Show that the components X^j of the position vector $\boldsymbol{r}$ associated with (6.22) are given by

$$X^j = i\mathrm{Tr}(\Phi^{-1}\Phi_{,\lambda}\sigma_j), \qquad j = 1, 2, 3.$$

   (c) Use the isomorphism $so(3) \cong su(2)$ to obtain the relation

$$\boldsymbol{r}\cdot\boldsymbol{L} = \hat{\Phi}^{-1}\hat{\Phi}_{,\lambda} \tag{6.62}$$

   where the $SO(3)$ matrix $\hat{\Phi}$ corresponding to Φ under the isomorphism is as set down in Appendix A.

   (d) Use the Sym-Tafel relation (6.22) to obtain the position vector $\boldsymbol{r}$ as given by (4.25) for the surface associated with the single soliton solution (4.23) of the NLS equation.

2. (a) If A and B are elements of $su(2)$, then show, using the decomposition notation (6.12), that the associated vectors $\boldsymbol{A}, \boldsymbol{B}$ satisfy the relations

$$\begin{aligned} &\text{(i)} \quad \boldsymbol{A}\cdot\boldsymbol{B} = -2\,\mathrm{Tr}(AB) \\ &\text{(ii)} \quad |\boldsymbol{A}|^2 = 4\det A \\ &\text{(iii)} \quad (\boldsymbol{A}\times\boldsymbol{B})\cdot\boldsymbol{e} = [A, B]. \end{aligned} \tag{6.63}$$

(b) Prove that the unit normal to a surface N in matrix notation takes the form

$$N = \boldsymbol{N} \cdot \boldsymbol{e} = \frac{[r_{,1}, r_{,2}]}{2 \det^{1/2}[r_{,1}, r_{,2}]}. \tag{6.64}$$

Derive the expression $(6.27)_2$ for $h_{\mu\nu}$. Apply the latter to obtain the 2nd fundamental form (6.28) for pseudospherical surfaces.

3. (a) Establish the relations (6.42) and (6.43).
   (b) Show that, for $N = 2$, $A_N = -1$ the system (6.43) generates the NLS equation

$$iq_t + q_{xx} + \frac{1}{2}|q|^2 q = 0.$$

4. Use the compatibility condition (6.49) to obtain the relations (6.50), (6.51).
5. Verify that, in the case $m = 1$, $n = 2$ with the identification $t_1 = x$, $t_2 = t$, the condition (6.54) holds modulo the canonical NLS equation

$$\begin{pmatrix} q \\ -\bar{q} \end{pmatrix}_t = -iL^2 \begin{pmatrix} q \\ \bar{q} \end{pmatrix}.$$

6. (a) Rederive the expressions (6.59) for the fundamental forms I and II associated with NLS surfaces by using Theorem 19.
   (b) Use the expressions (4.26) and (4.28) for I and II together with the invariance (4.80) to obtain the fundamental forms (6.59) with $s = x$.

6.2 NLS Eigenfunction Hierarchies. Geometric Properties. The Miura Transformation

In this section, it is established that the position vector of a soliton surface associated with the NLS hierarchy itself satisfies an integrable hierarchy, namely the so-called potential NLS eigenfunction hierarchy. The NLS eigenfunction hierarchy is then, in turn, shown to be generated by the action of a gauge transformation on the AKNS linear representation for the NLS hierarchy. To conclude, a classical representation for the curvature and torsion of a curve in terms of its intrinsic derivatives is interpreted as a Miura transformation linking the NLS hierarchy with its eigenfunction hierarchy.

6.2.1 *Soliton Surface Position Vectors as Solutions of Eigenfunction Equations*

The discovery of a link between the KdV and mKdV equations in 1968 is due to Miura [262]. The existence of this Miura transformation reflects the fact that the mKdV equation may be regarded as an eigenfunction equation of the KdV equation. Thus, a standard derivation of the Miura transformation is based on the observation that the KdV equation

$$u_t = u_{xxx} - 6uu_x \tag{6.65}$$

may be obtained as the compatibility condition for the *Lax pair* [228]

$$\begin{aligned} \lambda\psi &= \psi_{xx} - u\psi, \\ \psi_t &= 4\psi_{xxx} - 6u\psi_x - 3u_x\psi. \end{aligned} \tag{6.66}$$

If we express u in terms of the eigenfunction ψ for $\lambda = 0$, viz.

$$u = v_x + v^2, \tag{6.67}$$

where $v = \psi_x/\psi$, then, on insertion of u into $(6.66)_2$, the mKdV equation

$$v_t = v_{xxx} - 6v^2 v_x \tag{6.68}$$

results. Since (6.68) encapsulates an equation for the eigenfunction ψ, it is commonly termed the *eigenfunction equation* of the KdV equation. If v is a solution of this mKdV equation, then u as given by the *Miura transformation* (6.67) satisfies the KdV equation (6.65).

The Miura transformation may be exploited to generate a hierarchy of conservation laws for the KdV equation. Extensions of Miura-type transformations together with their interpretation as the inverse of gauge transformations have been widely studied [15, 16, 280]. On the other hand, general aspects of the integrability properties of eigenfunction equations associated with solitonic equations have been investigated by Konopelchenko [206].

Here, we are concerned with a geometric interpretation of Miura-type transformations in the context of the NLS hierarchy (6.43). The starting point is the elimination of the function q from the linear representation (6.33), (6.34) to obtain a nonlinear equation for a quantity which depends on the eigenfunction Φ alone. As in the derivation of the classical Miura transformation, we set $\lambda = 0$ and use the abbreviation $\Phi_0 = \Phi|_{\lambda=0}$. It proves convenient to introduce the matrices

$$S = \Phi_0^{-1}\sigma_3\Phi_0, \quad V_k = \Phi_0^{-1}\begin{pmatrix} A_k & -iB_k \\ i\bar{B}_k & -A_k \end{pmatrix}\Phi_0 \tag{6.69}$$

for $k = 0, \ldots, N$, where it is noted that

$$S^2 = \mathbb{1}, \quad S^\dagger = S, \tag{6.70}$$

since $\Phi_0^\dagger \Phi_0 = \mathbb{1}$, i.e., $\Phi_0 \in SU(2)$ and

$$V_N = A_N(t) S. \tag{6.71}$$

Here, $S^\dagger$ designates complex transposition of S, that is if $S = (S_{ij})$ then $S^\dagger = (\bar{S}_{ji})$.

Recursion relations for the matrices V_k may now be obtained in the following manner. Relation $(6.69)_2$ yields

$$\begin{aligned} V_{kx} &= \Phi_0^{-1} \left\{ \begin{pmatrix} A_{kx} & -iB_{kx} \\ i\bar{B}_{kx} & -A_{kx} \end{pmatrix} + \frac{1}{2} \left[\begin{pmatrix} A_k & -iB_k \\ i\bar{B}_k & -A_k \end{pmatrix}, \begin{pmatrix} 0 & q \\ -\bar{q} & 0 \end{pmatrix} \right] \right\} \Phi_0 \\ &= \Phi_0^{-1} \begin{pmatrix} 0 & B_{k-1} \\ \bar{B}_{k-1} & 0 \end{pmatrix} \Phi_0, \qquad k = 1, \ldots, N-1 \end{aligned} \tag{6.72}$$

by virtue of the scattering problem (6.33) and the recursion relations (6.38). The right-hand side of (6.72) expressed in terms of V_{k-1} now produces

$$V_{k-1} = -iSV_{kx} + \frac{1}{2} \operatorname{Tr}(SV_{k-1})\, S, \qquad k = 1, \ldots, N-1 \tag{6.73}$$

and the traces of (6.73) and (6.73) multiplied by S_x yield, in turn,

$$\begin{aligned} &\operatorname{Tr}(SV_{kx}) = 0 \\ &\operatorname{Tr}(S_x V_{k-1}) = -i \operatorname{Tr}(S_x S V_{kx}) = i \operatorname{Tr}(S S_x V_{kx}). \end{aligned} \tag{6.74}$$

Accordingly,

$$[\operatorname{Tr}(SV_{k-1})]_x = i \operatorname{Tr}(SS_x V_{kx}), \qquad k = 2, \ldots, N-1 \tag{6.75}$$

from which we conclude that

$$V_{k-1} = R V_k, \tag{6.76}$$

where the linear operator R is defined by

$$R = -iS \left[\partial_x - \frac{1}{2} \partial_x^{-1} \operatorname{Tr}(SS_x \partial_x \,\cdot\,) \right]. \tag{6.77}$$

On the other hand, the definition $(6.69)_1$ of S delivers the conservation law

$$S_t = \Phi_0^{-1} \begin{pmatrix} 0 & B_0 \\ \bar{B}_0 & 0 \end{pmatrix} \Phi_0 = V_{1x}, \tag{6.78}$$

where, by virtue of (6.76),

$$V_1 = R^{N-1} V_N = R^{N-1} A_N(t) S. \tag{6.79}$$

Thus, we obtain a nonlinear matrix equation for S, namely

$$S_t = A_N (R^{N-1} S)_x . \tag{6.80}$$

By construction, the constituent equations of the latter system are the eigenfunction equations of members of the NLS hierarchy (6.43). Hence, (6.80) is termed the *NLS eigenfunction hierarchy*. The first two members with ($A_2 = -1$, $A_3 = 1$) adopt the form

$$\begin{aligned} &iS_t + (SS_x)_x = 0, \\ &S_t + S_{xxx} + \frac{3}{2}\left(SS_x^2\right)_x = 0. \end{aligned} \tag{6.81}$$

An alternative representation of (6.81) may be given in terms of the vector notation

$$S = \boldsymbol{S} \cdot \boldsymbol{\sigma}, \quad \boldsymbol{S}^2 = 1, \tag{6.82}$$

which produces the vector equations

$$\begin{aligned} &\boldsymbol{S}_t + \boldsymbol{S} \times \boldsymbol{S}_{xx} = \boldsymbol{0}, \\ &\boldsymbol{S}_t + \boldsymbol{S}_{xxx} + \frac{3}{2}\left(\boldsymbol{S}_x^2 \boldsymbol{S}\right)_x = \boldsymbol{0}. \end{aligned} \tag{6.83}$$

In the previous section, it was established that the 'soliton surfaces' associated with the NLS equation via Theorem 19 may be identified with the Hasimoto surfaces generated by the evolution of a curve with $\boldsymbol{r}_t = \kappa \boldsymbol{b}$. The tangent vector $\boldsymbol{t}$ to such a curve obeys the Heisenberg spin equation (4.38). Since the latter coincides, up to a sign, with the base NLS eigenfunction equation $(6.83)_1$, it is natural to enquire as to the relation between $\boldsymbol{S}$ and the position vector $\boldsymbol{r}$. To determine this connection, we introduce the notation $r_0 = r|_{\lambda=0}$ and derive from the definition of the position matrix (6.31) the relations

$$r_{0x} = \frac{1}{2} iS, \quad r_{0t} = \frac{1}{2} iV_1. \tag{6.84}$$

Hence, the position matrix is nothing but a potential associated with the conservation law (6.78). According to the decomposition (6.31) the relations (6.84) become, in vector notation

$$\boldsymbol{r}_{0x} = -\boldsymbol{S}, \quad \boldsymbol{r}_{0t} = -\boldsymbol{V}_1, \tag{6.85}$$

where

$$V_1 = \boldsymbol{V}_1 \cdot \boldsymbol{\sigma}. \tag{6.86}$$

The relation $(6.85)_1$ establishes the following important result:

Theorem 20. *The position vector $\boldsymbol{r}_0$ of the soliton surface associated with the NLS hierarchy (6.43) satisfies the potential NLS eigenfunction hierarchy.*

The first two members of the potential NLS eigenfunction hierarchy are

$$\begin{aligned} \boldsymbol{r}_{0t} &= \boldsymbol{r}_{0x} \times \boldsymbol{r}_{0xx}, \\ -\boldsymbol{r}_{0t} &= \boldsymbol{r}_{0xxx} + \frac{3}{2}\boldsymbol{r}_{0xx}^2 \boldsymbol{r}_{0x} \end{aligned} \tag{6.87}$$

subject to the constraint $\boldsymbol{r}_{0x} \cdot \boldsymbol{r}_{0x} = 1$.

6.2.2 *The Serret-Frenet Equations and the NLS Hierarchy*

In geometric terms, the relation $(6.85)_1$ shows that $-\boldsymbol{S}$ is the tangent vector to the coordinate lines t = const on the surface with the position vector $\boldsymbol{r}_0$. This result may be tied in with the $\{\boldsymbol{t}, \boldsymbol{n}, \boldsymbol{b}\}$ formalism introduced to derive the NLS equation via a binormal motion in Section 4.1. Thus, on use of Appendix A, if Φ_0 is a solution of the scattering problem (6.33) at $\lambda = 0$, then the $SO(3)$ matrix

$$\hat{\Phi}_0 = \begin{pmatrix} \boldsymbol{A} \\ \boldsymbol{B} \\ \boldsymbol{C} \end{pmatrix} = \left(-2\,\mathrm{Tr}\left(\Phi_0^{-1} e_i \Phi_0 e_j\right)\right)_{i,j=1,2,3} \tag{6.88}$$

obeys the linear equation

$$\begin{pmatrix} \boldsymbol{A} \\ \boldsymbol{B} \\ \boldsymbol{C} \end{pmatrix}_x = \begin{pmatrix} 0 & 0 & -\kappa\cos\sigma \\ 0 & 0 & \kappa\sin\sigma \\ \kappa\cos\sigma & -\kappa\sin\sigma & 0 \end{pmatrix} \begin{pmatrix} \boldsymbol{A} \\ \boldsymbol{B} \\ \boldsymbol{C} \end{pmatrix}, \tag{6.89}$$

where we have introduced the orthonormal triad $\{\boldsymbol{A}, \boldsymbol{B}, \boldsymbol{C}\}$ consisting of the row vectors of $\hat{\Phi}_0$ and

$$q = \kappa \exp(i\sigma). \tag{6.90}$$

On application of the rotation

$$\begin{pmatrix} \boldsymbol{A}' \\ \boldsymbol{B}' \\ \boldsymbol{C}' \end{pmatrix} = \begin{pmatrix} 0 & 0 & -1 \\ -\cos\sigma & \sin\sigma & 0 \\ \sin\sigma & \cos\sigma & 0 \end{pmatrix} \begin{pmatrix} \boldsymbol{A} \\ \boldsymbol{B} \\ \boldsymbol{C} \end{pmatrix}, \tag{6.91}$$

where $\sigma_x = \tau$, we recover the usual Serret-Frenet relations if we identify the orthonormal triads $\{\boldsymbol{t}, \boldsymbol{n}, \boldsymbol{b}\} = \{\boldsymbol{A}', \boldsymbol{B}', \boldsymbol{C}'\}$ and take x as arc length. Comparison of the definition $(6.69)_1$ of S and the transformation (6.88) confirms that $\boldsymbol{C} = \boldsymbol{S} = -\boldsymbol{t}$.

It is remarked that the NLS eigenfunction hierarchy $(6.80)_{N=1,2,\dots}$ may also be obtained via a *gauge* transformation applied to the AKNS linear representation (6.33), (6.34) for the NLS hierarchy. In this connection, a new eigenfunction matrix Φ' is introduced via the gauge transformation

$$\Phi = \Phi_0 \Phi' \tag{6.92}$$

which takes (6.33), (6.34) to

$$\Phi'_x = \frac{1}{2} i\lambda S \Phi', \quad \Phi'_t = \frac{1}{2} i \sum_{k=1}^{N} \lambda^k V_k \Phi'. \tag{6.93}$$

Consequently, the NLS eigenfunction hierarchy (6.80) may be generated as the compatibility condition for the linear representation (6.93). It is noted that, since the gauge matrix Φ_0 does not depend on the spectral parameter λ, the position vector is unchanged by the gauge transformation (6.92), that is

$$r' = \Phi'^{-1} \Phi'_{,\lambda} = \Phi^{-1} \Phi_{,\lambda} = r. \tag{6.94}$$

Hence, Theorem 19 applied to the linear representation (6.93) leads to the same fundamental forms I and II, but now parametrised in terms of S rather than q. In fact, on use of the Serret-Frenet relations, it is easily verified that

$$\kappa = |\boldsymbol{S}_x|, \quad \tau = \frac{\boldsymbol{S}_x \cdot (\boldsymbol{S} \times \boldsymbol{S}_{xx})}{\boldsymbol{S}_x^2}. \tag{6.95}$$

These relations are nothing but the definition of the curvature and torsion of curves in $\mathbb{R}^3$. Here, in view of the decomposition (6.60), they represent a Miura-type transformation between the NLS hierarchy and the NLS eigenfunction hierarchy.

6.3 Reciprocal Transformations. Loop Solitons

Here, reciprocal transformations are introduced and their geometric importance illustrated in connection with loop solitons and the generators of soliton surfaces. In the previous section, it has been seen that gauge transformations may be exploited to give an interpretation of the position vector to a soliton surface as a solution of an eigenfunction equation. Reciprocal transformations, by contrast, represent a natural change of coordinate system on the soliton surface. In these reciprocal coordinates, it will be seen that loop solitons constitute both solutions of eigenfunction equations and generators of associated soliton surfaces. It is recalled that looped-curve phenomena have already been encountered in the generation of pseudospherical soliton surfaces associated with the sine-Gordon equation. A more general discussion of loop solitons in the context of soliton surfaces has been given by Sym [355]. Here, it will be seen that, reciprocal transformations provide a natural parametrisation of soliton surfaces associated with the NLS hierarchy. In contrast to the gauge transformations (6.92), a reciprocal change of coordinates transforms the coefficients $\{E, F, G\}$ and $\{e, f, g\}$ of the fundamental forms I and II, but leaves invariant the surfaces themselves. Subsequently, in Section 6.4, it will be shown that loop soliton and Dym hierarchies are generically invariant under certain reciprocal transformations. This reciprocal invariance may be used to induce auto-Bäcklund transformations in associated integrable hierarchies.

6.3.1 Reciprocal Transformations and the Loop Soliton Equation

It has been established that the position vector $\boldsymbol{r}_0$ of the soliton surfaces Σ associated with the NLS hierarchy is governed by the potential NLS eigenfunction hierarchy. The surfaces Σ were in that context parametrised in terms of independent variables x and t, that is

$$\boldsymbol{r}_0 = \begin{pmatrix} X^1 \\ X^2 \\ X^3 \end{pmatrix} (x, t), \tag{6.96}$$

where $\boldsymbol{r}_0$ satisfies a member of the potential NLS eigenfunction hierarchy. However, to understand the way the soliton surfaces Σ are generated by the motion of loop solitons, it proves convenient to replace x as an independent variable by one of the components X^i of $\boldsymbol{r}_0$. This leads to the notion of reciprocal transformations and the subsequent derivation thereby of the loop soliton equation.

Thus, use of the Pauli decomposition (6.82) together with

$$V_1 = \boldsymbol{V}_1 \cdot \boldsymbol{\sigma} \tag{6.97}$$

leads, via the conservation laws implicit in (6.85), to the relations

$$dX^i = -S_i dx - V_{1i} dt, \qquad i = 1, 2, 3. \tag{6.98}$$

A change of independent variables $(x, t) \to (X^i, T)$ of the type

$$dX^i = -S_i dx - V_{1i} dt, \quad dT = dt \tag{6.99}$$

or equivalently,

$$\partial_x = -S_i \partial_{X^i}, \quad \partial_t = \partial_T - V_{1i} \partial_{X^i} \tag{6.100}$$

for one of $i = 1, 2, 3$ is commonly called a *reciprocal transformation* [295]. In the present context of the NLS hierarchy, we select $Z = X^3$ and $T = t$ as new independent variables. It is emphasised that such a change of variable does not alter the corresponding soliton surface Σ. Even though the coefficients $\{E, F, G\}$ and $\{e, f, g\}$ of the fundamental forms I and II are transformed, I and II themselves are invariant.

Now, under the reciprocal transformation

$$dZ = -S_3 dx - V_{13} dt, \quad dT = dt, \tag{6.101}$$

the relations (6.85) that encapsulate as integrability conditions the NLS eigenfunction hierarchy in $\mathbf{S}$, become

$$\frac{\partial \boldsymbol{r}_0}{\partial Z} = \frac{\mathbf{S}}{S_3}, \quad \frac{\partial \boldsymbol{r}_0}{\partial T} = -\boldsymbol{V}_1 + V_{13} \frac{\partial \boldsymbol{r}_0}{\partial Z} = -\boldsymbol{V}_1 + \frac{V_{13}}{S_3} \mathbf{S}. \tag{6.102}$$

By construction, the Z-components of the relations (6.102) are identically satisfied. The residual X, Y-components or, equivalently, the reciprocal relations

$$dX = dX^1 = -S_1 dx - V_{11} dt, \quad dY = dX^2 = -S_2 dx - V_{12} dt \tag{6.103}$$

show that

$$\begin{aligned} \frac{\partial X}{\partial Z} &= \frac{S_1}{S_3}, \quad \frac{\partial X}{\partial T} = \frac{S_1}{S_3} V_{13} - V_{11}, \\ \frac{\partial Y}{\partial Z} &= \frac{S_2}{S_3}, \quad \frac{\partial Y}{\partial T} = \frac{S_2}{S_3} V_{13} - V_{12}, \end{aligned} \tag{6.104}$$

where it is recalled that

$$\mathbf{S}^2 = S_1^2 + S_2^2 + S_3^2 = 1. \tag{6.105}$$

Thus,

$$S_1 = \frac{\partial X}{\partial Z} S_3, \quad S_2 = \frac{\partial Y}{\partial Z} S_3, \tag{6.106}$$

where

$$S_3 = \pm 1 \Big/ \sqrt{1 + \left(\frac{\partial X}{\partial Z}\right)^2 + \left(\frac{\partial Y}{\partial Z}\right)^2}. \tag{6.107}$$

Moreover, $\boldsymbol{V}_1$ can be expressed in terms of $\boldsymbol{S}$ and derivatives of $\boldsymbol{S}$ with respect to Z on use of (6.79) and the relation $\partial_x = -S_3\partial_Z$. Accordingly, the relations (6.104) generate evolution equations for X and Y in the form

$$\frac{\partial X}{\partial T} = f(X, Y, X_Z, Y_Z, \ldots), \quad \frac{\partial Y}{\partial T} = G(X, Y, X_Z, Y_Z, \ldots). \tag{6.108}$$

This class of nonlinear equations reciprocally associated with the NLS eigenfunction hierarchy (6.80) represents a subclass of the 'new integrable systems' introduced by Wadati et al. [376]. The role of composite reciprocal and gauge transformations in linking the general AKNS and WKI systems has been described in [316].

A geometric interpretation of the WKI subclass (6.108) is readily given. Thus, for fixed 'time' T, any solution of (6.108) is associated with a curve (X, Y, Z) in $\mathbb{R}^3$ parametrised in terms of its Z-component. As time T evolves, the curve sweeps out the surface Σ associated with the corresponding member of the NLS hierarchy (6.43) as given by Theorem 19. Further, since (6.108) provides evolution equations for the components of the surface vector $\boldsymbol{r}_0$ of Σ, any solitonic property of these nonlinear equations may be 'seen' on the surface. This is best illustrated by the nonlinear equation which is reciprocally related via the above procedure to the potential mKdV equation. In this connection, it is recalled that the mKdV hierarchy is given by the odd members of the NLS hierarchy for real q. Accordingly, it is consistent to assume that S is a real matrix with the parametrisation

$$\boldsymbol{S} = \begin{pmatrix} \sin\theta \\ 0 \\ \cos\theta \end{pmatrix}. \tag{6.109}$$

Insertion into the NLS eigenfunction equation $(6.83)_2$ produces the potential mKdV equation

$$\theta_t + \theta_{xxx} + \frac{1}{2}\theta_x^3 = 0, \tag{6.110}$$

while the reciprocal relations (6.98) yield

$$\begin{aligned} dX^1 &= dX = -\sin\theta dx + \left(\cos\theta\,\theta_{xx} + \frac{\theta_x^2}{2}\sin\theta\right) dt, \\ dX^3 &= dZ = -\cos\theta dx + \left(-\sin\theta\,\theta_{xx} + \frac{\theta_x^2}{2}\cos\theta\right) dt. \end{aligned} \tag{6.111}$$

Under the reciprocal transformation (6.101), namely

$$dZ = -\cos\theta dx - \left(\theta_{xx}\sin\theta - \frac{1}{2}\theta_x^2\cos\theta\right) dt, \quad dT = dt, \tag{6.112}$$

the relations $(6.104)_{1,2}$ read

$$X_Z = \tan\theta, \quad X_T = (\sin\theta)_{ZZ}, \tag{6.113}$$

whence we obtain the *loop soliton* equation [376]

$$X_T = \pm\left(\frac{X_Z}{\sqrt{1+X_Z^2}}\right)_{ZZ}. \tag{6.114}$$

The relations $(6.104)_{3,4}$ yield $Y = Y_0$, where Y_0 is a constant of integration. Accordingly, the soliton surfaces associated with the mKdV equation are given by

$$\boldsymbol{r}_0 = \begin{pmatrix} X(Z,T) \\ Y_0 \\ Z \end{pmatrix}, \tag{6.115}$$

where $X = X(Z,T)$ is a solution of the loop soliton equation. Hence, it is deduced that the soliton surfaces $\Sigma : \boldsymbol{r}_0 = \boldsymbol{r}_0(Z,T)$ for the parametrisation (6.109) associated with the mKdV equation for vanishing spectral parameter λ are planes. Indeed, it will be seen in Section 6.4 that the entire mKdV hierarchy may be generated by a particular motion of a curve in the plane (cf. Section 2.4).

6.3.2 Loop Solitons

The loop soliton equation (6.114) is so called because of the solitonic behaviour of its loop-like solutions. The single-loop soliton solution of (6.114) is readily derived from the single soliton solution of the potential mKdV equation (6.110), namely

$$\theta = 4 \arctan e^{\beta x - \beta^3 t}. \tag{6.116}$$

Insertion of (6.116) into (6.111) and integration produces

$$\boldsymbol{r}_0 = \begin{pmatrix} -(2/\beta)\,\mathrm{sech}(\beta x - \beta^3 t) \\ 0 \\ -x + (2/\beta)\tanh(\beta x - \beta^3 t) \end{pmatrix}, \tag{6.117}$$

where constants of integration have been neglected. The corresponding solution of the loop soliton equation (6.114) is therefore given parametrically by

$$\begin{aligned} X &= -(2/\beta)\,\mathrm{sech}\,\sigma, \\ Z &= -\beta^2 T - \sigma/\beta + (2/\beta)\tanh\sigma, \end{aligned} \tag{6.118}$$

where $\sigma = \beta x - \beta^3 t$.

At $T = \text{const}$, the relations (6.118) determine a planar curve containing a loop. The curve $T = 0$ in the XZ-plane is depicted in Figure 6.1. As T evolves, the soliton surface (6.117) is swept out by the loop soliton curve travelling in the negative Z-direction with speed β^2. Since the 'amplitude' of the loop is $2/\beta$, we note the interesting feature that the smaller loop soliton travels faster than the larger.

Interaction properties of N-loop solitons have been investigated by Konno and Jeffrey [180, 202] via the inverse scattering transform. Reciprocal transformations have been used to obtain parametric representations of N-loop solitons by Dmitrieva [106]. Iterated application of the Bäcklund transformation at the

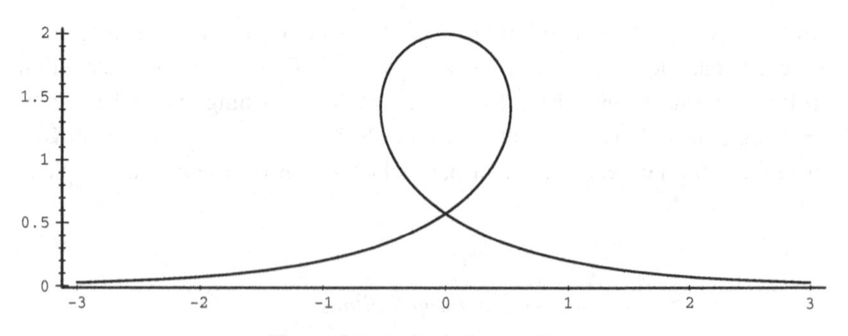

Figure 6.1. A single loop soliton.

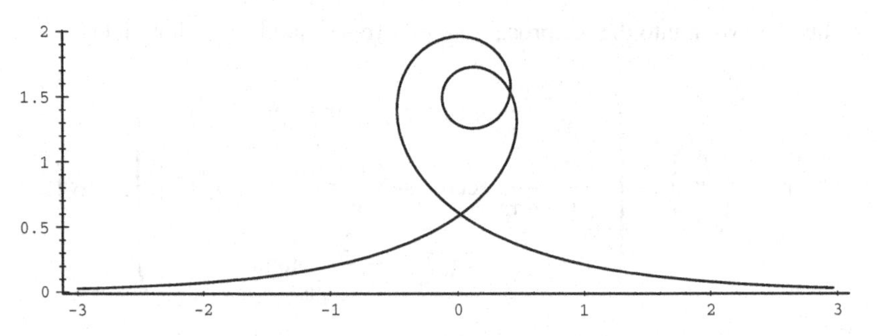

Figure 6.2. Loop soliton interaction.

soliton surface level has been used to construct N-loop solitons by Sym [355]. Loop soliton interaction is displayed in Figure 6.2.

It is natural to enquire as to whether loop solitons are also present in the case of the complex NLS hierarchy. To investigate this matter, we consider the member of the WKI system reciprocally related to the base member of the NLS eigenfunction hierarchy, namely the Heisenberg spin equation $(6.83)_1$, given via the conservation law

$$\boldsymbol{S}_t + (\boldsymbol{S} \times \boldsymbol{S}_x)_x = \boldsymbol{0} \tag{6.119}$$

and obtained as the integrability condition for the relations

$$\boldsymbol{r}_{0x} = -\boldsymbol{S}, \quad \boldsymbol{r}_{0t} = -\boldsymbol{V}_1 = \boldsymbol{S} \times \boldsymbol{S}_x. \tag{6.120}$$

On introduction of the decomposition $U = X + iY$, the relations (6.104), (6.105) deliver as the member of the WKI system reciprocally associated with the Heisenberg spin equation (6.119) the single complex equation

$$U_T = \mp i \left(\frac{U_Z}{\sqrt{1 + |U_Z|^2}} \right)_Z . \tag{6.121}$$

The solutions $U = U(Z, T)$ of (6.121) parametrise the soliton surface of the NLS equation according to

$$\boldsymbol{r}_0 = \begin{pmatrix} \Re(U) \\ \Im(U) \\ Z \end{pmatrix}. \tag{6.122}$$

A solution of the WKI equation (6.121) is readily obtained corresponding to the single soliton solution (4.23) of the NLS equation derived in Section 4.1.

Thus, insertion into the reciprocal relations (6.98) and integration yields

$$\boldsymbol{r}_0 = \begin{pmatrix} X \\ Y \\ Z \end{pmatrix} = \begin{pmatrix} -\dfrac{2\nu}{\nu^2+\tau_0^2}\operatorname{sech}\sigma\cos\left[\dfrac{\tau_0}{\nu}\sigma + (\nu^2+\tau_0^2)T\right] \\ -\dfrac{2\nu}{\nu^2+\tau_0^2}\operatorname{sech}\sigma\sin\left[\dfrac{\tau_0}{\nu}\sigma + (\nu^2+\tau_0^2)T\right] \\ \dfrac{\sigma}{\nu} + 2\tau_0 T - \dfrac{2\nu}{\nu^2+\tau_0^2}\tanh\sigma \end{pmatrix}, \tag{6.123}$$

where $\xi = s - ct$, $\sigma = \epsilon\xi$, $c = 2\tau_0$.

It is observed that the radial distance of the soliton surface from the Z-axis is given by

$$|U| = \sqrt{X^2+Y^2} = \frac{2|\nu|}{\nu^2+\tau_0^2}\operatorname{sech}\sigma.$$

Comparison with the loop soliton equation (6.118) shows that, at constant T, in the $(|U|, Z)$-plane, the relation (6.123) delivers a parametric representation of a loop soliton solution of the nonlinear equation (6.121). This loop soliton curve may be tracked on the one-soliton Hasimoto surface by measuring the radial distance of the coordinate lines T = const from the Z-axis.

In our preceding discussion of parametrisation of soliton surfaces, the notions of eigenfunction equations, gauge and reciprocal transformations have been introduced in a geometric context. It has been shown via the notion of a reciprocal transformation how loop solitons are naturally associated with the generation of corresponding soliton surfaces. In the next section, it is seen that the integrable hierarchy of Dym type is *invariant* under a reciprocal-type transformation.

Exercises

1. (a) Show that the first two members of the NLS eigenfunction hierarchy (6.80) are given by (6.81).
   (b) Use the vector notation (6.82) to derive the vector equations (6.83).
2. Establish the relations (6.84).
3. (a) Establish the linear equation (6.89) for $\begin{pmatrix} \boldsymbol{A} \\ \boldsymbol{B} \\ \boldsymbol{C} \end{pmatrix}$.
   (b) Show that $\boldsymbol{C} = \boldsymbol{S}$.

4. (a) Use the definition $(6.69)_1$ of the matrix S and the scattering problem (6.33) to obtain the Miura-type transformation (6.95) linking the NLS hierarchy (6.43) and the NLS eigenfunction hierarchy (6.80).
   (b) Derive the fundamental forms I and II for the Heisenberg spin equation $(6.83)_1$ via Theorem 19. Show that the Miura-type transformation (6.95) delivers the fundamental forms (6.59) for the NLS equation.
5. Verify the expression (6.117) for the position vector $\boldsymbol{r}_0$.
6. (a) Verify that the base WKI equation (6.121) is reciprocally related to the Heisenberg spin equation (6.119).
   (b) Under what circumstances does the planar curve

$$\sigma \to (\,\operatorname{sech}\sigma,\ \alpha\sigma + \beta\tanh\sigma)$$

   contain a loop? Deduce the constraints on the parameters ν and τ_0 in (6.123) for the underlying solution of the nonlinear equation (6.121) to represent a loop soliton.

6.4 The Dym, mKdV, and KdV Hierarchies. Connections

Reciprocal transformations have a long history and a wide range of applications. Thus, in 1928, Haar [156], in a paper devoted to adjoint variational problems, established a reciprocal-type invariance of a plane potential gasdynamic system. A decade later, Bateman [25] introduced an associated but less restricted class of invariant transformations that subsequently were termed reciprocal relations. Application of invariance properties to approximation theory in subsonic gasdynamics had been noted as early as 1939 by Tsien [367].

That the reciprocal relations of Bateman constitute a Bäcklund transformation was established in [22] via the Martin formulation of the gasdynamic equations as a Monge-Ampère system [242]. Indeed, both the adjoint and reciprocal relations may be shown to be induced via specialisations of an important class of matrix Bäcklund transformations, introduced in a gasdynamic context by Loewner [237].

Invariance of nonlinear gasdynamic and magnetogasdynamic systems under reciprocal-type transformations has been extensively studied in [286, 295–297, 304, 306, 307]. The application of reciprocal transformations to provide analytic solutions to both stationary and moving boundary value problems of practical interest, in particular, in soil mechanics and nonlinear heat conduction has likewise been the subject of much research [56, 57, 266, 298, 299, 301, 302, 310, 315, 317]. An extensive account of reciprocal-type transformations and their applications in continuum mechanics is presented in a monograph by Meirmanov et al. [258].

The importance of reciprocal transformations in the classification of so-called hydrodynamic systems has been established in a series of papers by Ferapontov [124–126, 132]. In terms of geometry, the classification of weakly nonlinear Hamiltonian systems of hydrodynamic type up to reciprocal transformations has been shown to be completely equivalent to the classification of Dupin hypersurfaces up to Lie sphere transformations [127, 130]. Interesting connections between the classical theory of congruences and systems of conservation laws of the Temple class [362] invariant under reciprocal transformations have also been discovered recently [5].

Here, it is the role of reciprocal transformations in soliton theory that will engage our attention. In this connection, it has already been seen in the previous section how reciprocal transformations arise naturally in links between important integrable equations such as the potential mKdV and loop soliton equations or the Heisenberg spin equation and the base member of the WKI system. Moreover, it has recently been shown that an integrable Camassa-Holm equation is connected via a reciprocal transformation to the first negative flow of the KdV hierarchy [174]. In this section, invariance of loop soliton and Dym hierarchies under reciprocal-type transformations is established.

6.4.1 Invariance under Reciprocal Transformations. A Class of Planar Curve Motions

In the sequel, we shall routinely call upon the following basic result [195]:

Theorem 21. *The conservation law*

$$\frac{\partial}{\partial t}\{T(\partial/\partial x;\partial/\partial t;u)\}+\frac{\partial}{\partial x}\{F(\partial/\partial x;\partial/\partial t;u)\}=0 \tag{6.124}$$

is transformed to the associated conservation law

$$\frac{\partial T'}{\partial t'}+\frac{\partial F'}{\partial x'}=0 \tag{6.125}$$

via the reciprocal transformation

$$\left.\begin{array}{c} dx'=T\,dx-F\,dt,\quad dt'=dt,\\ T'=\dfrac{1}{T},\quad F'=-\dfrac{F}{T},\\ 0<|T|<\infty \end{array}\right\}\mathcal{R}. \tag{6.126}$$

The reciprocal nature of the transformation (6.126) resides in the fact that $\mathcal{R}^2 = \text{id}$.

A general result on invariance of 1+1-dimensional evolution equations under reciprocal transformations was presented in [196]. Here, we restrict our attention to invariance of certain important soliton equations under reciprocal transformations.

Let us consider a plane curve C which propagates in the (x, y)-plane in such a way that

$$\boldsymbol{v} \cdot \boldsymbol{n} = \alpha \frac{\partial^n \kappa}{\partial s^n}, \qquad n = 0, 1, \dots, \tag{6.127}$$

where $\boldsymbol{v} \cdot \boldsymbol{n}$ is the (principal) normal component of the velocity of propagation and α is an arbitrary constant. Here, $\kappa = \theta_s$ denotes the curvature of a generic point P on C and s denotes arc length. It is recalled that a particular such planar motion with $n = 1$ has been seen to generate the mKdV equation in Section 2.4.[2] In terms of the 'Eulerian' coordinates x and y, the curve may be parametrised according to

$$C : y = \Omega(x, t) \tag{6.128}$$

so that $\Omega_x = \tan\theta$ and the normal $\boldsymbol{n}$ to C is given by

$$\boldsymbol{n} = \frac{1}{\sqrt{1 + \Omega_x^2}} \begin{pmatrix} \Omega_x \\ -1 \end{pmatrix}. \tag{6.129}$$

However, to calculate the velocity $\boldsymbol{v}$, it is necessary to adopt a 'Lagrangian' description. Thus, if the position vector $\boldsymbol{r} = (x, y)^{\mathrm{T}}$ to the curve C is parametrised in terms of a Lagrangian variable σ, that is $\boldsymbol{r} = \boldsymbol{r}(\sigma, t)$ or, explicitly,

$$x = x(\sigma, t), \quad y = y(\sigma, t), \tag{6.130}$$

then

$$\boldsymbol{v} = \begin{pmatrix} \left.\dfrac{\partial x}{\partial t}\right|_\sigma \\ \left.\dfrac{\partial y}{\partial t}\right|_\sigma \end{pmatrix} = \begin{pmatrix} \left.\dfrac{\partial x}{\partial t}\right|_\sigma \\ \Omega_x \left.\dfrac{\partial x}{\partial t}\right|_\sigma + \Omega_t \end{pmatrix}. \tag{6.131}$$

[2] Interestingly, motions with $\boldsymbol{v} \cdot \boldsymbol{n} = \alpha\kappa_s$ arise physically in the dynamics of long front waves in a quasi-geostrophic flow [279].

Accordingly, the normal speed reads

$$\boldsymbol{v}\cdot\boldsymbol{n} = -\frac{\Omega_t}{\sqrt{1+\Omega_x^2}}. \tag{6.132}$$

The relation (6.127) now yields

$$\frac{\partial}{\partial t}(\tan\theta) + \alpha\frac{\partial}{\partial x}\left(\sec\theta\,\frac{\partial^{n+1}\theta}{\partial s^{n+1}}\right) = 0 \tag{6.133}$$

in terms of θ. The specialisations $n=0$ and $n=2$ in the latter equation lead to 1+1-dimensional models of the propagation of surface grooves due to evaporation-condensation and surface diffusion in heated polycrystals [58, 265, 366]. In the case $n=1$, (6.133) reduces to

$$\frac{\partial}{\partial t}(\tan\theta) + \alpha\frac{\partial^3}{\partial x^3}(\sin\theta) = 0, \tag{6.134}$$

namely the loop soliton equation implicit in (6.113) on making the replacements $Z \to -x$, $T \to t$ and setting $\alpha = 1$.

The nonlinear evolution equation (6.133) is readily seen, on application of Theorem 21, to be invariant under the reciprocal transformation

$$\begin{aligned} dx' &= \tan\theta\,dx - \alpha\sec\theta\,\frac{\partial^{n+1}\theta}{\partial s^{n+1}}\,dt, \quad dt' = dt \\ \theta' &= -\theta + \pi/2. \end{aligned} \tag{6.135}$$

This corresponds to invariance of (6.127) under the *isotherm* transformation

$$x' = \Omega, \quad \Omega' = x, \quad t' = t. \tag{6.136}$$

Isotherm transformations have been exploited in the numerical treatment of nonlinear moving boundary problems in Crank [90].

It is recalled that the loop soliton equation (6.134) is reciprocally linked to the potential mKdV equation. This connection is encoded in the reciprocal transformation

$$\begin{aligned} dx^* &= \sec\theta\,dx + \alpha\left[\frac{1}{2}\cos^2\theta\,\theta_x^2 - \sin\theta\frac{\partial}{\partial x}(\cos\theta\,\theta_x)\right]dt \\ dt^* &= dt \end{aligned} \tag{6.137}$$

with integrability condition the conservation law

$$\frac{\partial}{\partial t}(\sec\theta) - \alpha\frac{\partial}{\partial x}\left[\frac{1}{2}\theta_x^2\cos^2\theta - \sin\theta\frac{\partial}{\partial x}(\cos\theta\,\theta_x)\right] = 0 \tag{6.138}$$

admitted by the loop soliton equation (6.134). Under the reciprocal transformation (6.137), the loop soliton equation becomes, on use of Theorem 21,

$$\frac{\partial}{\partial t^*}(\cos\theta) + \alpha\frac{\partial}{\partial x^*}\left[\frac{1}{2}\cos\theta\,\theta_{x^*}^2 - \sin\theta\,\theta_{x^*x^*}\right] = 0.$$

The relation $dx^* = ds$ shows that the reciprocal variable x^* is, in fact, nothing but arc length, and we obtain

$$\theta_{t^*} + \alpha\left(\theta_{sss} + \frac{\theta_s^3}{2}\right) = 0, \tag{6.139}$$

namely the potential mKdV equation. The curvature $\kappa = \theta_s$ evolves according to the mKdV equation

$$\kappa_t + \alpha\left(\kappa_{sss} + \frac{3}{2}\kappa^2\kappa_s\right) = 0. \tag{6.140}$$

6.4.2 The Dym, mKdV, and KdV Hierarchies

Invariance of the Dym equation

$$\rho_t = \rho^{-1}(\rho^{-1})_{xxx} \tag{6.141}$$

under the reciprocal transformation

$$\begin{aligned} dx' &= \rho^2 dx + 2(\rho^{-1})_{xx}dt, \quad dt' = dt, \\ \rho' &= \rho^{-1} \end{aligned} \tag{6.142}$$

was originally noted in [316] in the context of the connection between the AKNS and WKI integrable systems. Indeed, it was subsequently shown in [308] that, more generally, the Dym hierarchy as set down in Calogero and Degasperis [65], namely

$$\rho_t = \rho^{-1}(-D^3 r I r)^n \rho\rho_x, \quad n = 1, 2, \ldots, \tag{6.143}$$

where the operators D and I are defined by

$$D\phi := \phi_x, \quad I\phi := \int_x^\infty \phi(\sigma, t)\,d\sigma \tag{6.144}$$

and $r = 1/\rho$, is invariant under the reciprocal transformation

$$\left.\begin{aligned} dx' &= \rho^2 dx - 2(\rho^{-1}\mathcal{E}_{n-1})_{xx}dt, \quad dt' = dt \\ \rho' &= \rho^{-1} \end{aligned}\right\}\mathcal{R}', \tag{6.145}$$

where

$$\mathcal{E}_n = -\int_x^\infty \rho^{-1}(\rho^{-1}\mathcal{E}_{n-1})_{yyy}dy, \quad \mathcal{E}_0 = 1, \qquad n = 1, 2, \ldots . \tag{6.146}$$

The reciprocal invariance of the Dym equation and its hierarchy has been exploited by Dmitrieva [105, 106] to construct both N-gap and N-loop solutions. Ibragimov [177] demonstrated via Lie-Bäcklund arguments that the Dym equation (6.141) is but another avatar of the celebrated KdV equation. This result may be readily established in the more general context of the Dym and KdV hierarchies by a combination of reciprocal and gauge transformations. Thus, under the reciprocal transformation

$$\left.\begin{aligned} d\tilde{x} = \rho\, dx - \mathcal{E}_n dt, \quad d\tilde{t} = dt, \\ \tilde{\rho} = \frac{1}{\rho} \end{aligned}\right\} \tilde{\mathcal{R}}, \tag{6.147}$$

the Dym hierarchy (6.143) is taken to the so-called *Krichever-Novikov* hierarchy in $\phi(\tilde{x}, \tilde{t})$ where $\tilde{\rho} = \phi_{\tilde{x}}$. The above invariance of the Dym hierarchy induces invariance under $\phi_{\tilde{x}} \to 1/\phi_{\tilde{x}}$ of the Krichever-Novikov hierarchy. The latter arises naturally out of the application of the WTC expansion procedure.[3]

The substitution

$$v = \left(\tilde{\rho}^{-1/2}\right)_{\tilde{x}} / \tilde{\rho}^{-1/2}, \tag{6.148}$$

into the Krichever-Novikov hierarchy generates the mKdV hierarchy

$$v_{\tilde{t}} = M^n v_{\tilde{x}}, \qquad n = 1, 2, \ldots, \tag{6.149}$$

where

$$M := \frac{\partial^2}{\partial \tilde{x}^2} - 4v^2 + 4v_{\tilde{x}} \int_{\tilde{x}}^\infty v(y, \tilde{t}) \cdot dy \tag{6.150}$$

as obtained via the specialisation of the NLS hierarchy for odd N in the previous section.

The alternative substitution

$$u = \left(\tilde{\rho}^{-1/2}\right)_{\tilde{x}\tilde{x}} / \tilde{\rho}^{-1/2} \tag{6.151}$$

[3] The Weiss-Tabor-Carnavale (WTC) procedure represents an important version of the Painlevé integrability test based on a series expansion about the singularity manifold $S : \phi = \phi(\tilde{x}, \tilde{t})$. A survey of this method is given by Tabor in [357].

into the Krichever-Novikov hierarchy produces the *KdV hierarchy*

$$u_{\tilde{t}} = K^n u_{\tilde{x}}, \qquad n = 1, 2, \ldots, \tag{6.152}$$

where

$$K := \frac{\partial^2}{\partial \tilde{x}^2} - 4u + 2u_{\tilde{x}} \int_{\tilde{x}}^{\infty} \cdot \, dy. \tag{6.153}$$

It is seen that the relations (6.148) and (6.151) combine to produce the Miura transformation

$$u = v_{\tilde{x}} + v^2, \tag{6.154}$$

which links the mKdV and KdV hierarchies. The relation (6.151) shows that $\tilde{\rho}^{-1/2}$ plays the role of the eigenfunction at $\lambda = 0$ in the linear problem $(6.66)_1$ for the KdV hierarchy.

The invariance of the Dym hierarchy (6.143) under the reciprocal transformation (6.145) when taken together with the relations (6.147), (6.148) induces the simple symmetry $v \to -v$ in the mKdV hierarchy. This, in turn, together with the Miura transformation (6.154) and the invariance of the KdV hierarchy under the Galilean transformation

$$u' = u + \frac{\beta}{2}, \quad x' = x - 3\beta t, \quad t' = t \tag{6.155}$$

induces the spatial part of the auto-Bäcklund transformation generic to the KdV hierarchy. This is given by

$$\mathbb{B}_\beta : \quad (\Lambda^+ + \Lambda^-)_x = \beta - \frac{1}{2}[\Lambda^+ - \Lambda^-]^2, \tag{6.156}$$

where $u^\pm = -\Lambda_x^\pm$ are the solutions of the KdV hierarchy corresponding to solutions $\pm v$ of the linked mKdV hierarchy. Thus, as in the case of the classical sine-Gordon equation, a Lie symmetry is used to inject a parameter β into the Bäcklund transformation.

6.4.3 A Permutability Theorem

Here, a permutability theorem is established via (6.156) for the potential KdV equation. This is embodied in the following:

Theorem 22. *Let Λ_0 be a seed solution of the* potential *KdV equation*

$$\Lambda_t = \Lambda_{xxx} - 3\Lambda_x^2 \tag{6.157}$$

and Λ_1, Λ_2 *be the Bäcklund transforms of* Λ_0 *via* $\mathbb{B}_{\beta_1}$ *and* $\mathbb{B}_{\beta_2}$ *so that* $\Lambda_1 = \mathbb{B}_{\beta_1}(\Lambda_0)$ *and* $\Lambda_2 = \mathbb{B}_{\beta_2}(\Lambda_0)$*, where* $\mathbb{B}_\beta$ *designates the spatial part of the Bäcklund transformation as given by (6.156). Then, a new solution of the potential KdV equation is given by*

$$\Lambda = \Lambda_0 + 2(\beta_1 - \beta_2)/(\Lambda_1 - \Lambda_2), \tag{6.158}$$

where $\Lambda = \Lambda_{12} = \Lambda_{21}$ *in the notation* $\Lambda_{12} = \mathbb{B}_{\beta_2}(\Lambda_1)$, $\Lambda_{21} = \mathbb{B}_{\beta_1}(\Lambda_2)$.

The above result is readily established. Thus, the Bäcklund transformation $\mathbb{B}_\beta$ yields

$$\begin{aligned} \Lambda_{0,x} + \Lambda_{1,x} &= \beta_1 - \frac{1}{2}(\Lambda_0 - \Lambda_1)^2, \\ \Lambda_{0,x} + \Lambda_{2,x} &= \beta_2 - \frac{1}{2}(\Lambda_0 - \Lambda_2)^2, \\ \Lambda_{1,x} + \Lambda_{12,x} &= \beta_2 - \frac{1}{2}(\Lambda_1 - \Lambda_{12})^2, \\ \Lambda_{2,x} + \Lambda_{21,x} &= \beta_1 - \frac{1}{2}(\Lambda_2 - \Lambda_{21})^2. \end{aligned} \tag{6.159}$$

If we now postulate the existence of Λ with

$$\Lambda_{12} = \Lambda_{21} = \Lambda$$

then the operations $(6.159)_1$–$(6.159)_2$ and $(6.159)_3$–$(6.159)_4$ produce, in turn, the relations

$$\begin{aligned} (\Lambda_1 - \Lambda_2)_x &= \beta_1 - \beta_2 + \frac{1}{2}(\Lambda_1 - \Lambda_2)(2\Lambda_0 - \Lambda_1 - \Lambda_2), \\ (\Lambda_1 - \Lambda_2)_x &= \beta_2 - \beta_1 + \frac{1}{2}(\Lambda_2 - \Lambda_1)(\Lambda_1 + \Lambda_2 - 2\Lambda). \end{aligned}$$

Subtraction now produces the nonlinear superposition principle as set down in Theorem 22. It is readily verified that the Bäcklund relations $(6.159)_{3,4}$ are indeed satisfied with $\Lambda_{12} = \Lambda_{21} = \Lambda$ as given by (6.158), modulo the Bäcklund relations $(6.159)_{1,2}$.

The permutability theorem (6.158) has been used to construct multi-soliton solutions of the KdV equation by Wahlquist and Estabrook [377]. Grammaticos et al. [153] have recently observed that a version of the KdV permutability theorem also makes a remarkable appearance in numerical analysis. Thus, the

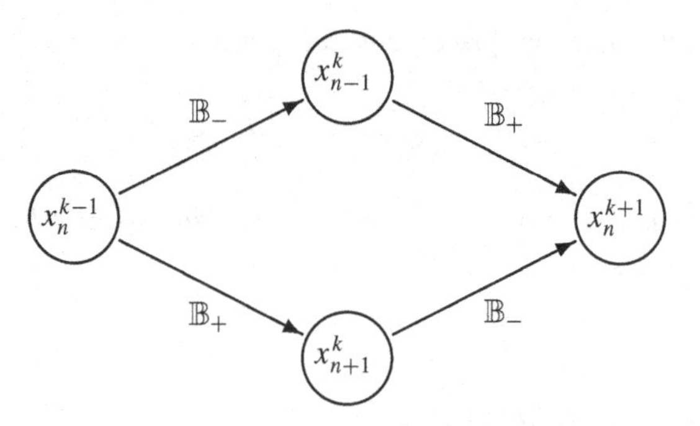

Figure 6.3. A Bianchi diagram associated with the ϵ-algorithm.

so-called ϵ-algorithm as described by Brezinski [55] adopts the form

$$x_n^{k+1} = x_n^{k-1} + \frac{1}{x_{n+1}^k - x_{n-1}^k}, \tag{6.160}$$

where x_n denotes a member of a sequence and x_n^k denotes its kth iteration in an expansion. Herein, the initial data are the x_n^1 with the x_n^0 set to be zero. It is seen that (6.160) is, 'mutatis mutandis', nothing but a version of the nonlinear superposition principle (6.158) for the potential KdV equation. Therein, the Bäcklund parameters β_1, β_2 are chosen so that $\beta_1 - \beta_2 = 1/2$. Thus, the algorithm (6.160) is equivalent to the permutability theorem encoded in the Bianchi diagram as set down in Figure 6.3. Here, the action of $\mathbb{B}^+$ is to augment the lower index by 1, while the action of $\mathbb{B}^-$ is to decrease the lower index by 1. The upper index corresponds to the level of the iteration.

The above novel aspect of the permutability theorem foreshadows a key role played by Bäcklund transformations and their associated nonlinear superposition principles in integrable discretisation (see [45, 205, 229, 230, 272–274, 291, 325, 327, 331]).

6.4.4 A Geometric Derivation of the mKdV Hierarchy

It has been seen that the mKdV equation may be derived in a purely geometric manner via a particular motion of a curve in the plane. This result is here extended to the mKdV hierarchy as follows:

Theorem 23. *If a curve $C : y = \Lambda(x, t)$ moves in the xy-plane in such a way that*

$$\boldsymbol{v} \cdot \boldsymbol{n} = M^{n-1} \kappa_s, \tag{6.161}$$

where M is the mKdV integro-differential operator

$$M := \frac{\partial^2}{\partial s^2} + \kappa^2 - \kappa_s \int_s^\infty \kappa \cdot ds, \tag{6.162}$$

then the curvature κ of the curve C evolves according to the nth member of the mKdV hierarchy, namely

$$\kappa_t + M^n \kappa_s = 0. \tag{6.163}$$

Thus, the condition (6.161) yields

$$-\Lambda_t / \sqrt{1 + \Lambda_x^2} = M^{n-1} \kappa_s, \tag{6.164}$$

whence, since $\Lambda_x = \tan\theta$,

$$\left[\frac{\partial^2}{\partial s^2} + \kappa^2 - \kappa_s \int_s^\infty \kappa \cdot ds \right] (\Lambda_t \cos\theta) = -M^n \kappa_s. \tag{6.165}$$

To show that (6.165) is equivalent to the mKdV hierarchy (6.163), it proves convenient to introduce a generalisation of the reciprocal transformation (6.137) which links the loop soliton equation and the mKdV equation. Thus, (6.164) yields

$$\frac{\partial}{\partial t}(\tan\theta) + \frac{\partial}{\partial x}(\sec\theta\, M^{n-1} \kappa_s) = 0, \tag{6.166}$$

whence

$$\frac{\partial}{\partial t}(\sec\theta) - \frac{\partial}{\partial x} \int_x^\infty \sin\theta \frac{\partial}{\partial x}(\sec\theta M^{n-1} \kappa_s) dx = 0. \tag{6.167}$$

Under the reciprocal transformation

$$\begin{aligned} dx^* &= \sec\theta\, dx + \left(\int_x^\infty \sin\theta \frac{\partial}{\partial x}(\sec\theta M^{n-1} \kappa_s)\, dx \right) dt \\ dt^* &= dt, \end{aligned} \tag{6.168}$$

it is seen that, since $dx^* = ds$,

$$\left.\frac{\partial}{\partial t}\right|_x = -\left(\int_x^\infty \Lambda_{xt} \sin\theta\, dx \right) \frac{\partial}{\partial s} + \left.\frac{\partial}{\partial t}\right|_s \tag{6.169}$$

Use of (6.169) in the relation (6.165) leads to the mKdV hierarchy (6.163). The integrable hierarchy (6.166) contains the loop soliton equation (6.114) as its base member. Its invariance properties were discussed in [309].

Reciprocal transformations have other applications in 1+1-dimensional soliton theory. Thus, novel integrable systems analogous to the Dym hierarchy may be constructed which are related by a combination of reciprocal and gauge transformations to the Caudrey-Dodd-Gibbon and Kaup-Kuperschmidt integrable hierarchies [303, 382].

In terms of physical applications, it has been shown by Kadanoff [187] that a version of the Dym equation arises in an analysis of the Saffman-Taylor problem with surface tension. On the other hand, Camassa and Holm [67], in a study of a dispersive shallow water system, have derived the equation

$$\mu_t = -\left(\partial_x - \partial_x^3\right)\frac{1}{\sqrt{\mu}}, \tag{6.170}$$

akin to the Dym equation (6.141). This itself represents the base member of an integrable hierarchy with interesting geometric properties [7, 8]. The geometry of yet another extension of the Dym-type equation (6.170) is the subject of the following section.

In [280], it has been shown that reciprocal transformations may also be constructed for certain 2+1-dimensional integrable systems. These were used to link the modified Kadomtsev-Petviashvili (mKP) and Kadomtsev-Petviashvili (KP) hierarchies with a 2+1-dimensional Dym hierarchy. Invariance of the latter under reciprocal-type transformations was established. Multi-soliton solutions of the 2+1-dimensional Dym equation have been derived via reciprocal transformations by Dmitrieva and Khlabystova [107].

Exercises

1. (a) Establish the connection between the conservation laws (6.124) and (6.125) via the reciprocal transformation $\mathcal{R}$.
   (b) Show that $\mathcal{R}^2 = \text{id}$.
2. Show that the system of nonlinear evolution equations (6.133) is invariant under the reciprocal transformation (6.135).
3. Use the permutability theorem to construct the two-soliton solution of the KdV equation.
4. Show that the hierarchy (6.166) is invariant under the reciprocal transformation

$$\begin{gathered} dx' = \tan\theta\, dx - \sec\theta\, M^{n-1}\kappa_s\, dt, \quad dt' = dt, \\ \theta' = -\theta + \pi/2. \end{gathered} \tag{6.171}$$

5. (a) Use a reciprocal transformation to link equation (6.170) to the

integrable equation [137]

$$\theta_{\tilde{t}} + \theta_{\tilde{x}\tilde{x}\tilde{x}} - \frac{1}{2}\theta_{\tilde{x}}^3 - \frac{3}{2}e^{2\theta}\theta_{\tilde{x}} = 0 \tag{6.172}$$

in reciprocal variables $\tilde{x}, \tilde{t}$.

(b) Show that the class of equations

$$\phi_t + \phi_{xxx} + \frac{1}{2}\phi_x^3 + 6\Phi^2(\phi)\phi_x = 0, \tag{6.173}$$

with

$$\Phi'' + \Phi = 0, \tag{6.174}$$

is connected to the potential mKdV equation (6.110) via the transformation

$$\theta_x = \phi_x - 2\Phi. \tag{6.175}$$

6. Use the reciprocal relation (6.168) to show that the evolution (6.164) produces the mKdV hierarchy (6.163).

6.5 The Binormal Motion of Curves of Constant Curvature. Extended Dym Surfaces

Here, we return to the geometric formulation introduced in Chapter 4 in connection with the NLS equation as generated by the binormal motion of a curve travelling with velocity $\boldsymbol{v} = \kappa \boldsymbol{b}$. This representation will be exploited here to derive a novel solitonic equation associated with the purely binormal motion of a curve of constant curvature [339].

Let us examine the implications of purely binormal motion. If a curve C moving in space has generic position vector $\boldsymbol{r} = \boldsymbol{r}(\sigma, t)$, where time t and the parameter σ are independent, then the arc length of C in its configuration at constant time is given by

$$s(\sigma, t) = \int_0^\sigma \sqrt{h(\sigma^*, t)}\, d\sigma^* + s(0, t), \tag{6.176}$$

where

$$h = \frac{\partial \boldsymbol{r}}{\partial \sigma} \cdot \frac{\partial \boldsymbol{r}}{\partial \sigma}. \tag{6.177}$$

For non-vanishing curvature κ and torsion τ,

$$\frac{\partial \boldsymbol{r}}{\partial \sigma} \times \boldsymbol{t} = \boldsymbol{0}, \quad \frac{\partial \boldsymbol{b}}{\partial \sigma} \times \boldsymbol{n} = \boldsymbol{0} \tag{6.178}$$

so that the assumption of a purely binormal motion

$$\frac{\partial \boldsymbol{r}}{\partial t} = \Xi \boldsymbol{b} \tag{6.179}$$

yields

$$\frac{\partial h}{\partial t} = 2\frac{\partial \boldsymbol{r}}{\partial \sigma} \cdot \frac{\partial^2 \boldsymbol{r}}{\partial \sigma \partial t} = 2\frac{\partial \boldsymbol{r}}{\partial \sigma} \cdot \left(\frac{\partial \Xi}{\partial \sigma}\boldsymbol{b} + \Xi \frac{\partial \boldsymbol{b}}{\partial \sigma} \right) = 0 \tag{6.180}$$

by virtue of (6.178). This implies, in turn, that $\partial(s - s(0, t))/\partial t = 0$ and hence purely binormal motions are only possible for *inextensible* curves.

Here, as in our previous discussion of the NLS equation, we restrict our attention to geometries with the vanishing abnormality constraint $\Omega_n = 0$ so that there exists a one-parameter family of surfaces Σ which contain the s-lines and b-lines. It is recalled that if the s-lines and b-lines are taken as coordinate curves on Σ, then the surface metric may be reduced to the form

$$\mathrm{I}_\Sigma = ds^2 + g(s, b)db^2 \tag{6.181}$$

so that

$$\frac{\partial \boldsymbol{r}}{\partial b} = g^{1/2}\boldsymbol{b}. \tag{6.182}$$

The s-lines are geodesics and the b-lines are parallels on individual surfaces Σ. If the variable b is interpreted as a measure of time, then the relation (6.182) gives the velocity in the purely binormal motion. The Gauss-Mainardi-Codazzi equations for an individual surface Σ become

$$\begin{gathered} \frac{\partial}{\partial s}(g\tau) + g^{1/2}\frac{\partial \kappa}{\partial b} = 0, \\ \frac{\partial \tau}{\partial b} = \frac{\partial}{\partial s}\left[g^{1/2}(\kappa + \operatorname{div}\boldsymbol{n})\right] + \kappa\frac{\partial g^{1/2}}{\partial s}, \\ [\kappa(\kappa + \operatorname{div}\boldsymbol{n}) + \tau^2]g^{1/2} = \frac{\partial^2 g^{1/2}}{\partial s^2} \end{gathered} \tag{6.183}$$

and elimination of $\kappa + \operatorname{div}\boldsymbol{n}$ in the latter pair of relations yields (cf. (4.144))

$$\frac{\partial \tau}{\partial b} = \frac{\partial}{\partial s}\left[\frac{1}{\kappa}\left(\frac{\partial^2 g^{1/2}}{\partial s^2} - \tau^2 g^{1/2}\right)\right] + \kappa\frac{\partial g^{1/2}}{\partial s}. \tag{6.184}$$

6.5.1 Curves of Constant Curvature

If we restrict our attention to the purely binormal motion of curves of constant curvature κ, then $(6.183)_1$ shows that

$$g = \alpha(b)\tau^{-1} \tag{6.185}$$

whence on elimination of g in (6.184) and reparametrisation $\alpha^{-1/2}\partial/\partial b \to \partial/\partial b$, we obtain

$$\boxed{\tau_b = \left[\frac{1}{\kappa}\left(\frac{1}{\tau^{1/2}}\right)_{ss} - \tau^{3/2} + \kappa\frac{1}{\tau^{1/2}}\right]_s .} \tag{6.186}$$

This will subsequently be termed the *extended Dym* (ED) equation [339]. On an appropriate scaling and limiting process in (6.186), it leads to the Dym equation

$$\tau_b = \left(\frac{1}{\tau^{1/2}}\right)_{sss} \tag{6.187}$$

as a specialisation. It is important to remark that it is the ED equation rather than the Dym equation (6.187) that admits a simple geometric derivation. Thus, it is generated by the purely binormal motion of an inextensible curve of constant curvature κ moving with velocity

$$\boldsymbol{r}_b = \tau^{-1/2}\boldsymbol{b}. \tag{6.188}$$

The Dym equation has been seen to be solitonic and to be invariant under a reciprocal transformation. It is natural to enquire as to whether the nonlinear evolution equation (6.186) shares these properties.

That the ED equation (6.186) is solitonic is readily established by a reciprocal link to the m^2KdV equation. Thus, it is noted that (6.186) admits the conservation law

$$(V^{-1})_b = \frac{1}{2}\left[\frac{1}{\kappa}\left(VV_{ss} - \frac{1}{2}V_s^2\right) - \frac{3}{2}V^{-2} + \kappa\frac{V^2}{2}\right]_s . \tag{6.189}$$

where $V = \tau^{-1/2}$ denotes the speed of the binormal motion. On application of the reciprocal transformation

$$\begin{aligned} dx &= \sqrt{2}\,V^{-1}ds + \frac{1}{\sqrt{2}}\left[\frac{1}{\kappa}\left(VV_{ss} - \frac{1}{2}V_s^2\right) - \frac{3}{2}V^{-2} + \kappa\frac{V^2}{2}\right]db \\ dt &= -\sqrt{2}\,db, \end{aligned} \tag{6.190}$$

the ED equation delivers an evolution equation for the speed V, namely

$$V_t = \left\{ \frac{V}{\kappa} \left[\left(\frac{V_x}{V} \right)_x - \frac{1}{2} \left(\frac{V_x}{V} \right)^2 \right] - \frac{3}{4} V^{-1} + \frac{\kappa}{4} V^3 \right\}_x . \tag{6.191}$$

The insertion of $V = e^{\varphi}$ into the latter now produces the so-called *m^2KdV equation* [66, 137].

$$\varphi_t = \varphi_{xxx} - \frac{1}{2}\varphi_x^3 + \frac{3}{2}\varphi_x \cosh 2\varphi. \tag{6.192}$$

An auto-Bäcklund transformation for the latter was constructed by Calogero and Degasperis [66]. Here, we shall present auto-Bäcklund transformations both for the ED equation and the reciprocally related m^{2}KdV equation. Soliton surfaces swept out by the binormal motion of curves of constant curvature will then be constructed by means of the Sym-Tafel formula. It is noted that the ED equation sits in a wide class of nonlinear evolution equations which are invariant under reciprocal transformations and which were delimited in [196].

Under the reciprocal transformation

$$\begin{gathered} ds' = \tau ds + \left[\frac{1}{\kappa} \left(\frac{1}{\tau^{1/2}} \right)_{ss} - \tau^{3/2} + \kappa \left(\frac{1}{\tau^{1/2}} \right) \right] db, \quad db' = db \\ \tau' = \frac{1}{\tau}, \end{gathered} \tag{6.193}$$

(6.186) becomes

$$\tau'_{b'} = \left[\frac{1}{\kappa} \left(\frac{1}{\tau'^{1/2}} \right)_{s's'} - \kappa \tau'^{3/2} + \frac{1}{\tau'^{1/2}} \right]_{s'} \tag{6.194}$$

leading to invariance when, without loss of generality, we set $\kappa = 1$. In this case, the ED equation (6.186) adopts the form

$$\tau_b = X_s, \quad X = \left(\frac{1}{\tau^{1/2}} \right)_{ss} - \tau^{3/2} + \frac{1}{\tau^{1/2}}, \tag{6.195}$$

while the corresponding Serret-Frenet equations and the 'time' evolution of the triad $\{\boldsymbol{t}, \boldsymbol{n}, \boldsymbol{b}\}$ read

$$\begin{pmatrix} \boldsymbol{t} \\ \boldsymbol{n} \\ \boldsymbol{b} \end{pmatrix}_s = \begin{pmatrix} 0 & 1 & 0 \\ -1 & 0 & \tau \\ 0 & -\tau & 0 \end{pmatrix} \begin{pmatrix} \boldsymbol{t} \\ \boldsymbol{n} \\ \boldsymbol{b} \end{pmatrix},$$

$$\begin{pmatrix} \boldsymbol{t} \\ \boldsymbol{n} \\ \boldsymbol{b} \end{pmatrix}_b = \begin{pmatrix} 0 & -\tau^{1/2} & -\dfrac{1}{2}\dfrac{\tau_s}{\tau^{3/2}} \\ \tau^{1/2} & 0 & X - \dfrac{1}{\tau^{1/2}} \\ \dfrac{1}{2}\dfrac{\tau_s}{\tau^{3/2}} & -X + \dfrac{1}{\tau^{1/2}} & 0 \end{pmatrix} \begin{pmatrix} \boldsymbol{t} \\ \boldsymbol{n} \\ \boldsymbol{b} \end{pmatrix}. \tag{6.196}$$

In this formulation, it is readily verified that the scalar conservation law (6.195) possesses the companion vector-valued conservation law

$$(\tau \boldsymbol{b})_b = \left(X\boldsymbol{b} + \tau^{1/2}\boldsymbol{t}\right)_s, \tag{6.197}$$

whence a vector-valued function $\boldsymbol{r}'$ may be introduced such that

$$\boldsymbol{r}'_s = \tau \boldsymbol{b}, \quad \boldsymbol{r}'_b = X\boldsymbol{b} + \tau^{1/2}\boldsymbol{t}. \tag{6.198}$$

Under the reciprocal transformation (6.193) with $\kappa = 1$, that is

$$ds' = \tau ds + X db, \quad db' = db, \tag{6.199}$$

these relations become

$$\boldsymbol{r}'_{s'} = \boldsymbol{b}, \quad \boldsymbol{r}'_{b'} = \tau'^{-1/2}\boldsymbol{t}. \tag{6.200}$$

It is therefore natural to introduce the right-handed orthonormal triad

$$\{\boldsymbol{t}', \boldsymbol{n}', \boldsymbol{b}'\} = \{\boldsymbol{b}, -\boldsymbol{n}, \boldsymbol{t}\} \tag{6.201}$$

which satisfies the primed version of the Serret-Frenet equations $(6.196)_1$ with the new curvature and torsion given by

$$\kappa' = 1, \quad \tau' = \frac{1}{\tau}. \tag{6.202}$$

Thus, the unit vectors $\boldsymbol{t}'$, $\boldsymbol{n}'$ and $\boldsymbol{b}'$ indeed constitute the unit tangent, principal normal and binormal associated with the curves $\boldsymbol{r}'(s', t' = \text{const})$. Moreover, as the latter propagate in a purely binormal direction, the surface swept out as b' evolves has the same character as the original ED surface with generic position vector $\boldsymbol{r}$. This is reflected at the nonlinear level by the fact that the ED equation is form-invariant under the reciprocal transformation $(\tau, s, b) \to (\tau', s', b')$.

Remarkably, the relations (6.198) may be integrated explicitly. Indeed, it is readily shown that

$$\boldsymbol{r}' = \boldsymbol{r} + \boldsymbol{n} \tag{6.203}$$

without loss of generality. This implies that the ED surfaces Σ and Σ' are parallel since $\boldsymbol{N} = -\boldsymbol{n}$. Accordingly, we have the following result:

Theorem 24 (The dual ED surface). *Every ED surface Σ obtained by integration of the system (6.196) has an associated parallel dual ED surface Σ' with position vector $\boldsymbol{r}' = \boldsymbol{r} - \boldsymbol{N}$ generated by the reciprocal transformation (6.198), (6.199). The latter leaves invariant the ED equation determined by (6.195). At corresponding points under this reciprocal transformation, the tangent vectors to the coordinate lines on the ED surface and its dual are interchanged, that is $\boldsymbol{t}' = \boldsymbol{b}$, $\boldsymbol{b}' = \boldsymbol{t}$.*

It is interesting to note that the geodesics b = const on an ED surface and their associated geodesics on the dual ED surface constitute a family of pairs of *Bertrand curves* [380]. In fact, for b = const, the reciprocal relation (6.199) is classical. In general, Bertrand curves are characterised by

$$\alpha\kappa + \beta\tau = 1, \tag{6.204}$$

where α, β are constants. Their conjugate curves (dual Bertrand curves) represent examples of *offset curves* [278] which are used in computer-aided design (CAD) and computer-aided manufacture (CAM). It has been shown by Razzaboni [293] that surfaces on which there exist a one-parameter family of geodesic Bertrand curves admit a Bäcklund transformation and hence are solitonic [334]. ED surfaces and the extended sine-Gordon surfaces as discussed in the next section are particular cases of such surfaces.

The reciprocal invariance of the ED equation (6.195) implies that, if $\tau(s, b)$ is a seed solution then a solution of its primed counterpart is given by

$$\begin{gathered} \tau' = \frac{1}{\tau(s, b)}, \\ s' = s'(s, b), \quad b' = b, \end{gathered} \tag{6.205}$$

where s' is obtained by insertion of the seed solution into the reciprocal relation (6.199) and subsequent integration.

If we now introduce the reciprocal counterparts x', t' of the independent variables x, t in the m^2KdV equation via

$$
\begin{aligned}
dx' &= \sqrt{2}\, V'^{-1} ds' + \frac{1}{\sqrt{2}}\left[V' V'_{s's'} - \frac{1}{2} V'^2_{s'} - \frac{3}{2} V'^{-2} + \frac{V'^2}{2} \right] db' \\
dt' &= -\sqrt{2}\, db'
\end{aligned}
\tag{6.206}
$$

with $V' = \tau'^{-1/2}$, then it is seen that

$$dx' = dx, \quad dt' = dt. \tag{6.207}$$

Accordingly, the ED surface and its dual may be parametrised in terms of the same coordinates. The transition from the ED surface to its dual is induced by the simple discrete symmetry $\varphi \to \varphi' = -\varphi$ in the m^2KdV equation.

6.5.2 Extended Dym Surfaces. The su(2) Linear Representation

The 1st and 2nd fundamental forms associated with the ED surfaces are readily obtained from the linear system (6.196) and the relations

$$\boldsymbol{r}_b = \frac{1}{\tau^{1/2}} \boldsymbol{b}, \quad \boldsymbol{N} = \frac{\boldsymbol{r}_s \times \boldsymbol{r}_b}{|\boldsymbol{r}_s \times \boldsymbol{r}_b|} = -\boldsymbol{n}. \tag{6.208}$$

They read

$$
\begin{aligned}
\mathrm{I} &= \quad d\boldsymbol{r} \cdot d\boldsymbol{r} = ds^2 + \frac{1}{\tau} db^2 \\
\mathrm{II} &= -d\boldsymbol{r} \cdot d\boldsymbol{N} = -ds^2 + 2\tau^{1/2} ds\, db + \left(\frac{X}{\tau^{1/2}} - \frac{1}{\tau} \right) db^2.
\end{aligned}
\tag{6.209}
$$

According to Bonnet's theorem, a surface is uniquely determined up to its position in space by its fundamental forms. Thus, any position vector $\boldsymbol{r} = \boldsymbol{r}(s, b)$ which gives rise to the fundamental forms (6.209) defines an ED surface. In what follows, it is shown how such position vectors may be constructed explicitly in terms of eigenfunctions of the ED equation. It is noted that the relation $\boldsymbol{N} = -\boldsymbol{n}$ implies that the linear system (6.196) represents the $\{\boldsymbol{t}, \boldsymbol{b}, \boldsymbol{N}\}$ version of the Gauss-Weingarten equations for ED surfaces.

Now, consider the linear system

$$\Phi_s = \frac{1}{2}\begin{pmatrix} 0 & -\tau - \Lambda \\ \tau + \dfrac{1}{\Lambda} & 0 \end{pmatrix}\Phi,$$

$$\Phi_b = \frac{1}{2}\begin{pmatrix} \dfrac{1}{4}\left(\Lambda - \dfrac{1}{\Lambda}\right)\dfrac{\tau_s}{\tau^{3/2}} & -X + \dfrac{1}{2}\left(\Lambda - \dfrac{1}{\Lambda}\right)\tau^{1/2} + \dfrac{1}{2}(1 - \Lambda^2)\dfrac{1}{\tau^{1/2}} \\ X + \dfrac{1}{2}\left(\Lambda - \dfrac{1}{\Lambda}\right)\tau^{1/2} - \dfrac{1}{2}\left(1 - \dfrac{1}{\Lambda^2}\right)\dfrac{1}{\tau^{1/2}} & -\dfrac{1}{4}\left(\Lambda - \dfrac{1}{\Lambda}\right)\dfrac{\tau_s}{\tau^{3/2}} \end{pmatrix}\Phi, \tag{6.210}$$

where Λ is an arbitrary constant (complex) parameter. One may directly verify that the above Lax pair is compatible modulo the ED equation (6.195). At $\Lambda = i$, it takes the form

$$\begin{aligned} \Phi_s &= [e_1 + \tau e_2]\Phi \\ \Phi_b &= \left[-\tau^{1/2} e_1 + \left(X - \frac{1}{\tau^{1/2}}\right) e_2 - \frac{1}{2}\frac{\tau_s}{\tau^{3/2}} e_3\right]\Phi \end{aligned} \tag{6.211}$$

with the skew-symmetric matrices e_i defined by (cf. Section 2.2)

$$e_1 = \frac{1}{2i}\begin{pmatrix} 0 & 1 \\ 1 & 0 \end{pmatrix}, \quad e_2 = \frac{1}{2i}\begin{pmatrix} 0 & -i \\ i & 0 \end{pmatrix}, \quad e_3 = \frac{1}{2i}\begin{pmatrix} 1 & 0 \\ 0 & -1 \end{pmatrix}. \tag{6.212}$$

It is recalled that the e_i obey the $so(3)$ commutator relations

$$[e_1, e_2] = e_3, \quad [e_2, e_3] = e_1, \quad [e_3, e_1] = e_2. \tag{6.213}$$

On the other hand, the Gauss-Weingarten equations (6.196) may be formulated as

$$\begin{aligned} \Psi_s &= [D_1 + \tau D_2]\Psi \\ \Psi_b &= \left[-\tau^{1/2} D_1 + \left(X - \frac{1}{\tau^{1/2}}\right) D_2 - \frac{1}{2}\frac{\tau_s}{\tau^{3/2}} D_3\right]\Psi, \end{aligned} \tag{6.214}$$

where

$$\Psi = \begin{pmatrix} \boldsymbol{t} \\ \boldsymbol{n} \\ \boldsymbol{b} \end{pmatrix} \tag{6.215}$$

and the anti-symmetric matrices D_i are given by

$$D_1 = \begin{pmatrix} 0 & 1 & 0 \\ -1 & 0 & 0 \\ 0 & 0 & 0 \end{pmatrix}, \quad D_2 = \begin{pmatrix} 0 & 0 & 0 \\ 0 & 0 & 1 \\ 0 & -1 & 0 \end{pmatrix}, \quad D_3 = \begin{pmatrix} 0 & 0 & 1 \\ 0 & 0 & 0 \\ -1 & 0 & 0 \end{pmatrix}. \tag{6.216}$$

Since the matrices D_i are nothing but the generators of $so(3)$ which satisfy the commutator relations

$$[D_1, D_2] = D_3, \quad [D_2, D_3] = D_1, \quad [D_3, D_1] = D_2, \tag{6.217}$$

the correspondence between the linear representations (6.211) and (6.214) reflects the isomorphism between the Lie algebras $so(3)$ and $su(2)$. Accordingly, the linear representation (6.211) may be regarded as an $su(2)$ version of the Gauss-Weingarten equations (6.196) for ED surfaces.

If we now require that the parameter Λ be confined to the unit circle in the complex plane, it is natural to introduce the parametrisation

$$\Lambda = ie^{-i\lambda}, \quad \lambda \in \mathbb{R} \tag{6.218}$$

so that the linear representation (6.210) takes the form

$$\Phi_{,\mu} = g_\mu(\lambda)\Phi, \quad \mu = 1, 2 \tag{6.219}$$

with the notation $_{,1} = \partial/\partial s$ and $_{,2} = \partial/\partial b$. Since λ is real, the matrices g_μ are skew-symmetric and hence the eigenfunction Φ may be assumed to be an element of $SU(2)$, that is

$$\Phi^\dagger \Phi = \mathbb{1}. \tag{6.220}$$

Consequently, the matrix

$$\hat{r} = \Phi^{-1}\Phi_\lambda \tag{6.221}$$

provided by the Sym-Tafel formula (6.31) is trace-free and may be decomposed into the generators e_i according to

$$\hat{r} = \hat{\boldsymbol{r}} \cdot \boldsymbol{e}, \quad \boldsymbol{e} = \begin{pmatrix} e_1 \\ e_2 \\ e_3 \end{pmatrix}. \tag{6.222}$$

Thus, the vector-valued function $\hat{\boldsymbol{r}} = \hat{\boldsymbol{r}}(\lambda)$ parametrises a one-parameter family of surfaces in $\mathbb{R}^3$.

To establish the connection between ED surfaces and the surfaces represented by $\hat{\boldsymbol{r}}$, it is necessary to investigate the geometry of the latter. In this connection, the tangent vectors $\hat{\boldsymbol{r}}_{,\mu}$ may be obtained from the decomposition

$$\hat{\boldsymbol{r}}_{,\mu} \cdot \boldsymbol{e} = \hat{r}_{,\mu} = \Phi^{-1} g_{\mu\lambda} \Phi \tag{6.223}$$

which implies that

$$\hat{\boldsymbol{r}}_{,\mu} \cdot \hat{\boldsymbol{r}}_{,\nu} = -2\,\mathrm{Tr}(\hat{r}_{,\mu}\hat{r}_{,\nu}) = -2\,\mathrm{Tr}(g_{\mu\lambda} g_{\nu\lambda}) \tag{6.224}$$

by virtue of the relation

$$-2\,\mathrm{Tr}(e_i e_j) = \delta_{ij}. \tag{6.225}$$

Hence, the one-parameter family of 1st fundamental forms associated with the position vectors $\hat{\boldsymbol{r}}(\lambda)$ reads

$$\hat{\mathrm{I}}(\lambda) = -2\,\mathrm{Tr}(g_{\mu\lambda} g_{\nu\lambda})\, dx^{\mu} dx^{\nu}, \tag{6.226}$$

where $(x^1, x^2) = (s, b)$ and Einstein's summation convention has been adopted.

It is straightforward to derive the family of 2nd fundamental forms. They are (cf. Section 6.1)

$$\hat{\mathrm{II}}(\lambda) = -\frac{1}{\det^{1/2}[g_{1\lambda}, g_{2\lambda}]} \mathrm{Tr}([g_{1\lambda}, g_{2\lambda}](g_{\mu,\nu\lambda} + [g_{\mu\lambda}, g_{\nu}]))\, dx^{\mu} dx^{\nu}. \tag{6.227}$$

At $\lambda = 0$, the fundamental forms (6.209) are retrieved, that is

$$\hat{\mathrm{I}}(0) = \mathrm{I}, \quad \hat{\mathrm{II}}(0) = \mathrm{II}. \tag{6.228}$$

Thus, the position vector

$$\boldsymbol{r} = \hat{\boldsymbol{r}}(0) = -2\,\mathrm{Tr}(\Phi^{-1} \Phi_{\lambda}|_{\lambda=0} \boldsymbol{e}) \tag{6.229}$$

generates ED surfaces.

6.5.3 A CC-Ideal Formulation

The cc-ideal formulation [122, 158] is valuable, in particular, for isolating Miura and reciprocal-type links between 1+1-dimensional solitonic equations [169–171, 173, 323, 336, 339]. Importantly, it also simultaneously reduces the construction of a matrix Darboux transformation for the spatial and temporal parts of a Lax pair to consideration of invariance of a single one-form equation [173, 323, 336, 339]. The cc-ideal procedure is summarised in Appendix B.

Here, the cc-ideal for the ED and m^2KdV equations is found by rewriting the linear representation (6.210) as a single matrix one-form equation, namely

$$d\Phi = \mathcal{X}(\Lambda)\Phi = \sum_{i=1}^{5} X_i \xi^i \Phi, \tag{6.230}$$

where the matrices X_i are given by

$$X_1 = \begin{pmatrix} 0 & -1 \\ 1 & 0 \end{pmatrix}, \quad X_2 = \begin{pmatrix} 0 & -\Lambda \\ \Lambda^{-1} & 0 \end{pmatrix}, \quad X_3 = \begin{pmatrix} 0 & -\Lambda^{-1} \\ \Lambda & 0 \end{pmatrix}$$
$$X_4 = \begin{pmatrix} \Lambda - \Lambda^{-1} & 0 \\ 0 & -\Lambda + \Lambda^{-1} \end{pmatrix}, \quad X_5 = \begin{pmatrix} 0 & -\Lambda^2 \\ \Lambda^{-2} & 0 \end{pmatrix} \tag{6.231}$$

and the one-forms $\xi^i = \alpha^i ds + \beta^i db$ are obtained by expanding (6.210) in terms of the matrices X_i. The latter may be regarded as linearly independent generators of a subalgebra of the loop algebra $sl(2) \otimes R(\Lambda, \Lambda^{-1})$ [186]. Since not all commutators $[X_i, X_j]$ may be expressed in terms of the generators X_k, we have the following incomplete commutator table:

$$\begin{aligned} &[X_1, X_2] = X_4, && [X_2, X_4] = 2X_1 - 2X_5 \\ &[X_1, X_3] = -X_4, && [X_2, X_5] = X_4 \\ &[X_1, X_4] = -2X_2 + 2X_3, && [X_3, X_4] = ? \\ &[X_1, X_5] = -[X_2, X_3], && [X_3, X_5] = ? \\ & && [X_4, X_5] = ?. \end{aligned} \tag{6.232}$$

Therefore, evaluation of the integrability condition

$$0 = d^2\Phi = (d\mathcal{X} - \mathcal{X} \wedge \mathcal{X})\Phi \tag{6.233}$$

produces the cc-ideal

$$\begin{aligned} &d\xi^1 = 2\xi^2\xi^4, && \xi^1\xi^5 = \xi^2\xi^3 \\ &d\xi^2 = -2\xi^1\xi^4, && \xi^3\xi^4 = 0 \\ &d\xi^3 = 2\xi^1\xi^4, && \xi^3\xi^5 = 0 \\ &d\xi^4 = \xi^1\xi^2 - \xi^1\xi^3 + \xi^2\xi^5, && \xi^4\xi^5 = 0 \\ &d\xi^5 = -2\xi^2\xi^4. && \end{aligned} \tag{6.234}$$

In the above, the wedge between differential forms has been suppressed.

One may investigate different parametrisations of the integral manifolds of the above cc-ideal. Thus, the algebraic part of (6.234) is satisfied identically by setting

$$\xi^2 = v\xi^1 + p\xi^3, \quad \xi^4 = u\xi^3, \quad \xi^5 = v\xi^3 \tag{6.235}$$

with functions u, v and p as yet unspecified. Consequently, the number of independent one-forms, namely the genus of the cc-ideal, is $g = 2$. However, if one inserts the parametrisation (6.235) into the differential part of the cc-ideal, then the following four exact one-forms result:

$$\begin{aligned} d\left(v^{1/2}\xi^3\right) = 0 &\Rightarrow dB = v^{1/2}\xi^3 \\ d\left(\xi^2 + \xi^3\right) = 0 &\Rightarrow dS = \xi^2 + \xi^3 \\ d\left(\xi^1 + \xi^5\right) = 0 &\Rightarrow dS' = \xi^1 + \xi^5 \\ d\left(v^{1/2}\xi^1 + \frac{1}{v^{1/2}}\xi^2 - \frac{u}{v^{1/2}}\xi^4\right) = 0 &\Rightarrow dy = v^{1/2}\xi^1 + \frac{1}{v^{1/2}}\xi^2 - \frac{u}{v^{1/2}}\xi^4. \end{aligned} \tag{6.236}$$

Any pair of functions S, B, S' and y may be used as independent variables while the remaining pair constitute potentials. For instance, if we single out the functions S and B as independent variables, then the one-forms read

$$\begin{aligned} \xi^1 &= \frac{dS}{v} + q\,dB, \quad \xi^2 = dS - \frac{dB}{v^{1/2}} \\ \xi^3 &= \frac{dB}{v^{1/2}}, \quad \xi^4 = w\,dB, \quad \xi^5 = v^{1/2}dB \end{aligned} \tag{6.237}$$

while the potentials S' and y are defined by

$$\begin{aligned} dS' &= \frac{dS}{v} + \left(q + v^{1/2}\right)dB \\ dy &= 2\frac{dS}{v^{1/2}} + \left(v^{1/2}q - \frac{1}{v} - w^2\right)dB, \end{aligned} \tag{6.238}$$

where, for convenience, we have set $w = u/v^{1/2}$ and $q = -(p+1)/v^{3/2}$.

Now, the differential part of the cc-ideal delivers the relations

$$\begin{aligned} -\left(\frac{1}{v}\right)_B + q_S &= 2w \\ \left(\frac{1}{v^{1/2}}\right)_S = 2\frac{w}{v}, \quad w_S &= -q - \frac{2}{v^{3/2}} + v^{1/2}. \end{aligned} \tag{6.239}$$

The latter two relations serve to define w and q in terms of v, whence

$$w = -\frac{1}{2}\left(v^{1/2}\right)_S, \quad q = \frac{1}{2}\left(v^{1/2}\right)_{SS} + v^{1/2} - \frac{2}{v^{3/2}}, \tag{6.240}$$

so that $(6.239)_1$ reduces to the third-order equation

$$\left(\frac{1}{v}\right)_B = \frac{1}{2}\left(v^{1/2}\right)_{SSS} + 2\left(v^{1/2}\right)_S - 2\left(\frac{1}{v^{3/2}}\right)S. \tag{6.241}$$

The substitutions

$$v = \frac{1}{\tau}, \quad S = \frac{s}{2}, \quad B = \frac{b}{4} \tag{6.242}$$

now show that (6.241) is nothing but the ED equation (6.195).

Alternatively, one could choose S' and B say, as the independent variables and derive the corresponding nonlinear equation in a similar manner. However, inspection of $(6.238)_1$ shows that S' is related to the reciprocal variable s' introduced in (6.199) by

$$S' = s'/2 \tag{6.243}$$

and hence the ED equation reciprocally related to (6.241) is obtained. Moreover, the potential y may be identified with

$$y = x/\sqrt{2} \tag{6.244}$$

so that the pair of independent variables (y, B) gives rise to the m^2KdV equation (6.192). Accordingly, the cc-ideal (6.234) is seen to provide a convenient encapsulation of both the ED equation and the reciprocally related m^2KdV equation. In what follows, the cc-ideal formulation will be shown to be also well-adapted to the construction of both matrix Darboux transformations and associated Bäcklund transformations.

6.5.4 A Matrix Darboux Transformation. A Bäcklund Transformation for the Extended Dym and m²KdV Equations

To construct a transformation which leaves form-invariant the cc-ideal (6.234), it is necessary to give an algebraic characterisation of the linear representation (6.230). In fact, therein, it is readily verified that the generators X_i, $i = 1, \ldots, 5$ lie in the subalgebra

$$\mathfrak{g} = \operatorname{span}\{X^n, Y^n; n \in \mathbb{Z}\},$$

$$X^n = \begin{pmatrix} \Lambda^n - \Lambda^{-n} & 0 \\ 0 & -\Lambda^n + \Lambda^{-n} \end{pmatrix}, \quad Y^n = \begin{pmatrix} 0 & -\Lambda^{-n} \\ \Lambda^n & 0 \end{pmatrix} \tag{6.245}$$

of the loop algebra $sl(2) \otimes R(\Lambda, \Lambda^{-1})$. It is noted that $\mathfrak{g}$ is equivalently characterised by

$$\mathcal{X}(\Lambda) \in \mathfrak{g} \quad \Leftrightarrow \quad \mathcal{X}(\Lambda^{-1}) = M\mathcal{X}(\Lambda)M^{-1}, \quad M = \begin{pmatrix} 0 & -1 \\ 1 & 0 \end{pmatrix} \tag{6.246}$$

if we assume that $\mathcal{X}(\Lambda)$ is trace-free. Hence, the following preliminary result obtains:

Lemma 1 (Characterisation of the ED-m²KdV cc-ideal). *Let $\mathcal{X}(\Lambda)$ be a one-parameter family of trace-free 2×2 matrix-valued one-forms. Then, the linear equation*

$$d\Phi = \mathcal{X}(\Lambda)\Phi \tag{6.247}$$

coincides with the linear representation (6.230) of the ED-m²KdV cc-ideal if and only if (6.247) is compatible and

- *(i)* $\mathcal{X}(\Lambda)$ *is a quadratic polynomial in Λ and Λ^{-1}*
- *(ii)* $\mathcal{X}(\Lambda) \in \mathfrak{g}$
- *(iii)* $\mathcal{X}(\Lambda)$ *does* not *contain the generators X^2 and Y^2.*

The description given in this lemma is typical for linear representations of cc-ideals. In fact, it allows the construction of a matrix Darboux transformation in a purely algebraic manner. The result given below [339] constitutes an extension to cc-ideals of the matrix Darboux formalism to be discussed extensively in geometric terms in the next chapter.

Theorem 25 (A matrix Darboux transformation for the ED-m²KdV cc-ideal). *If $\phi = (\phi^1 \ \phi^2)^{\mathsf{T}}$ is a vector-valued solution of the linear representation (6.230) of the ED-m²KdV cc-ideal (6.234) with parameter $\Lambda_0 = -\mu$, then (6.230) is form-invariant under the matrix Darboux transformation*

$$\begin{aligned} \Phi \to \tilde{\Phi} &= P(\Lambda)\Phi \\ \mathcal{X}(\Lambda) \to \tilde{\mathcal{X}}(\Lambda) &= P(\Lambda)\mathcal{X}(\Lambda)P^{-1}(\Lambda) + dP(\Lambda)P^{-1}(\Lambda), \end{aligned} \tag{6.248}$$

where the Darboux matrix $P(\Lambda)$ is given by

$$P(\Lambda) = \begin{pmatrix} A & B \\ C & D \end{pmatrix} + \Lambda \begin{pmatrix} D & -C \\ -B & A \end{pmatrix} \tag{6.249}$$

with

$$\frac{D}{A} = \frac{\mu^2 + \zeta^2}{\mu(1+\zeta^2)}, \quad \zeta = \frac{\phi^1}{\phi^2} \tag{6.250}$$
$$AD = 1, \quad B = 0, \quad C = (D - \mu^{-1}A)\zeta.$$

Proof. Here, it is shown that the particular form (6.250) of the functions A, B, C, D is a consequence of algebraic constraints which, in turn, guarantee that (6.248) preserves the form of the linear representation (6.230). Thus, the symmetry embodied in (6.246) implies that if $\phi_1 = \phi$ is a vector-valued eigenfunction with parameter $\Lambda_1 = \Lambda_0$, then $\phi_2 = M\phi$ is another eigenfunction corresponding to the parameter $\Lambda_2 = \Lambda_0^{-1}$. Furthermore, the identity

$$\Lambda P(\Lambda^{-1}) = M P(\Lambda) M^{-1} \tag{6.251}$$

shows that if $P(\Lambda_1)\phi_1 = 0$ then $P(\Lambda_2)\phi_2 = 0$. Hence, imposition of the linear constraints

$$P(\Lambda_i)\phi_i = 0 \tag{6.252}$$

reduces the number of unknown functions in $P(\Lambda)$ by two. The conditions (6.252) are standard for matrix Darboux transformations and guarantee that the polynomial structure and degree of $\mathcal{X}(\Lambda)$ are preserved provided that $\det P(\Lambda) = \text{const}$ [173, 269, 270, 323, 336]. The latter condition is the third restriction on the Darboux matrix $P(\Lambda)$. Consequently, the new one-form $\tilde{\mathcal{X}}(\Lambda)$ obeys condition (i) in Lemma 1 while condition (ii) is satisfied by virtue of the symmetry (6.251). Now, evaluation of cubic terms in the transformation law

$$\tilde{\mathcal{X}}(\Lambda)P(\Lambda) = P(\Lambda)\mathcal{X}(\Lambda) + dP(\Lambda) \tag{6.253}$$

reveals that the fourth condition $B = 0$ ensures that the generators X^2 and Y^2 do not appear in $\tilde{\mathcal{X}}(\Lambda)$, that is, condition (iii) is satisfied. The determinant condition then reduces to $AD = \text{const}$. This completes the proof. □

Corollary 5 (A Bäcklund transformation for the ED and m²KdV equations). *The ED equation (6.195), its dual and the m^2KdV equation (6.192) are form-invariant under* $(\tau, s, b, s', x) \to (\tilde{\tau}, \tilde{s}, \tilde{b}, \tilde{s}', \tilde{x})$, *where*

$$\tilde{\tau} = \frac{A^2}{D^2}\tau, \quad \begin{array}{ll} \tilde{s} = s' + 2\arctan\zeta, & \tilde{b} = b \\ \tilde{s}' = s - 2\arctan(\zeta/\mu), & \tilde{x} = x. \end{array} \tag{6.254}$$

The independent variables of the m^2KdV equation are preserved.

Proof. Evaluation of the transformation law (6.253) leads to expressions for the new one-forms $\tilde{\xi}^i$ which are linear combinations of ξ^k with ζ-dependent coefficients. For instance, terms cubic in Λ and Λ^{-1} yield

$$\tilde{\xi}^3 = \frac{A}{D}\xi^3, \quad \tilde{\xi}^5 = \frac{D}{A}\xi^5 \tag{6.255}$$

so that insertion of the parametrisation (6.237) yields

$$\tilde{\upsilon}^{1/2} = \frac{D}{A}\upsilon^{1/2}, \quad d\tilde{B} = dB. \tag{6.256}$$

The expression for $\tilde{\xi}^4$ may then be used to determine $\tilde{u}$. It turns out that

$$\begin{aligned} d\tilde{y} &= \tilde{\upsilon}^{1/2}\tilde{\xi}^1 + \frac{1}{\tilde{\upsilon}^{1/2}}\tilde{\xi}^2 - \frac{\tilde{u}}{\tilde{\upsilon}^{1/2}}\tilde{\xi}^4 \\ &= \upsilon^{1/2}\xi^1 + \frac{1}{\upsilon^{1/2}}\xi^2 - \frac{u}{\upsilon^{1/2}}\xi^4 = dy \end{aligned} \tag{6.257}$$

and

$$\begin{aligned} \tilde{\xi}^2 + \tilde{\xi}^3 &= \xi^1 + \xi^5 + d[\arctan\zeta] \\ \tilde{\xi}^1 + \tilde{\xi}^5 &= \xi^2 + \xi^3 - d[\arctan(\zeta/\mu)]. \end{aligned} \tag{6.258}$$

To establish the latter two relations, use has been made of the expression for the differential $d\zeta = d(\phi^1/\phi^2)$ provided by the linear representation (6.230). □

6.5.5 *Soliton Surfaces*

It has been shown in Section 6.5 that if $\Phi \in SU(2)$ is a vector-valued eigenfunction of the ED equation, then

$$\boldsymbol{r} = -2\,\mathrm{Tr}(r\boldsymbol{e}), \quad r = \Phi^{-1}\Phi_\Lambda|_{\Lambda=i} \tag{6.259}$$

parametrises an ED surface since $\Lambda_\lambda|_{\lambda=0} = 1$. However, for computational purposes, it is sufficient to assume that $\Phi^\dagger\Phi = f(\Lambda)\mathbb{1}$ since the function $f(\Lambda)$ merely generates trace terms in r which do not contribute to the position vector $\boldsymbol{r}$. Accordingly, the new position vector $\tilde{\boldsymbol{r}}$ induced by the matrix Darboux transformation (6.248) is given by

$$\tilde{\boldsymbol{r}} = -2\,\mathrm{Tr}(\tilde{r}\boldsymbol{e}), \quad \tilde{r} = r + \Phi^{-1}P^{-1}P_\Lambda\Phi. \tag{6.260}$$

Here and in the following, it is to be understood that functions depending on Λ are evaluated at $\Lambda = i$. The Darboux matrix P as given by (6.249), (6.250) now yields

$$P^{-1}P_{\Lambda} = i\frac{\mu^2 - 1}{(\mu^2+1)(\zeta^2+1)}\begin{pmatrix} \frac{1}{2}(\zeta^2 - 1) & \zeta \\ \zeta & -\frac{1}{2}(\zeta^2-1) \end{pmatrix} \tag{6.261}$$

which implies the *constant length property* $|\Delta \boldsymbol{r}| = |\tilde{\boldsymbol{r}} - \boldsymbol{r}| = \text{const}$ (cf. Section 7.2).

The simplest ED surface is generated by the binormal evolution of a circular helix. Indeed, if we choose the seed solution $\tau = 1$, then the linear representation (6.210) reduces to

$$\begin{aligned} \Phi_x &= \frac{1}{2\sqrt{2}}\begin{pmatrix} 0 & -\Lambda - 1 \\ \Lambda^{-1} + 1 & 0 \end{pmatrix}\Phi, \\ \Phi_t &= \frac{1}{4\sqrt{2}}\begin{pmatrix} 0 & \Lambda^{-1} + \Lambda^2 \\ -\Lambda - \Lambda^{-2} & 0 \end{pmatrix}\Phi \end{aligned} \tag{6.262}$$

in the m^{2}KdV coordinates

$$x = \sqrt{2}s - b/\sqrt{2}, \quad t = -\sqrt{2}b. \tag{6.263}$$

Its fundamental solution is given by

$$\Phi = \begin{pmatrix} (\Lambda+1)e^{\alpha} & (\Lambda+1)e^{-\alpha} \\ -2\sqrt{2}ke^{\alpha} & 2\sqrt{2}ke^{-\alpha} \end{pmatrix}, \tag{6.264}$$

where

$$\alpha = kx + \tilde{k}t, \quad \begin{aligned} k &= \frac{i}{2\sqrt{2}}(\Lambda^{1/2} + \Lambda^{-1/2}) \\ \tilde{k} &= -\frac{i}{4\sqrt{2}}(\Lambda^{3/2} + \Lambda^{-3/2}), \end{aligned} \tag{6.265}$$

so that the generic position vector (6.259) for the cylindrical seed surface Σ becomes

$$\boldsymbol{r} = \begin{pmatrix} \frac{1}{2}\cos\tilde{\alpha} \\ \frac{1}{2}\sin\tilde{\alpha} \\ t - \frac{1}{2}\tilde{\alpha} \end{pmatrix}, \quad \tilde{\alpha} = x + \frac{1}{2}t. \tag{6.266}$$

For real parameter $\Lambda_0 = -\mu$, the general vector-valued solution $\phi = (\phi^1\ \phi^2)$ of (6.262) leads to

$$\zeta = \frac{\phi^1}{\phi^2} = \frac{\mu - 1}{2\sqrt{2}\,c}\tanh\beta, \tag{6.267}$$

where

$$\beta = cx + \tilde{c}t, \qquad \begin{aligned} c &= \frac{1}{2\sqrt{2}}\left(\mu^{1/2} - \mu^{-1/2}\right) \\ \tilde{c} &= \frac{1}{4\sqrt{2}}\left(\mu^{3/2} - \mu^{-3/2}\right). \end{aligned} \tag{6.268}$$

Insertion into (6.260), (6.261) produces the new ED surface $\tilde{\Sigma}$ represented by

$$\tilde{\boldsymbol{r}} = \boldsymbol{r} + \hat{c}\begin{pmatrix} \dfrac{-[(\mu-1)\cosh 2\beta - \mu - 1]\cos\tilde{\alpha} - \sqrt{2}\sqrt{\mu}\,\sinh 2\beta\,\sin\tilde{\alpha}}{(\mu+1)\cosh 2\beta - \mu + 1} \\ \dfrac{-[(\mu-1)\cosh 2\beta - \mu - 1]\sin\tilde{\alpha} + \sqrt{2}\sqrt{\mu}\,\sinh 2\beta\,\cos\tilde{\alpha}}{(\mu+1)\cosh 2\beta - \mu + 1} \\ \dfrac{\sqrt{2}\sqrt{\mu}\,\sinh 2\beta}{(\mu+1)\cosh 2\beta - \mu + 1} \end{pmatrix}$$

$$\hat{c} = \frac{\mu^2 - 1}{\mu^2 + 1}. \tag{6.269}$$

The curve of constant curvature which generates the surface $\tilde{\Sigma}$ is given by $b = \text{const}$ or, equivalently, $t = \text{const}$.

For $\mu > 0$, the evolution of the generator may be interpreted as the motion at constant speed of a soliton on a circular helix. This is reflected by the soliton solutions

$$\begin{gathered} \tilde{\tau} = e^{-2\tilde{\varphi}}, \quad e^{\tilde{\varphi}} = \frac{\mu + \tanh^2\beta}{\mu\tanh^2\beta + 1} \\ \tilde{s} = s + 2\arctan\zeta, \quad \tilde{b} = b \end{gathered} \tag{6.270}$$

of the ED and m^2KdV equations. It is noted that the former solution is given parametrically in terms of s and b by virtue of (6.263). A typical generator is displayed in Figure 6.4 along with the surfaces Σ and $\tilde{\Sigma}$. Interestingly, the expression

$$(\tilde{x}^1)^2 + (\tilde{x}^2)^2 = \frac{1}{4} + 8\mu^2\frac{\mu^2 - 1}{(\mu^2+1)^2}\frac{\cosh 2\beta}{[(\mu+1)\cosh 2\beta - \mu + 1]^2}, \tag{6.271}$$

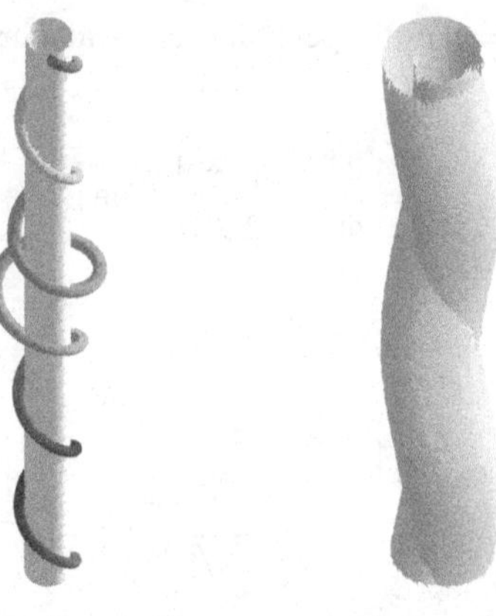

Figure 6.4. An ED surface and its generator of constant curvature ($\mu = 2$).

where $\tilde{\boldsymbol{r}} = (\tilde{x}^1, \tilde{x}^2, \tilde{x}^3)^{\mathsf{T}}$, shows that for $\mu > 1$ the surface $\tilde{\Sigma}$ is confined to the exterior of the seed cylinder Σ while for $\mu < 1$ it lies inside Σ.

On the other hand, for $\mu < 0$, the solutions (6.270) become periodic in both independent variables. In fact, inspection of the position vector (6.269) reveals that if

$$\frac{t - \frac{1}{2}\tilde{\alpha}}{\beta} = \text{const}, \tag{6.272}$$

then $\tilde{\Sigma}$ constitutes a surface of revolution. It turns out that the above condition may indeed be satisfied. One obtains two solutions, namely

$$\mu = -2 \pm \sqrt{3}. \tag{6.273}$$

An ED surface of revolution and its generator is depicted in Figure 6.5. In contrast to the case $\mu > 0$, it may be shown that the generator does intersect the seed cylinder Σ.

Exercise

1. Use Theorem 21 to establish invariance of the extended Dym equation (6.186) under the reciprocal transformation (6.193).

6.6 The Binormal Motion of Curves of Constant Torsion. The Extended Sine-Gordon System

It is recalled that the classical Bäcklund transformation for the construction of pseudospherical surfaces is characterised by remarkably simple geometric

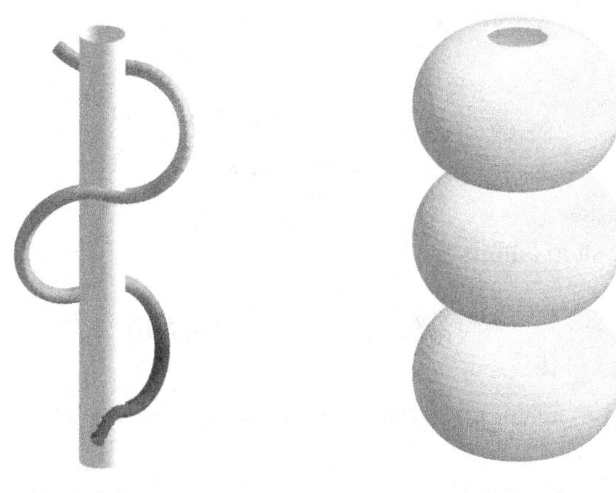

Figure 6.5. An ED surface of revolution and its generator ($\mu = -2 - \sqrt{3}$).

properties. Thus, if both a pseudospherical surface $\Sigma : \boldsymbol{r} = \boldsymbol{r}(u, v)$ and its Bäcklund transform $\Sigma' : \boldsymbol{r}' = \boldsymbol{r}'(u, v)$ are parametrised in terms of asymptotic coordinates u, v, then the vector $\boldsymbol{r}' - \boldsymbol{r}$ joining corresponding points P and P' is tangential to both Σ and Σ'. Moreover, the distance $|\boldsymbol{r}' - \boldsymbol{r}|$ is constant as is the angle between the corresponding unit normals $\boldsymbol{N}$ and $\boldsymbol{N}'$. Here, it is shown that the Bäcklund transformation for surfaces swept out by curves of constant torsion when they are subject to a purely binormal motion is of equal simplicity.[4]

6.6.1 The Extended Sine-Gordon System

The Gauss-Mainardi-Codazzi equations (6.183) for the surfaces swept out by the purely binormal motion of curves of constant torsion τ yield

$$\begin{gathered}\left(2g^{1/2}\tau\right)_s + \kappa_b = 0 \\ \left(\frac{\left(g^{1/2}\right)_{ss} - \tau^2 g^{1/2}}{\kappa}\right)_s + \kappa\left(g^{1/2}\right)_s = 0.\end{gathered} \tag{6.274}$$

If we now set

$$q = -2g^{1/2}\tau, \quad \rho = \frac{q_{ss} - \tau^2 q}{\kappa} \tag{6.275}$$

[4] Pseudospherical surfaces have already been seen to be generated by inextensible curves of constant torsion in motion without a binormal component (cf. Chapter 2).

then the system

$$\boxed{\begin{aligned} q_s = \kappa_b, \quad \rho_s + \kappa q_s = 0, \\ q_{ss} = \tau^2 q + \rho\kappa \end{aligned}} \tag{6.276}$$

results. This system admits the integral

$$q_s^2 + \rho^2 - \tau^2 q^2 = \tilde{c}, \quad \tilde{c} = \tilde{c}(b) \tag{6.277}$$

which, if $\tilde{c} = 1$ is taken, gives rise to the parametrisation

$$q_s = \sin\sigma \cosh\xi, \quad \rho = \cos\sigma \cosh\xi, \quad \tau q = \sinh\xi \tag{6.278}$$

so that the system (6.276) becomes

$$\begin{aligned} \xi_s &= \tau \sin\sigma \\ \sigma_{sb} - \tau(\cos\sigma \tanh\xi)_b &= \sin\sigma \cosh\xi. \end{aligned} \tag{6.279}$$

The system (6.279) or equivalently (6.276) we term the *extended sine-Gordon system* [339]. Indeed, it reduces to the classical sine-Gordon equation in the formal limit $\tau = \xi = 0$. The system is associated with the purely binormal motion of a curve travelling with velocity

$$\boldsymbol{r}_b = -\frac{q}{2\tau}\boldsymbol{b} \tag{6.280}$$

and curvature

$$\kappa = \sigma_s - \tau \cos\sigma \tanh\xi. \tag{6.281}$$

6.6.2 Fundamental Forms. An su(2) Linear Representation

The surfaces swept out by the bilinear motion (6.280) and associated with the extended sine-Gordon system with $\tau = 1$ have fundamental forms

$$\begin{aligned} \mathrm{I} &= ds^2 + \frac{1}{4}q^2 db^2, \\ \mathrm{II} &= -\kappa ds^2 - q\, ds\, db + \frac{1}{4}q\rho\, db^2 \end{aligned} \tag{6.282}$$

while the Gauss-Weingarten equations yield

$$\begin{pmatrix} \boldsymbol{t} \\ \boldsymbol{n} \\ \boldsymbol{b} \end{pmatrix}_s = \begin{pmatrix} 0 & \kappa & 0 \\ -\kappa & 0 & 1 \\ 0 & -1 & 0 \end{pmatrix} \begin{pmatrix} \boldsymbol{t} \\ \boldsymbol{n} \\ \boldsymbol{b} \end{pmatrix},$$
$$\begin{pmatrix} \boldsymbol{t} \\ \boldsymbol{n} \\ \boldsymbol{b} \end{pmatrix}_b = \frac{1}{2} \begin{pmatrix} 0 & q & -q_s \\ -q & 0 & -\rho \\ q_s & \rho & 0 \end{pmatrix} \begin{pmatrix} \boldsymbol{t} \\ \boldsymbol{n} \\ \boldsymbol{b} \end{pmatrix}. \tag{6.283}$$

A linear representation for the extended sine-Gordon system which contains an arbitrary parameter may be obtained via the *prolongation procedure* due to Estabrook and Wahlquist [123, 378]. The latter is based on Cartan's calculus of differential forms [68]. It provides a semi-algorithmic way of finding Lax pairs for 1+1-dimensional integrable systems. Here, we merely state that, in the present context, it leads to the linear representation [339]

$$\Phi_s = \frac{1}{2}\left[\begin{pmatrix} 0 & \kappa \\ -\kappa & 0 \end{pmatrix} + i\lambda \begin{pmatrix} 1 & 0 \\ 0 & -1 \end{pmatrix}\right]\Phi,$$
$$\Phi_b = \frac{1}{2(1+\lambda^2)}\left[i\lambda \begin{pmatrix} -\rho & q_s \\ q_s & \rho \end{pmatrix} + \begin{pmatrix} 0 & q \\ -q & 0 \end{pmatrix}\right]\Phi, \tag{6.284}$$

compatible modulo the extended sine-Gordon system. In fact, at $\lambda = 1$, the $su(2)$ Lax pair (6.284) is seen to be equivalent to the Gauss-Weingarten equations (6.283) if we identify

$$D_1 \simeq -e_2, \quad D_2 \simeq -e_3, \quad D_3 \simeq e_1. \tag{6.285}$$

The 'spatial' part of the Lax pair (6.284) constitutes the scattering problem for an important subclass of the AKNS integrable system studied earlier, namely the mKdV hierarchy with prototypical member

$$\theta_t = \theta_{sss} + \frac{1}{2}\theta_s^3. \tag{6.286}$$

This connection is not coincidental. Thus, the angle θ is a potential associated with the conservation law $(6.276)_1$, that is

$$\kappa = \theta_s, \quad q = \theta_b \tag{6.287}$$

and it turns out that the mKdV equation (6.286) is a symmetry of the extended sine-Gordon system written as

$$\left(\frac{\theta_{bss} - \theta_b}{\theta_s}\right)_s + \theta_s \theta_{bs} = 0. \tag{6.288}$$

6.6.3 A Bäcklund Transformation

The fundamental forms (6.282) are retrievable from the linear representation (6.284) using the Sym-Tafel formula. It emerges that the action of an associated matrix Darboux transformation at the surface level is of remarkable simplicity. Indeed, the following theorem may be verified directly.

Theorem 26 (An analogue of Bäcklund's classical transformation). *The Gauss-Weingarten equations (6.283) and the extended sine-Gordon equation (6.288) are form-invariant under*

$$\begin{aligned} \boldsymbol{r} \to \tilde{\boldsymbol{r}} &= \boldsymbol{r} + \frac{2\mu}{1+\mu^2}(\cos\varphi\, \boldsymbol{t} - \sin\varphi\, \boldsymbol{n}) \\ \theta \to \tilde{\theta} &= 2\varphi - \theta, \end{aligned} \tag{6.289}$$

where the function φ is a solution of the compatible system

$$\begin{aligned} \varphi_s &= \kappa + \mu \sin\varphi \\ \varphi_b &= \frac{1}{1-\mu^2}(\mu q_s \cos\varphi - \mu p \sin\varphi + q) \end{aligned} \tag{6.290}$$

and μ is an arbitrary constant (Bäcklund) parameter. The vector $\tilde{\boldsymbol{r}} - \boldsymbol{r}$ joining corresponding points on the surfaces Σ and $\tilde{\Sigma}$ lies in both the $(\boldsymbol{t}, \boldsymbol{n})$-plane and the $(\tilde{\boldsymbol{t}}, \tilde{\boldsymbol{n}})$-plane and the distance $|\tilde{\boldsymbol{r}} - \boldsymbol{r}|$ is constant. The angle between the binormals $\boldsymbol{b}$ and $\tilde{\boldsymbol{b}}$ is constant, that is

$$\tilde{\boldsymbol{b}} \cdot \boldsymbol{b} = \frac{1-\mu^2}{1+\mu^2}. \tag{6.291}$$

The compatible system (6.290) may be linearised by setting

$$\varphi = 2\arctan\left(\frac{\phi^1}{\phi^2}\right), \tag{6.292}$$

where $\phi = (\phi^1\ \phi^2)^{\mathrm{T}}$ is a vector-valued solution of (6.284) with parameter $\lambda = -i\mu$. Symmetry dictates that $\tilde{\boldsymbol{r}} - \boldsymbol{r}$ is orthogonal to both $\boldsymbol{b}$ and $\tilde{\boldsymbol{b}}$, and a short calculation produces the relation (6.291). It is also noted that the transformation law $(6.289)_2$ may be used to eliminate the pseudopotential φ from the nonlinear Lax pair (6.290). In this way, one obtains the analogue of the classical

Bäcklund relations. In particular, the spatial part of the Bäcklund transformation becomes

$$\left(\frac{\tilde{\theta}-\theta}{2}\right)_s = \mu \sin\left(\frac{\tilde{\theta}+\theta}{2}\right) \tag{6.293}$$

and this is generic to the entire mKdV hierarchy.

In conclusion, it is observed that application of the above Bäcklund transformation to a straight line (degenerate surface) produces the simplest Hasimoto surface (cf. Chapter 4). If one chooses a cylinder generated by a circular helix as the seed surface Σ, then the new surface $\tilde{\Sigma}$ is, as in the case of the ED equation, swept out by a soliton which propagates on a circular helix.

6.6.4 An Analogue of the Bianchi Transformation. Dual Surfaces

In Chapter 1, it has been pointed out that Bäcklund's classical transformation for pseudospherical surfaces may be regarded as a generalisation of a transformation due to Bianchi who assumed that the normals $\boldsymbol{N}$ and $\boldsymbol{N}'$ to the pseudospherical surface Σ and its transform Σ' at corresponding points are perpendicular. In the case of the surfaces governed by the extended sine-Gordon system, the natural analogue of Bianchi's assumption is orthogonality of the binormals $\boldsymbol{b}$ and $\boldsymbol{b}'$ associated with two surfaces Σ and Σ', that is

$$\boldsymbol{b}' \cdot \boldsymbol{b} = 0. \tag{6.294}$$

The relation (6.291) shows that the transformation given in Theorem 26 would have to be specialised to $\mu = \pm 1$ to meet this assumption. Even though the transformation formulae (6.289) and $(6.290)_1$ formally allow the specialisation $\mu = \pm 1$, the b-evolution $(6.290)_2$ of φ is *not* defined. However, the latter may be replaced in such a way that Theorem 26 retains its validity for $\mu = \pm 1$. This observation proves to be important in the investigation of integrable discrete models of the extended sine-Gordon surfaces [328]. It leads to a system

$$\begin{aligned} &\left(\frac{\theta'-\theta}{2}\right)_s = \epsilon \sin\left(\frac{\theta'+\theta}{2}\right), \qquad \epsilon = \pm 1 \\ &\theta_{bs} = c \sin\left(\frac{\theta'+\theta}{2}\right) - \epsilon \theta_b \cos\left(\frac{\theta'+\theta}{2}\right) \end{aligned} \tag{6.295}$$

which is shown below to be equivalent to the extended sine-Gordon equation (6.288). It is noted that $(6.295)_1$ coincides with the Bäcklund relation (6.293) for $\mu = \epsilon$.

Let θ and θ' be solutions of the system (6.295). Then, differentiation of $(6.295)_2$ with respect to s and substitution for θ'_s from $(6.295)_1$ yields

$$\left(\frac{\theta_{bss} - \theta_b}{\theta_s}\right)^2 + \theta_{bs}^2 - \theta_b^2 = c^2, \tag{6.296}$$

namely the first integral (6.277) written in terms of θ. Moreover, differentiation of $(6.295)_1$ reveals that

$$\theta'_{bs} = c \sin\left(\frac{\theta' + \theta}{2}\right) + \epsilon \theta'_b \cos\left(\frac{\theta' + \theta}{2}\right). \tag{6.297}$$

This implies that the system (6.295) is invariant under the discrete transformation $(\theta, \theta', \epsilon) \to (\theta', \theta, -\epsilon)$. Consequently, both θ and θ' are solutions of the extended sine-Gordon equation (6.288). It should be noted that the equations $(6.295)_2$ and (6.297) are equivalent to the system (6.295) modulo an appropriate transformation of the form $\theta \to \theta + f(s)$, $\theta' \to \theta' + f(s)$.

Conversely, for a solution θ of the extended sine-Gordon equation with associated first integral (6.296), we define a function φ according to

$$\cos\varphi = \frac{c\rho - \epsilon q q_s}{c^2 + q^2}, \quad \sin\varphi = \frac{c q_s + \epsilon \rho q}{c^2 + q^2}. \tag{6.298}$$

It is then readily seen that

$$\begin{aligned} \varphi_s &= \kappa + \epsilon \sin\varphi, \\ q_s &= c \sin\varphi - \epsilon q \cos\varphi \end{aligned} \tag{6.299}$$

which is exactly the system (6.295) if the function θ' is defined by

$$\theta' = 2\varphi - \theta. \tag{6.300}$$

This implies, in turn, that θ' is another solution of the extended sine-Gordon equation. The following theorem may now be directly verified:

Theorem 27 (An analogue of Bianchi's classical transformation). *Let θ be a solution of the extended sine-Gordon equation (6.288) with associated first integral (6.296) and $\Sigma : \boldsymbol{r} = \boldsymbol{r}(s, b)$ the corresponding surface. Then, the position vector $\boldsymbol{r}'$ of a dual surface Σ' is given by*

$$\boldsymbol{r}' = \boldsymbol{r} + \epsilon(\cos\varphi\, \boldsymbol{t} - \sin\varphi\, \boldsymbol{n}), \tag{6.301}$$

where φ is defined by (6.298). The corresponding solution θ' of the extended sine-Gordon equation reads

$$\theta' = 2\varphi - \theta \tag{6.302}$$

and the binormal $\boldsymbol{b}'$ has the form

$$\epsilon \boldsymbol{b}' = \sin\varphi\, \boldsymbol{t} + \cos\varphi\, \boldsymbol{n} \tag{6.303}$$

so that

$$\boldsymbol{b}' \cdot \boldsymbol{b} = 0. \tag{6.304}$$

We therefore conclude that Theorem 27 coincides with Theorem 26 for $\mu = \pm 1$ if the b-evolution $(6.290)_2$ is replaced by the explicit formulae (6.298). The remaining relation $(6.290)_1$ is satisfied identically.

Exercise

1. Show that the avatar (6.288) of the extended sine-Gordon system is invariant under the Bäcklund transformation

$$\theta \rightarrow \theta' = 2\varphi - \theta,$$

where φ is given by the relations (6.298).

7

Bäcklund Transformation and Darboux Matrix Connections

In 1882, Darboux introduced his celebrated invariance of what has come to be called the one-dimensional Schrödinger equation [92]. This result, which is but a special case of the Moutard transformation obtained in 1878 [264], allows new solutions of the linear Schrödinger equation to be generated via solutions corresponding to a seed potential.

The spectral properties of the classical Darboux transformation were investigated much later in 1955 by the Oxford mathematician Crum [91]. Twenty years on, Wadati et al. [375], in pioneering work, showed that not only Bäcklund transformations but also conservation laws could be generated for canonical 1+1-dimensional soliton equations by what they termed the Crum transformation. The role of the classical Darboux transformation in soliton theory is detailed in an appendix to the monograph on the spectral transform by Calogero and Degasperis [65].

The Darboux matrix formalism was introduced in connection with the dressing method as originated by Zakharov and Shabat in [395] and described in the text [393]. Important work on the Darboux matrix method was conducted, on the one hand, by Matveev and Salle (detailed in the monograph [251]) and, on the other hand, by Neugebauer and Meinel [270]. Links between Bäcklund transformations and the dressing method, as well as between the latter and the classical Darboux transformation, have been elucidated by Levi et al. [232, 233]. Therein, the Darboux matrix method plays a key role. An excellent recent account of the Darboux matrix method with particular attention to non-isospectral problems has been given by Cieśliński [82]. Darboux transformations and their geometric applications have been the subject of a recent monograph by Gu et al. [155].

This chapter deals, in a geometric setting, with the important connections between matrix Darboux and Bäcklund transformations.

7.1 The Connection for Pseudospherical and Nonlinear Schrödinger Surfaces

It has been shown how Bäcklund transformations for soliton equations are induced by an invariance of their linear representation. In geometric terms, a Bäcklund transformation for a soliton equation encapsulated in a Gauss-Mainardi-Codazzi system is induced by a transformation of the associated linear Gauss-Weingarten equations. This property of generating an invariance of a soliton equation via invariance of its linear representations is one that is shared with matrix Darboux transformations. Here, it is established that the auto-Bäcklund transformations for pseudospherical and NLS surfaces are nothing but disguised matrix Darboux transformations. In the next section, this equivalence between Bäcklund transformations acting at the surface level and matrix Darboux transformations will be extended to a geometrically important subclass of the AKNS system.

7.1.1 Pseudospherical Surfaces

To establish the important connection between the classical Bäcklund transformation (1.52) for pseudospherical surfaces and a matrix Darboux transformation, we shall exploit the Sym-Tafel relation (6.22). It is recalled that the latter provides a compact expression for the generic position vector to a soliton surface in terms of the λ-dependent eigenfunction matrix of the associated AKNS representation.

The standard AKNS representation for pseudospherical surfaces adopts the form (2.32), that is, if $\rho = 1$ and the tilde is dropped therein,

$$\begin{aligned} \Phi_u &= \frac{1}{2}\begin{pmatrix} i\lambda & \omega_u \\ -\omega_u & -i\lambda \end{pmatrix}\Phi = g_1\Phi \\ \Phi_v &= \frac{i}{2\lambda}\begin{pmatrix} -\cos\omega & \sin\omega \\ \sin\omega & \cos\omega \end{pmatrix}\Phi = g_2\Phi \end{aligned} \tag{7.1}$$

with compatibility condition the sine-Gordon equation

$$\omega_{uv} = \sin\omega. \tag{7.2}$$

The associated fundamental forms may be obtained from the relations (6.32) whence

$$\begin{aligned} \mathrm{I} &= du^2 + 2\lambda^{-2}\cos\omega\, du dv + \lambda^{-4} dv^2 \\ \mathrm{II} &= 2\lambda^{-1}\sin\omega\, du dv. \end{aligned} \tag{7.3}$$

These determine pseudospherical surfaces with Gaussian curvature

$$\mathcal{K} = -\lambda^2. \tag{7.4}$$

The fundamental forms given by (1.24) with $\rho = 1$ for pseudospherical surfaces of Gaussian curvature $\mathcal{K} = -1$ may be retrieved on introduction of the scalings

$$u|_{\mathcal{K}=-1} = \lambda u, \quad v|_{\mathcal{K}=-1} = \lambda^{-1} v, \quad \boldsymbol{r}|_{\mathcal{K}=-1} = \lambda \boldsymbol{r}. \tag{7.5}$$

Use of the these relations in combination with (1.52) implies that the classical Bäcklund transformation acting on pseudospherical surfaces with the fundamental forms (7.3) reads

$$\boldsymbol{r}' = \boldsymbol{r} + 2\frac{\mu}{\lambda^2 + \mu^2}\left[\frac{\sin(\omega - \varphi)}{\sin\omega}\boldsymbol{r}_u + \lambda^2 \frac{\sin\varphi}{\sin\omega}\boldsymbol{r}_v\right], \tag{7.6}$$

where the 'swivel angle' φ is, by virtue of (1.36), (1.37), a solution of the compatible system

$$\varphi_u = \omega_u + \mu \sin\varphi, \quad \varphi_v = \mu^{-1}\sin(\varphi - \omega) \tag{7.7}$$

with $\mu = \beta\lambda$.

We proceed in terms of the matrix triad $\{e_1, e_2, e_3\}$, where

$$e_i = \frac{\sigma_i}{2i} \tag{7.8}$$

as introduced in Section 2.2. This triad has the property that it is orthonormal with respect to the usual $su(2)$ metric

$$g(a, b) = -2\,\mathrm{Tr}(ab), \tag{7.9}$$

that is

$$g(e_i, e_j) = \delta_{ij}. \tag{7.10}$$

By virtue of the isomorphism between $su(2)$ and $so(3)$ (cf. Appendix A), it is evident that this triad $\{e_i\}$ of 2×2 matrices corresponds to an orthonormal triad of vectors $\{\boldsymbol{e}_i\}$ in the ambient space $\mathbb{R}^3$.

The relation (6.19) yields

$$r_{,i} = \Phi^{-1} g_{i,\lambda} \Phi, \quad i = 1, 2, \tag{7.11}$$

where $r = \boldsymbol{r} \cdot \boldsymbol{e} = x^1(u, v)e_1 + x^2(u, v)e_2 + x^3(u, v)e_3$ is the generic position matrix. Insertion of the g_i as given by (7.1) into (7.11) yields

$$r_u = -t_3, \quad r_v = \lambda^{-2}(\sin \omega\, t_1 - \cos \omega\, t_3), \tag{7.12}$$

wherein the 'moving' orthonormal triad $\{t_i\}$ of matrices

$$t_i = \Phi^{-1} e_i \Phi \tag{7.13}$$

has been introduced. If we adopt the usual decomposition

$$t_i = \boldsymbol{t}_i \cdot \boldsymbol{e}, \quad \boldsymbol{e} = \begin{pmatrix} e_1 \\ e_2 \\ e_3 \end{pmatrix} \tag{7.14}$$

then the $su(2)$-$so(3)$ isomorphism implies that $\{\boldsymbol{t}_1, \boldsymbol{t}_2, \boldsymbol{t}_3\}$ constitutes an orthonormal triad in $\mathbb{R}^3$. The relations (1.28) of Section 1.2 for the orthonormal set $\{\boldsymbol{A}, \boldsymbol{B}, \boldsymbol{C}\}$, in view of (7.12), indicate that

$$\boldsymbol{A} = -\boldsymbol{t}_3, \quad \boldsymbol{B} = \boldsymbol{t}_1, \quad \boldsymbol{C} = \boldsymbol{A} \times \boldsymbol{B} = -\boldsymbol{t}_3 \times \boldsymbol{t}_1 = -\boldsymbol{t}_2. \tag{7.15}$$

The preceding gives rise to a simple geometric interpretation of the eigenfunction Φ. Thus, the relation (7.13) implies that the orthonormal triad $\{\boldsymbol{t}_i\}$ is related to the rigid triad $\{\boldsymbol{e}_i\}$ by the rotation

$$\boldsymbol{t}_i = \Phi^{(3)^{\mathrm{T}}} \boldsymbol{e}_i, \tag{7.16}$$

where $\Phi^{(3)}$ is the usual $SO(3)$ eigenfunction of the $so(3)$ linear representation.

The matrix version of the Bäcklund transformation (7.6) reads

$$r' = r + 2\frac{\mu}{\lambda^2 + \mu^2}(\sin \varphi\, t_1 - \cos \varphi\, t_3). \tag{7.17}$$

On use of (7.13) and the Sym-Tafel formula, the Bäcklund transformation may be rewritten as

$$\begin{aligned} r' &= r - \frac{i\mu}{\lambda^2 + \mu^2} \Phi^{-1} Q_0 \Phi \\ &= \Phi^{-1}\Phi_\lambda + \Phi^{-1} Q^{-1} \Phi - \frac{\lambda}{\lambda^2 + \mu^2} \mathbb{1} \\ &= (pQ\Phi)^{-1} (pQ\Phi)_\lambda, \end{aligned} \tag{7.18}$$

where

$$Q_0 = \begin{pmatrix} -\cos \varphi & \sin \varphi \\ \sin \varphi & \cos \varphi \end{pmatrix}, \quad Q = \lambda \mathbb{1} + i\mu Q_0, \quad p = \frac{1}{\sqrt{\lambda^2 + \mu^2}}. \tag{7.19}$$

Substitution of the Sym-Tafel relation $r' = \Phi'^{-1}\Phi'_\lambda$ into (7.18) and integration now yields

$$\Phi' = pQ_1 Q\Phi, \tag{7.20}$$

where $Q_1(u, v) \in SU(2)$ is an arbitrary matrix independent of λ. By construction, (7.20) constitutes a form-invariance of the linear representation for the sine-Gordon equation if Q_1 is chosen appropriately. Thus, insertion of Φ' as given by (7.20) into the primed version of the linear representation (7.1) leads to

$$g_i' P = P g_i + P_{,i}, \quad i = 1, 2, \tag{7.21}$$

where $P = Q_1 Q$. Since the solutions ω and ω' of the sine-Gordon equation are independent of λ, it is admissible to sort the relations (7.21) with respect to various powers of λ. In particular, for $i = 1$, the terms proportional to λ^2 provide the vanishing commutator relation

$$[\sigma_3, Q_1] = 0, \tag{7.22}$$

while, for $i = 2$, the terms proportional to λ^{-1} yield

$$Q_1 = \mathbb{1} \tag{7.23}$$

since $\omega' = 2\varphi - \omega$.

Thus, the λ-dependent gauge transformation (7.20) reduces to

$$\Phi \to \Phi' = p(\lambda\mathbb{1} + i\mu Q_0)\Phi, \tag{7.24}$$

where the presence of the factor $p(\lambda)$ guarantees that $\det \Phi' = 1$, as required for the validity of the primed Sym-Tafel formula. Hence, $p = p(\lambda)$ serves merely as a normalisation factor and does not appear in the relation (7.21).

The invariance $\Phi \to \Phi'$ embodied in the gauge transformation (7.24) represents the prototype of a *matrix Darboux* transformation [82, 231–233, 251, 269, 270, 319, 393, 395]. The gauge matrix P is called a *Darboux* matrix. The rationale for this terminology is that the invariance (7.24) may be regarded as a matrix version of the classical Darboux transformation.

For the general theory to be developed subsequently in Section 7.2, it proves convenient to present here an alternative expression for the matrix Q_0 dependent on the swivel angle φ. To this end, it is observed that the system (7.7) may be transformed into a pair of compatible Riccati equations via the change of variable

$$\varphi = 2 \arctan \xi. \tag{7.25}$$

This implies that the swivel angle equations (7.7) are linearisable. Indeed, ξ may be identified with the ratio

$$\xi = \frac{\phi_1}{\phi_2}, \tag{7.26}$$

where $\phi = (\phi_1 \ \phi_2)^{\mathrm{T}}$ is a solution of the Lax pair (7.1) with parameter $\lambda = -i\mu$.

In terms of ξ, the matrix Q_0 is parametrised according to

$$Q_0 = \frac{1}{\xi^2 + 1} \begin{pmatrix} \xi^2 - 1 & 2\xi \\ 2\xi & -(\xi^2 - 1) \end{pmatrix}, \tag{7.27}$$

where, on use of (7.26), we obtain the relation[1]

$$Q_0 = \Psi \sigma_3 \Psi^{-1}, \quad \Psi = \begin{pmatrix} \phi_1 & -\phi_2 \\ \phi_2 & \phi_1 \end{pmatrix}. \tag{7.28}$$

Thus, we have established the important fact that the classical Bäcklund transformation for pseudospherical surfaces may be formulated entirely in terms of eigenfunctions of the associated $su(2)$ linear representation according to, on use of (7.18),

$$r' = r - \frac{i\mu}{\lambda^2 + \mu^2} \Phi^{-1} \Psi \sigma_3 \Psi^{-1} \Phi. \tag{7.29}$$

7.1.2 NLS Surfaces

In Chapter 4, the NLS equation was generated in connection with the motion of a curve $\boldsymbol{r} = \boldsymbol{r}(s, t)$ with time evolution of the s-lines given by

$$v = \boldsymbol{r}_t = \kappa \boldsymbol{b}. \tag{7.30}$$

The 1st and 2nd second fundamental forms of the NLS soliton surfaces swept out by such a motion were shown to be given by (4.26) and (4.28) where, as indicated in Section 4.3, $t \leftrightarrow b$ so that

$$\mathrm{I} = ds^2 + \kappa^2 db^2 \tag{7.31}$$

$$\mathrm{II} = -\kappa\, ds^2 + 2\kappa\tau\, ds db + (\kappa_{ss} - \kappa\tau^2) db^2. \tag{7.32}$$

[1] It is emphasised that Ψ is not itself a matrix-valued eigenfunction. The significance of Ψ will be elaborated upon in Section 7.2.

It is recalled that the associated Gauss-Mainardi-Codazzi equations lead, on use of the Hasimoto transformation

$$q = \kappa e^{i\int \tau\, ds}, \tag{7.33}$$

to the NLS equation

$$iq_b + q_{ss} + \frac{1}{2}|q|^2 q = 0. \tag{7.34}$$

In Section 4.3, the Bäcklund transformation for NLS surfaces was constructed via the requirement of form-invariance of the fundamental forms (7.31) and (7.32) under a transformation of the position vector of the type $\boldsymbol{r}' = \boldsymbol{r} + \alpha\boldsymbol{t} + \beta\boldsymbol{n} + \gamma\boldsymbol{b}$ subject to the constraint that the length $|\boldsymbol{r}' - \boldsymbol{r}|$ be constant. Thus, it was established that, if $\Sigma : \boldsymbol{r} = \boldsymbol{r}(s, b)$ is a seed NLS surface, then the position vector $\boldsymbol{r}'$ of a second NLS surface is given by[2]

$$\boldsymbol{r}' = \boldsymbol{r} + 2\frac{\Im(\lambda_0)}{|\lambda_0|^2}\left(\frac{|\tilde{\xi}|^2 - 1}{|\tilde{\xi}|^2 + 1}\boldsymbol{t} + \frac{2\Re(\tilde{\xi})}{|\tilde{\xi}|^2 + 1}\boldsymbol{n} + \frac{2\Im(\tilde{\xi})}{|\tilde{\xi}|^2 + 1}\boldsymbol{b}\right), \tag{7.35}$$

where

$$\tilde{\xi} = \xi e^{-i\int \tau ds}, \quad \xi = \frac{\phi_1}{\phi_2} \tag{7.36}$$

and the eigenfunction $\Phi|_{\lambda=\lambda_0} = (\phi_1\ \phi_2)^{\mathsf{T}}$ satisfies the standard linear representation for the NLS equation, namely (cf. (4.167)),

$$\begin{aligned} \Phi_s &= \frac{1}{2}\begin{pmatrix} i\lambda & q \\ -\bar{q} & -i\lambda \end{pmatrix}\Phi = g_1\Phi \\ \Phi_b &- \frac{1}{2}\begin{pmatrix} i\left(\frac{1}{2}|q|^2 - \lambda^2\right) & iq_s - \lambda q \\ i\bar{q}_s + \lambda\bar{q} & -i\left(\frac{1}{2}|q|^2 - \lambda^2\right) \end{pmatrix}\Phi = g_2\Phi \end{aligned} \tag{7.37}$$

with $\lambda = \lambda_0$.

To derive the $su(2)$ representation r' of the position vector $\boldsymbol{r}'$, it is recalled that the fundamental forms (7.31) and (7.32) may be obtained from the linear representation (7.37) by evaluation of the eigenfunction Φ at $\lambda = 0$.

The relation (6.19) shows that the $su(2)$ analogues of the tangent vectors $\boldsymbol{r}_s$ and $\boldsymbol{r}_b$ to the NLS surface corresponding to the fundamental forms (7.31) and

[2] For notational convenience, we have, interchanged ξ and $\tilde{\xi}$.

(7.32), wherein $\lambda = 0$ are given by

$$\begin{aligned} r_s &= \Phi^{-1} g_{1,\lambda} \Phi = -t_3 \\ r_b &= \Phi^{-1} g_{2,\lambda} \Phi = \Re(q) t_2 + \Im(q) t_1, \end{aligned} \tag{7.38}$$

where the matrices t_i are again introduced according to

$$t_i = \Phi^{-1} e_i \Phi \tag{7.39}$$

and in (7.38) are understood to be evaluated at $\lambda = 0$.

The relations[3] $r_s = \mathfrak{t}$ and $r_b = \kappa \mathfrak{b}$ imply that the $su(2)$ versions of the unit tangent and the unit binormal are

$$\mathfrak{t} = -t_3, \quad \mathfrak{b} = \cos(\textstyle\int \tau ds)\, t_2 + \sin(\textstyle\int \tau ds)\, t_1, \tag{7.40}$$

while that of the principal normal $\boldsymbol{n}$ reads

$$\mathfrak{n} = \sin(\textstyle\int \tau ds)\, t_2 - \cos(\textstyle\int \tau ds)\, t_1. \tag{7.41}$$

It is readily verified that $(\mathfrak{t}, \mathfrak{n}, \mathfrak{b})$ as given by (7.40), (7.41) constitutes a right-handed orthonormal triad with respect to (7.9) and the usual matrix commutator. Insertion of these $\mathfrak{t}, \mathfrak{n}, \mathfrak{b}$ into the $su(2)$ version of the Bäcklund transformation (7.35) yields

$$r' = r - 2\frac{\Im(\lambda_0)}{|\lambda_0|^2}\left(\frac{2\Re(\xi)}{|\xi|^2+1} t_1 - \frac{2\Im(\xi)}{|\xi|^2+1} t_2 + \frac{|\xi|^2-1}{|\xi|^2+1} t_3\right). \tag{7.42}$$

It remains to obtain the $su(2)$ version of the Bäcklund transformation for NLS surfaces associated with the linear representation (7.37) with $\lambda \neq 0$. To this end, it is observed that this linear representation may be generated by a Lie point symmetry applied to $(7.37)_{\lambda=0}$. Indeed, the NLS equation is invariant under the change of variables (cf. Section 4.2).

$$s \to s^* = s + 2\lambda b, \quad b \to b^* = b, \quad q \to q^* = q e^{i\lambda(s+\lambda b)}, \tag{7.43}$$

where λ is a real constant. This invariance is now supplemented by the gauge transformation

$$\Phi^* = G\Phi|_{\lambda=0}, \quad G = \begin{pmatrix} e^{\frac{1}{2}i\lambda(s+\lambda b)} & 0 \\ 0 & e^{-\frac{1}{2}i\lambda(s+\lambda b)} \end{pmatrix} \tag{7.44}$$

[3] Here, we use gothic symbols for the $su(2)$ analogues of the unit vectors $\boldsymbol{t}$, $\boldsymbol{n}$ and $\boldsymbol{b}$.

which generates precisely the starred version of (7.37), namely

$$\begin{aligned}\Phi^*_{s^*} &= \frac{1}{2}\begin{pmatrix} i\lambda & q^* \\ -\bar{q}^* & -i\lambda \end{pmatrix}\Phi^* = g_1^*\Phi^* \\ \Phi^*_{b^*} &= \frac{1}{2}\begin{pmatrix} i(\frac{1}{2}|q^*|^2 - \lambda^2) & iq^*_{s^*} - \lambda q^* \\ i\bar{q}^*_{s^*} + \lambda\bar{q}^* & -i(\frac{1}{2}|q^*|^2 - \lambda^2) \end{pmatrix}\Phi^* = g_2^*\Phi^*.\end{aligned} \tag{7.45}$$

It is readily shown that $\phi^* = G\Phi|_{\lambda=\lambda_0}$ is a solution of (7.45) with $\lambda \to \lambda_0^* = \lambda + \lambda_0$.

The transition from unstarred to starred variables merely corresponds to a change of the coordinate system on the same surfaces, that is $\boldsymbol{r}^*(s^*, b^*) = \boldsymbol{r}(s, b)$. The new velocity condition (4.81), namely

$$v^* = \boldsymbol{r}_{b^*} = \kappa\boldsymbol{b} - 2\lambda\boldsymbol{t}, \tag{7.46}$$

has, due to the introduction of the parameter λ, a constant component tangential to the s^*-parametric lines. It is emphasised that the s- and s^*-parametric lines coincide. Hence, the orthonormal triad $\{\boldsymbol{t}, \boldsymbol{n}, \boldsymbol{b}\}$ is unaffected by the transformation (7.43).

Now, the starred version of (7.39), namely

$$t_i^* = \Phi^{*-1}e_i\Phi^* = \Phi^{-1}|_{\lambda=0}G^{-1}e_iG\,\Phi|_{\lambda=0} \tag{7.47}$$

shows that the orthonormal triads $\{t_i\}$ and $\{t_i^*\}$ are related by

$$\begin{aligned} t_1^* &= \sin z\, t_2 + \cos z\, t_1 \\ t_2^* &= \cos z\, t_2 - \sin z\, t_1 \\ t_3^* &= t_3, \end{aligned} \tag{7.48}$$

where $z = \lambda(s + \lambda b)$. Accordingly, the Bäcklund transformation (7.42) becomes

$$r' = r - 2\frac{\Im(\lambda_0^*)}{|\lambda - \lambda_0^*|^2}\left(\frac{2\Re(\xi^*)}{|\xi^*|^2+1}t_1^* - \frac{2\Im(\xi^*)}{|\xi^*|^2+1}t_2^* + \frac{|\xi^*|^2-1}{|\xi^*|^2+1}t_3^*\right), \tag{7.49}$$

where $\xi^* = e^{iz}\xi$ and we have taken into account that λ is real.

On introduction of the matrix

$$\Psi = \begin{pmatrix} \phi_1^* & -\bar{\phi}_2^* \\ \phi_2^* & \bar{\phi}_1^* \end{pmatrix}, \tag{7.50}$$

it is seen that the Bäcklund transformation (7.49) for NLS surfaces may be rewritten compactly as

$$r' = r + \frac{i\Im(\lambda_0^*)}{|\lambda - \lambda_0^*|^2} \Phi^{*-1} \Psi \sigma_3 \Psi^{-1} \Phi^* \tag{7.51}$$

in close analogy with the form of the Bäcklund transformation (7.29) for pseudospherical surfaces. In fact, if Ψ is real and λ_0^* is purely imaginary, then (7.29) and (7.51) are identical in form. Indeed, in the next section, it is established that the Bäcklund transformation for surfaces associated with the entire AKNS class $r = -\bar{q}$ is generically of the form (7.51) and (7.29) is a canonical reduction thereof corresponding to $\bar{q} = q$.

The preceding suggests that the Bäcklund transformation of the eigenfunction Φ^* of (7.45) should be a 'complexified' version of (7.24). This is indeed the case. Thus, if we set, as in (7.28),

$$Q_0 = \Psi \sigma_3 \Psi^{-1} \tag{7.52}$$

but where Ψ is now given by (7.50), then (7.51) together with the Sym-Tafel formula yields

$$\begin{aligned} r' &= r + \frac{i\Im(\lambda_0^*)}{|\lambda - \lambda_0^*|^2} \Phi^{*-1} Q_0 \Phi^* \\ &= \Phi^{*-1}\Phi^*_\lambda + \Phi^{*-1}[(\lambda - \Re(\lambda_0^*))\mathbb{1} - i\Im(\lambda_0^*)Q_0]^{-1}\Phi^* - \frac{\lambda - \Re(\lambda_0^*)}{|\lambda - \lambda_0^*|^2}\mathbb{1} \\ &= (pQ\Phi^*)^{-1}(pQ\Phi^*)_\lambda, \end{aligned} \tag{7.53}$$

where

$$Q = (\lambda - \Re(\lambda_0^*))\mathbb{1} - i\Im(\lambda_0^*)Q_0, \quad p = \frac{1}{|\lambda - \lambda_0^*|}. \tag{7.54}$$

Insertion of the Sym-Tafel relation $r' = \Phi^{*\prime -1}\Phi^{*\prime}_\lambda$ into (7.53) and integration gives, in analogy with (7.20),

$$\Phi^{*\prime} = pQ_1 Q \Phi^*. \tag{7.55}$$

Substitution of $\Phi^{*\prime}$ as given by (7.55) into the primed version of the linear representation (7.45) produces, in analogy with (7.21),

$$g_i^{*\prime} P = P g_i^* + P_{,i}, \quad i = 1, 2, \tag{7.56}$$

where $P = Q_1 Q$. The relations (7.56) show that, as in the pseudospherical case, $Q_1 = \mathbb{1}$. Accordingly, the action of the Bäcklund transformation at the eigenfunction level for NLS surfaces can be represented by a matrix Darboux transformation, viz.

$$\Phi^{*\prime} = p[(\lambda - \Re(\lambda_0^*))\mathbb{1} - i\Im(\lambda_0^*)Q_0]\Phi^*. \tag{7.57}$$

To summarise, we have established that the Bäcklund transformations for both pseudospherical and NLS surfaces may be interpreted as matrix Darboux transformations acting on the eigenfunctions of the underlying $su(2)$ representations. In the following section, we present a set of algebraic conditions which define uniquely the elementary matrix Darboux transformation for the important AKNS class $r = -\bar{q}$. It is shown that these conditions admit solutions which give rise to the particular forms (7.24) and (7.57).

Exercises

1. Let the two $su(2)$ matrices S and T be related by the similarity transformation

$$T = \Phi^{-1} S \Phi, \quad \Phi \in SU(2).$$

Show that the vectors $\boldsymbol{S}$ and $\boldsymbol{T}$ obtained from the decompositions

$$S = \boldsymbol{S} \cdot \boldsymbol{e}, \quad T = \boldsymbol{T} \cdot \boldsymbol{e}$$

are connected by the rotation

$$\boldsymbol{T} = \Phi^{(3)^{\mathsf{T}}} \boldsymbol{S}$$

with rotation matrix (cf. Appendix A)

$$\Phi^{(3)}_{ik} = -2\,\mathrm{Tr}(\Phi^{-1} e_i \Phi e_k) \in SO(3).$$

2. (a) Show that the NLS equation (7.34) is invariant under

$$s \to s - 2cb, \quad b \to b, \quad q \to q e^{-ic(s-cb)},$$

where c is a real constant.

(b) Verify that the NLS linear representation (7.37) is invariant under the preceding change of variables augmented by the gauge transformation

$$\Phi \to \begin{pmatrix} e^{-\frac{1}{2}ic(s-cb)} & 0 \\ 0 & e^{\frac{1}{2}ic(s-cb)} \end{pmatrix} \Phi$$

and the change of parameter $\lambda \to \lambda + c$.

7.2 Darboux Matrix and Induced Bäcklund Transformations for the AKNS System. The Constant Length Property

It has been shown that the Bäcklund transformations for pseudospherical and NLS surfaces induce a gauge-invariance of their linear representations of the form

$$\begin{aligned}\Phi \to \Phi' &= pP(\lambda)\Phi \\ &= p(\lambda\mathbb{1} + P_0)\Phi \\ &= p[(\lambda - \Re(\lambda_0))\mathbb{1} - i\Im(\lambda_0)Q_0]\Phi,\end{aligned} \tag{7.58}$$

where

$$Q_0 = \Psi\sigma_3\Psi^{-1}, \quad \Psi = \begin{pmatrix} \phi_1 & -\bar{\phi}_2 \\ \phi_2 & \bar{\phi}_1 \end{pmatrix}, \quad p = \frac{1}{|\lambda - \lambda_0|} \tag{7.59}$$

and Φ, $\Phi|_{\lambda=\lambda_0} = (\phi_1\ \phi_2)^{\mathrm{T}}$ are solutions of the AKNS scattering problem

$$\Phi_s = \frac{1}{2}\begin{pmatrix} i\lambda & q \\ -\bar{q} & -i\lambda \end{pmatrix}\Phi \tag{7.60}$$

with parameters λ, λ_0, respectively. In the pseudospherical case, $q = \omega_s$ and ϕ are real while λ_0 is purely imaginary. The form of the transformation matrix P may, in fact, be generated by algebraic constraints. These constraints are stated in Theorem 28. They are subsequently shown to be consonant with the construction of a matrix Darboux transformation valid for the entire AKNS class $r = -\bar{q}$. The Sym-Tafel formula may be adduced to translate the matrix Darboux transformation into a Bäcklund transformation which acts on the associated AKNS surfaces.

7.2.1 An Elementary Matrix Darboux Transformation

The structure of the transformation matrix P is described in the following result [269, 270]:

Theorem 28 (An Elementary Matrix Darboux Transformation). *Consider the linear 2×2 matrix equation*

$$\Phi_s = g(\lambda)\Phi, \qquad \det\Phi = \text{const} \neq 0, \tag{7.61}$$

where the matrix $g(\lambda) \in sl(2)$ has entries polynomials of degree n in a (complex) parameter λ. Let $\phi_{[1]}$ and $\phi_{[2]}$ be two known vector-valued 'eigenfunctions' *of the equation (7.61) corresponding to the parameters* $\lambda_1 \neq \lambda_2$ *and* $P(\lambda) = \lambda \mathbb{1} + P_0$, P_0 *independent of* λ, *be the* 2×2 *matrix uniquely defined by the linear algebraic system*

$$\begin{aligned} P(\lambda_1)\phi_{[1]} &= (\lambda_1 \mathbb{1} + P_0)\phi_{[1]} = 0 \\ P(\lambda_2)\phi_{[2]} &= (\lambda_2 \mathbb{1} + P_0)\phi_{[2]} = 0. \end{aligned} \tag{7.62}$$

Then, the transformation

$$\begin{aligned} \Phi &\to \Phi' = P(\lambda)\Phi \\ g(\lambda) &\to g'(\lambda) = P(\lambda)g(\lambda)P^{-1}(\lambda) + P_s(\lambda)P^{-1}(\lambda) \end{aligned} \tag{7.63}$$

is such that

(i) $g'(\lambda) \in sl(2)$

(ii) *the polynomial structure of (7.61) is preserved.*

Proof. (i) It is noted that

$$\mathrm{Tr}(g) = \mathrm{Tr}(\Phi_s \Phi^{-1}) = [\ln(\det \Phi)]_s = 0, \tag{7.64}$$

since $g \in sl(2) \Leftrightarrow \mathrm{Tr}(g) = 0$. Thus, $\det \Phi = \text{const}$. In the following, we assume that $\det \Phi = 1$ without loss of generality.

Now,

$$\mathrm{Tr}(g') = \mathrm{Tr}(\Phi'_s \Phi'^{-1}) = [\ln(\det \Phi')]_s = [\ln(\det P)]_s, \tag{7.65}$$

where

$$\det P(\lambda) = (\lambda - \lambda_1)(\lambda - \lambda_2) = \text{const}. \tag{7.66}$$

Accordingly, $\mathrm{Tr}(g') = 0$ whence $g'(\lambda) \in sl(2)$.

(ii) If $\Phi(\lambda)$ is a fundamental solution of the linear equation (7.61), then there exist constant vectors v_1 and v_2 such that

$$\phi_{[i]} = \Phi(\lambda_i)v_i, \qquad i = 1, 2, \tag{7.67}$$

where $\phi_{[i]}$ are the eigenfunctions of (7.61) corresponding to the parameters λ_i which determine the matrix P via (7.62). Now,

$$\Phi'(\lambda_i)v_i = P(\lambda_i)\Phi(\lambda_i)v_i = P(\lambda_i)\phi_{[i]} = 0 \tag{7.68}$$

which implies that

$$\Phi'_s(\lambda_i)v_i = [\Phi'(\lambda_i)v_i]_s = 0, \tag{7.69}$$

whence (cf. Exercise 4)

$$\Phi'_s(\lambda_i)\hat{\Phi}'(\lambda_i) = 0, \tag{7.70}$$

where $\hat{\Phi}'$ is the adjoint of Φ' obeying $\Phi'\hat{\Phi}' = \det \Phi' \mathbb{1}$. On the other hand,

$$\Phi'_s\hat{\Phi}' = Pg\hat{P} + P_s\hat{P} \tag{7.71}$$

constitutes a polynomial in λ of degree $n+2$ with $\Phi'_s(\lambda)\hat{\Phi}'(\lambda)|_{\lambda=\lambda_i} = 0$ and hence may be decomposed according to

$$\Phi'_s\hat{\Phi}' = (\lambda - \lambda_1)(\lambda - \lambda_2)g'(\lambda) = \det P\, g' \tag{7.72}$$

so that

$$\Phi'_s\Phi'^{-1} = \frac{\Phi'_s\hat{\Phi}'}{\det \Phi'} = g' \tag{7.73}$$

is, as required, a matrix-valued polynomial of degree n in λ. □

It is important to note that the matrix Darboux transformation of Theorem 28 may be conjugated with a λ-independent gauge transformation

$$\Phi' \to \Phi'' = P_1\Phi', \qquad \frac{\partial P_1}{\partial \lambda} = 0, \quad \det P_1 = \text{const.} \tag{7.74}$$

Such a gauge-matrix P_1 may be shown to play a crucial role in connection with matrix Darboux transformations for the NLS eigenfunction hierarchy.

Theorem 28 is readily extended to the case when $g(\lambda)$ is a Laurent polynomial, that is a polynomial in λ and λ^{-1}, and may be generalised for a transformation matrix P which is a polynomial of arbitrary degree N in λ. In the latter case, P is determined by $2N$ eigenfunctions which define the kernel of P at $\lambda = \lambda_i$, $i = 1, \ldots, 2N$. Matrix Darboux transformations of this type give rise to the multi-soliton solutions of the AKNS hierarchy [270]. Particular choices of the parameters λ_i also lead to breather solutions and their associated surfaces (cf. Section 1.4).

7.2.2 *Invariance of a su(2) Constraint*

It is evident from Theorem 19, that from the point of view of Euclidean geometry, we are primarily interested in linear equations of the form

$$\Phi_s = g(\lambda)\Phi, \tag{7.75}$$

where λ is real and $g(\lambda) \in su(2)$. However, on occasion, it is necessary to relax this condition and subsume it in the requirement that

$$g(\lambda)|_{\lambda=\bar{\lambda}} \in su(2). \tag{7.76}$$

Thus, if $g(\lambda)$ is a Laurent polynomial in λ, then (7.76) requires that it is of the form

$$g(\lambda) = \sum_{i=-m}^{n} \lambda^i g_i, \quad g_i \in su(2). \tag{7.77}$$

In general, the matrix Darboux transformation (7.63) does not preserve the requirement (7.76). However, it is readily shown that there exists a constant matrix C such that

$$\Phi(\bar{\lambda}) = A\overline{\Phi(\lambda)}C, \qquad iA = \sigma_2. \tag{7.78}$$

Conversely, the constraint (7.78) on the linear matrix equation (7.75) ensures the condition (7.77).

If ϕ is a vector-valued solution of (7.75) with complex parameter λ_0, then a natural choice for the parameters and eigenfunctions in Theorem 28 is

$$\begin{aligned} \phi_{[1]} &= \phi, & \lambda_1 &= \lambda_0 \\ \phi_{[2]} &= A\bar{\phi}, & \lambda_2 &= \bar{\lambda}_0. \end{aligned} \tag{7.79}$$

The solution of the algebraic equations $P(\lambda_i)\phi_{[i]} = 0$ then reads

$$P(\lambda) = (\lambda - \lambda_0)\mathbb{1} + (\bar{\lambda}_0 - \lambda_0)\frac{A\bar{\phi}\phi^{\mathsf{T}}A}{\bar{\phi}^{\mathsf{T}}\phi}, \tag{7.80}$$

whence, on introduction of the matrices

$$Q_0 = \Psi\sigma_3\Psi^{-1}, \quad \Psi = \left(\phi_{[1]}\ \phi_{[2]}\right) = \begin{pmatrix} \phi_1 & -\bar{\phi}_2 \\ \phi_2 & \bar{\phi}_1 \end{pmatrix}, \tag{7.81}$$

we arrive at the representation

$$P(\lambda) = (\lambda - \Re(\lambda_0))\mathbb{1} - i\Im(\lambda_0)Q_0. \tag{7.82}$$

This is precisely the form of the transformation matrix P for pseudospherical and NLS surfaces. It is noted that even though Ψ is not a matrix-valued eigenfunction, its columns are vector-valued eigenfunctions for $\lambda = \lambda_1$ and $\lambda = \lambda_2$, respectively.

It remains to be proven that the matrix Darboux transformation generated by (7.82) does indeed preserve the constraint (7.78). Here, we use the identity

$$P(\bar{\lambda}) = -\,A\overline{P(\lambda)}A \tag{7.83}$$

which follows directly from (7.80). This relation together with (7.78) yields

$$\Phi'(\bar{\lambda}) = P(\bar{\lambda})\Phi(\bar{\lambda}) = [-A\overline{P(\lambda)}A][A\overline{\phi(\lambda)}C] = A\overline{\Phi'(\lambda)}C \tag{7.84}$$

so that Φ' satisfies the primed version of (7.78) and hence $g'(\lambda)$ is of the required form.

In the simplest case, if we consider the action of the matrix Darboux transform $P(\lambda) = \lambda\mathbb{1} + P_0$ on

$$\Phi_s = g(\lambda)\Phi = (g_1\lambda + g_0)\Phi, \tag{7.85}$$

then the terms proportional to λ^2 and λ in

$$g'(\lambda)P(\lambda) = P(\lambda)g(\lambda) + P_s(\lambda) \tag{7.86}$$

deliver the transformation formulae

$$g_0' = g_0 + [P_0, g_1], \quad g_1' = g_1. \tag{7.87}$$

Hence, it is consistent to assume that g_1 is constant and g_0 lies in the image of g_1 under the commutation operation so that g_0 is of the form $[\,\cdot\,, g_1]$. In particular, the AKNS scattering problem

$$\Phi_s = \frac{1}{2}\begin{pmatrix} i\lambda & q \\ -\bar{q} & -i\lambda \end{pmatrix}\Phi \tag{7.88}$$

satisfies these basic criteria and accordingly, with regard to its invariance under an elementary matrix Darboux transformation, may be considered canonical.

7.2.3 The AKNS Class $r = -\bar{q}$ and Its Elementary Bäcklund Transformation

We now return to the AKNS class $r = -\bar{q}$ associated with the NLS hierarchy as introduced in Section 6.1, namely (cf. (6.43))

$$\begin{pmatrix} q \\ -\bar{q} \end{pmatrix}_{t_N} = (-1)^{N+1} i L^N \begin{pmatrix} q \\ \bar{q} \end{pmatrix}, \tag{7.89}$$

where L is the NLS recursion operator defined by

$$L = i \begin{pmatrix} -\partial_s - \frac{1}{2} q \partial_s^{-1} \bar{q} & \frac{1}{2} q \partial_s^{-1} q \\ -\frac{1}{2} \bar{q} \partial_s^{-1} \bar{q} & \partial_s + \frac{1}{2} \bar{q} \partial_s^{-1} q \end{pmatrix}. \tag{7.90}$$

There, the NLS hierarchy was derived via the compatibility condition between the scattering problem (7.88) and the adjoined time evolution

$$\Phi_{t_N} = \frac{1}{2} \begin{pmatrix} iA(\lambda) & B(\lambda) \\ -\bar{B}(\lambda) & -iA(\lambda) \end{pmatrix} \Phi, \tag{7.91}$$

where A and B are polynomials of degree N in λ and $A_N = (-1)^{N+1}$. It is crucial to note that the polynomials A and B are determined by this compatibility condition. Thus, any transformation which leaves invariant both the NLS scattering problem (7.88) and the time evolution (7.91) induces an invariance of the NLS hierarchy.

Now, it has already been established that the elementary matrix Darboux transformation (7.63) with $P(\lambda)$ given by (7.82) preserves (7.88). Moreover, the time evolution (7.91) constitutes the most general form of a linear matrix equation of the type (7.75) subject to the constraint (7.83), except for the particular choice of A_N. It may be verified that the latter is preserved under the matrix Darboux transformation, that is, $A'_N = A_N$. This result together with the induced invariance for the NLS hierarchy is incorporated in the following:

Theorem 29 (An Elementary Matrix Darboux Transformation for the NLS Hierarchy). *The matrix Darboux transformation (7.63), (7.82) leaves form-invariant the linear representation (7.88), (7.91) of the NLS hierarchy. The induced invariance of the NLS hierarchy itself is given by*

$$q \to q' = q - 4\Im(\lambda_0) \frac{\phi_1 \bar{\phi}_2}{\phi_1 \bar{\phi}_1 + \phi_2 \bar{\phi}_2}, \tag{7.92}$$

where $\phi = (\phi_1 \ \phi_2)^{\mathsf{T}}$ *is a vector-valued solution of the linear representation with parameter* λ_0.

Proof. It only remains to establish the validity of (7.92). This is a consequence of the relation $(7.87)_1$ which yields

$$q' = q - 2i(P_0)_{12} \tag{7.93}$$

and hence (7.92). □

To obtain the auto-Bäcklund transformation for the NLS hierarchy explicitly in terms of just q and q', it is observed that q' as given by (7.92) depends only on q and the ratio

$$\xi = \frac{\phi_1}{\phi_2} \tag{7.94}$$

via

$$q' = q - 4\Im(\lambda_0)\frac{\xi}{\xi\bar{\xi} + 1}. \tag{7.95}$$

The quantity ξ satisfies the Riccati equation

$$\xi_s = \frac{q}{2} + i\lambda_0\xi + \frac{\bar{q}}{2}\xi^2, \tag{7.96}$$

while inversion of (7.95) yields

$$\xi = -\frac{q' - q}{2\Im(\lambda_0) \mp \sqrt{4\Im(\lambda_0)^2 - |q' - q|^2}}. \tag{7.97}$$

Insertion of the latter expression into (7.96) produces

$$q'_s - q_s = i\Re(\lambda_0)(q' - q) \pm \frac{q' + q}{2}\sqrt{4\Im(\lambda_0)^2 - |q' - q|^2}. \tag{7.98}$$

This is the spatial part of the auto-Bäcklund transformation associated with the elementary matrix Darboux transformation for the NLS hierarchy. It is the same for all members of the hierarchy. However, the temporal part of the Bäcklund transformation depends on the particular member of the hierarchy under investigation. For instance, the time evolution (6.58) with $x \to s$ for the NLS equation

$$iq_b + q_{ss} + \frac{1}{2}|q|^2 q = 0, \qquad b = t_2 \tag{7.99}$$

gives rise to the Riccati equation

$$\xi_b = \tfrac{1}{2}(iq_s - \lambda_0 q) + i(\tfrac{1}{2}|q|^2 - \lambda_0^2)\xi - \tfrac{1}{2}(i\bar{q}_s + \lambda_0\bar{q})\xi^2, \tag{7.100}$$

which, on substitution of ξ from (7.97), may be written as

$$\begin{aligned} q'_b - q_b = -\Re(\lambda_0)(q'_s - q_s) + i\frac{q' - q}{4}(|q'|^2 + |q|^2) \\ \pm\, i\frac{q'_s + q_s}{2}\sqrt{4\Im(\lambda_0)^2 - |q' - q|^2}. \end{aligned} \tag{7.101}$$

It may be verified directly that the relations (7.98) and (7.101) are compatible if and only if q satisfies the NLS equation (7.99). This is a consequence of the fact that the Riccati equations (7.96) and (7.100) may be regarded as a nonlinear Lax pair for the NLS equation. Compatibility guarantees, in turn, that q' is likewise a solution of the NLS equation so that the pair of equations (7.98), (7.101) constitute a *strong* auto-Bäcklund transformation.

In conclusion, we recall that the NLS hierarchy (7.89) reduces to the mKdV hierarchy if N is odd and q is real. In that case, it is consistent to assume that ϕ is real if $\lambda_0 = -i\mu$ is taken to be purely imaginary. The transformation (7.92) reveals that these reality constraints are preserved by the matrix Darboux transformation. Now, the mKdV hierarchy may be represented in a compact conservative form, namely

$$q_{t_N} = [R(q)]_s, \tag{7.102}$$

where $R(q)$ is a polynomial in q and its s-derivatives. Hence, if we set

$$q = \omega_s \tag{7.103}$$

the evolution equations (7.102) may be integrated to obtain the potential mKdV hierarchy

$$\omega_{t_N} = R(\omega_s). \tag{7.104}$$

The spatial part of the auto-Bäcklund transformation, that is

$$(\omega' - \omega)_{ss} = \pm\frac{(\omega' + \omega)_s}{2}\sqrt{4\mu^2 - (\omega' - \omega)_s^2}, \tag{7.105}$$

may also be integrated once to produce

$$\left(\frac{\omega' - \omega}{2}\right)_s = \pm\mu\sin\left(\frac{\omega' + \omega}{2}\right). \tag{7.106}$$

In the above, time-dependent 'constants' of integration have been neglected.

The relation (7.106) provides the spatial part of the auto-Bäcklund transformation for the potential mKdV hierarchy. By construction, it coincides with one of the classical Bäcklund relations for the sine-Gordon equation. This is because the latter is nothing but the first 'hyperbolic' member of the potential mKdV hierarchy. It is recalled that the potential mKdV equation arises in connection with a compatible motion of pseudospherical surfaces described by the sine-Gordon equation (cf. Section 2.4).

7.2.4 The Constant Length Property

It remains to analyse the geometric implications of the elementary matrix Darboux transformation. Thus, let us consider compatible linear equations of the form

$$\Phi_{,\nu} = g_\nu(\lambda)\Phi, \quad \nu = 1, 2, \tag{7.107}$$

where the matrices $g_\nu \in su(2)$ are Laurent polynomials in the real parameter λ. If we choose $\det \Phi = 1$, then the above Lax pair is associated with a surface Σ in $\mathbb{R}^3$ through the Sym-Tafel formula

$$r = \Phi^{-1}\Phi_\lambda \tag{7.108}$$

as set down in Theorem 19. The normalised matrix Darboux transformation

$$\Phi' = \frac{1}{|\lambda - \lambda_0|} P(\lambda)\Phi = p(\lambda)P(\lambda)\Phi, \tag{7.109}$$

where $P(\lambda)$ is given by (7.82) and

$$\det P = p^{-2} = (\lambda - \lambda_0)(\lambda - \bar{\lambda}_0), \tag{7.110}$$

leaves invariant the Lax pair (7.107) and, accordingly, preserves the structure of the fundamental forms I, II associated with the surface Σ. The position matrix of the new surface Σ' is given by

$$r' = \Phi'^{-1}\Phi'_\lambda = \Phi^{-1}\Phi_\lambda + p^{-1}p_\lambda \mathbb{1} + \Phi^{-1}P^{-1}P_\lambda\Phi, \tag{7.111}$$

whence, on insertion of p and P, we obtain the compact representation

$$r' = r + i\frac{\Im(\lambda_0)}{|\lambda - \lambda_0|^2}\Phi^{-1}\Psi\sigma_3\Psi^{-1}\Phi \tag{7.112}$$

for the Bäcklund transformation linking the surfaces Σ and Σ'. The components of the generic position vector $\boldsymbol{r}' = x^{1'}\boldsymbol{e}_1 + x^{2'}\boldsymbol{e}_2 + x^{3'}\boldsymbol{e}_3$ to Σ' are given by

$$x^{j'} = i\,\mathrm{Tr}(\Phi'_\lambda \sigma_j \Phi'^{-1}) \tag{7.113}$$

so that, in terms of the orthonormal triad $\boldsymbol{t}_i$ determined by

$$t_i = \boldsymbol{t}_i \cdot \boldsymbol{e} = \Phi^{-1} e_i \Phi, \tag{7.114}$$

the Bäcklund transformation becomes

$$\boldsymbol{r}' = \boldsymbol{r} - 2\frac{\Im(\lambda_0)}{|\lambda - \lambda_0|^2}\left(\frac{2\Re(\xi)}{|\xi|^2+1}\boldsymbol{t}_1 - \frac{2\Im(\xi)}{|\xi|^2+1}\boldsymbol{t}_2 + \frac{|\xi|^2-1}{|\xi|^2+1}\boldsymbol{t}_3\right). \tag{7.115}$$

Remarkably, it is seen that the magnitude of the distance vector $\boldsymbol{r}' - \boldsymbol{r}$, namely

$$|\boldsymbol{r}' - \boldsymbol{r}| = 2\frac{|\Im(\lambda_0)|}{|\lambda - \lambda_0|^2} \tag{7.116}$$

is constant. This result is embodied in the following theorem [354]:

Theorem 30 (The Constant Length Property). *The Bäcklund transformation (7.112) associated with the elementary matrix Darboux transformation (7.109) for the Lax pair (7.107) is such that the distance between corresponding points on the surfaces Σ and Σ' is constant.*

It should be emphasised that the constant length property holds for any class of surfaces associated with a Lax pair of the form (7.107) and is not restricted to surfaces linked to the AKNS class $r = -\bar{q}$. The Bäcklund transformation (7.115) generalises that obtained for the canonical NLS equation via an alternative geometric approach in Chapter 4.

Exercises

1. Prove that

$$\det P(\lambda) = (\lambda - \lambda_1)(\lambda - \lambda_2),$$

where the matrix $P(\lambda) = \lambda \mathbb{1} + P_0$ is determined by the linear algebraic system (7.62).

2. Prove that, if $\Phi(\lambda)$ is a fundamental solution of the linear equation (7.61), then there exist constant vectors v_i, $i = 1, 2$ such that

$$\phi_{[i]} = \Phi(\lambda_i) v_i, \qquad i = 1, 2,$$

where the $\phi_{[i]}$ are eigenfunctions of equation (7.61) corresponding to the parameters λ_i.

3. The adjoint of a matrix $B \in \mathbb{R}^{m,m}$ is defined by

$$\hat{B}_{ij} = (-1)^{i+j} \det \begin{pmatrix} B_{11} & \cdots & B_{1i-1} & B_{1i+1} & \cdots & B_{1m} \\ \vdots & & \vdots & \vdots & & \vdots \\ B_{j-11} & \cdots & B_{j-1i-1} & B_{j-1i+1} & \cdots & B_{j-1m} \\ B_{j+11} & \cdots & B_{j+1i-1} & B_{j+1i+1} & \cdots & B_{j+1m} \\ \vdots & & \vdots & \vdots & & \vdots \\ B_{m1} & \cdots & B_{mi-1} & B_{mi+1} & \cdots & B_{mm} \end{pmatrix}.$$

Show that

$$B\hat{B} = \det B\, \mathbb{1}.$$

4. If $A, B \in \mathbb{R}^{m,m}$ and $v \in \mathbb{R}^m$ then show that

$$\left.\begin{aligned} Av &= 0 \\ Bv &= 0 \end{aligned}\right\} \Rightarrow A\hat{B} = 0.$$

5. Show that an equivalent characterisation of the condition (7.77) is that

$$g(\bar{\lambda}) = -A\overline{g(\lambda)}A, \quad iA = \sigma_2$$

and that this condition is, in turn, equivalent to (7.78). Verify the identity

$$P(\bar{\lambda}) = -A\overline{P(\lambda)}A.$$

7.3 Iteration of Matrix Darboux Transformations. Generic Permutability Theorems

In this section, we show that matrix Darboux transformations can be iterated in a purely algebraic manner and that N iterations of the elementary matrix Darboux transformation may be interpreted as a nonlinear superposition of N elementary matrix Darboux transformations applied to the same seed solution. This observation is exploited to give new solutions $q^{(N)}$ in terms of a seed solution q together with N Bäcklund transforms $q_1^{(1)}, \ldots, q_N^{(1)}$ and the corresponding Bäcklund parameters $\lambda_1, \ldots, \lambda_N$, that is

$$q^{(N)} = q^{(N)}\left[q, q_i^{(1)}, \lambda_i\right]. \tag{7.117}$$

The simplest such superposition principle is obtained for $N = 2$ and gives rise to a permutability theorem generic to the AKNS hierarchy $r = -\bar{q}$. If q is real then this reduces to the classical permutability theorem for the sine-Gordon equation. In geometric terms, at the surface level, the iterated matrix Darboux transformation will be seen to generate a suite of surfaces such that each pair of neighbouring surfaces possesses the constant length property.

7.3.1 Iteration of Matrix Darboux Transformations

Here, we confine ourselves to the geometrically relevant case (cf. Subsection 7.2.2)

$$\Phi_s = g(\lambda)\Phi, \quad g(\lambda) \in su(2) \tag{7.118}$$

which admits the elementary matrix Darboux transformation

$$\Phi \to \Phi^1 = P^1(\lambda)\Phi, \tag{7.119}$$

where $P^1 = \lambda \mathbb{1} + P_0^1$ is defined by

$$P^1(\lambda_1)\phi_1 = 0, \quad P^1(\bar{\lambda}_1)A\bar{\phi}_1 = 0. \tag{7.120}$$

The superscript ^{1} indicates that the Darboux transformation (7.119) is generated by the eigenfunction ϕ_1 with parameter λ_1. A second application of the matrix Darboux transformation requires the knowledge of a vector-valued solution ϕ_{12} of

$$\Phi_s^1 = g^1(\lambda)\Phi^1 \tag{7.121}$$

with parameter λ_2, where g^1 is the Darboux transform of g under (7.119). The second Darboux matrix is then given by

$$P^{12}(\lambda_2)\phi_{12} = 0, \quad P^{12}(\bar{\lambda}_2)A\bar{\phi}_{12} = 0 \tag{7.122}$$

which defines the corresponding transformation according to

$$\Phi \to \Phi^{12} = P^{12}(\lambda)\Phi^1 = P^{12}(\lambda)P^1(\lambda)\Phi. \tag{7.123}$$

Now, the key observation is that for $\lambda_2 \neq \lambda_1, \bar{\lambda}_1$, the Darboux matrix $P^1(\lambda_2)$ is regular, that is

$$\det P^1(\lambda_2) = (\lambda_2 - \lambda_1)(\lambda_2 - \bar{\lambda}_1) \neq 0, \tag{7.124}$$

and hence $P^1(\lambda_2)\Phi(\lambda_2)$ is a fundamental solution of (7.121) if $\Phi(\lambda_2)$ is a fundamental solution of the seed equation (7.118). Consequently, ϕ_{12} may be regarded as the image

$$\phi_{12} = P^1(\lambda_2)\phi_2 \tag{7.125}$$

of an eigenfunction ϕ_2 while the Darboux matrix

$$P(\lambda) = P^{12}(\lambda)P^1(\lambda) \tag{7.126}$$

satisfies the relations

$$P(\lambda_i)\phi_i = 0, \quad P(\bar{\lambda}_i)A\bar{\phi}_i = 0, \qquad i = 1, 2 \tag{7.127}$$

by virtue of the definitions (7.120) and (7.122). On the other hand, since $P(\lambda)$ is of the form

$$P(\lambda) = \lambda^2 + \lambda P_1 + P_0, \tag{7.128}$$

the property (7.127) determines $P(\lambda)$ uniquely. If we take into account that ϕ_2 also generates a matrix Darboux transformation of the form (7.119), (7.120), viz.

$$\Phi \to \Phi^2 = P^2(\lambda)\Phi, \tag{7.129}$$

then we conclude that conjugation of two elementary Darboux transformations leads to a *nonlinear superposition* of two elementary matrix Darboux transformations applied to the *same* seed solution.

The above result may be readily generalised to yield:

Theorem 31 (The Iterated Matrix Darboux Transformation). *N applications of the matrix Darboux transformation (7.63) with parameters $\lambda_1, \ldots, \lambda_N$ take the form of the matrix Darboux transformation*

$$\begin{aligned} \Phi \to \ \Phi^{(N)} &= P(\lambda)\Phi \\ g(\lambda) \to g^{(N)}(\lambda) &= P(\lambda)g(\lambda)P^{-1}(\lambda) + P_s(\lambda)P^{-1}(\lambda), \end{aligned} \tag{7.130}$$

where the Darboux matrix

$$P(\lambda) = \lambda^N + \sum_{i=0}^{N-1} \lambda^i P_i \tag{7.131}$$

is defined by the solution of the linear algebraic system

$$P(\lambda_i)\phi_i = 0, \quad P(\bar{\lambda}_i)A\bar{\phi}_i = 0. \tag{7.132}$$

The elementary matrix Darboux transformations commute in the sense that the iterated matrix Darboux transformation is independent of the order of application.

Proof. The Darboux matrix associated with N successive applications of the elementary matrix Darboux transformation reads

$$P(\lambda) = P^{1\cdots N}(\lambda)\cdots P^{12}(\lambda)P^{1}(\lambda) \tag{7.133}$$

with the definitions

$$P^{1\cdots i}(\lambda_i)\phi_{1\cdots i} = 0, \quad P^{1\cdots i}(\bar{\lambda}_i)A\bar{\phi}_{1\cdots i} = 0. \tag{7.134}$$

Since the eigenfunctions $\phi_{1\cdots i+1}$ may be regarded as the images of eigenfunctions ϕ_{i+1} under i elementary matrix Darboux transformations, that is

$$\phi_{1\cdots i+1} = P^{1\cdots i}(\lambda_{i+1})\cdots P^{1}(\lambda_{i+1})\phi_{i+1}, \tag{7.135}$$

the Darboux matrix $P(\lambda)$, which is of the form (7.131), indeed satisfies the conditions (7.132). The latter are symmetric in $1, \ldots, N$ and hence the order in which the elementary matrix Darboux transformations are applied is immaterial. Moreover, it is noted that both $P(\lambda)$ and $-A\overline{P(\bar{\lambda})}A$ satisfy the defining relations (7.131), (7.132). Consequently, $P(\lambda) = -A\overline{P(\bar{\lambda})}A$ so that the condition $g(\lambda) \in sl(2)$ is preserved.[4] □

The iterated matrix Darboux transformation acting on Lax pairs of the form (7.107) induces iterated Bäcklund transformations for the associated soliton surfaces. Thus, in principle, it is first necessary to normalise the matrix Darboux transformation in such a way that $\Phi^{(N)} \in SU(2)$ and then apply the Sym-Tafel formula. However, since we are only interested in a decomposition of the new position matrix

$$r^{(N)} = \boldsymbol{r}^{(N)} \cdot \boldsymbol{e} \tag{7.136}$$

[4] This is evident since the building blocks of the iterated matrix Darboux transformation leave invariant the geometric constraint $g(\lambda) \in sl(2)$.

into Pauli matrices which are trace-free, it is convenient to introduce the trace-free part of a matrix Q according to

$$Q|^{\mathrm{nt}} = Q - \frac{1}{2}\,\mathrm{Tr}(Q)\mathbb{1}. \tag{7.137}$$

The new position matrix $r^{(N)}$ then reads

$$r^{(N)} = r + \Phi^{-1}P^{-1}P_{\lambda}\Phi\big|^{\mathrm{nt}}. \tag{7.138}$$

By construction, the difference matrix $r^{(N)} - r$ may be written as a sum of matrices which obey the constant length condition, viz.

$$r^{(N)} - r = \sum_{i=1}^{N} s_i, \quad -2\,\mathrm{Tr}\big(s_i^2\big) = 4\frac{[\Im(\lambda_i)]^2}{|\lambda - \lambda_i|^4}, \tag{7.139}$$

since at each step of the iteration procedure

$$\Sigma^{(0)} \to \Sigma^{(1)} \to \cdots \to \Sigma^{(N)} \tag{7.140}$$

the new surface $\Sigma^{(i)}$ and the old surface $\Sigma^{(i-1)}$ possess the constant length property. Hence, one can think of the surfaces $\Sigma^{(1)}, \ldots, \Sigma^{(N)}$ as being generated by the vertices $\boldsymbol{r}_i$ of a polygon

$$[\boldsymbol{r}_0, \boldsymbol{r}_1, \ldots, \boldsymbol{r}_N], \qquad |\boldsymbol{r}_i - \boldsymbol{r}_{i-1}| = 2\frac{|\Im(\lambda_i)|}{|\lambda - \lambda_i|^2} \tag{7.141}$$

as the intial point $\boldsymbol{r}_0$ moves along the seed surface $\Sigma^{(0)}$ and the vertices $\boldsymbol{r}_i - \boldsymbol{r}_{i-1}$ swivel appropriately. The matrices s_i are best written in terms of the matrix Darboux transforms

$$\Phi^{(i)} = P^{1\cdots i}(\lambda)\cdots P^{1}(\lambda)\Phi, \quad \Phi^{(0)} = \Phi \tag{7.142}$$

associated with the surfaces $\Sigma^{(i)}$. Thus, the transformation formula (7.138) reads

$$\begin{aligned} r^{(N)} &= r \\ &+ \Phi^{-1}\left[\sum_{i=1}^{N}(P^{1\cdots N}\cdots P^{1})^{-1}P^{1\cdots N}\cdots P^{1\cdots i+1}P_{\lambda}^{1\cdots i}P^{1\cdots i-1}\cdots P^{1}\right]\Phi\Bigg|^{\mathrm{nt}} \end{aligned} \tag{7.143}$$

from which we conclude that

$$s_i = \big[\Phi^{(i-1)}\big]^{-1}\big[P^{1\cdots i}\big]^{-1}\Phi^{(i-1)}\big|^{\mathrm{nt}}. \tag{7.144}$$

The above decomposition has been discussed by Sym [354] in the context of a class of generalised AKNS scattering problems. It is emphasised, however, that the decomposition formulae (7.139), (7.144) are generic in the sense outlined below.

The general form of a Darboux matrix associated with a Lax pair which is polynomial in a constant parameter λ and its inverse λ^{-1} is given by

$$\tilde{P}(\lambda) = \sum_{i=0}^{N} \lambda^i \tilde{P}_i. \tag{7.145}$$

If $\tilde{P}_N$ is regular,[5] then this may be factorised into a product of a gauge matrix and a Darboux matrix of standard form, that is

$$\tilde{P}(\lambda) = P_N P(\lambda) = P_N \left(\lambda^N + \sum_{i=0}^{N-1} \lambda^i P_i \right). \tag{7.146}$$

Even though the solutions of the underlying nonlinear equations are affected by gauge matrices, the surface geometry is independent of P_N since

$$\tilde{r}^{(N)} = r + \Phi^{-1} \tilde{P}^{-1} \tilde{P}_\lambda \Phi \Big|^{\mathrm{nt}} = r + \Phi^{-1} P^{-1} P_\lambda \Phi \Big|^{\mathrm{nt}} = r^{(N)}. \tag{7.147}$$

This underlines the generic nature of the preceding analysis.

7.3.2 *Generic Permutability Theorems*

The fact that the iterated matrix Darboux transformation is symmetric in all eigenfunctions $\phi_1, \ldots, \phi_N$ has the important consequence that the Bianchi diagram associated with the corresponding Bäcklund transformation commutes. Here, we return to the AKNS class $r = -\bar{q}$ and establish a generic permutability theorem. The classical permutability theorem as originally obtained by Bianchi for the sine-Gordon equation is recovered as a specialisation.

It is readily seen that the analogue of the transformation formula (7.93) is given by

$$q^{(N)} = q - 2i(P_{N-1})_{12}. \tag{7.148}$$

[5] Note that singular matrices $\tilde{P}_N$ are relevant in connection with the Korteweg-de Vries hierarchy which governs surfaces in 2+1-dimensional Minkowski space.

The solution of the linear algebraic system (7.132)

$$\begin{pmatrix} \xi_1 & 1 & \cdots & \lambda_1^{N-1}\xi_1 & \lambda_1^{N-1} \\ -1 & \bar{\xi}_1 & \cdots & -\bar{\lambda}_1^{N-1} & \bar{\lambda}_1^{N-1}\bar{\xi}_1 \\ \vdots & \vdots & & \vdots & \vdots \\ \xi_N & 1 & \cdots & \lambda_N^{N-1}\xi_N & \lambda_N^{N-1} \\ -1 & \bar{\xi}_N & \cdots & -\bar{\lambda}_N^{N-1} & \bar{\lambda}_N^{N-1}\bar{\xi}_N \end{pmatrix} \begin{pmatrix} (P_0)_{11} \\ (P_0)_{12} \\ \vdots \\ (P_{N-1})_{11} \\ (P_{N-1})_{12} \end{pmatrix} = \begin{pmatrix} -\lambda_1^N\xi_1 \\ \bar{\lambda}_1^N \\ \vdots \\ -\lambda_N^N\xi_N \\ \bar{\lambda}_N^N \end{pmatrix} \tag{7.149}$$

$$\begin{pmatrix} \xi_1 & 1 & \cdots & \lambda_1^{N-1}\xi_1 & \lambda_1^{N-1} \\ -1 & \bar{\xi}_1 & \cdots & -\bar{\lambda}_1^{N-1} & \bar{\lambda}_1^{N-1}\bar{\xi}_1 \\ \vdots & \vdots & & \vdots & \vdots \\ \xi_N & 1 & \cdots & \lambda_N^{N-1}\xi_N & \lambda_N^{N-1} \\ -1 & \bar{\xi}_N & \cdots & -\bar{\lambda}_N^{N-1} & \bar{\lambda}_N^{N-1}\bar{\xi}_N \end{pmatrix} \begin{pmatrix} (P_0)_{21} \\ (P_0)_{22} \\ \vdots \\ (P_{N-1})_{21} \\ (P_{N-1})_{22} \end{pmatrix} = \begin{pmatrix} -\lambda_1^N \\ -\bar{\lambda}_1^N\bar{\xi}_1 \\ \vdots \\ -\lambda_N^N \\ -\bar{\lambda}_N^N\bar{\xi}_N \end{pmatrix}, \tag{7.150}$$

where $\xi_i = \phi_i^1/\phi_i^2$ is readily obtained by means of Cramer's rule. Thus,

$$(P_{N-1})_{12} = \frac{\begin{vmatrix} \xi_1 & 1 & \cdots & \lambda_1^{N-1}\xi_1 & -\lambda_1^N\xi_1 \\ -1 & \bar{\xi}_1 & \cdots & -\bar{\lambda}_1^{N-1} & \bar{\lambda}_1^N \\ \vdots & \vdots & & \vdots & \vdots \\ \xi_N & 1 & \cdots & \lambda_N^{N-1}\xi_N & -\lambda_N^N\xi_N \\ -1 & \bar{\xi}_N & \cdots & -\bar{\lambda}_N^{N-1} & \bar{\lambda}_N^N \end{vmatrix}}{\begin{vmatrix} \xi_1 & 1 & \cdots & \lambda_1^{N-1}\xi_1 & \lambda_1^{N-1} \\ -1 & \bar{\xi}_1 & \cdots & -\bar{\lambda}_1^{N-1} & \bar{\lambda}_1^{N-1}\bar{\xi}_1 \\ \vdots & \vdots & & \vdots & \vdots \\ \xi_N & 1 & \cdots & \lambda_N^{N-1}\xi_N & \lambda_N^{N-1} \\ -1 & \bar{\xi}_N & \cdots & -\bar{\lambda}_N^{N-1} & \bar{\lambda}_N^{N-1}\bar{\xi}_N \end{vmatrix}}. \tag{7.151}$$

For $N = 1$, the new solution of the AKNS hierarchy is therefore given by

$$q_1^{(1)} = q - 4\Im(\lambda_1)\frac{\xi_1}{1 + |\xi_1|^2}, \tag{7.152}$$

which is in agreement with (7.92). At this stage the subscript on $q^{(1)}$ is redundant, but will be useful in the sequel. In the case of two eigenfunctions, the new solution reads

$$q^{(2)} = q - 2i\frac{Q}{R}, \tag{7.153}$$

where

$$\begin{aligned}
Q &= (\lambda_1 - \bar{\lambda}_1)(\lambda_2 - \bar{\lambda}_2)(\lambda_1 + \bar{\lambda}_1 - \lambda_2 - \bar{\lambda}_2)(\bar{\xi}_1 - \bar{\xi}_2)\xi_1\xi_2 \\
&\quad + (\lambda_1 - \bar{\lambda}_1)(\lambda_1 - \bar{\lambda}_2)(\bar{\lambda}_1 - \bar{\lambda}_2)(1 + |\xi_2|^2)\xi_1 \\
&\quad - (\lambda_2 - \bar{\lambda}_2)(\lambda_2 - \bar{\lambda}_1)(\bar{\lambda}_1 - \bar{\lambda}_2)(1 + |\xi_1|^2)\xi_2 \qquad (7.154) \\
R &= (\lambda_1 - \bar{\lambda}_1)(\lambda_2 - \bar{\lambda}_2)|\xi_1 - \xi_2|^2 \\
&\quad - |\lambda_1 - \lambda_2|^2(1 + |\xi_1|^2)(1 + |\xi_2|^2).
\end{aligned}$$

It is noted that $q^{(2)} = q_1^{(1)}$ if $\lambda_2 = 0$. In general, $q^{(N)}$ may be formulated as

$$q^{(N)} = q^{(N)}[q, \xi_1, \ldots, \xi_N, \lambda_1, \ldots, \lambda_N]. \qquad (7.155)$$

To eliminate the eigenfunctions from this expression, it is observed that each eigenfunction ϕ_i generates a new solution

$$q_i^{(1)} = q - 4\Im(\lambda_i)\frac{\xi_i}{1 + |\xi_i|^2}. \qquad (7.156)$$

Hence, if we solve for the eigenfunctions to obtain

$$\xi_i = -\frac{q_i^{(1)} - q}{2\Im(\lambda_i) \mp \sqrt{4\Im(\lambda_i)^2 - \left|q_i^{(1)} - q\right|^2}}, \qquad (7.157)$$

a *nonlinear superposition principle* is retrieved in the form

$$q^{(N)} = q^{(N)}\left[q, q_i^{(1)}, \lambda_i\right]. \qquad (7.158)$$

Accordingly, we have the following result:

Theorem 32 (Superposition Principles for the AKNS Hierarchy). *The iterated matrix Darboux transformation (7.130)–(7.132) for the AKNS hierarchy $r = -\bar{q}$ is associated with explicit nonlinear superposition principles which express a new solution $q^{(N)}$ in terms of a seed solution q and N Bäcklund transforms $q_1^{(1)}, \ldots, q_N^{(1)}$ generated by the Bäcklund transformation (7.98) and its temporal extensions (7.101), ... with parameters $\lambda_1, \ldots, \lambda_N$.*

It is remarked that for $N = 2$, the above superposition principle represents a generalisation of the classical permutability theorem for the sine-Gordon equation. To justify this assertion, it is convenient to turn to the potential mKdV hierarchy (7.104) for $\omega = \int q\, ds$. Instead of integrating the superposition formula (7.158) to obtain an explicit superposition principle for ω, we exploit the fact that the sine-Gordon equation is the first 'negative' member of this hierarchy. We therefore focus on the transformation properties of the corresponding 'v-evolution'

$$\Phi_v = \frac{i}{2\lambda} S(\omega)\Phi, \quad S(\omega) = \begin{pmatrix} -\cos\omega & \sin\omega \\ \sin\omega & \cos\omega \end{pmatrix}. \tag{7.159}$$

The action of the iterated matrix Darboux transformation on $S(\omega)$ is given by

$$S\big(\omega^{(N)}\big) = P_0 S(\omega) P_0^{-1}. \tag{7.160}$$

By definition, the matrix P_0 may be decomposed into

$$P_0 = P_0^{1\cdots N} \cdots P_0^1. \tag{7.161}$$

According to (7.19), each matrix $P^{1\cdots i}$ has the structure

$$P^{1\cdots i} \sim S(\varphi_{1\cdots i}), \quad \varphi_{1\cdots i} = 2\arctan \xi_{1\cdots i} \tag{7.162}$$

so that

$$S\big(\omega_i^{(1)}\big) = S(\varphi_i) S(\omega) S^{-1}(\varphi_i), \qquad i = 1, 2 \tag{7.163}$$

and

$$S\big(\omega^{(2)}\big) = \tilde{S}(\chi) S(\omega) \tilde{S}^{-1}(\chi), \tag{7.164}$$

where

$$\tilde{S}(\chi) = \begin{pmatrix} \cos\chi & \sin\chi \\ -\sin\chi & \cos\chi \end{pmatrix}, \quad \chi = \varphi_{12} - \varphi_1. \tag{7.165}$$

Evaluation of the above yields

$$\omega_i^{(1)} = 2\varphi_i - \omega, \quad \omega^{(2)} = \omega + 2\chi \tag{7.166}$$

or, equivalently,

$$\xi_i = \tan\left(\frac{\omega_i^{(1)} + \omega}{4}\right), \quad \xi = \tan\left(\frac{\omega^{(2)} - \omega}{4}\right) \tag{7.167}$$

if we set

$$\chi = 2\arctan\xi. \tag{7.168}$$

Under this change of variable, P_0 becomes

$$P_0 \sim \tilde{S}(\chi) = \frac{1}{1+\xi^2}\begin{pmatrix} 1-\xi^2 & 2\xi \\ -2\xi & 1-\xi^2 \end{pmatrix} \tag{7.169}$$

which, on comparison with the solution of the linear system (7.149), (7.150), reveals that

$$\xi = \left(\frac{\mu_2+\mu_1}{\mu_2-\mu_1}\right)\frac{\xi_2-\xi_1}{1+\xi_1\xi_2} \tag{7.170}$$

with $\lambda_i = -i\mu_i$. Combination of (7.167) and (7.170) now produces the classical permutability theorem for the sine-Gordon equation, viz.

$$\tan\left(\frac{\omega^{(2)}-\omega}{4}\right) = \frac{\mu_2+\mu_1}{\mu_2-\mu_1}\tan\left(\frac{\omega_2^{(1)}-\omega_1^{(1)}}{4}\right). \tag{7.171}$$

Exercise

1. Show that for real eigenfunctions ξ_1, ξ_2 and purely imaginary parameters $\lambda_i = -i\mu_i$, the Darboux matrix

$$P(\lambda) = \lambda^2 + \lambda P_1 + P_0$$

is such that

$$P_0 \sim \frac{1}{1+\xi^2}\begin{pmatrix} 1-\xi^2 & 2\xi \\ -2\xi & 1-\xi^2 \end{pmatrix}, \quad \xi = \left(\frac{\mu_2+\mu_1}{\mu_2-\mu_1}\right)\frac{\xi_2-\xi_1}{1+\xi_1\xi_2}.$$

8

Bianchi and Ernst Systems. Bäcklund Transformations and Permutability Theorems

This chapter is concerned with the construction of Bäcklund-Darboux transformations for the generation of exact solutions to Einstein's equations for axially symmetric gravitational fields. That important connections exist between soliton theory and certain areas of general relativity was first established around 1978. Thus, in that year, Maison [241] constructed a Lax pair for the stationary, axially symmetric Einstein equations while Belinsky and Zakharov [27] applied the inverse scattering method to isolate simple soliton-type solutions of these reduced gravitation equations. In the same year, Harrison [157] derived a Bäcklund transformation for the Ernst equation of general relativity [121,217] by using the Wahlquist-Estabrook procedure. In 1979, Neugebauer [267] independently established a Bäcklund transformation for Ernst's equation. This allowed the iterative generation of multi-parameter solutions from a starting 'seed' solution. There has since been extensive research on the application of Bäcklund transformations in general relativity [172,221]. In particular, Cosgrove [88] established important connections between group-theoretic and soliton-theoretic methods for generating not only well-known, but also new stationary axially symmetric solutions of Einstein's equations.

Here, a remarkable analogy is described between the Bianchi system of classical differential geometry as discussed in Chapter 1 and the Ernst equation of general relativity. Moreover, the Harrison transformation [157] is shown to be an 'elliptic' equivalent of the classical Bäcklund transformation for Bianchi surfaces as derived as long ago as 1890 [33]. The Neugebauer transformations [267] emerge as the basic building blocks for the known auto-Bäcklund transformations for the Ernst equation. For instance, the Harrison transformation may be decomposed into two Neugebauer transformations. This motivates the adoption of Neugebauer's elementary Bäcklund transformations in the construction of matrix Darboux transformations for the Ernst equation. A fundamental permutability theorem for the Ernst equation and its dual are thereby established from which solutions may be generated in an iterative manner. Its vectorial form [81] is then related to a permutability theorem for the unit normal to Bianchi surfaces.

8.1 Bianchi Surfaces. Application of the Sym-Tafel Formula

In the previous chapter, it has been shown that if a class of surfaces is associated with a polynomial *su*(2) Lax pair through the Sym-Tafel formula and admits a corresponding elementary matrix Darboux transformation, then the distance between the position vector $\boldsymbol{r}$ and its Bäcklund transform $\boldsymbol{r}'$ is constant. To prove the constant length property, the matrices $g_\nu(\lambda)$ in the Lax pair (7.107) were assumed to be Laurent polynomials in a constant parameter λ. However, one may regard the magnitude of the distance vector $\boldsymbol{r}'-\boldsymbol{r}$ as a function of the 'spectral parameters' λ and λ_0. This interpretation is important when one considers matrix Darboux transformations for 'non-isospectral' Lax pairs, namely those which contain a *non-constant* parameter. Thus, we shall demonstrate that, whereas the Bäcklund transformation (1.187) for Bianchi surfaces does not possess the constant length property, nevertheless the distance $|\boldsymbol{r}' - \boldsymbol{r}|$ may be expressed entirely in terms of non-constant Bäcklund parameters. Indeed, this property is proved to be generic to elementary matrix Darboux transformations for a broad class of non-isospectral linear representations. The proof is based on the Sym-Tafel formula as described subsequently.

In Section 1.6, the derivation of the Bäcklund transformation for Bianchi surfaces governed by

$$\boxed{\begin{gathered} \omega_{uv} + \frac{1}{2}\left(\frac{\rho_u}{\rho}\frac{b}{a}\sin\omega\right)_u + \frac{1}{2}\left(\frac{\rho_v}{\rho}\frac{a}{b}\sin\omega\right)_v - ab\sin\omega = 0 \\ a_v + \frac{1}{2}\frac{\rho_v}{\rho}a - \frac{1}{2}\frac{\rho_u}{\rho}b\cos\omega = 0 \\ b_u + \frac{1}{2}\frac{\rho_u}{\rho}b - \frac{1}{2}\frac{\rho_v}{\rho}a\cos\omega = 0 \\ \rho_{uv} = 0 \end{gathered}} \tag{8.1}$$

led to a non-isospectral version of the Gauss-Weingarten equations, namely

$$\begin{aligned} \Phi_u &= \frac{i}{2}\left[\lambda a\left(\sin\frac{\omega}{2}\sigma_1 + \cos\frac{\omega}{2}\sigma_3\right) + \frac{1}{2}\left(\omega_u + \frac{a}{b}\frac{\rho_v}{\rho}\sin\omega\right)\sigma_2\right]\Phi \\ \Phi_v &= \frac{i}{2}\left[\frac{1}{\lambda}b\left(\sin\frac{\omega}{2}\sigma_1 - \cos\frac{\omega}{2}\sigma_3\right) - \frac{1}{2}\left(\omega_v + \frac{b}{a}\frac{\rho_u}{\rho}\sin\omega\right)\sigma_2\right]\Phi, \end{aligned} \tag{8.2}$$

where the 'spectral parameter' λ is given by

$$\lambda = \pm\sqrt{\frac{K - V(v)}{K + U(u)}}, \qquad \rho = U(u) + V(v) \tag{8.3}$$

and K is an arbitrary real constant. It is noted that in the limit $K \to \infty$ or, equivalently, $\lambda = \pm 1$, the linear system (8.2) is nothing but an $su(2)$ version of the Gauss-Weingarten equations in the form (1.156). The system (8.1) was set down by Bianchi in connection with the isometric deformation of conjugate nets [34, 89]. In that context, the constant K may be identified as the deformation parameter. Purely geometric considerations led Bianchi to the invariance

$$a \to \lambda a, \quad b \to \lambda^{-1} b, \quad \rho \to \frac{(\lambda - \lambda^{-1})^2}{\rho} \tag{8.4}$$

of the system (8.1). This invariance constitutes a generalisation of Lie's transformation for pseudospherical surfaces (cf. Section 1.2). It is remarked that the admittance of a Lie point symmetry by the $su(2)$ form of the Gauss-Weingarten equations for the general Bianchi system has been used by Levi and Sym [234] to determine the integrability constraint $\rho_{uv} = 0$. In the context of the Ernst equation of general relativity, the transformation (8.4) constitutes a Neugebauer transformation (cf. (8.78)) to be discussed in Section 8.6.

It is natural to enquire as to whether it is possible to retrieve the fundamental forms for Bianchi surfaces from the linear representation (8.2). Thus, if we regard $K = K(k)$ as a function of a constant k with respect to which the Sym-Tafel formula is to be evaluated, the position matrix associated with (8.2) reads

$$r = \Phi^{-1}\Phi_k = \lambda_k \Phi^{-1}\Phi_\lambda \tag{8.5}$$

and the corresponding 1st fundamental form may be written as

$$\mathrm{I} = -2\lambda_k^2 \,\mathrm{Tr}(g_{\mu,\lambda} g_{\nu,\lambda}) dx^\mu dx^\nu. \tag{8.6}$$

Evaluation of the latter results in

$$\mathrm{I} = \lambda_k^2 (a^2 du^2 + 2\lambda^{-2} ab \cos\omega \, du dv + \lambda^{-4} b^2 dv^2). \tag{8.7}$$

Now,

$$\lambda_k = \frac{\rho}{2\lambda} \frac{K_k}{(K + U)^2}, \tag{8.8}$$

so that the choice

$$K = -\frac{1}{2k} \tag{8.9}$$

produces

$$\lambda_k = \frac{\rho}{\lambda}\frac{1}{(1-2kU)^2}. \tag{8.10}$$

Accordingly, at $k=0$, so that $\lambda = \pm 1$, the 1st fundamental form (8.7) reduces to that for Bianchi surfaces, namely (cf. $(1.126)_1$)

$$\mathrm{I} = \rho^2(a^2du^2 + 2ab\cos\omega\, du dv + b^2dv^2). \tag{8.11}$$

Similarly,

$$\mathrm{II} = -\frac{\lambda_k}{\det^{1/2}[g_{1,\lambda}, g_{2,\lambda}]}\operatorname{Tr}([g_{1,\lambda}, g_{2,\lambda}](g_{\mu,\nu,\lambda} + [g_{\mu,\lambda}, g_\mu]))dx^\mu dx^\nu \tag{8.12}$$

evaluated at $\lambda = 1$ coincides with the 2nd fundamental form $(1.126)_2$ for Bianchi surfaces.

8.2 Matrix Darboux Transformations for Non-Isospectral Linear Representations

Here, a generalisation of the notion of matrix Darboux transformations valid for a wide class of non-isospectral Lax pairs is presented. In this connection, let us recall the form of the Bäcklund transformation for Bianchi surfaces. Thus, if $\boldsymbol{r}$ is the position vector of a Bianchi surface and $\phi = (\phi^1\ \phi^2)^{\mathsf{T}}$ is a solution of the linear representation (8.2) with parameter $\lambda_0 = i\mu(k_0)$, then a second Bianchi surface Σ' is represented by

$$\boldsymbol{r}' = \boldsymbol{r} + 2\frac{\mu}{1+\mu^2}\frac{1}{\sin\omega}\left[\sin\left(\theta - \frac{\omega}{2}\right)\frac{\boldsymbol{r}_u}{a} - \sin\left(\theta + \frac{\omega}{2}\right)\frac{\boldsymbol{r}_v}{b}\right], \tag{8.13}$$

where

$$\theta = 2\arctan\frac{\phi^1}{\phi^2} \tag{8.14}$$

and μ is assumed to be real in a certain (u, v)-domain. The distance between the surfaces Σ and Σ' is therefore given by

$$|\boldsymbol{r}' - \boldsymbol{r}| = 2|\rho|\frac{|\mu|}{1+\mu^2} \tag{8.15}$$

as a function of the coordinates u and v. On use of the relation (8.10), this may be cast into the form

$$|\boldsymbol{r}' - \boldsymbol{r}| = 2|\lambda_k|\frac{|\Im(\lambda_0)|}{|\lambda - \lambda_0|^2}\Bigg|_{\lambda=1} \tag{8.16}$$

which shows that, although $|\boldsymbol{r}' - \boldsymbol{r}|$ is non-constant, it nevertheless may be expressed in terms of the one-parameter family $\lambda = \lambda(k)$.

To determine a class of non-isospectral Lax pairs for which a generalised elementary matrix Darboux transformation generically admits a distance property of the kind (8.16), it is observed that the compatible system

$$\lambda_u = \frac{1}{2}(\lambda^3 - \lambda)\frac{\rho_u}{\rho}, \quad \lambda_v = \frac{1}{2}(\lambda - \lambda^{-1})\frac{\rho_v}{\rho} \tag{8.17}$$

equally defines the parameter λ, where the K of (8.9) constitutes the constant of integration. Hence, λ itself satisfies (Laurent) polynomial equations. This observation suggests that a matrix Darboux transformation may exist for non-isospectral linear representations if the parameter λ is defined via a compatible system of differential equations which constitute Laurent polynomials of appropriate degrees. This indeed proves to be the case. The result is as follows:

Theorem 33 (A Generalised Elementary Matrix Darboux Transformation). *Consider the linear 2×2 matrix equation*

$$\Phi_u = g(\lambda)\Phi, \qquad \det \Phi = \text{const} \neq 0, \tag{8.18}$$

where the matrix $g(\lambda)$ has entries which are polynomials of degree n in a (complex) 'parameter' λ and degree m in λ^{-1}. The function $f(\lambda)$ in the scalar companion equation

$$\lambda_u = f(\lambda) \tag{8.19}$$

is assumed to be a polynomial of degree $n+2$ in λ and degree m in λ^{-1}. Let $\phi_{[1]}$ and $\phi_{[2]}$ be two known vector-valued 'eigenfunctions' of the equation (8.18) corresponding to the parameters $\lambda_1 \neq \lambda_2$ and

$$P(\lambda) = \lambda P_1 + P_0, \qquad \det P_1 = 1, \tag{8.20}$$

P_0, P_1 independent of λ, be a 2×2 matrix which obeys the linear algebraic system

$$P(\lambda_1)\phi_{[1]} = 0, \quad P(\lambda_2)\phi_{[2]} = 0. \tag{8.21}$$

Then, the transformation

$$\begin{aligned} \Phi \to \quad \Phi' &= p(\lambda)P(\lambda)\Phi \\ g(\lambda) \to g'(\lambda) &= P(\lambda)g(\lambda)P^{-1}(\lambda) + \frac{dP(\lambda)}{du}P^{-1}(\lambda) + \frac{dp(\lambda)}{du}p^{-1}(\lambda), \end{aligned} \tag{8.22}$$

where $p^{-2} = \det P = (\lambda - \lambda_1)(\lambda - \lambda_2)$, *is such that*

(i) $g'(\lambda) \in sl(2)$

(ii) the polynomial structure of (8.18) is preserved.

Proof. (i) As in the proof of Theorem 28, we deduce from

$$\operatorname{Tr}(g') = \operatorname{Tr}\left(\frac{d(pP)}{du}P^{-1}p^{-1}\right) = \frac{d}{du}[\ln\det(pP)] = 0 \qquad (8.23)$$

that $g' \in sl(2)$.

(ii) In analogy with the proof of Theorem 28, it is readily shown that the first two terms in $g'(\lambda)$ form a Laurent polynomial in λ. The remaining term

$$\frac{dp(\lambda)}{du}p^{-1}(\lambda) = -\frac{1}{2}\frac{f(\lambda) - f(\lambda_1)}{\lambda - \lambda_1} - \frac{1}{2}\frac{f(\lambda) - f(\lambda_2)}{\lambda - \lambda_2} \qquad (8.24)$$

is regular at the zeros λ_1 and λ_2, since $f(\lambda)$ constitutes a Laurent polynomial in λ. It therefore remains to be proven that $g'(\lambda)$ is of the same polynomial degree as $g(\lambda)$. Now, on the one hand, it is evident that $g'(\lambda)$ is of degree m in λ^{-1}. On the other hand, on writing the transformation law $(8.22)_2$ in the form

$$g'(\lambda)P(\lambda) = P(\lambda)g(\lambda) + \frac{dP(\lambda)}{du} + \frac{dp(\lambda)}{du}p^{-1}(\lambda)P(\lambda), \qquad (8.25)$$

it is readily verified that terms proportional to λ^{n+2} in (8.25) cancel out so that $g'(\lambda)$ is indeed a polynomial in λ of degree n. □

8.3 Invariance of the $su(2)$ Constraint. A Distance Property

Here, we return to the geometrically relevant case

$$g(\lambda)|_{\lambda=\bar{\lambda}} \in su(2), \quad f(\lambda)|_{\lambda=\bar{\lambda}} \text{ real.} \qquad (8.26)$$

As in Section 7.2, the constraint $(8.26)_1$ is preserved by the generalised matrix Darboux transformation if, for a given solution ϕ of (8.18) with parameter λ_0, the choice

$$\begin{aligned} \phi_{[1]} &= \phi, & \lambda_1 &= \lambda_0 \\ \phi_{[2]} &= A\bar{\phi}, & \lambda_2 &= \bar{\lambda}_0 \end{aligned} \qquad (8.27)$$

is made with $iA = \sigma_2$. For the Bianchi system (8.1), it may be verified that the matrices and functions in (8.2) and (8.17), respectively, satisfy the conditions of Theorem 33 and the geometric constraints (8.26). Furthermore, by choosing the matrix-valued function P_1 appropriately, it is possible to ensure that the particular structure of the linear representation (8.2) is preserved. In fact, such P_1 may be chosen in a purely algebraic manner. Here, we suppress the details of this procedure and refer to the next subsection in which a similar statement is established for the Ernst equation of general relativity. We note parenthetically that, by construction, the function $f(\lambda)$ does not change under the generalised matrix Darboux transformation. This confirms the known result that ρ and therefore the Gaussian curvature $\mathcal{K}$ are preserved by the Bäcklund transformation for Bianchi surfaces.

We conclude this subsection with the remark that since the Sym-Tafel formula (8.5) involves only differentiation with respect to the parameter k, the Bäcklund transformation at the surface level is identical with that for isospectral Lax pairs (7.115), except for a factor of λ_k, namely

$$\boldsymbol{r}' = \boldsymbol{r} - 2\lambda_k \frac{\Im(\lambda_0)}{|\lambda - \lambda_0|^2}\left(\frac{2\Re(\xi)}{|\xi|^2+1}\boldsymbol{t}_1 - \frac{2\Im(\xi)}{|\xi|^2+1}\boldsymbol{t}_2 + \frac{|\xi|^2-1}{|\xi|^2+1}\boldsymbol{t}_3\right). \tag{8.28}$$

As usual, the orthonormal triad $\{\boldsymbol{t}_1, \boldsymbol{t}_2, \boldsymbol{t}_3\}$ is defined by the decomposition

$$\boldsymbol{t}_i \cdot \boldsymbol{e} = \Phi^{-1} e_i \Phi \tag{8.29}$$

and $\xi = \phi^1/\phi^2$. Accordingly, the magnitude of the distance vector $\boldsymbol{r}' - \boldsymbol{r}$ reads

$$|\boldsymbol{r}' - \boldsymbol{r}| = 2|\lambda_k| \frac{|\Im(\lambda_0)|}{|\lambda - \lambda_0|^2}, \tag{8.30}$$

which is consistent with the expression (8.16) for Bianchi surfaces.

8.4 The Ernst Equation of General Relativity

In what follows, a remarkable analogy is noted between the Bianchi system of classical differential geometry and an important equation due to Ernst resulting from Einstein's theory of relativity. Thus, it has been demonstrated in Section 1.6 that the normal $\boldsymbol{N}$ to Bianchi surfaces of total curvature $\mathcal{K} = -1/\rho^2$ with $\rho_{uv} = 0$ satisfies the vectorial equation

$$(\rho \boldsymbol{N} \times \boldsymbol{N}_u)_v + (\rho \boldsymbol{N} \times \boldsymbol{N}_v)_u = 0 \tag{8.31}$$

and that, in fact, the surfaces may be reconstructed from the solutions of this equation. It has also been remarked that the parametrisations

$$N = \frac{1}{|\mathcal{E}|^2+1}\begin{pmatrix} \mathcal{E}+\bar{\mathcal{E}} \\ -i(\mathcal{E}-\bar{\mathcal{E}}) \\ |\mathcal{E}|^2-1 \end{pmatrix}, \quad N = \boldsymbol{N}\cdot\boldsymbol{\sigma} \tag{8.32}$$

produce the alternative form

$$\mathcal{E}_{uv} + \frac{1}{2}\frac{\rho_v}{\rho}\mathcal{E}_u + \frac{1}{2}\frac{\rho_u}{\rho}\mathcal{E}_v = 2\frac{\mathcal{E}_u\mathcal{E}_v\bar{\mathcal{E}}}{|\mathcal{E}|^2+1}, \quad \rho_{uv} = 0 \tag{8.33}$$

and the characterisation

$$(\rho N N_u)_v + (\rho N N_v)_u = 0, \quad N^2 = \mathbb{1}, \quad N^\dagger = N. \tag{8.34}$$

In the latter, it is understood that the degenerate case $N = \pm\mathbb{1}$ is excluded so that N may be assumed to be trace-free.

It is recalled that Sym's approach allows one to associate in a canonical manner two-dimensional submanifolds (surfaces) with any simple Lie algebra [356]. In this monograph, we have restricted our attention, in the main, to the $su(2)$ Lie algebra which gives rise to $\mathbb{R}^3$ as the ambient space of surfaces. However, it should be noted that the Lie algebra $sl(2) \cong so(2,1)$ which corresponds to the embedding of surfaces in a three-dimensional Minkowski space $\mathbb{M}^3$ also arises in soliton theory. Here, we merely state that there exist 'elliptic' surfaces ($\mathcal{K} > 0$) of Bianchi type in $\mathbb{M}^3$ [358] which are represented by an elliptic analogue of the complex equation (8.33), namely

$$\xi_{z\bar{z}} + \frac{1}{2}\frac{\rho_{\bar{z}}}{\rho}\xi_z + \frac{1}{2}\frac{\rho_z}{\rho}\xi_{\bar{z}} = 2\frac{\xi_z\xi_{\bar{z}}\bar{\xi}}{|\xi|^2-1}, \quad \rho_{z\bar{z}} = 0. \tag{8.35}$$

As in the Euclidean case, ξ and ρ are complex and real, respectively, while z an $\bar{z}$ are complex conjugate variables.

The above equation is well-known in general relativity. It provides particular solutions of Einstein's vacuum equations with ξ being the 'gravitational' potential. Indeed, on introduction of the *Ernst potential*

$$\mathcal{E} = \frac{1-\xi}{1+\xi}, \tag{8.36}$$

the celebrated *Ernst equation* [121]

$$\boxed{\mathcal{E}_{z\bar{z}} + \frac{1}{2}\frac{\rho_{\bar{z}}}{\rho}\mathcal{E}_z + \frac{1}{2}\frac{\rho_z}{\rho}\mathcal{E}_{\bar{z}} = \frac{\mathcal{E}_z\mathcal{E}_{\bar{z}}}{\Re(\mathcal{E})}, \quad \rho_{z\bar{z}} = 0} \tag{8.37}$$

results.

8.4.1 Linear Representations

The Ernst equation may also be formulated in terms of a non-isospectral extension of the principal chiral field model equation associated with the $O(2, 1)$ Lie group. Thus, if one parametrises a real 2×2 matrix S subject to the constraints

$$S^2 = -\mathbb{1}, \quad \operatorname{Tr} S = 0 \tag{8.38}$$

according to

$$S = \frac{1}{\mathcal{E} + \bar{\mathcal{E}}} \begin{pmatrix} i(\mathcal{E} - \bar{\mathcal{E}}) & -2\mathcal{E}\bar{\mathcal{E}} \\ 2 & -i(\mathcal{E} - \bar{\mathcal{E}}) \end{pmatrix}, \tag{8.39}$$

then the Ernst equation adopts the matrix form

$$(\rho S S_z)_{\bar{z}} + (\rho S S_{\bar{z}})_z = 0. \tag{8.40}$$

The corresponding linear representation reads (see [220, 393])

$$\Psi_z = \frac{1}{2}(1 - \lambda) S S_z \Psi, \qquad \Psi_{\bar{z}} = \frac{1}{2}(1 - \lambda^{-1}) S S_{\bar{z}} \Psi, \tag{8.41}$$

where the non-constant parameter λ is given by

$$\lambda = \pm\sqrt{\frac{k - i\bar{Z}(\bar{z})}{k + iZ(z)}} = \bar{\lambda}^{-1}, \qquad \rho = Z(z) + \bar{Z}(\bar{z}). \tag{8.42}$$

It is readily verified that cross-differentation of (8.41) results in the matrix equation (8.40).

The Lax pair (8.41) is but one of many linear representations for the Ernst equation that have appeared in the literature. Important amongst these are representations due to Belinsky and Zakharov [27], Harrison [157], Hauser and Ernst [159], Kinnersley and Chitre [199] and Maison [241]. A survey of these linear representations and their gauge-equivalence has been given both by Cosgrove [88] and Kramer [216]. Here, we focus on a formulation proposed by Neugebauer in [267] which gives rise to a polynomial form of the Ernst equation. Therein, the constraints (8.38) are satisfied identically by setting

$$S = \phi^{-1} \begin{pmatrix} -i & 0 \\ 0 & i \end{pmatrix} \phi. \tag{8.43}$$

The parametrisation (8.39) is obtained by choosing

$$\phi = \begin{pmatrix} i & \mathcal{E} \\ -i & \bar{\mathcal{E}} \end{pmatrix} \tag{8.44}$$

and the gauge transformation

$$\Phi = \phi\Psi \tag{8.45}$$

then implies that ϕ may be identified with $\Phi|_{\lambda=1}$ if we make the choice $\Psi|_{\lambda=1} = \mathbb{1}$. Indeed, it is evident that ϕ is a particular solution of the transformed Lax pair

$$\begin{aligned}
\Phi_z &= \left[\begin{pmatrix} \frac{\mathcal{E}_z}{\mathcal{E}+\bar{\mathcal{E}}} & 0 \\ 0 & \frac{\bar{\mathcal{E}}_z}{\mathcal{E}+\bar{\mathcal{E}}} \end{pmatrix} + \lambda \begin{pmatrix} 0 & \frac{\mathcal{E}_z}{\mathcal{E}+\bar{\mathcal{E}}} \\ \frac{\bar{\mathcal{E}}_z}{\mathcal{E}+\bar{\mathcal{E}}} & 0 \end{pmatrix}\right]\Phi \\
\phi_{\bar{z}} &= \left[\begin{pmatrix} \frac{\mathcal{E}_{\bar{z}}}{\mathcal{E}+\bar{\mathcal{E}}} & 0 \\ 0 & \frac{\bar{\mathcal{E}}_{\bar{z}}}{\mathcal{E}+\bar{\mathcal{E}}} \end{pmatrix} + \frac{1}{\lambda} \begin{pmatrix} 0 & \frac{\mathcal{E}_{\bar{z}}}{\mathcal{E}+\bar{\mathcal{E}}} \\ \frac{\bar{\mathcal{E}}_{\bar{z}}}{\mathcal{E}+\bar{\mathcal{E}}} & 0 \end{pmatrix}\right]\Phi
\end{aligned} \tag{8.46}$$

evaluated at $\lambda = 1$. Hence, it has been established that the eigenfunction Φ encodes explicitly the Ernst potential $\mathcal{E}$. Since Darboux transformations act on eigenfunctions, this result is crucial to the direct construction of the associated transformed Ernst potentials.

8.4.2 The Dual 'Ernst Equation'

We now investigate the algebraic structure of the linear representation (8.46). It is observed that, as in the case of Bianchi surfaces, the parameter λ obeys the pair of equations

$$\lambda_z = \frac{1}{2}(\lambda^3 - \lambda)C, \quad \lambda_{\bar{z}} = \frac{1}{2}(\lambda - \lambda^{-1})C^*, \tag{8.47}$$

where the functions C and C^* are given by

$$C = \frac{\rho_z}{\rho}, \quad C^* = \frac{\rho_{\bar{z}}}{\rho}. \tag{8.48}$$

Hence, (8.46) is of the algebraic form

$$\begin{aligned}
\Phi_z &= \left[\begin{pmatrix} A & 0 \\ 0 & B \end{pmatrix} + \lambda \begin{pmatrix} 0 & A \\ B & 0 \end{pmatrix}\right]\Phi \\
\phi_{\bar{z}} &= \left[\begin{pmatrix} B^* & 0 \\ 0 & A^* \end{pmatrix} + \frac{1}{\lambda} \begin{pmatrix} 0 & B^* \\ A^* & 0 \end{pmatrix}\right]\Phi,
\end{aligned} \tag{8.49}$$

where λ is governed by the system (8.47). It is therefore natural to consider Lax pairs of the type (8.47), (8.49) with as yet unspecified (complex) functions A, B, C, A^*, B^* and C^*. Thus, the compatibility conditions $\phi_{z\bar{z}} = \phi_{\bar{z}z}$ and $\lambda_{z\bar{z}} = \lambda_{\bar{z}z}$ yield

$$\begin{aligned} A_{\bar{z}} + AA^* - AB^* + \frac{1}{2}C^*A + \frac{1}{2}CB^* &= 0 \\ B_{\bar{z}} + BB^* - A^*B + \frac{1}{2}C^*B + \frac{1}{2}CA^* &= 0 \\ A^*_z + AA^* - A^*B + \frac{1}{2}CA^* + \frac{1}{2}C^*B &= 0 \\ B^*_z + BB^* - AB^* + \frac{1}{2}CB^* + \frac{1}{2}C^*A &= 0 \\ C_{\bar{z}} + CC^* &= 0 \\ C^*_z + CC^* &= 0 \end{aligned} \tag{8.50}$$

and the relations

$$\begin{aligned} A_{\bar{z}} + AA^* &= B^*_z + BB^* \\ A^*_z + AA^* &= B_{\bar{z}} + BB^* \\ C_{\bar{z}} &= C^*_z, \end{aligned} \tag{8.51}$$

which are readily seen to be a consequence of (8.50). The subsystem (8.51) guarantees the existence of functions $\mathcal{F}$, $\mathcal{F}^*$ and ρ obeying the linear equations

$$\begin{aligned} &\mathcal{F}_z = A(\mathcal{F} + \mathcal{F}^*), \quad &\mathcal{F}_{\bar{z}} = B^*(\mathcal{F} + \mathcal{F}^*) \\ &\mathcal{F}^*_z = B(\mathcal{F} + \mathcal{F}^*), \quad &\mathcal{F}^*_{\bar{z}} = A^*(\mathcal{F} + \mathcal{F}^*) \\ &\rho_z = C\rho, &\rho_{\bar{z}} = C^*\rho \end{aligned} \tag{8.52}$$

which may be used to parametrise the functions A, B, C, A^*, B^*, C^* in terms of $\mathcal{F}$, $\mathcal{F}^*$, ρ. The system (8.50) then reduces to

$$\begin{aligned} \mathcal{F}_{z\bar{z}} + \frac{1}{2}\frac{\rho_{\bar{z}}}{\rho}\mathcal{F}_z + \frac{1}{2}\frac{\rho_z}{\rho}\mathcal{F}_{\bar{z}} &= 2\frac{\mathcal{F}_z\mathcal{F}_{\bar{z}}}{\mathcal{F} + \mathcal{F}^*} \\ \mathcal{F}^*_{z\bar{z}} + \frac{1}{2}\frac{\rho_{\bar{z}}}{\rho}\mathcal{F}^*_z + \frac{1}{2}\frac{\rho_z}{\rho}\mathcal{F}^*_{\bar{z}} &= 2\frac{\mathcal{F}^*_z\mathcal{F}^*_{\bar{z}}}{\mathcal{F} + \mathcal{F}^*} \end{aligned} \tag{8.53}$$

together with the harmonic condition

$$\rho_{z\bar{z}} = 0. \tag{8.54}$$

The particular case

$$A^* = \bar{A}, \quad B^* = \bar{B}, \quad C^* = \bar{C} \tag{8.55}$$

implies, without loss of generality, that

$$\mathcal{F}^* = \bar{\mathcal{F}}, \quad \bar{\rho} = \rho \tag{8.56}$$

and the Ernst equation (8.37) is retrieved if we identify $\mathcal{F}$ with the Ernst potential $\mathcal{E}$. Thus, the Ernst equation is encapsulated in the linear representation (8.49) with (8.47) and subject to the constraint (8.55).

A different specialisation of the Ernst-type system (8.53), (8.54) is given by

$$\bar{\mathcal{F}} = \mathcal{F}, \quad \bar{\mathcal{F}}^* = \mathcal{F}^*, \quad \bar{\rho} = \rho \tag{8.57}$$

and is directly linked to the space-time metric under consideration. This is seen as follows. It is readily verified that the linear substitution

$$\begin{aligned} \hat{A} &= -B + \frac{1}{2}C, \quad & \hat{A}^* &= -A^* + \frac{1}{2}C^* \\ \hat{B} &= -A + \frac{1}{2}C, \quad & \hat{B}^* &= -B^* + \frac{1}{2}C^* \\ \hat{C} &= C, & \hat{C}^* &= C^* \end{aligned} \tag{8.58}$$

preserves the form of (8.50). If $A, \ldots, C^*$ are now associated with the Ernst equation, i.e., they satisfy the constraint (8.55), then this corresponds to the specialisation

$$\hat{B}^* = \bar{\hat{A}}, \quad \hat{A}^* = \bar{\hat{B}}, \quad \hat{C}^* = \bar{\hat{C}}. \tag{8.59}$$

This implies, in turn, that $\hat{A}$, $\hat{A}^*$, $\hat{B}$, $\hat{B}^*$ may be parametrised in terms of two real functions $\mathcal{F}$, $\mathcal{F}^*$ according to (8.52). Thus, the Ernst equation (8.37) may be mapped to the real-valued *'dual Ernst equation'* (8.53) and vice versa. It is noted that the dual Ernst equation is formally obtained from the Ernst equation by replacing complex conjugation with the star operation. Evaluation of the linear transformation (8.58) in terms of the Ernst potential $\mathcal{E}$ and $\mathcal{F}$, $\mathcal{F}^*$ leads to the following [217]:

Theorem 34. *The solutions of the Ernst equation (8.37) and the (real-valued) dual Ernst equation (8.53) are related by the contact transformation*

$$(\mathfrak{S}) \qquad \mathcal{F} = \frac{\rho}{f} + \omega, \quad \mathcal{F}^* = \frac{\rho}{f} - \omega, \tag{8.60}$$

where $f = \Re(\mathcal{E})$ and the real potential ω is defined by the compatible system

$$\omega_z = i\rho \frac{[\Im(\mathcal{E})]_z}{\Re(\mathcal{E})^2}, \quad \omega_{\bar{z}} = -i\rho \frac{[\Im(\mathcal{E})]_{\bar{z}}}{\Re(\mathcal{E})^2}. \tag{8.61}$$

The system (8.50) possesses another important property. It admits the 'conservation law'

$$(ABC^{-1})_{\bar{z}} = (A^* B^* C^{*-1})_z. \tag{8.62}$$

For the Ernst equation, this implies that there exists a real potential γ which satisfies

$$\gamma_z = \frac{1}{4}\frac{\rho}{\rho_z}\frac{\mathcal{E}_z \bar{\mathcal{E}}_z}{\Re(\mathcal{E})^2}, \quad \gamma_{\bar{z}} = \frac{1}{4}\frac{\rho}{\rho_{\bar{z}}}\frac{\mathcal{E}_{\bar{z}} \bar{\mathcal{E}}_{\bar{z}}}{\Re(\mathcal{E})^2}. \tag{8.63}$$

It is emphasized that these defining relations may also be expressed in terms of $\mathcal{F}$ and $\mathcal{F}^*$ by virtue of (8.58). If we now define the four-dimensional pseudo-Riemannian metric [236]

$$ds^2 = \frac{1}{f}(e^{2\gamma} dz d\bar{z} + \rho^2 d\varphi^2) - f(dt - \omega d\varphi)^2 \tag{8.64}$$

then it may be shown that Einstein's vacuum equations

$$R_{ij} = 0, \tag{8.65}$$

where R_{ij} is the usual Ricci tensor [117], are satisfied modulo the dual Ernst equation and the Frobenius system (8.63). The main features of the metric (8.64) are that it is of block-diagonal form and does not depend on the ignorable coordinates t and φ. In the terminology of general relativity, the space-time metric (8.64) is said to admit two commuting time- and space-like Killing vectors ∂_t and ∂_φ, respectively, which are hypersurface-orthogonal [221]. In physical terms, t and φ may be regarded as temporal and angular coordinates, respectively. Accordingly, the dual Ernst equation, like the Ernst equation itself, governs stationary and axi-symmetric gravitational fields in a vacuum [121].

8.5 The Ehlers and Matzner-Misner Transformations

The Kramer-Neugebauer transformation (8.60), (8.61) plays a pivotal role in the construction of solutions of the Ernst equation and its dual. This is due to the existence of two Möbius transformations associated with the Ernst equation and its dual with which it may be conjugated. Thus, it is evident that the matrix equation (8.40) is preserved by the similarity transformation

$$S \to C^{-1} S C, \tag{8.66}$$

where C is an arbitrary constant real matrix. This invariance is reflected by the Möbius transformation

$$(\mathfrak{E}) \qquad \mathcal{E} \to \frac{\alpha\mathcal{E} + i\beta}{i\gamma\mathcal{E} + \delta}, \qquad \alpha, \beta, \gamma, \delta \in \mathbb{R}, \tag{8.67}$$

which is known as *Ehlers transformation* [115]. The dual Ernst equation may also be cast in the form

$$(\rho F F_z)_{\bar{z}} + (\rho F F_{\bar{z}})_z = 0 \tag{8.68}$$

in terms of the matrix

$$F = \frac{1}{\mathcal{F} + \mathcal{F}^*} \begin{pmatrix} \mathcal{F} - \mathcal{F}^* & 2\mathcal{F}\mathcal{F}^* \\ 2 & \mathcal{F}^* - \mathcal{F} \end{pmatrix}. \tag{8.69}$$

In this case, the associated Möbius invariance reads

$$(\mathfrak{M}) \qquad \mathcal{F} \to \frac{a\mathcal{F} + b}{c\mathcal{F} + d}, \quad \mathcal{F}^* \to \frac{a\mathcal{F}^* - b}{-c\mathcal{F}^* + d} \tag{8.70}$$

with constants $a, b, c, d \in \mathbb{R}$. By construction, the *Matzner-Misner transformation* [252] (8.70) acts directly on the metric coefficients f, ω, ρ and may be compensated for by a linear transformation of the ignorable coordinates t and φ. Hence, the Möbius transformation acting on the dual Ernst equation is generated by trivial linear superpositions of the Killing vectors ∂_t and ∂_φ.

The transformations $\mathfrak{S}$, $\mathfrak{E}$, $\mathfrak{M}$ can be exploited in the following way. Given a solution $(\mathcal{F}, \mathcal{F}^*)$ of the dual Ernst equation (or, equivalently, a space-time metric of the Lewis-Papapetrou form (8.64)), one first applies the Matzner-Misner transformation $\mathfrak{M}$ and then maps the new solution of the dual Ernst equation to a solution of the Ernst equation by means of the inverse of the Kramer-Neugebauer transformation $\mathfrak{S}^{-1}$. The Ehlers transformation $\mathfrak{E}$ is then applied, and the result is mapped to another solution of the dual Ernst equation via the Kramer-Neugebauer transformation $\mathfrak{S}$. The composite *Geroch transformation* $\mathfrak{G}$ obtained thereby is given by

$$(\mathfrak{G}) \qquad (\mathcal{F}, \mathcal{F}^*) \to (\mathfrak{S} \circ \mathfrak{E} \circ \mathfrak{S}^{-1} \circ \mathfrak{M})(\mathcal{F}, \mathcal{F}^*). \tag{8.71}$$

In two remarkable papers, Geroch [149] showed that, in principle, one can generate solutions of the (dual) Ernst equation containing an arbitrary number of parameters by successive application of the Geroch transformation (8.71), that is via a conjugation of the form

$$\cdots \circ \mathfrak{G} \circ \mathfrak{G} \circ \mathfrak{G} \circ \mathfrak{G}. \tag{8.72}$$

This is possible because the transformation $\mathfrak{G}$ contains arbitrary parameters to be chosen differently at each step of the iteration process. Geroch conjectured that thereby it would be possible to generate any stationary axi-symmetric space-time by acting on flat space. Hauser and Ernst [160] subsequently confirmed a restricted version of this conjecture. However, in practice, the explicit representation of the compound Geroch transformations (8.72) has proven to be highly non-trivial and requires the introduction of an infinite hierarchy of so-called *Kinnersley-Chitre potentials* [197, 198]. These potentials have been used to show that the collection of Geroch transformations forms an infinite-dimensional Banach Lie group [341]. The finite *Hoenselaers-Kinnersley-Xanthopoulos (HKX) transformations* [168] constitute an important subset of this group. An application of these HKX transformations has confirmed, by means of an exact solution, the conjecture of the existence of gravitational spin-spin repulsion [104].

8.6 The Neugebauer and Harrison Bäcklund Transformations

In 1978, Harrison [157], using the Wahlquist-Estabrook procedure [123, 378] and building upon earlier work of Maison [241], constructed an auto-Bäcklund transformation for the Ernst equation. This Bäcklund transformation is here shown to constitute the 'elliptic' equivalent of the classical Bäcklund transformation for Bianchi surfaces.

Independently, in 1979, Neugebauer [267] derived Bäcklund transformations which prove to be the elementary building blocks for all known Bäcklund transformations for the Ernst equation. For instance, the Harrison transformation may be decomposed into two Neugebauer transformations while the Darboux transformations used by Belinsky and Zakharov [27] to generate multi-soliton solutions of the Ernst equation may also be formulated in terms of Neugebauer transformations. It is therefore natural to adopt Neugebauer's elementary Bäcklund transformations in the construction of matrix Darboux-type transformations for the Ernst equation of general relativity.

Consider a vector-valued solution ϕ of the linear system (8.47), (8.49) for the coupled Ernst-type system (8.53), (8.54). It is readily verified that the ratio

$$q = -\frac{\phi^2}{\phi^1}, \quad \phi = \begin{pmatrix} \phi^1 \\ \phi^2 \end{pmatrix} \tag{8.73}$$

satisfies the Riccati equations

$$\begin{aligned} q_z &= (\lambda q^2 - q)A + (q - \lambda)B \\ q_{\bar{z}} &= (\lambda^{-1} q^2 - q)B^* + (q - \lambda^{-1})A^*, \end{aligned} \tag{8.74}$$

where λ is a solution of (8.47), viz.

$$\lambda_z = \frac{1}{2}(\lambda^3 - \lambda)C, \quad \lambda_{\bar{z}} = \frac{1}{2}(\lambda - \lambda^{-1})C^*. \tag{8.75}$$

It is consistent to impose the following natural constraints in the Ernst and dual Ernst case corresponding, in turn, to $\mathcal{F} = \mathcal{E}$, $\mathcal{F}^* = \bar{\mathcal{E}}$ (*Ernst picture*) and $\bar{\mathcal{F}} = \mathcal{F}$, $\bar{\mathcal{F}}^* = \mathcal{F}^*$ (*metric picture*), respectively:

$$\begin{aligned} &\text{ERNST PICTURE:} \quad && \bar{q} = 1/q, \quad \bar{\lambda} = 1/\lambda \\ &\text{METRIC PICTURE:} \quad && \bar{q} = q, \quad \bar{\lambda} = 1/\lambda. \end{aligned} \tag{8.76}$$

Now, the Neugebauer Bäcklund transformation for the sextuplet $\Omega = (A, A^*, B, B^*, C, C^*)$ may be expressed entirely in terms of the pair of *pseudopotentials* [123, 378] $\beta = (q, \lambda)$. In what follows, a superscript (i) on transformed objects refers to the pair $\beta_i = (q_i, \lambda_i)$ which 'drive' the corresponding transformation:

Theorem 35 (The Neugebauer Bäcklund Transformation $\mathfrak{I}$). *Let $\Omega = (A, A^*, B, B^*, C, C^*)$ be a solution of the Ernst-type system in polynomial form (8.50) and $\beta = (q, \lambda)$, $\beta_1 = (q_1, \lambda_1)$ corresponding solutions of the compatible system (8.74), (8.75). Then, a second solution $\Omega^{(1)}$, $\beta^{(1)}$ of the Ernst-type system and its associated Riccati system is given by*

$$(\mathfrak{I}) \begin{cases} A^{(1)} = \lambda_1 q_1 A, & B^{(1)} = \dfrac{\lambda_1}{q_1} B, \quad C^{(1)} = \lambda_1^2 C \\ A^{*(1)} = \dfrac{1}{q_1 \lambda_1} A^*, & B^{*(1)} = \dfrac{q_1}{\lambda_1} B^*, \quad C^{*(1)} = \dfrac{1}{\lambda_1^2} C^* \\ q^{(1)} = \dfrac{q}{q_1}, & \lambda^{(1)} = \dfrac{\lambda}{\lambda_1}. \end{cases} \tag{8.77}$$

Both Ernst and metric pictures are preserved by $\mathfrak{I}$.

The above theorem may be verified directly. In particular, if the constraints (8.76) are imposed, then the new solution $\Omega^{(1)}$ again satisfies the 'reality conditions' (8.55) and (8.59), respectively. Consequently, the Neugebauer transformation maps within solutions of the Ernst equation or the dual Ernst equation. It is noted that the Neugebauer transformation changes the harmonic function ρ which is defined by $(8.52)_{5,6}$. In fact, one obtains

$$\rho^{(1)} = \hat{c}\frac{\left(\lambda_1 - \lambda_1^{-1}\right)^2}{\rho}, \tag{8.78}$$

where $\hat{c}$ is an arbitrary constant. This constitutes the equivalent of Bianchi's invariance (8.4). However, the products ABC^{-1} and $A^*B^*C^{*-1}$ in the conservation law (8.62) are invariant so that the metric function γ is unaffected by $\mathfrak{I}$.

Interestingly, at $\lambda_1 = 1$, the solution of the Riccati system (8.74) is given by

$$q_1 = -\frac{\mathcal{F}^* - \kappa}{\mathcal{F} + \kappa}, \tag{8.79}$$

where $\kappa \in \mathbb{C}$ is an arbitary constant of integration. In this case, the transformation laws (8.77), namely

$$\begin{aligned} \mathcal{F}^{(1)}_z &= q_1\frac{\mathcal{F}^{(1)} + \mathcal{F}^{*(1)}}{\mathcal{F} + \mathcal{F}^*}\mathcal{F}_z, \qquad & \mathcal{F}^{*(1)}_z &= \frac{1}{q_1}\frac{\mathcal{F}^{(1)} + \mathcal{F}^{*(1)}}{\mathcal{F} + \mathcal{F}^*}\mathcal{F}^*_z \\ \mathcal{F}^{(1)}_{\bar{z}} &= q_1\frac{\mathcal{F}^{(1)} + \mathcal{F}^{*(1)}}{\mathcal{F} + \mathcal{F}^*}\mathcal{F}_{\bar{z}}, \qquad & \mathcal{F}^{*(1)}_{\bar{z}} &= \frac{1}{q_1}\frac{\mathcal{F}^{(1)} + \mathcal{F}^{*(1)}}{\mathcal{F} + \mathcal{F}^*}\mathcal{F}^*_{\bar{z}}, \end{aligned} \tag{8.80}$$

may be integrated explicitly for the new potentials $\mathcal{F}^{(1)}$ and $\mathcal{F}^{*(1)}$. Indeed, it is readily verified that

$$\mathcal{F}^{(1)} = \frac{a\mathcal{F} + b}{c\mathcal{F} + d}, \quad \mathcal{F}^{*(1)} = \frac{a\mathcal{F}^* - b}{-c\mathcal{F}^* + d}, \qquad d = c\kappa, \tag{8.81}$$

which constitutes nothing but a Möbius invariance of the general Ernst-type system (8.53). In the case of the Ernst equation and its dual, the Neugebauer transformation $\mathfrak{I}$ therefore represents, in turn, a generalisation of the Ehlers (8.67) and Matzner-Misner (8.70) transformations.

A second application of the Neugebauer transformation acts trivially on the solution space. This is seen as follows. If $\beta_2 = (q_2, \lambda_2)$ is another solution associated with Ω then, according to Theorem 35,

$$q_2^{(1)} = \frac{q_2}{q_1}, \quad \lambda_2^{(1)} = \frac{\lambda_2}{\lambda_1} \tag{8.82}$$

constitutes a solution corresponding to $\Omega^{(1)}$. This solution may be used in a second application of the Neugebauer transformation which produces

$$\begin{aligned} q^{(12)} &= \frac{q^{(1)}}{q_2^{(1)}} = \frac{q}{q_2} =: q^{(2)} \\ \lambda^{(12)} &= \frac{\lambda^{(1)}}{\lambda_2^{(1)}} = \frac{\lambda}{\lambda_2} =: \lambda^{(2)}. \end{aligned} \tag{8.83}$$

Thus, $\Omega^{(12)}$ may be generated directly by means of one Neugebauer transformation if one chooses β_2 instead of β_1 in Theorem 35. Consequently, Neugebauer transformations do not commute so that the order of application is crucial. The composition laws

$$\begin{aligned} \mathfrak{I}\left(\beta_2^{(1)}\right) \circ \mathfrak{I}(\beta_1) &= \mathfrak{I}(\beta_2) \\ \mathfrak{I}\left(\beta_1^{(2)}\right) \circ \mathfrak{I}(\beta_2) &= \mathfrak{I}(\beta_1) \end{aligned} \tag{8.84}$$

are illustrated in Figure 8.1.

Neugebauer transformations may be iterated if one links any two successive transformations $\mathfrak{I}$ by a Kramer-Neugebauer transformation $\mathfrak{S}$. To do this, one needs to know how $\beta = (q, \lambda)$ transforms under $\mathfrak{S}$. This information is encapsulated in the following:

Theorem 36 (The Kramer-Neugebauer Transformation $\mathfrak{S}$). *Let $\Omega = (A, A^*, B, B^*, C, C^*)$ be a solution of the Ernst-type system in polynomial form (8.50) and let $\beta = (q, \lambda)$ be a corresponding solution of the compatible system (8.74), (8.75). Then, a second solution $\hat{\Omega}$, $\hat{\beta}$ of the Ernst-type system and its associated*

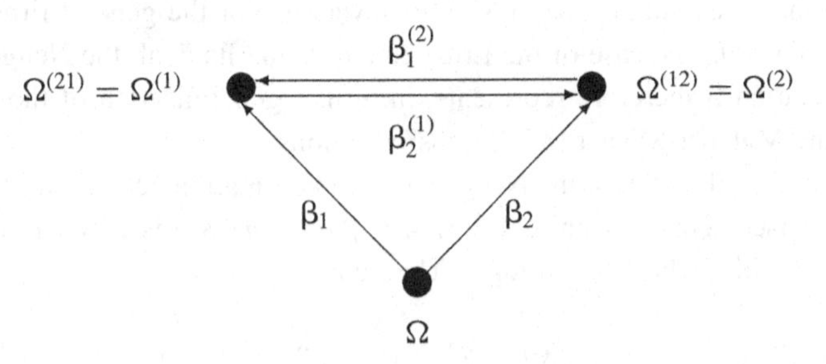

Figure 8.1. A group property of the Neugebauer transformation $\mathfrak{I}$.

Riccati system is given by

$$(\mathfrak{S})\left\{\begin{array}{lll} \hat{A}=-B+\frac{1}{2}C, & \hat{A}^*=-A^*+\frac{1}{2}C^*, & \hat{C}=C \\ \hat{B}=-A+\frac{1}{2}C, & \hat{B}^*=-B^*+\frac{1}{2}C^*, & \hat{C}^*=C^* \\ \hat{q}=\dfrac{q-\lambda}{\lambda q-1}, & \hat{\lambda}=\lambda. & \end{array}\right. \tag{8.85}$$

The transformation $\mathfrak{S}$ is involutive, that is $\mathfrak{S}^2 =$ id. In particular, the Ernst picture is mapped to the metric picture and vice versa.

It is noted that the Kramer-Neugebauer transformation interchanges the reality constraints (8.76). The harmonic function ρ remains unchanged while

$$\begin{aligned} \hat{A}\hat{B}\hat{C}^{-1} &= ABC^{-1} - \frac{1}{2}(A+B) + \frac{1}{4}C \\ &= \gamma_z - \frac{1}{2}\frac{\mathcal{F}_z + \mathcal{F}_z^*}{\mathcal{F}+\mathcal{F}^*} + \frac{1}{4}\frac{\rho_z}{\rho}. \end{aligned} \tag{8.86}$$

A similar expression holds for the product $\hat{A}^*\hat{B}^*\hat{C}^{*-1}$ which implies that the solution of the 'hat' analogue of (8.63) reads

$$\hat{\gamma} = \gamma + \frac{1}{4}\ln\left[\frac{\rho}{(\mathcal{F}+\mathcal{F}^*)^2}\right] \tag{8.87}$$

up to an irrelevant additive constant. Hence, the new metric coefficient $\hat{\gamma}$ in (8.64) may be given without quadrature.

Let us now consider a transformation of the form

$$\tilde{\mathfrak{I}} = \mathfrak{S} \circ \mathfrak{I} \circ \mathfrak{S}. \tag{8.88}$$

If β and β_2 are two pairs of pseudopotentials associated with the solution Ω then, according to Theorem 36,

$$\begin{aligned} \hat{q} &= \frac{q-\lambda}{\lambda q - 1}, & \hat{\lambda} &= \lambda \\ \hat{q}_2 &= \frac{q_2-\lambda_2}{\lambda_2 q_2 - 1}, & \hat{\lambda}_2 &= \lambda_2 \end{aligned} \tag{8.89}$$

are pseudopotentials corresponding to the solution $\hat{\Omega}$. Subsequent application of the Neugebauer transformation $\mathfrak{I}$ to $\hat{\beta}$ then produces a solution $\hat{\Omega}^{(2)}$ with associated pseudopotentials

$$\hat{q}^{(2)} = \frac{\hat{q}}{\hat{q}_2} = \frac{(q-\lambda)(\lambda_2 q_2 - 1)}{(q_2-\lambda_2)(\lambda q - 1)}, \qquad \hat{\lambda}^{(2)} = \frac{\hat{\lambda}}{\hat{\lambda}_2} = \frac{\lambda}{\lambda_2}. \tag{8.90}$$

Finally, a second application of the Kramer-Neugebauer transformation $\mathfrak{S}$ gives rise to a solution $\tilde{\Omega}^{(2)}$ with corresponding $\tilde{\beta}^{(2)}$ given in the following:

Theorem 37 (The Neugebauer Bäcklund Transformation $\tilde{\mathfrak{I}}$). *Let $\Omega = (A, A^*, B, B^*, C, C^*)$ be a solution of the Ernst-type system in polynomial form (8.50) and $\beta = (q, \lambda)$, $\beta_2 = (q_2, \lambda_2)$ corresponding solutions of the compatible system (8.74), (8.75). Then, a second solution $\tilde{\Omega}^{(2)}$, $\tilde{\beta}^{(2)}$ of the Ernst-type system and its associated Riccati system obtained by means of the composite transformation $\tilde{\mathfrak{I}} = \mathfrak{S} \circ \mathfrak{I} \circ \mathfrak{S}$ is given by*

$$(\tilde{\mathfrak{I}}) \left\{ \begin{aligned} \tilde{A}^{(2)} &= \lambda_2 \frac{\lambda_2 q_2 - 1}{q_2 - \lambda_2} A - \frac{1}{2}\frac{\lambda_2^3 - \lambda_2}{q_2 - \lambda_2} C \\ \tilde{A}^{*(2)} &= \frac{1}{\lambda_2}\frac{\lambda_2 q_2 - 1}{q_2 - \lambda_2} A^* - \frac{q_2}{2\lambda_2^2}\frac{\lambda_2^2 - 1}{q_2 - \lambda_2} C^* \\ \tilde{B}^{(2)} &= \lambda_2 \frac{q_2 - \lambda_2}{\lambda_2 q_2 - 1} B + \frac{q_2}{2}\frac{\lambda_2^3 - \lambda_2}{\lambda_2 q_2 - 1} C \\ \tilde{B}^{*(2)} &= \frac{1}{\lambda_2}\frac{q_2 - \lambda_2}{\lambda_2 q_2 - 1} B^* + \frac{1}{2\lambda_2^2}\frac{\lambda_2^2 - 1}{\lambda_2 q_2 - 1} C^* \\ \tilde{C}^{(2)} &= \lambda_2^2 C, \quad \tilde{C}^{*(2)} = \frac{1}{\lambda_2^2} C^* \\ \tilde{q}^{(2)} &= \frac{\lambda_2(q-\lambda)(\lambda_2 q_2 - 1) - \lambda(q_2 - \lambda_2)(\lambda q - 1)}{\lambda(q-\lambda)(\lambda_2 q_2 - 1) - \lambda_2(q_2 - \lambda_2)(\lambda q - 1)}, \quad \tilde{\lambda}^{(2)} = \frac{\lambda}{\lambda_2}. \end{aligned} \right. \tag{8.91}$$

Both Ernst and metric pictures are preserved by $\tilde{\mathfrak{I}}$.

The Neugebauer Bäcklund transformations $\mathfrak{I}$ and $\tilde{\mathfrak{I}}$ may be iterated in the following manner. Let $\beta_1, \beta_2, \beta_3, \ldots$ be pairs of pseudopotentials associated with the seed solution Ω. Then, β_1 gives rise to a second solution $\Omega^{(1)}$ via the Neugebauer transformation $\mathfrak{I}(\beta_1)$. The corresponding pseudopotentials are generated in accordance with Theorem 35 as follows:

$$\beta_2, \beta_3, \ldots \xrightarrow{\mathfrak{I}(\beta_1)} \beta_2^{(1)}, \beta_3^{(1)}, \ldots \tag{8.92}$$

The pair $\beta_2^{(1)}$, in turn, may be used to produce a third solution $\tilde{\Omega}^{(2)}$ by means of the Neugebauer transformation $\tilde{\mathfrak{I}}(\beta_2^{(1)})$, and Theorem 37 provides associated pseudopotentials via the mapping

$$\beta_3^{(1)}, \ldots \xrightarrow{\tilde{\mathfrak{I}}(\beta_2^{(1)})} \tilde{\beta}_3^{(12)}, \ldots. \tag{8.93}$$

Another application of the Neugebauer transformation $\mathfrak{I}$ induced by $\beta_3^{(12)}$ results in a fourth solution $\tilde{\Omega}^{(123)}$. This process may be repeated *ad infinitum* to generate new solutions of the system (8.50) of arbitrary complexity. The underlying composite transformation may be written as

$$\cdots \circ \tilde{\mathfrak{I}}(\beta_4^{(123)}) \circ \mathfrak{I}(\tilde{\beta}_3^{(12)}) \circ \tilde{\mathfrak{I}}(\beta_2^{(1)}) \circ \mathfrak{I}(\beta_1). \tag{8.94}$$

Alternatively, one may choose $\tilde{\mathfrak{I}}$ as the first transformation in the suite of Neugebauer transformations

$$\cdots \circ \mathfrak{I}(\tilde{\tilde{\beta}}_4^{(123)}) \circ \tilde{\mathfrak{I}}(\tilde{\beta}_3^{(12)}) \circ \mathfrak{I}(\beta_2^{(1)}) \circ \tilde{\mathfrak{I}}(\beta_1). \tag{8.95}$$

In general, the classes of solutions generated by the transformations (8.94) and (8.95) are not identical. In particular, explicit application of the transformation formulae (8.77) and (8.91) shows that, in the generic case,

$$\tilde{\mathfrak{I}}(\beta_2^{(1)}) \circ \mathfrak{I}(\beta_1) \neq \mathfrak{I}(\tilde{\beta}_2^{(1)}) \circ \tilde{\mathfrak{I}}(\beta_1). \tag{8.96}$$

However, in the case

$$\lambda_2 = 1, \quad q_2 = 1, \tag{8.97}$$

the Neugebauer transformations $\mathfrak{I}$ and $\tilde{\mathfrak{I}}$ produce the pseudopotentials

$$q_2^{(1)} = \frac{1}{q_1}, \quad \lambda_2^{(1)} = \frac{1}{\lambda_1}, \quad \tilde{q}_2^{(1)} = q_1, \quad \tilde{\lambda}_2^{(1)} = \frac{1}{\lambda_1}, \tag{8.98}$$

respectively, and

$$\tilde{\mathfrak{I}}(\beta_2^{(1)}) \circ \mathfrak{I}(\beta_1) = \mathfrak{I}(\tilde{\beta}_2^{(1)}) \circ \tilde{\mathfrak{I}}(\beta_1). \tag{8.99}$$

This 'commutation theorem' was set down by Neugebauer in [267] and is illustrated in Figure 8.2.

For the Ernst equation ($\mathcal{F} = \mathcal{E}, \mathcal{F}^* = \bar{\mathcal{E}}$), the conjugated transformations in the commutation theorem (8.99) are equivalent to a single transformation derived by Harrison using a different but albeit algebraically related set of pseudopotentials [157]. It is therefore natural to refer to (8.99) as the *Harrison transformation*. Its explicit action on the seed solution Ω is the content of the following result:

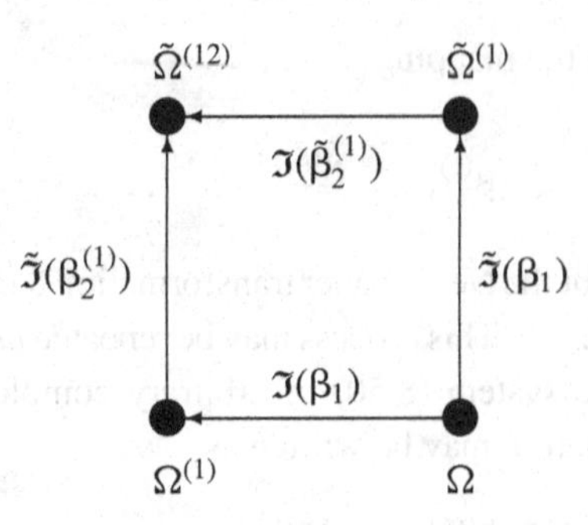

Figure 8.2. The Neugebauer commutation theorem. The pseudopotentials $\beta_2^{(1)}$ and $\tilde{\beta}_2^{(1)}$ are given by (8.98).

Theorem 38 (The Harrison Bäcklund Transformation). *Let $\Omega = (A, A^*, B, B^*, C, C^*)$ be a solution of the Ernst-type system in polynomial form (8.50) and $\beta_1 = (q_1, \lambda_1)$ a corresponding solution of the compatible system (8.74), (8.75). Then, the action of the Harrison transformation*

$$\tilde{\mathfrak{I}}\left(\frac{1}{q_1}, \frac{1}{\lambda_1}\right) \circ \mathfrak{I}(q_1, \lambda_1) = \mathfrak{I}\left(q_1, \frac{1}{\lambda_1}\right) \circ \tilde{\mathfrak{I}}(q_1, \lambda_1) \tag{8.100}$$

is given by

$$(\mathfrak{H}) \left\{ \begin{aligned} \tilde{A}^{(1\cdot)} &= q_1 \frac{\lambda_1 q_1 - 1}{q_1 - \lambda_1} A - \frac{q_1}{2} \frac{\lambda_1^2 - 1}{q_1 - \lambda_1} C \\ \tilde{A}^{*(1\cdot)} &= \frac{1}{q_1} \frac{\lambda_1 q_1 - 1}{q_1 - \lambda_1} A^* - \frac{1}{2} \frac{\lambda_1 - \lambda_1^{-1}}{q_1 - \lambda_1} C^* \\ \tilde{B}^{(1\cdot)} &= \frac{1}{q_1} \frac{q_1 - \lambda_1}{\lambda_1 q_1 - 1} B + \frac{1}{2} \frac{\lambda_1^2 - 1}{\lambda_1 q_1 - 1} C \\ \tilde{B}^{*(1\cdot)} &= q_1 \frac{q_1 - \lambda_1}{\lambda_1 q_1 - 1} B^* + \frac{q_1}{2} \frac{\lambda_1 - \lambda_1^{-1}}{\lambda_1 q_1 - 1} C^* \\ \tilde{C}^{(1\cdot)} &= C, \quad \tilde{C}^{*(1\cdot)} = C^*. \end{aligned} \right. \tag{8.101}$$

Both Ernst and metric pictures are preserved by $\mathfrak{H}$.

In the above theorem, the superscript $(1\cdot)$ indicates that Harrison transforms do not depend on β_2 by virtue of the particular choice $\lambda_2 = 1$. The transformation laws $(8.101)_{5,6}$ show that, in contrast to the Neugebauer transformations $\mathfrak{I}$ and $\tilde{\mathfrak{I}}$, the Harrison transformation $\mathfrak{H}$ preserves C and C^* and therefore the harmonic function ρ. It turns out that the new potentials $\mathcal{F}^{(1\cdot)}$ and $\mathcal{F}^{*(1\cdot)}$, which are obtained from the analogue of the linear system (8.52), cannot be expressed explicitly in terms of the seed potentials $\mathcal{F}$ and $\mathcal{F}^*$ and the pseudopotentials β_1. However, expressions for the potentials associated with the transformation

$$\mathfrak{D} = \mathfrak{S} \circ \mathfrak{H}, \tag{8.102}$$

which implies the decomposition

$$\mathfrak{H} = \mathfrak{S} \circ \mathfrak{D}, \tag{8.103}$$

do exist. These can be obtained algebraically by means of the matrix Darboux transformation to be discussed in the next section. Indeed, the transformation $\mathfrak{D}$ may be identified with the simplest matrix Darboux transformation for the Ernst-type system (8.53). In terms of solutions of Einstein's equations, this means that the metric (8.64) associated with a Harrison transformation may be given explicitly in terms of the seed Ernst potential $\mathcal{E}$ and corresponding pseudopotentials β_1. For instance, the Ernst potential of flat space-time leads to the Kerr black hole metric. In fact, N applications of the Harrison transformation with this seed solution gives rise to a nonlinear superposition of N Kerr-NUT fields [218]. Another natural class of seed Ernst potentials constitutes the Weyl class [383]

$$\mathcal{E} = \bar{\mathcal{E}} = e^{2U}, \quad U_{z\bar{z}} + \frac{1}{2}\frac{\rho_{\bar{z}}}{\rho}U_z + \frac{1}{2}\frac{\rho_z}{\rho}U_{\bar{z}} = 0. \tag{8.104}$$

In this case, the application of N Harrison transformations leads to a nonlinear superposition of N Kerr-NUT fields on the Weyl background [268]. The Papapetrou class

$$\mathcal{E} = (\cosh U)^{-1} + i \tanh U \tag{8.105}$$

is obtained from the Weyl class by means of the Ehlers transformation $\mathfrak{E}$. A single application of the Harrison transformation produces the Schwarzschild solution on the Papapetrou background [215].

The above list of solutions is merely an indication of the wide variety of important solutions of Einstein's equations which may be generated via Bäcklund transformations for the Ernst equation and its dual. For details, reference may be made to the review articles [109, 219].

8.7 A Matrix Darboux Transformation for the Ernst Equation

To derive a matrix Darboux transformation for the Ernst-type system (8.53) and, in particular, the Ernst equation and its dual, it is necessary to give a unique algebraic characterisation of the linear representation (8.49):

Theorem 39. *The linear representation (8.49) is uniquely characterised by its polynomial structure*

$$\begin{aligned} \Phi_z &= F(\lambda)\Phi = (F_0 + \lambda F_1)\Phi, \qquad \lambda_z = \frac{1}{2}(\lambda^3 - \lambda)C \\ \phi_{\bar{z}} &= G(\lambda)\Phi = (G_0 + \lambda^{-1}G_1)\Phi, \quad \lambda_{\bar{z}} = \frac{1}{2}(\lambda - \lambda^{-1})C^* \end{aligned} \tag{8.106}$$

and the linear constraints

$$\begin{aligned} &\text{(i)} \quad F(-\lambda) = \sigma_3 F(\lambda)\sigma_3, \quad G(-\lambda) = \sigma_3 G(\lambda)\sigma_3 \\ &\text{(ii)} \quad F(1)\begin{pmatrix} 1 \\ -1 \end{pmatrix} = \mathbf{0}, \qquad G(1)\begin{pmatrix} 1 \\ -1 \end{pmatrix} = \mathbf{0} \end{aligned} \tag{8.107}$$

with the usual Pauli matrix σ_3.

The constraint (i) guarantees that $F(\lambda)$ and $G(\lambda)$ are of the form

$$\begin{pmatrix} * & 0 \\ 0 & * \end{pmatrix} + \lambda^{\pm 1}\begin{pmatrix} 0 & * \\ * & 0 \end{pmatrix}, \tag{8.108}$$

respectively, where the asterisks represent some, in general, non-zero functions. Hence, the condition (ii) gives rise to the particular form (8.49). Now, if $\phi_{[1]}$ and $\phi_{[2]}$ are two vector-valued eigenfunctions of (8.106) with parameters λ_1 and λ_2, then Theorem 33 states that there exists an elementary matrix Darboux transformation

$$\Phi' = \mathcal{P}(\lambda)\Phi = p(\lambda)P(\lambda)\Phi \tag{8.109}$$

with

$$P(\lambda) = \lambda P_1 + P_0, \quad p(\lambda) = \frac{1}{\sqrt{(\lambda - \lambda_1)(\lambda - \lambda_2)}}. \tag{8.110}$$

It is noted that the normalisation $\det P_1 = 1$ does not apply since no trace conditions are imposed on $F(\lambda)$ and $G(\lambda)$.

In analogy with the geometric $su(2)$ constraints (8.26), (8.27), we now make the choice

$$\begin{aligned} \phi_{[1]} &= \phi_0, \qquad \lambda_1 = \lambda_0 \\ \phi_{[2]} &= \sigma_3\phi_0, \quad \lambda_2 = -\lambda_0, \end{aligned} \tag{8.111}$$

which guarantees that P_1 may be chosen in such a way that

$$P(\lambda) = \begin{pmatrix} * & 0 \\ 0 & * \end{pmatrix} + \lambda \begin{pmatrix} 0 & * \\ * & 0 \end{pmatrix} \tag{8.112}$$

or, equivalently,

$$P(-\lambda) = \sigma_3 P(\lambda)\sigma_3. \tag{8.113}$$

The latter implies that $F'(\lambda)$ and $G'(\lambda)$ as given by $(8.22)_2$ obey the primed version of the constraint (i). Thus, it only remains to show that the condition (ii) is preserved by the above transformation. To this end, it is observed that if

$$P(1)\begin{pmatrix} 1 \\ -1 \end{pmatrix} = \hat{c}\begin{pmatrix} 1 \\ -1 \end{pmatrix}, \qquad \hat{c} = \text{const}, \tag{8.114}$$

then the particular eigenfunction $(1 \quad -1)^{\mathsf{T}}$ of (8.106) at $\lambda = 1$ is mapped to itself modulo the irrelevant constant factor $\hat{c}$. Consequently,

$$F'(1)\begin{pmatrix} 1 \\ -1 \end{pmatrix} = \mathbf{0}, \quad G'(1)\begin{pmatrix} 1 \\ -1 \end{pmatrix} = \mathbf{0}. \tag{8.115}$$

Conditions (8.112) and (8.114) determine the Darboux matrix $\mathcal{P}(\lambda)$ uniquely. Hence, we have the following result:

Theorem 40 (An Elementary Matrix Darboux Transformation for the Ernst-Type System). *Let $\Omega = (A, A^*, B, B^*, C, C^*)$ be a solution of the Ernst-type system in polynomial form (8.50). Consider two matrix- and vector-valued solutions Φ, ϕ_0 of the linear representation (8.49) with parameters λ and λ_0, respectively. Then, in terms of the pseudopotential $q_0 = -\phi_0^2/\phi_0^1$, the corresponding elementary matrix Darboux transformation takes the form*

$$\Phi' = \mathcal{P}(\lambda)\Phi = i\sqrt{\frac{1-\lambda_0^2}{\lambda^2-\lambda_0^2}}\begin{pmatrix} \dfrac{\lambda_0 q_0}{\lambda_0 q_0 - 1} & \dfrac{\lambda}{\lambda_0 q_0 - 1} \\ \dfrac{\lambda q_0}{\lambda_0 - q_0} & \dfrac{\lambda_0}{\lambda_0 - q_0} \end{pmatrix}\Phi. \tag{8.116}$$

The new solution Ω' is given by

$$(\mathfrak{D})\left\{\begin{aligned} A' &= \frac{1}{q_0}\frac{\lambda_0 - q_0}{\lambda_0 q_0 - 1}B - \frac{\lambda_0}{2}\frac{\lambda_0 - q_0}{\lambda_0 q_0 - 1}C \\ A^{*\prime} &= \frac{1}{q_0}\frac{\lambda_0 q_0 - 1}{\lambda_0 - q_0}A^* - \frac{1}{2\lambda_0}\frac{\lambda_0 q_0 - 1}{\lambda_0 - q_0}C^* \\ B' &= q_0\frac{\lambda_0 q_0 - 1}{\lambda_0 - q_0}A - \frac{\lambda_0}{2}\frac{\lambda_0 q_0 - 1}{\lambda_0 - q_0}C \\ B^{*\prime} &= q_0\frac{\lambda_0 - q_0}{\lambda_0 q_0 - 1}B^* - \frac{1}{2\lambda_0}\frac{\lambda_0 - q_0}{\lambda_0 q_0 - 1}C^* \\ C' &= C, \quad C^{*\prime} = C^*, \end{aligned}\right. \tag{8.117}$$

while the new potentials $\mathcal{F}'$ and $\mathcal{F}^{*\prime}$ read

$$\mathcal{F}' = i\frac{\lambda_0 q_0 \mathcal{F} + \mathcal{F}^*}{\lambda_0 q_0 - 1}, \quad \mathcal{F}^{*\prime} = i\frac{q_0 \mathcal{F} + \lambda_0 \mathcal{F}^*}{\lambda_0 - q_0} \tag{8.118}$$

without loss of generality. The matrix Darboux transformation $\mathfrak{D}$ and the Harrison Bäcklund transformation $\mathfrak{H}$ are related by

$$\mathfrak{H} = \mathfrak{S} \circ \mathfrak{D} = \mathfrak{D} \circ \mathfrak{S}. \tag{8.119}$$

Thus, the transformations $\mathfrak{S}$ and $\mathfrak{D}$ commute.

Proof. It may be directly verified that the Darboux matrix $\mathcal{P}(\lambda)$ given in (8.116) satisfies the linear algebraic system (8.21), (8.111) together with the constraints (8.113), (8.114). The transformation formulae (8.117) are obtained by comparison of like terms in λ in

$$\begin{aligned} F'(\lambda)\mathcal{P}(\lambda) &= \mathcal{P}(\lambda)F(\lambda) + [\mathcal{P}(\lambda)]_z \\ G'(\lambda)\mathcal{P}(\lambda) &= \mathcal{P}(\lambda)G(\lambda) + [\mathcal{P}(\lambda)]_{\bar{z}}. \end{aligned} \tag{8.120}$$

Moreover, the general solution of the linear representation (8.49) at $\lambda = 1$ is given by

$$\Phi|_{\lambda=1} = \begin{pmatrix} i & \mathcal{F} \\ -i & \mathcal{F}^* \end{pmatrix} \mathcal{C}, \tag{8.121}$$

where $\mathcal{C}$ is an arbitrary (complex) matrix. The latter merely corresponds to a subcase of the Möbius transformation (8.81) so that the identifications

$$\phi_{12}|_{\lambda=1} = \mathcal{F}, \quad \phi_{22}|_{\lambda=1} = \mathcal{F}^* \tag{8.122}$$

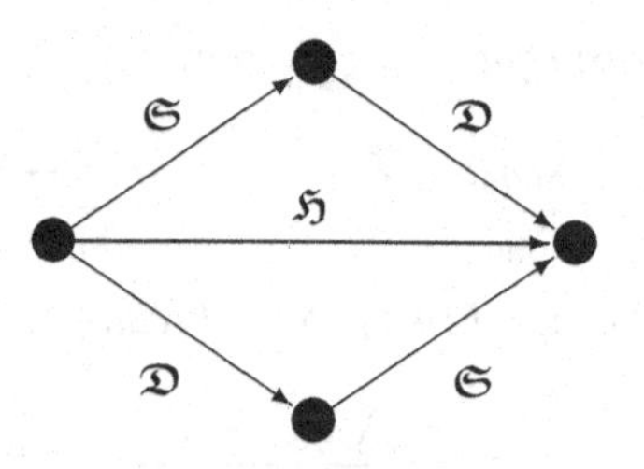

Figure 8.3. Decomposition of the Harrison transformation $\mathfrak{H}$.

are admissible. Evaluation of the primed counterpart of these relations then produces (8.118). Finally, if we set $q_0 = q_1$ and $\lambda_0 = \lambda_1$, then application of the Kramer-Neugebauer transformation to Ω' results in the Harrison transformation laws (8.101). Conversely, it is readily shown that the Kramer-Neugebauer transformation followed by the matrix Darboux transformation generate the same result. Thus, the Harrison transformation $\mathfrak{H}$ may indeed be decomposed into the Kramer-Neugebauer transformation $\mathfrak{S}$ and the elementary matrix Darboux transformation $\mathfrak{D}$ as illustrated in Figure 8.3. □

In the case of the Ernst equation and its dual, it is consistent to assume that

$$\begin{aligned} &\text{Ernst picture:} \quad &&\bar{\Phi} = \sigma_1 \Phi, \quad \bar{\lambda} = 1/\lambda \\ &\text{Metric picture:} \quad &&\bar{\Phi} = \Phi, \quad \bar{\lambda} = 1/\lambda \end{aligned} \tag{8.123}$$

which implies the pseudopotential conditions (8.76). Thus, if Φ, ϕ_0 and λ, λ_0 in Theorem 40 are constrained by (8.123), then the Darboux matrix $\mathcal{P}$ satisfies

$$\begin{aligned} &\text{Ernst picture:} \quad &&\bar{\mathcal{P}} = \mathcal{P}\sigma_1 \\ &\text{Metric picture:} \quad &&\bar{\mathcal{P}} = \sigma_1 \mathcal{P} \end{aligned} \tag{8.124}$$

and the new eigenfunction Φ' obeys the relations

$$\begin{aligned} &\text{Ernst picture:} \quad &&\bar{\Phi}' = \Phi' \\ &\text{Metric picture:} \quad &&\bar{\Phi}' = \sigma_1 \Phi'. \end{aligned} \tag{8.125}$$

Consequently, the following important result has been established:

Theorem 41 (The Elementary Matrix Darboux Transformation for the Ernst Equation and Its Dual). *The elementary matrix Darboux transformation (8.116) subject to the 'reality constraints' (8.123) maps solutions of the Ernst equation to solutions of its dual and vice versa. In the Ernst picture, the new* *solution of the dual Ernst equation is given by*

$$\mathcal{F}' = i\frac{\lambda_0 q_0 \mathcal{E} + \bar{\mathcal{E}}}{\lambda_0 q_0 - 1}, \quad \mathcal{F}^{*\prime} = i\frac{q_0 \mathcal{E} + \lambda_0 \bar{\mathcal{E}}}{\lambda_0 - q_0}. \tag{8.126}$$

In the metric picture, the new Ernst potential reads

$$\mathcal{E}' = i\frac{\lambda_0 q_0 \mathcal{F} + \mathcal{F}^*}{\lambda_0 q_0 - 1}. \tag{8.127}$$

8.8 A Permutability Theorem for the Ernst Equation and Its Dual. A Classical Bianchi Connection

In this section, the successive application of two matrix Darboux transformations is shown to lead to a permutability theorem associated with the Harrison transformation. This stems from the decomposition (8.119) of the Harrison transformation which implies that

$$\mathfrak{H} \circ \mathfrak{H} = \mathfrak{S} \circ \mathfrak{D} \circ \mathfrak{S} \circ \mathfrak{D} = \mathfrak{D} \circ \mathfrak{S} \circ \mathfrak{S} \circ \mathfrak{D} = \mathfrak{D} \circ \mathfrak{D} \tag{8.128}$$

since $\mathfrak{S}^2 = \mathrm{id}$. Thus, the action of two Harrison transformations $\mathfrak{H}$ is equivalent to that of two elementary matrix Darboux transformations $\mathfrak{D}$.

Consider two pairs of pseudopotentials $\beta_1 = (q_1, \lambda_1)$ and $\beta_2 = (q_2, \lambda_2)$ associated with a seed solution $(\mathcal{F}, \mathcal{F}^*)$ of the Ernst-type system (8.53). Application of the matrix Darboux transformation $\mathfrak{D}$ with respect to the pseudopotentials β_1 produces the new solution

$$\mathcal{F}' = i\frac{\lambda_1 q_1 \mathcal{F} + \mathcal{F}^*}{\lambda_1 q_1 - 1}, \quad \mathcal{F}^{*\prime} = i\frac{q_1 \mathcal{F} + \lambda_1 \mathcal{F}^*}{\lambda_1 - q_1}. \tag{8.129}$$

According to Theorem 40, an associated pair of pseudopotentials is given by

$$q_2' = \frac{(\lambda_2 q_1 - \lambda_1 q_2)(\lambda_1 q_1 - 1)}{(\lambda_2 q_2 - \lambda_1 q_1)(\lambda_1 - q_1)}, \quad \lambda_2' = \lambda_2. \tag{8.130}$$

Hence, a second application of the matrix Darboux transformation $\mathfrak{D}$ generates the solution

$$\mathcal{F}'' = i\frac{\lambda_2' q_2' \mathcal{F}' + \mathcal{F}^{*\prime}}{\lambda_2' q_2' - 1}, \quad \mathcal{F}^{*\prime\prime} = i\frac{q_2' \mathcal{F}' + \lambda_2' \mathcal{F}^{*\prime}}{\lambda_2' - q_2'}. \tag{8.131}$$

Theorem 42 (A Permutability Theorem for the (Dual) Ernst Equation). *In terms of two pairs of pseudopotentials β_1 and β_2 corresponding to a seed solution $(\mathcal{F}, \mathcal{F}^*)$ of the Ernst-type system (8.53), the action of two elementary* *matrix Darboux transformations is given by*

$$\mathcal{F}'' = -\mathcal{F} + \frac{(\lambda_2^2 - \lambda_1^2)(\mathcal{F} + \mathcal{F}^*)}{\lambda_2^2 - \lambda_1^2 + q_1\lambda_1(1 - \lambda_2^2) - q_2\lambda_2(1 - \lambda_1^2)}$$

$$\mathcal{F}^{*\prime\prime} = -\mathcal{F}^* + \frac{q_1 q_2(\lambda_2^2 - \lambda_1^2)(\mathcal{F} + \mathcal{F}^*)}{q_1 q_2(\lambda_2^2 - \lambda_1^2) + q_2\lambda_1(1 - \lambda_2^2) - q_1\lambda_2(1 - \lambda_1^2)}. \tag{8.132}$$

The two matrix Darboux transformations $\mathfrak{D}(\beta_1)$ and $\mathfrak{D}(\beta_2)$ commute, that is, the permutability theorem

$$\mathfrak{D}(\beta_2') \circ \mathfrak{D}(\beta_1) = \mathfrak{D}(\beta_1') \circ \mathfrak{D}(\beta_2) \tag{8.133}$$

holds. If either of the reality conditions (8.123) are imposed, then the corresponding Ernst or metric picture is preserved. For instance, if $\mathcal{F} = \mathcal{E}$, $\mathcal{F}^* = \bar{\mathcal{E}}$ then (8.132), that is

$$\mathcal{E}'' = -\mathcal{E} + \frac{(\lambda_2^2 - \lambda_1^2)(\mathcal{E} + \bar{\mathcal{E}})}{\lambda_2^2 - \lambda_1^2 + q_1\lambda_1(1 - \lambda_2^2) - q_2\lambda_2(1 - \lambda_1^2)}, \tag{8.134}$$

enshrines a permutability theorem for the Ernst equation.

Since the pairs β_1 and β_2 appear on equal footing in the transformation laws (8.132), it is evident that the permutability theorem (8.133) holds. In the case of the Ernst equation, Neugebauer [268] has shown that the superposition principle (8.134) is but a special case of a determinantal expression for Ernst potentials generated by means of $2N$ Harrison (or matrix Darboux) transformations. In principle, the relations (8.129) and their analogues associated with β_2 may be solved for the pseudopotentials so that (8.134) may be written entirely in terms of the Ernst potentials $\mathcal{E}$, $\mathcal{E}''$ and the intermediate solutions $(\mathcal{F}', \mathcal{F}^{*\prime})(\beta_1)$, $(\mathcal{F}', \mathcal{F}^{*\prime})(\beta_2)$ of the dual Ernst equation. The corresponding Bianchi diagram is depicted in Figure 8.4.

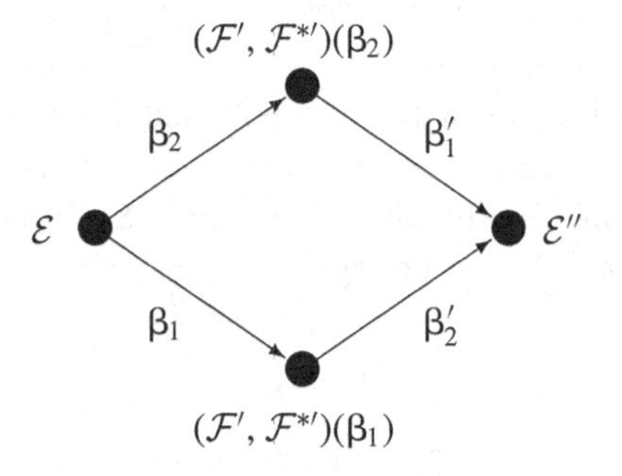

Figure 8.4. A permutability theorem for the Ernst equation.

A Classical Bianchi Connection

It has been noted that Bianchi surfaces of Gaussian curvature $\mathcal{K} = -1/\rho^2$ are governed by the vector equation (8.31) for the unit normal $\boldsymbol{N}$. This vector equation possesses natural counterparts associated with the Ernst equation and its dual. In both cases, the vector-valued functions involved may be interpreted as 'unit normals' in Minkowski space $\mathbb{M}^3$ which are parametrised in terms of the Ernst potential $\mathcal{E}$ and $\mathcal{F}$, $\mathcal{F}^*$, respectively. On use of these vector equations, Chinea [81] has derived a vectorial Bäcklund transformation for the Ernst equation without appeal to pseudopotentials. This vectorial transformation is, in fact, equivalent to the elementary matrix Darboux transformation $\mathfrak{D}$. Two applications of the vectorial Bäcklund transformation deliver a compact vectorial formulation of the nonlinear superposition principle (8.134). An analogue of the latter may be readily obtained for the classical Bianchi transformation (8.13). Thus, if $\boldsymbol{r}$ denotes the position vector of a Bianchi surface Σ, then the position vector $\boldsymbol{r}_1$ of the transformed Bianchi surface Σ_1 satisfies the relation (1.163), that is

$$\boldsymbol{r}_1 - \boldsymbol{r} = \boldsymbol{\nu}_1 \times \boldsymbol{\nu}, \tag{8.135}$$

where $\boldsymbol{\nu} = \sqrt{\rho}\boldsymbol{N}$ and $\boldsymbol{\nu}_1 = \sqrt{\rho_1}\boldsymbol{N}_1$. By virtue of (1.176), the angle σ_1 between the normals $\boldsymbol{N}_1$ and $\boldsymbol{N}$ is given in terms of the 'parameter' μ_1 by

$$\boldsymbol{N}_1 \cdot \boldsymbol{N} = \cos\sigma_1 = \frac{1+\mu_1^2}{1-\mu_1^2}. \tag{8.136}$$

A third Bianchi surface Σ_2 may be generated from Σ by choosing a different parameter μ_2. Hence, we have the relations

$$\boldsymbol{r}_2 - \boldsymbol{r} = \boldsymbol{\nu}_2 \times \boldsymbol{\nu} \tag{8.137}$$

and

$$\boldsymbol{N}_2 \cdot \boldsymbol{N} = \cos\sigma_2 = \frac{1+\mu_2^2}{1-\mu_2^2}, \tag{8.138}$$

where σ_2 is the angle between the normals $\boldsymbol{N}_2$ and $\boldsymbol{N}$. Since the Bianchi diagram associated with the Bianchi transformation is closed, the surfaces Σ_1 and Σ_2 may be mapped to the same fourth Bianchi surface Σ_{12} if one uses the parameters μ_2 and μ_1, respectively. Accordingly, the position vector $\boldsymbol{r}_{12}$ obeys the two relations

$$\begin{aligned} \boldsymbol{r}_{12} - \boldsymbol{r}_2 &= \boldsymbol{\nu}_{12} \times \boldsymbol{\nu}_2 \\ \boldsymbol{r}_{12} - \boldsymbol{r}_1 &= \boldsymbol{\nu}_{12} \times \boldsymbol{\nu}_1. \end{aligned} \tag{8.139}$$

An appropriate linear combination of the vector equations (8.135), (8.137) and (8.139) delivers

$$(\boldsymbol{\nu}_{12} - \boldsymbol{\nu}) \times (\boldsymbol{\nu}_1 - \boldsymbol{\nu}_2) = \mathbf{0}, \tag{8.140}$$

which implies that the vectors $\boldsymbol{\nu}_{12} - \boldsymbol{\nu}$ and $\boldsymbol{\nu}_1 - \boldsymbol{\nu}_2$ are parallel. Moreover, since the Bianchi transformation preserves the Gaussian curvature, that is $\rho = \rho_1 = \rho_2 = \rho_{12}$, the magnitude of $\boldsymbol{\nu}$ is invariant so that there exists a function H such that

$$\boldsymbol{N}_{12} - \boldsymbol{N} = H(\boldsymbol{N}_1 - \boldsymbol{N}_2), \tag{8.141}$$

whence

$$H = \frac{\boldsymbol{N}_2 \cdot \boldsymbol{N} - \boldsymbol{N}_1 \cdot \boldsymbol{N}}{1 - \boldsymbol{N}_1 \cdot \boldsymbol{N}_2}. \tag{8.142}$$

Accordingly,

$$\boxed{\boldsymbol{N}_{12} = \boldsymbol{N} + \frac{\cos\sigma_2 - \cos\sigma_1}{1 - \boldsymbol{N}_1 \cdot \boldsymbol{N}_2}(\boldsymbol{N}_1 - \boldsymbol{N}_2).} \tag{8.143}$$

This nonlinear superposition principle for the unit normals to Bianchi surfaces is the analogue of Chinea's vectorial superposition principle for the Ernst equation. In turn, the Harrison transformation $\mathfrak{H}$ for the Ernst equation is the counterpart of the Bäcklund transformation for the classical Bianchi system.

Exercises

1. Show that the expression $(8.12)_{\lambda=1}$ associated with the linear representation (8.2), (8.3), (8.9) coincides with the 2nd fundamental form $(1.126)_2$ for Bianchi surfaces.
2. Verify that the action of the Kramer-Neugebauer transformation (8.58) on the Ernst potential $\mathcal{E}$ is given by (8.60), (8.61).
3. (a) Verify directly that the pseudopotential equations (8.74), (8.75) are form-invariant under the Neugebauer transformation $\mathfrak{I}$.
   (b) Show that the corresponding transformation law for the harmonic function ρ is given by (8.78).
   (c) Prove that at $\lambda_1 = 1$, the Neugebauer transformation $\mathfrak{I}(\beta_1)$ reduces to the Möbius invariance (8.81).
4. Prove Theorems 36 and 37.

5. (a) Derive explicit expressions for the pseudopotentials associated with the transformations $\tilde{\mathfrak{I}}(\beta_2^{(1)}) \circ \mathfrak{I}(\beta_1)$ and $\mathfrak{I}(\tilde{\beta}_2^{(1)}) \circ \tilde{\mathfrak{I}}(\beta_1)$. Deduce that

$$\tilde{\mathfrak{I}}(\beta_2^{(1)}) \circ \mathfrak{I}(\beta_1) \neq \mathfrak{I}(\tilde{\beta}_2^{(1)}) \circ \tilde{\mathfrak{I}}(\beta_1)$$

   in the generic case.

   (b) Justify the choice of the pseudopotentials (8.98) and show that, for such, the transformations $\tilde{\mathfrak{I}}(\beta_2^{(1)}) \circ \mathfrak{I}(\beta_1)$ and $\mathfrak{I}(\tilde{\beta}_2^{(1)}) \circ \tilde{\mathfrak{I}}(\beta_1)$ coincide.

   (c) Derive the Harrison transformation laws (8.101).

6. Prove that the Harrison transformation $\mathfrak{H}$ decomposes into the Kramer-Neugebauer transformation $\mathfrak{S}$ and the elementary matrix Darboux transformation $\mathfrak{D}$ according to $\mathfrak{H} = \mathfrak{S} \circ \mathfrak{D} = \mathfrak{D} \circ \mathfrak{S}$.

9

Projective-Minimal and Isothermal-Asymptotic Surfaces

The study of the projective differential geometry of surfaces has roots in the work of Wilczynski [384] at the beginning of the last century. There have been a number of monographs on the subject, notably by Fubini and Ĉech [143], Bol [51], Finikov [135], Lane [227], Akivis and Goldberg [6]. It has been established recently that privileged classes of surfaces in classical projective differential geometry are, in fact, described by integrable systems that pertain to modern soliton theory. A lucid summary of these connections and their historical origins have been given by Ferapontov [131]. It will be with these integrable classes of surfaces of projective differential geometry that the present chapter will be concerned. The emphasis will be on projective-minimal and isothermal-asymptotic surfaces.

Projective-minimal surfaces arise out of the Euler-Lagrange equations associated with extremals of a projective area functional. These Euler-Lagrange equations were set down by Thomsen [365] in 1928 and later taken up by Sasaki [322]. Included in the class of projective-minimal surfaces are the surfaces of Godeaux-Rozet [51] and those of Demoulin [102]. The latter are governed by a coupled Tzitzeica system [135], which may be derived as a reduction of the two-dimensional Toda lattice system [259]. Bäcklund transformations and associated permutability theorems for Godeaux-Rozet and Demoulin surfaces were derived in a purely geometric manner by Demoulin [102]. It is remarked that projective-minimal and, in particular, Godeaux-Rozet and Demoulin surfaces arise in the theory of Lie quadrics [51, 135, 227].

The geometry of projective-minimal surfaces, in general, has been studied in the early papers of Thomsen [364] and Mayer [254]. Following upon Demoulin's early work in 1933, Godeaux-Rozet and Demoulin surfaces were studied extensively by Godeaux [152], Rozet [318] and most recently by Ferapontov and Schief [133].

The introduction of isothermal-asymptotic surfaces is attributed to Fubini (see [143]). They are discussed in the classical treatises of Lane [227], Bol [51] and Finikov [135, 136]. The class of isothermal-asymptotic surfaces includes arbitrary quadric and cubic surfaces, the quartics of Kummer [176], as well as projective transforms of surfaces of revolution [131]. It has been shown recently by Ferapontov [129] that isothermal-asymptotic surfaces are governed by the stationary modified Nizhnik-Veselov-Novikov equation [47]. The latter constitutes a symmetric 2+1-dimensional integrable version of the modified Korteweg-de Vries equation.

9.1 Analogues of the Gauss-Mainardi-Codazzi Equations in Projective Differential Geometry

Here, we identify the components of a four-dimensional vector

$$\boldsymbol{r} = (r^0, r^1, r^2, r^3) \in \mathbb{R}^4 \tag{9.1}$$

with the homogeneous coordinates of a point

$$\boldsymbol{r} = (r^0 : r^1 : r^2 : r^3) \in \mathbb{P}^3 \tag{9.2}$$

in a three-dimensional projective space. Thus, any straight line which passes through the origin of $\mathbb{R}^4$ is mapped to a point $\boldsymbol{r} \in \mathbb{P}^3$. If $r^0 \neq 0$, then a particular representation of that point is given by

$$\boldsymbol{r} = (1, \hat{\boldsymbol{r}}), \quad \hat{\boldsymbol{r}} = (r^1/r^0, r^2/r^0, r^3/r^0). \tag{9.3}$$

The latter gives rise to a natural mapping between $\mathbb{P}^3$ and $\mathbb{R}^3$, that is

$$\mathbb{P}^3 \to \mathbb{R}^3, \quad \boldsymbol{r} \mapsto \hat{\boldsymbol{r}}. \tag{9.4}$$

Consequently, the group of linear transformations

$$\boldsymbol{r} \to A\boldsymbol{r}, \quad A \in \mathbb{R}^{4,4}, \quad \det A \neq 0 \tag{9.5}$$

is represented by linear fractional transformations which act on $\hat{\boldsymbol{r}}$ and are known as *projective transformations*. In what follows, we consider surfaces $\Sigma \subset \mathbb{P}^3$ which are defined up to arbitrary linear transformations of the form (9.5) and may therefore be thought of as surfaces $\hat{\Sigma} \subset \mathbb{R}^3$ whose properties are invariant under projective transformations.

As in the case of Euclidean differential geometry, asymptotic coordinates are particularly useful in the discussion of surfaces in projective differential geometry. A surface $\Sigma : \boldsymbol{r} = \boldsymbol{r}(x, y) \in \mathbb{P}^3$ is parametrised in terms of asymptotic coordinates if the position vector $\boldsymbol{r}$ obeys two second-order equations of the form

$$\boldsymbol{r}_{xx} = s\boldsymbol{r}_x + p\boldsymbol{r}_y + \pi\boldsymbol{r}, \quad \boldsymbol{r}_{yy} = q\boldsymbol{r}_x + t\boldsymbol{r}_y + \chi\boldsymbol{r}. \tag{9.6}$$

The associated position vector $\hat{\boldsymbol{r}}$ of the surface $\hat{\Sigma}$ therefore satisfies the linear system

$$\hat{\boldsymbol{r}}_{xx} = a\hat{\boldsymbol{r}}_x + p\hat{\boldsymbol{r}}_y, \quad \hat{\boldsymbol{r}}_{yy} = b\hat{\boldsymbol{r}}_y + q\hat{\boldsymbol{r}}_x \tag{9.7}$$

with $a = s - 2(\ln r^0)_x, \ b = t - 2(\ln r^0)_y$ which implies that the coordinates x and y are indeed asymptotic on $\hat{\Sigma}$. The compatibility condition $\boldsymbol{r}_{xxyy} = \boldsymbol{r}_{yyxx}$ of the linear system (9.6) is readily shown to produce the condition $s_y = t_x$ provided that the vectors $\boldsymbol{r}, \boldsymbol{r}_x, \boldsymbol{r}_y$ and $\boldsymbol{r}_{xy}$ are linearly independent. We may set $s = t = 0$ without loss of generality since the position vector of a surface in projective space is only defined up to a multiplicative arbitrary function. Thus, application of a suitable gauge transformation of the form $\boldsymbol{r} \to g\boldsymbol{r}$ removes the coefficients s and t and brings the linear system (9.6) into canonical form. The latter was set down and extensively discussed by Wilczynski [384]. Its compatibility condition is readily shown to lead to the following theorem.

Theorem 43 ('Gauss-Weingarten' and 'Gauss-Mainardi-Codazzi' equations in projective differential geometry). *The position vector of a surface $\Sigma \subset \mathbb{P}^3$ parametrised in terms of asymptotic coordinates may be normalised in such a way that it obeys a linear system of the form*

$$\boldsymbol{r}_{xx} = p\boldsymbol{r}_y + \frac{1}{2}(V - p_y)\boldsymbol{r}, \quad \boldsymbol{r}_{yy} = q\boldsymbol{r}_x + \frac{1}{2}(W - q_x)\boldsymbol{r}. \tag{9.8}$$

The latter is compatible if and only if p, q and V, W constitute a solution of the underdetermined nonlinear system

$$\begin{aligned} p_{yyy} - 2p_yW - pW_y &= q_{xxx} - 2q_xV - qV_x \\ W_x &= 2qp_y + pq_y \\ V_y &= 2pq_x + qp_x. \end{aligned} \tag{9.9}$$

Particular integrable reductions of the above 'Gauss-Mainardi-Codazzi equations' of projective differential geometry and their associated classes of surfaces will be the subject of this chapter. We remark that in order to preserve the structure of the 'Gauss-Weingarten equations' (9.8) under an arbitrary reparametrisation

$$x^* = f(x), \quad y^* = g(y) \tag{9.10}$$

of the asymptotic coordinates, the position vector $\boldsymbol{r}$ must be renormalised according to

$$\boldsymbol{r}^* = \sqrt{f'(x)g'(y)}\,\boldsymbol{r}. \tag{9.11}$$

The associated coefficients p^*, q^* and V^*, W^* are related to the original ones by

$$\begin{aligned} p^* &= p\frac{g'}{f'^2}, \quad V^* = \frac{V + S(f)}{f'^2} \\ q^* &= q\frac{f'}{g'^2}, \quad W^* = \frac{W + S(g)}{g'^2}, \end{aligned} \tag{9.12}$$

where $S(\cdot)$ denotes the usual Schwarzian derivative, that is

$$S(f) = \frac{f'''}{f'} - \frac{3}{2}\left(\frac{f''}{f'}\right)^2. \tag{9.13}$$

It may be verified directly that (9.10), (9.12) constitute an invariance of the nonlinear system (9.9). Furthermore, the quadratic form

$$pq\,dxdy \tag{9.14}$$

and the class of cubic forms which are proportional ('conformally equivalent') to

$$p\,dx^3 + q\,dy^3 \tag{9.15}$$

are absolute projective invariants since they are preserved by the above class of transformations. The quantities (9.14) and (9.15) are known as the *projective metric* and *Darboux cubic form*, respectively, and play the role of the 1st and 2nd fundamental forms in projective differential geometry. In particular, they define a 'generic' surface uniquely up to projective equivalence [51].

Exercise

1. Verify that the projective Gauss-Weingarten equations (9.9) are form-invariant under the gauge transformation (9.10)–(9.12). Show that the quadratic form (9.14) is preserved as is the cubic form (9.15) up to a multiplicative factor.

9.2 Projective-Minimal, Godeaux-Rozet, and Demoulin Surfaces

One route to the isolation of integrable reductions of the projective Gauss-Mainardi-Codazzi equations involves seeking a Lie point symmetry of the latter which acts non-trivially on the projective Gauss-Weingarten equations. For instance, the scaling

$$p \to \lambda p, \quad q \to q/\lambda \tag{9.16}$$

takes the projective Gauss-Weingarten equations to

$$\begin{aligned} \boldsymbol{r}_{xx} &= \lambda p \boldsymbol{r}_y + \frac{1}{2}(V - \lambda p_y)\boldsymbol{r}, \\ \boldsymbol{r}_{yy} &= \frac{1}{\lambda} q \boldsymbol{r}_x + \frac{1}{2}(W - \frac{1}{\lambda} q_x)\boldsymbol{r}. \end{aligned} \tag{9.17}$$

The first of the projective Gauss-Mainardi-Codazzi equations becomes

$$\lambda^2(p_{yyy} - 2p_y W - pW_y) = q_{xxx} - 2q_x V - qV_x \tag{9.18}$$

while the remaining two are invariant. If we now require that the projective Gauss-Mainardi-Codazzi equations be independent of λ, then (9.18) has to be separated into two equations and the system (9.9) yields

$$\begin{aligned} p_{yyy} - 2p_y W - pW_y = 0, \quad W_x = 2qp_y + pq_y \\ q_{xxx} - 2q_x V - qV_x = 0, \quad V_y = 2pq_x + qp_x. \end{aligned} \tag{9.19}$$

In the terminology of soliton theory, the 'Lax pair' (9.17) with 'spectral parameter' λ is compatible if and only if p, q and V, W constitute a solution of the nonlinear system (9.19). In Section 9.5, it will be shown that an $so(3, 3)$ analogue of this linear representation may be exploited to construct a Bäcklund transformation for (9.19).

In geometric terms, it may be shown that (9.19) embody the Euler-Lagrange equations associated with the projective area functional

$$\iint pq \, dxdy. \tag{9.20}$$

Accordingly, the corresponding projective Gauss-Weingarten equations (9.8) are descriptive of *projective-minimal surfaces*. The latter arise naturally in the context of soliton theory and may be classified both geometrically and algebraically. In this connection, we first observe that multiplication of $(9.19)_{1,3}$ by

p and q, respectively, and subsequent integration results in

$$W = \frac{p_{yy}}{p} - \frac{1}{2}\left(\frac{p_y}{p}\right)^2 + \frac{\alpha(x)}{p^2}$$
$$V = \frac{q_{xx}}{q} - \frac{1}{2}\left(\frac{q_x}{q}\right)^2 + \frac{\beta(y)}{q^2}, \tag{9.21}$$

where α and β are functions of integration. The latter may be made constant by means of a transformation of the form (9.10), (9.12). Insertion of V and W as given by (9.21) into the remaining projective Gauss-Mainardi-Codazzi equations yields

$$[p(\ln p)_{xy} - p^2 q]_y = -p\left(\frac{\alpha}{p^2}\right)_x$$
$$[q(\ln q)_{xy} - q^2 p]_x = -q\left(\frac{\beta}{q^2}\right)_y \tag{9.22}$$

or, equivalently,

$$(\ln p)_{xy} = pq + \frac{A}{p}, \quad A_y = -p\left(\frac{\alpha}{p^2}\right)_x$$
$$(\ln q)_{xy} = pq + \frac{B}{q}, \quad B_x = -q\left(\frac{\beta}{q^2}\right)_y. \tag{9.23}$$

Case 1 (*General*). Both α and β are non-zero and hence may be normalised to ± 1. For instance, if $\alpha = \beta = 1$, then the projective Gauss-Mainardi-Codazzi equations reduce to

$$(\ln p)_{xy} = pq + \frac{A}{p}, \quad A_y = 2\frac{p_x}{p^2}$$
$$(\ln q)_{xy} = pq + \frac{B}{q}, \quad B_x = 2\frac{q_y}{q^2}. \tag{9.24}$$

Case 2 (*Surfaces of Godeaux-Rozet* [152,318]). Here, $\alpha = 0$, while β is non-zero (or vice versa). The normalisations $\beta = \pm 1$ and $A = 1$ may be adopted provided that A is non-zero. In particular, if $\beta = 1$, then we obtain the nonlinear system

$$\begin{aligned}(\ln p)_{xy} &= pq + \frac{1}{p}\\ (\ln q)_{xy} &= pq + \frac{B}{q}, \quad B_x = 2\frac{q_y}{q^2}.\end{aligned} \tag{9.25}$$

Case 3 (*Surfaces of Demoulin* [102]). Both α and β vanish. In this case, we may assume that $A = B = 1$ if A and B are non-zero and hence

$$(\ln p)_{xy} = pq + \frac{1}{p}, \quad (\ln q)_{xy} = pq + \frac{1}{q}. \tag{9.26}$$

Case 3a (*Surfaces of Tzitzeica*). If we assume that $p = q$, then it is readily shown that $\alpha = \beta = 0$ and the Tzitzeica equation

$$(\ln h)_{xy} = h - \frac{1}{h^2} \tag{9.27}$$

with $h = -1/p$ results. Thus, affine spheres constitute particular Demoulin surfaces and are characterised by the condition that they be projective-minimal and *isothermal-asymptotic*, that is $p = q$. We shall return to the class of isothermal-asymptotic surfaces in Section 9.7. The discovery that affine spheres constitute extrema of the area-minimising variational problem in projective differential geometry is due to Behnke (see [39]).

Exercise

1. Find a change of variables which normalises the functions α and β in (9.23) to $0, \pm 1$. In the case $\alpha = 0$, show that one may set $A = 0, 1$ without loss of generality.

9.3 Linear Representations

The Gauss-Mainardi-Codazzi equations associated with projective-minimal surfaces admit several linear representations which may be constructed geometrically and which are also meaningful in the context of soliton theory. We begin with the construction of an $sl(4)$ 4×4 linear representation which is based on the so-called Wilczynski moving tetrahedral. On use of the classical Plücker correspondence which encapsulates the isomorphism of the $sl(4)$ and $so(3, 3)$

Lie algebras, we derive a 6×6 linear representation. Its geometric significance becomes apparent in Section 9.4 in connection with the identification of the Demoulin system (9.26) as a periodic reduction of the two-dimensional Toda lattice.

9.3.1 The Wilczynski Tetrahedral and a 4 × 4 Linear Representation

In Chapter 1, it has been seen how the Gauss-Weingarten equations for surfaces in $\mathbb{R}^3$ may be rewritten as two compatible first-order linear matrix differential equations for an orthonormal triad (moving trihedral) consisting of two tangent vectors and the unit normal. In projective differential geometry, however, no canonically defined 'projective normal' exists. We here make use of Wilczynski's definition of a *moving tetrahedral* [51] whose vertices are defined in an invariant way by

$$\begin{gathered} \boldsymbol{r}, \quad \boldsymbol{r}_1 = \boldsymbol{r}_x - \frac{1}{2}\frac{q_x}{q}\boldsymbol{r}, \quad \boldsymbol{r}_2 = \boldsymbol{r}_y - \frac{1}{2}\frac{p_y}{p}\boldsymbol{r} \\ \boldsymbol{\eta} = \boldsymbol{r}_{xy} - \frac{1}{2}\frac{p_y}{p}\boldsymbol{r}_x - \frac{1}{2}\frac{q_x}{q}\boldsymbol{r}_y + \left(\frac{1}{4}\frac{p_y q_x}{pq} - \frac{1}{2}pq\right)\boldsymbol{r}. \end{gathered} \tag{9.28}$$

Under a transformation of the form (9.10), (9.12), the quantities $\boldsymbol{r}, \boldsymbol{r}_1, \boldsymbol{r}_2$ and $\boldsymbol{\eta}$ acquire non-zero multiplicative factors which do not change them as points in $\mathbb{P}^3$. The geometric interpretation of the edges $[\boldsymbol{r}, \boldsymbol{r}_1]$, $[\boldsymbol{r}, \boldsymbol{r}_2]$, $[\boldsymbol{r}_1, \boldsymbol{r}_2]$ and $[\boldsymbol{r}, \boldsymbol{\eta}]$ of the moving tetrahedral is as follows. In projective space, the line $(\boldsymbol{r}, \boldsymbol{r}_1)$ is represented by an arbitrary linear combination of $\boldsymbol{r}$ and $\boldsymbol{r}_1$ or, equivalently, the linear combination

$$f\boldsymbol{r} + g\boldsymbol{r}_x. \tag{9.29}$$

Its image under the mapping (9.4) is therefore given by

$$\hat{\boldsymbol{r}} + h\hat{\boldsymbol{r}}_x, \quad h = \frac{gr^0}{fr^0 + gr^0_x} \tag{9.30}$$

and this constitutes the tangent to the x-asymptotic line at $\hat{\boldsymbol{r}}$. Thus, the lines $(\boldsymbol{r}, \boldsymbol{r}_1)$ and $(\boldsymbol{r}, \boldsymbol{r}_2)$ are tangent to the x- and y-asymptotic lines on Σ, respectively, so that the points $\boldsymbol{r}, \boldsymbol{r}_1, \boldsymbol{r}_2$ span the tangent plane of Σ at $\boldsymbol{r}$. The line $(\boldsymbol{r}_1, \boldsymbol{r}_2)$ is known as the second directrix of Wilczynski and is tangent to Σ. The line $(\boldsymbol{r}, \boldsymbol{\eta})$ is transversal to Σ if the vectors $\boldsymbol{r}, \boldsymbol{r}_x, \boldsymbol{r}_y$ and $\boldsymbol{r}_{xy}$ are linearly independent. It plays the role of a projective normal and is termed the first directrix of Wilczynski.

Since $\boldsymbol{r}, \boldsymbol{r}_1, \boldsymbol{r}_2$ and $\boldsymbol{\eta}$ when regarded as vectors form a basis of $\mathbb{R}^4$, the Gauss-Weingarten equations for projective-minimal surfaces may be brought into a first-order matrix form. Differentiation of these vectors as defined by (9.28) and use of (9.8) yields

$$\begin{pmatrix} \boldsymbol{r} \\ \boldsymbol{r}_1 \\ \boldsymbol{r}_2 \\ \boldsymbol{\eta} \end{pmatrix}_x = \frac{1}{2} \begin{pmatrix} \frac{q_x}{q} & 2 & 0 & 0 \\ \frac{\beta}{q^2} & -\frac{q_x}{q} & 2p & 0 \\ -\frac{A}{p} & 0 & \frac{q_x}{q} & 2 \\ \frac{\alpha}{p} & -\frac{A}{p} & \frac{\beta}{q^2} & -\frac{q_x}{q} \end{pmatrix} \begin{pmatrix} \boldsymbol{r} \\ \boldsymbol{r}_1 \\ \boldsymbol{r}_2 \\ \boldsymbol{\eta} \end{pmatrix}$$

$$\begin{pmatrix} \boldsymbol{r} \\ \boldsymbol{r}_1 \\ \boldsymbol{r}_2 \\ \boldsymbol{\eta} \end{pmatrix}_y = \frac{1}{2} \begin{pmatrix} \frac{p_y}{p} & 0 & 2 & 0 \\ -\frac{B}{q} & \frac{p_y}{p} & 0 & 2 \\ \frac{\alpha}{p^2} & 2q & -\frac{p_y}{p} & 0 \\ \frac{\beta}{q} & \frac{\alpha}{p^2} & -\frac{B}{q} & -\frac{p_y}{p} \end{pmatrix} \begin{pmatrix} \boldsymbol{r} \\ \boldsymbol{r}_1 \\ \boldsymbol{r}_2 \\ \boldsymbol{\eta} \end{pmatrix}. \tag{9.31}$$

By construction, this linear system is compatible modulo (9.23) and a parameter may be injected via the scalings

$$\begin{aligned} p &\to \lambda p, \quad & A &\to \lambda A, \quad & \alpha &\to \lambda^2 \alpha \\ q &\to q/\lambda, \quad & B &\to B/\lambda, \quad & \beta &\to \beta/\lambda^2. \end{aligned} \tag{9.32}$$

It is noted that the matrices in the above linear representation are trace-free and hence (9.31) subject to (9.32) may be regarded as a Lax pair based on the Lie algebra $sl(4)$.

9.3.2 The Plücker Correspondence and a 6 × 6 Linear Representation

The exterior product of two vectors $\boldsymbol{a}, \boldsymbol{b} \in \mathbb{R}^4$ is defined as the six-dimensional vector

$$\boldsymbol{a} \wedge \boldsymbol{b} = (p_{01}, p_{02}, p_{03}, p_{23}, p_{31}, p_{12}), \tag{9.33}$$

where

$$p_{ij} = \det \begin{pmatrix} a^i & a^j \\ b^i & b^j \end{pmatrix} \tag{9.34}$$

denote the twelve sub-determinants of the 2×4 matrix

$$\begin{pmatrix} \boldsymbol{a} \\ \boldsymbol{b} \end{pmatrix} = \begin{pmatrix} a^0 & a^1 & a^2 & a^3 \\ b^0 & b^1 & b^2 & b^3 \end{pmatrix}. \tag{9.35}$$

It is noted that $p_{ji} = -p_{ij}$. The exterior product enjoys the usual properties

$$\begin{aligned} &\boldsymbol{a} \wedge \boldsymbol{b} = -\boldsymbol{b} \wedge \boldsymbol{a}, \quad \boldsymbol{a} \wedge \boldsymbol{a} = 0 && (\text{SKEW-SYMMETRY}) \\ &\kappa(\boldsymbol{a} \wedge \boldsymbol{b}) = (\kappa\boldsymbol{a}) \wedge \boldsymbol{b} = \boldsymbol{a} \wedge (\kappa\boldsymbol{b}) && (\text{ASSOCIATIVITY}) \\ &(\boldsymbol{a} + \boldsymbol{b}) \wedge \boldsymbol{c} = \boldsymbol{a} \wedge \boldsymbol{c} + \boldsymbol{b} \wedge \boldsymbol{c} && (\text{DISTRIBUTIVITY}) \\ &(\boldsymbol{a} \wedge \boldsymbol{b})' = \boldsymbol{a}' \wedge \boldsymbol{b} + \boldsymbol{a} \wedge \boldsymbol{b}' && (\text{LEIBNIZ RULE}), \end{aligned} \tag{9.36}$$

where $\boldsymbol{a}, \boldsymbol{b}, \boldsymbol{c} \in \mathbb{R}^4$, κ is a scalar and the prime denotes differentiation. The associativity law implies that

$$(\kappa\boldsymbol{a}) \wedge (\sigma\boldsymbol{b}) = \kappa\sigma(\boldsymbol{a} \wedge \boldsymbol{b}). \tag{9.37}$$

Consequently, if we regard $\boldsymbol{a}$ and $\boldsymbol{b}$ as points in $\mathbb{P}^3$, that is

$$\boldsymbol{a} = (a^0 : a^1 : a^2 : a^3) \in \mathbb{P}^3, \quad \boldsymbol{b} = (b^0 : b^1 : b^2 : b^3) \in \mathbb{P}^3, \tag{9.38}$$

then the exterior product $\boldsymbol{a} \wedge \boldsymbol{b}$ may be identified with a point in a five-dimensional projective space with homogeneous coordinates

$$\boldsymbol{a} \wedge \boldsymbol{b} = (p_{01} : p_{02} : p_{03} : p_{23} : p_{31} : p_{12}) \in \mathbb{P}^5. \tag{9.39}$$

Moreover, since any two points $\boldsymbol{a}, \boldsymbol{b} \in \mathbb{P}^3$ define a line $l(\boldsymbol{a}, \boldsymbol{b})$ in $\mathbb{P}^3$ and the exterior products of $\boldsymbol{a}$ and $\boldsymbol{b}$ on the one hand and arbitrary linear combinations of $\boldsymbol{a}$ and $\boldsymbol{b}$ on the other are identical as points in $\mathbb{P}^5$, the map

$$l(\boldsymbol{a}, \boldsymbol{b}) \mapsto \boldsymbol{a} \wedge \boldsymbol{b} \tag{9.40}$$

between lines in $\mathbb{P}^3$ and points in $\mathbb{P}^5$ is well-defined. In this way, the *Plücker correspondence* (9.40) provides homogeneous coordinates for lines in $\mathbb{P}^3$. It is emphasised that the image of $\mathbb{P}^3$ under the Plücker map is a quadric embedded in $\mathbb{P}^5$, that is the coordinates p_{ij} satisfy the quadratic Plücker relations

$$p_{01}p_{23} + p_{02}p_{31} + p_{03}p_{12} = 0. \tag{9.41}$$

We now introduce vector-valued functions $\varphi \in \mathbb{R}^6$ and $\psi \in \mathbb{R}^6$ which are defined by

$$\varphi = \frac{1}{2}(\boldsymbol{r}_1 \wedge \boldsymbol{r}_2 + \boldsymbol{r} \wedge \boldsymbol{\eta}), \quad \psi = \frac{1}{2}(\boldsymbol{r}_2 \wedge \boldsymbol{r}_1 + \boldsymbol{r} \wedge \boldsymbol{\eta}). \tag{9.42}$$

The linear system (9.31) then implies that

$$\begin{aligned} \varphi_x &= \boldsymbol{r}_1 \wedge \boldsymbol{\eta} + \frac{\beta}{2q^2}\boldsymbol{r} \wedge \boldsymbol{r}_2, \quad \varphi_y = \frac{B}{2q}\boldsymbol{r}_2 \wedge \boldsymbol{r} \\ \psi_y &= \boldsymbol{r}_2 \wedge \boldsymbol{\eta} + \frac{\alpha}{2p^2}\boldsymbol{r} \wedge \boldsymbol{r}_1, \quad \psi_x = \frac{A}{2p}\boldsymbol{r}_1 \wedge \boldsymbol{r} \end{aligned} \tag{9.43}$$

so that $\varphi, \varphi_x, \varphi_y, \psi, \psi_x, \psi_y$ are seen to form a basis of $\mathbb{R}^6$. Further differentiation delivers the closed system

$$\begin{aligned} \varphi_{xx} &= p\psi_y - \frac{q_x}{q}\varphi_x + 2\frac{\alpha}{A}\psi_x + 2\frac{\beta}{q^2}\varphi \\ \varphi_{xy} &= \frac{B_x}{B}\varphi_y - \frac{B}{q}\varphi \\ \varphi_{yy} &= \left(\ln\frac{B}{q}\right)_y \varphi_y + p\frac{B}{A}\psi_x \\ \psi_{xx} &= \left(\ln\frac{A}{p}\right)_x \psi_x + q\frac{A}{B}\varphi_y \\ \psi_{xy} &= \frac{A_y}{A}\psi_x - \frac{A}{p}\psi \\ \psi_{yy} &= q\varphi_x - \frac{p_y}{p}\psi_y + 2\frac{\beta}{B}\varphi_y + 2\frac{\alpha}{p^2}\psi. \end{aligned} \tag{9.44}$$

Its compatibility conditions are satisfied modulo the Gauss-Mainardi-Codazzi equations for projective-minimal surfaces. Once again, the scaling (9.32) produces a linear representation of (9.23) in the sense of soliton theory. It is observed, *en passant*, that in the case of Demoulin surfaces corresponding to $\alpha = \beta = 0$ and $A = B = 1$ the linear representation simplifies radically (cf. Section 9.4).

The Lie algebraic structure underlying the linear system (9.44) is revealed by introducing the vector

$$\Phi = (\phi^1, \phi^2, \phi^3, \phi^4, \phi^5, \phi^6)^{\mathsf{T}} \tag{9.45}$$

with

$$\begin{aligned} \phi^1 &= \psi_y + \frac{\alpha}{Ap}\psi_x, \quad \phi^2 = \psi, \quad \phi^3 = \frac{p}{A}\psi_x \\ \phi^6 &= \varphi_x + \frac{\beta}{Bq}\varphi_y, \quad \phi^5 = \varphi, \quad \phi^4 = \frac{q}{B}\varphi_y, \end{aligned} \tag{9.46}$$

in terms of which we obtain the first-order system

$$\Phi_x = F\Phi, \quad F = \begin{pmatrix} 0 & -\frac{A}{p} & 0 & \frac{\alpha}{p} & 0 & 0 \\ 0 & 0 & \frac{A}{p} & 0 & 0 & 0 \\ 0 & 0 & 0 & p & 0 & 0 \\ 0 & 0 & 0 & \frac{q_x}{q} & -1 & 0 \\ 0 & 0 & 0 & -\frac{\beta}{q^2} & 0 & 1 \\ p & 0 & \frac{\alpha}{p} & 0 & \frac{\beta}{q^2} & -\frac{q_x}{q} \end{pmatrix}$$

$$\Phi_y = G\Phi, \quad G = \begin{pmatrix} -\frac{p_y}{p} & \frac{\alpha}{p^2} & 0 & \frac{\beta}{q} & 0 & q \\ 1 & 0 & -\frac{\alpha}{p^2} & 0 & 0 & 0 \\ 0 & -1 & \frac{p_y}{p} & 0 & 0 & 0 \\ 0 & 0 & q & 0 & 0 & 0 \\ 0 & 0 & 0 & \frac{B}{q} & 0 & 0 \\ 0 & 0 & \frac{\beta}{q} & 0 & -\frac{B}{q} & 0 \end{pmatrix}. \tag{9.47}$$

The trace-free matrices F and G are readily shown to satisfy the relations

$$F^{\mathsf{T}}D = -DF, \quad G^{\mathsf{T}}D = -DG, \quad D = \begin{pmatrix} 0 & 0 & 1 & 0 & 0 & 0 \\ 0 & 1 & 0 & 0 & 0 & 0 \\ 1 & 0 & 0 & 0 & 0 & 0 \\ 0 & 0 & 0 & 0 & 0 & -1 \\ 0 & 0 & 0 & 0 & -1 & 0 \\ 0 & 0 & 0 & -1 & 0 & 0 \end{pmatrix}. \tag{9.48}$$

These, in turn, define the $so(3,3)$ Lie algebra since D may be transformed into $\operatorname{diag}(1, 1, 1, -1, -1, -1)$ by means of a similarity transformation. Thus, we have retrieved the well-known result that the Plücker embedding (9.40) encapsulates the $sl(4)$–$so(3, 3)$ isomorphism. The $so(3, 3)$ linear representation constitutes the starting point for the construction of a Bäcklund transformation for projective-minimal surfaces to be discussed in the following section.

Exercises

1. Show that the Wilczynski tetrahedral (9.28) is projective-invariant under the gauge transformation (9.10)–(9.12).
2. Verify the discrete symmetry (9.48) of the $so(3, 3)$ linear representation of projective-minimal surfaces and deduce that

$$\Phi^{\top} D \Phi = \text{const.}$$

9.4 The Demoulin System as a Periodic Toda Lattice

The relations $(9.43)_{2,4}$ and $(9.46)_{3,6}$ imply that the points $\phi^3 \in \mathbb{P}^5$ and $\phi^4 \in \mathbb{P}^5$ represent via the Plücker correspondence the tangents to the x- and y-asymptotic lines on $\Sigma \subset \mathbb{P}^3$, respectively. However, ϕ^3 and ϕ^4 may also be interpreted as the position vectors of two surfaces Σ_3 and Σ_4 in $\mathbb{P}^5$, respectively. Since these two position vectors are related by

$$\phi^3_x = p\phi^4, \quad \phi^4_y = q\phi^3, \tag{9.49}$$

the surfaces Σ_3 and Σ_4 are Laplace-Darboux transforms of each other (cf. Section 3.3). Indeed, elimination of either ϕ^3 or ϕ^4 leads to the conjugate net equations in projective space

$$\phi^3_{xy} = \frac{p_y}{p}\phi^3_x + pq\phi^3, \quad \phi^4_{xy} = \frac{q_x}{q}\phi^4_y + pq\phi^4. \tag{9.50}$$

Accordingly, the coordinates x and y are conjugate on Σ_3 and Σ_4. Continuation of the Laplace-Darboux sequence in both directions leads to the *Godeaux sequence* [51] of surfaces in $\mathbb{P}^5$

$$\cdots \longleftrightarrow \Sigma_2 \longleftrightarrow \Sigma_3 \longleftrightarrow \Sigma_4 \longleftrightarrow \Sigma_5 \longleftrightarrow \cdots. \tag{9.51}$$

Periodic Godeaux sequences are of particular interest. The only surfaces Σ for which the associated Godeaux sequences are of period 6 (in fact the smallest possible period) are those of Demoulin. This may be interpreted as an equivalent geometric description of Demoulin surfaces. In this case, the relations

$$\begin{aligned} \phi^1_x &= -\frac{1}{p}\phi^2, \quad \phi^2_y = \phi^1 \\ \phi^6_y &= -\frac{1}{q}\phi^5, \quad \phi^5_x = \phi^6 \end{aligned} \tag{9.52}$$

obtain so that both ϕ^1, ϕ^2 and ϕ^5, ϕ^6 are related by Laplace-Darboux transformations with corresponding conjugate net equations

$$\begin{aligned} \phi^1_{xy} &= -\frac{p_y}{p}\phi^1_x - \frac{1}{p}\phi^1, \quad \phi^2_{xy} = -\frac{1}{p}\phi^2 \\ \phi^6_{xy} &= -\frac{q_x}{q}\phi^6_y - \frac{1}{q}\phi^6, \quad \phi^5_{xy} = -\frac{1}{q}\phi^5. \end{aligned} \tag{9.53}$$

In fact, the points $\phi^1, \ldots, \phi^6 \in \mathbb{P}^5$ constitute the position vectors of the surfaces $\Sigma_1, \ldots, \Sigma_6$ of the periodic Godeaux sequence. This follows from the fact that the Laplace-Darboux invariants h and k label the equivalence classes of conjugate net equations related by gauge transformations (cf. Section 3.3) and hence are identical in $\mathbb{P}^5$. The Laplace-Darboux invariants associated with (9.50), (9.53) are given by

$$\begin{aligned} h_1 &= h_4 = k_3 = k_6 = l \\ h_2 &= h_3 = k_1 = k_2 = k \\ h_5 &= h_6 = k_4 = k_5 = h \end{aligned} \tag{9.54}$$

with the definitions

$$h = -\frac{1}{q}, \quad k = -\frac{1}{p}, \quad l = \frac{1}{hk} = pq \tag{9.55}$$

so that the points ϕ^i are indeed related by the Laplace-Darboux transformations which generate the periodic Godeaux sequence.

In modern terminology, the classical Laplace-Darboux sequence of conjugate nets is governed by the two-dimensional Toda lattice (3.105). If we impose periodicity 6, then it reduces to the finite system

$$\begin{aligned} (\ln h_1)_{xy} &= -h_6 + 2h_1 - h_2, \quad (\ln h_2)_{xy} = -h_1 + 2h_2 - h_3 \\ (\ln h_3)_{xy} &= -h_2 + 2h_3 - h_4, \quad (\ln h_4)_{xy} = -h_3 + 2h_4 - h_5 \\ (\ln h_5)_{xy} &= -h_4 + 2h_5 - h_6, \quad (\ln h_6)_{xy} = -h_5 + 2h_6 - h_1 \end{aligned} \tag{9.56}$$

which is associated with the affine Lie algebra $A_5^{(1)}$. The latter contains the affine Lie algebra $D_3^{(2)}$ as a subalgebra which corresponds to the reduction (9.54). The periodic Toda lattice now specialises to

$$(\ln h)_{xy} = h - l, \quad (\ln k)_{xy} = k - l, \quad (\ln l)_{xy} = -h + 2l - k \qquad (9.57)$$

which implies that $(\ln hkl)_{xy} = 0$. Without loss of generality, we may set $hkl = 1$ and retrieve the *Demoulin system*

$$\boxed{(\ln h)_{xy} = h - \frac{1}{hk}, \quad (\ln k)_{xy} = k - \frac{1}{hk}} \qquad (9.58)$$

written in terms of the variables $(9.55)_{1,2}$. A linear representation is obtained by setting $A = -\lambda$, $B = -1/\lambda$ and applying the scaling

$$h \to -\lambda h, \quad k \to -k/\lambda \qquad (9.59)$$

in (9.44). Indeed, it is readily shown that the linear system

$$\begin{aligned}
\varphi_{xx} &= \frac{h_x}{h}\varphi_x + \lambda\frac{1}{k}\psi_y, & \psi_{xx} &= \frac{k_x}{k}\psi_x + \lambda\frac{1}{h}\varphi_y \\
\varphi_{xy} &= h\varphi, & \psi_{xy} &= k\psi \\
\varphi_{yy} &= \frac{h_y}{h}\varphi_y + \frac{1}{\lambda}\frac{1}{k}\psi_x, & \psi_{yy} &= \frac{k_y}{k}\psi_y + \frac{1}{\lambda}\frac{1}{h}\varphi_x
\end{aligned} \qquad (9.60)$$

is compatible modulo the Demoulin system (9.58). In the case $h = k$ corresponding to Tzitzeica surfaces, the identification $\varphi = \psi$ is admissible and the standard linear representation (3.15) for the Tzitzeica equation is recovered.

Exercise

1. In the case of Demoulin surfaces, show that, up to gauge transformations, the quantities $\phi^1, \ldots, \phi^6$ are related by Laplace-Darboux transformations in the sense of Section 3.3.

9.5 A Bäcklund Transformation for Projective-Minimal Surfaces

A Bäcklund transformation for projective-minimal surfaces may be derived in a systematic manner by imposition of suitable constraints on the Fundamental

Transformation. Here, we assume that α and β are constant and make the canonical change of variables

$$h = -\frac{1}{q}, \quad k = -\frac{1}{p}. \tag{9.61}$$

Furthermore, application of the scaling

$$\begin{aligned} h &\to -\lambda h, \quad A \to -\lambda A, \quad \alpha \to \lambda^2 \alpha \\ k &\to -k/\lambda, \quad B \to -B/\lambda, \quad \beta \to \beta/\lambda^2 \end{aligned} \tag{9.62}$$

takes the linear representation (9.44) to

$$\begin{aligned} \varphi_{xx} &= \frac{h_x}{h}\varphi_x + \lambda\frac{1}{k}\psi_y - 2\lambda\frac{\alpha}{A}\psi_x + 2\beta h^2\varphi \\ \varphi_{xy} &= Bh\varphi + \frac{B_x}{B}\varphi_y \\ \varphi_{yy} &= (\ln Bh)_y\varphi_y + \frac{1}{\lambda}\frac{B}{Ak}\psi_x \\ \psi_{xx} &= (\ln Ak)_x\psi_x + \lambda\frac{A}{Bh}\varphi_y \\ \psi_{xy} &= Ak\psi + \frac{A_y}{A}\psi_x \\ \psi_{yy} &= \frac{k_y}{k}\psi_y + \frac{1}{\lambda}\frac{1}{h}\varphi_x - 2\frac{1}{\lambda}\frac{\beta}{B}\varphi_y + 2\alpha k^2\psi \end{aligned} \tag{9.63}$$

with compatibility conditions

$$\begin{aligned} (\ln h)_{xy} &= Bh - \frac{1}{hk}, \quad B_x = 2\beta h_y \\ (\ln k)_{xy} &= Ak - \frac{1}{hk}, \quad A_y = 2\alpha k_x. \end{aligned} \tag{9.64}$$

We now seek an invariance of the above linear system which may then be formulated via the Plücker correspondence as a Bäcklund transformation for projective-minimal surfaces.

9.5.1 Invariance of the so(3, 3) Linear Representation

It is readily seen that the compatibility conditions for the linear system

$$\begin{aligned} \varphi_{xy} &= Bh\varphi + \frac{B_x}{B}\varphi_y, \quad \varphi_{yy} = (\ln Bh)_y\varphi_y + \frac{1}{\lambda}\frac{B}{Ak}\psi_x \\ \psi_{xy} &= Ak\psi + \frac{A_y}{A}\psi_x, \quad \psi_{xx} = (\ln Ak)_x\psi_x + \lambda\frac{A}{Bh}\varphi_y, \end{aligned} \tag{9.65}$$

which forms a subsystem of the Lax pair (9.63), yield

$$(\ln h)_{xy} = Bh - \frac{1}{hk}, \quad (\ln k)_{xy} = Ak - \frac{1}{hk}. \tag{9.66}$$

The latter is symmetric in the independent variables and may be regarded as a system for h and k with arbitrary functions A and B. It also guarantees the compatibility of the 'adjoint' system

$$\begin{aligned} \tilde{\varphi}_{xy} &= Bh\tilde{\varphi} + \frac{B_y}{B}\tilde{\varphi}_x, \quad \tilde{\varphi}_{xx} = (\ln Bh)_x\tilde{\varphi}_x + \lambda\frac{B}{Ak}\tilde{\psi}_y \\ \tilde{\psi}_{xy} &= Ak\tilde{\psi} + \frac{A_x}{A}\tilde{\psi}_y, \quad \tilde{\psi}_{yy} = (\ln Ak)_y\tilde{\psi}_y + \frac{1}{\lambda}\frac{A}{Bh}\tilde{\varphi}_x \end{aligned} \tag{9.67}$$

which is obtained from (9.65) by interchange of x and y and the transposition $\lambda \to 1/\lambda$. We first focus on the equations $(9.65)_{1,3}$ which may be viewed as normalised conjugate net equations in projective space and are therefore preserved by the Fundamental Transformation. The analogues of the parallel nets introduced in Section 5.4 are given by the bilinear potentials M and N which are defined by the compatible equations

$$\begin{aligned} M_x &= \frac{\tilde{\varphi}^\circ_x\varphi}{B}, \quad M_y = \frac{\tilde{\varphi}^\circ\varphi_y}{B} \\ N_x &= \frac{\tilde{\psi}^\circ\psi_x}{A}, \quad N_y = \frac{\tilde{\psi}^\circ_y\psi}{A}, \end{aligned} \tag{9.68}$$

where $(\tilde{\varphi}^\circ, \tilde{\psi}^\circ)$ is a solution of the adjoint system (9.67) with parameter $\lambda_\circ$. In addition, if φ° and ψ° are 'eigenfunctions' satisfying the linear system (9.65) with parameter $\lambda_\circ$, then the corresponding bilinear potentials M° and N° obey the relations

$$\begin{aligned} M^\circ_x &= \frac{\tilde{\varphi}^\circ_x\varphi^\circ}{B}, \quad M^\circ_y = \frac{\tilde{\varphi}^\circ\varphi^\circ_y}{B} \\ N^\circ_x &= \frac{\tilde{\psi}^\circ\psi^\circ_x}{A}, \quad N^\circ_y = \frac{\tilde{\psi}^\circ_y\psi^\circ}{A}. \end{aligned} \tag{9.69}$$

Differentiation of the ansätze

$$\varphi' = \varphi - \varphi^\circ \frac{M}{M^\circ}, \quad \psi' = \psi - \psi^\circ \frac{N}{N^\circ} \tag{9.70}$$

for the Fundamental Transforms of φ and ψ leads to

$$\varphi'_y = \Gamma\left(\varphi_y - \varphi^\circ_y \frac{M}{M^\circ}\right), \quad \psi'_x = \Lambda\left(\psi_x - \psi^\circ_x \frac{N}{N^\circ}\right), \tag{9.71}$$

where Γ and Λ are defined by

$$\Gamma = 1 - \frac{\tilde{\varphi}^\circ \varphi^\circ}{BM^\circ}, \quad \Lambda = 1 - \frac{\tilde{\psi}^\circ \psi^\circ}{AN^\circ}. \tag{9.72}$$

Accordingly, the mixed derivative of φ' may be cast in the form

$$\varphi'_{xy} = \Gamma\left(Bh - \frac{\tilde{\varphi}^\circ_x \varphi^\circ_y}{BM^\circ}\right)\varphi' + \left(\frac{\Gamma_x}{\Gamma} + \frac{B_x}{B}\right)\varphi'_y, \tag{9.73}$$

which shows that the hyperbolic equation $(9.65)_1$ is indeed invariant under the Fundamental Transformation $(9.70)_1$ with

$$B' = g\Gamma B, \quad gh' = h - \frac{\tilde{\varphi}^\circ_x \varphi^\circ_y}{B^2 M^\circ}, \tag{9.74}$$

where $g = g(y)$ is a function of integration. For symmetry reasons, the second hyperbolic equation $(9.65)_3$ is preserved by the Fundamental Transformation $(9.70)_2$ with

$$A' = f\Lambda A, \quad fk' = k - \frac{\tilde{\psi}^\circ_y \psi^\circ_x}{A^2 N^\circ} \tag{9.75}$$

and $f = f(x)$.

Preservation of the equations $(9.65)_{2,4}$ may now be guaranteed by appropriate specification of f and g and certain constants of integration. To this end, we observe that

$$\lambda M - \lambda_\circ N - \lambda \frac{\tilde{\varphi}^\circ_x \varphi_y}{B^2 h} + \lambda_\circ \frac{\tilde{\psi}^\circ_y \psi_x}{A^2 k} = c \tag{9.76}$$

constitutes a first integral, a particular first integral of which yields

$$M^\circ - N^\circ - \frac{\tilde{\varphi}^\circ_x \varphi^\circ_y}{B^2 h} + \frac{\tilde{\psi}^\circ_y \psi^\circ_x}{A^2 k} = c^\circ. \tag{9.77}$$

In terms of primed quantities, these relations may be written as

$$\lambda g\frac{h'}{h}M - \lambda_\circ f\frac{k'}{k}N - Q = c, \tag{9.78}$$

where

$$Q = \lambda g\frac{\tilde{\varphi}_x^\circ\varphi_y'}{B'Bh} - \lambda_\circ f\frac{\tilde{\psi}_y^\circ\psi_x'}{A'Ak}, \tag{9.79}$$

and

$$g\frac{h'}{h}M^\circ - f\frac{k'}{k}N^\circ = c^\circ. \tag{9.80}$$

Accordingly, the relation

$$\lambda\frac{M}{M^\circ} - \lambda_\circ\frac{N}{N^\circ} = \frac{1}{M^\circ N^\circ}\frac{h}{gh'}[(c+Q)N^\circ - c^\circ N] \tag{9.81}$$

obtains.

Now, differentiation of φ_y' as given by $(9.71)_1$ produces

$$\begin{aligned}\varphi_{yy}' = {}& \left(\frac{\Gamma_y}{\Gamma} + (\ln Bh)_y - \frac{\tilde{\varphi}^\circ\varphi_y^\circ}{BM^\circ}\right)\varphi_y' \\ &+ \frac{1}{\lambda}\frac{\Gamma B}{Ak}\left[\frac{\psi_x'}{\Lambda} - \frac{\psi_x^\circ}{\lambda_\circ}\left(\lambda\frac{M}{M^\circ} - \lambda_\circ\frac{N}{N^\circ}\right)\right]\end{aligned} \tag{9.82}$$

which is of the form

$$\varphi_{yy}' = Q_1\varphi_y' + \frac{1}{\lambda}Q_2\psi_x' \tag{9.83}$$

provided that, by virtue of (9.79) and (9.81),

$$c = c^\circ = 0. \tag{9.84}$$

It is emphasized that the above constraints are admissible since the bilinear potentials M, N and M°, N° are only defined up to arbitrary additive constants. Thus, if the choice (9.84) is made, then the coefficients Q_1 and Q_2 read

$$\begin{aligned}Q_1 &= \frac{\Gamma_y}{\Gamma} + (\ln Bh)_y - \frac{\tilde{\varphi}^\circ\varphi_y^\circ}{BM^\circ} - \frac{\Gamma}{\lambda_\circ}\frac{\tilde{\varphi}_x^\circ\psi_x^\circ}{AkB'h'M^\circ} = (\ln B'h')_y \\ Q_2 &= \frac{\Gamma B}{Ak}\left(\frac{1}{\Lambda} + \frac{\tilde{\psi}_y^\circ\psi_x^\circ}{AA'k'N^\circ}\right) = \frac{B'}{gA'k'}\end{aligned} \tag{9.85}$$

so that preservation of $(9.65)_2$ requires that

$$g = 1. \tag{9.86}$$

Analogously, the specialisation

$$f = 1 \tag{9.87}$$

ensures that $(9.65)_4$ is invariant under the Fundamental Transformation. Consequently, (9.74), (9.75) with $f = g = 1$ and $c^\circ = 0$ constitute another solution of the nonlinear system (9.66).

To obtain necessary conditions for the invariance of the remaining *linear* equations $(9.63)_{1,6}$, we now deal directly with the residual *nonlinear* equations $(9.64)_{2,4}$. Invariance of the latter implies certain eigenfunction-adjoint eigenfunction constraints which may indeed be satisfied. Thus, if φ and ψ are eigenfunctions which are solutions of the complete set of linear equations (9.63), then the quantities

$$\begin{aligned}\tilde{\varphi} &= B\varphi - 2\beta\frac{h_y}{Bh}\varphi_y - 2\frac{1}{\lambda}\frac{\beta}{Ak}\psi_x\\ \tilde{\psi} &= A\psi - 2\alpha\frac{k_x}{Ak}\psi_x - 2\lambda\frac{\alpha}{Bh}\varphi_y\end{aligned} \tag{9.88}$$

are particular adjoint eigenfunctions. On use of the relations

$$\tilde{\varphi}_x = B\varphi_x - 2\beta h\varphi_y, \quad \tilde{\psi}_y = A\psi_y - 2\alpha k\psi_x, \tag{9.89}$$

it is readily verified that $\tilde{\varphi}$ and $\tilde{\psi}$ indeed satisfy the adjoint system (9.67). With this choice of adjoint eigenfunctions, the general solution of the defining relations (9.69) is given by

$$M^\circ = \frac{1}{2}\varphi^{\circ 2} - \beta\frac{\varphi_y^{\circ 2}}{B^2} + c_1, \quad N^\circ = \frac{1}{2}\psi^{\circ 2} - \alpha\frac{\psi_x^{\circ 2}}{A^2} + c_2. \tag{9.90}$$

Insertion into the first integral $(9.77)_{c^\circ=0}$ produces the quadratic constraint

$$\varphi^{\circ 2} - 2\frac{\varphi_y^\circ}{Bh}\left(\varphi_x^\circ - \beta\frac{h}{B}\varphi_y^\circ\right) - \psi^{\circ 2} + 2\frac{\psi_x^\circ}{Ak}\left(\psi_y^\circ - \alpha\frac{k}{A}\psi_x^\circ\right) = c_2 - c_1 \tag{9.91}$$

or, equivalently,

$$\Phi^{\circ\mathsf{T}} D\Phi^\circ = c_1 - c_2, \tag{9.92}$$

where Φ° and D are defined by (9.45), (9.46) and $(9.48)_3$, respectively. The latter expresses the fact that the norm of the vector Φ with respect to the Killing-Cartan metric D associated with the $so(3,3)$ Lie algebra is constant. This is a direct consequence of the identities (9.48).

Evaluation of the condition

$$B'_x = 2\beta h'_y \tag{9.93}$$

yields

$$2\beta h_y - \left(\frac{\tilde{\varphi}^\circ \varphi^\circ}{M^\circ}\right)_x = 2\beta\left[\left(\frac{h_y}{h} - \frac{\tilde{\varphi}^\circ \varphi^\circ_y}{BM^\circ}\right)h' - \frac{1}{\lambda_\circ}\frac{\tilde{\varphi}^\circ_x \psi^\circ_x}{AkBM^\circ}\right] \tag{9.94}$$

and further simplification results in

$$\left(M^\circ - \frac{1}{2}\varphi^{\circ 2} + \beta\frac{\varphi^{\circ 2}_y}{B^2}\right)\tilde{\varphi}^\circ \tilde{\varphi}^\circ_x = 0. \tag{9.95}$$

A similar result is obtained from the relation

$$A'_y = 2\alpha k'_x. \tag{9.96}$$

Thus, the choice

$$c_1 = c_2 = 0 \tag{9.97}$$

guarantees that the Gauss-Mainardi-Codazzi equations underlying projective-minimal surfaces are preserved by the Fundamental Transformation.

The bilinear potentials M and N may also be expressed explicitly in terms of eigenfunctions. Thus, one may directly verify that the expressions

$$\begin{aligned}
M &= -\kappa_0\varphi^\circ\varphi + \kappa_2\frac{\varphi^\circ_x\varphi_y}{Bh} + \kappa_0\frac{\varphi^\circ_y\varphi_x}{Bh} - 2\kappa_2\beta\frac{\varphi^\circ_y\varphi_y}{B^2}\\
&\quad + \kappa_1\psi^\circ\psi - \kappa_1\frac{\psi^\circ_y\psi_x}{Ak} - \kappa_1\frac{\psi^\circ_x\psi_y}{Ak} + 2\kappa_1\alpha\frac{\psi^\circ_x\psi_x}{A^2} + c_3\\
N &= -\kappa_1\varphi^\circ\varphi + \kappa_1\frac{\varphi^\circ_x\varphi_y}{Bh} + \kappa_1\frac{\varphi^\circ_y\varphi_x}{Bh} - 2\kappa_1\beta\frac{\varphi^\circ_y\varphi_y}{B^2}\\
&\quad + \kappa_2\psi^\circ\psi - \kappa_0\frac{\psi^\circ_y\psi_x}{Ak} - \kappa_2\frac{\psi^\circ_x\psi_y}{Ak} + 2\kappa_0\alpha\frac{\psi^\circ_x\psi_x}{A^2} + c_4,
\end{aligned} \tag{9.98}$$

where the constants κ_i are given by

$$\kappa_0 = \frac{\lambda^2_\circ}{\lambda^2 - \lambda^2_\circ}, \quad \kappa_1 = \frac{\lambda\lambda_\circ}{\lambda^2 - \lambda^2_\circ}, \quad \kappa_2 = \frac{\lambda^2}{\lambda^2 - \lambda^2_\circ}, \tag{9.99}$$

satisfy the defining relations (9.68). Evaluation (by means of a computer algebra program) of the primed versions of the remaining linear equations $(9.63)_{1,6}$ now shows that

$$c_3 = c_4 = 0. \tag{9.100}$$

The above result is summarised in the following:

Theorem 44 (An invariance of the so(3, 3) linear representation for projective-minimal surfaces). *Let $(\varphi, \psi, h, k, A, B)$ be a solution of the linear representation (9.63) and the nonlinear system (9.64). If $(\varphi^\circ, \psi^\circ)$ is another pair of eigenfunctions with parameter $\lambda_\circ$ subject to the admissible quadratic constraint*

$$\varphi^{\circ 2} - 2\frac{\varphi_y^\circ \varphi_x^\circ}{Bh} + 2\beta\frac{\varphi_y^{\circ 2}}{B^2} - \psi^{\circ 2} + 2\frac{\psi_x^\circ \psi_y^\circ}{Ak} - 2\alpha\frac{\psi_x^{\circ 2}}{A^2} = 0 \tag{9.101}$$

and the bilinear potentials M, N and M°, N° are defined by $(9.98)_{c_3=c_4=0}$ and

$$M^\circ = \frac{1}{2}\varphi^{\circ 2} - \beta\frac{\varphi_y^{\circ 2}}{B^2}, \quad N^\circ = \frac{1}{2}\psi^{\circ 2} - \alpha\frac{\psi_x^{\circ 2}}{A^2}, \tag{9.102}$$

respectively, then a second solution of (9.63), (9.64) is given by

$$\mathbb{B}: \begin{cases} \varphi' = \varphi - \varphi^\circ \dfrac{M}{M^\circ}, & \psi' = \psi - \psi^\circ \dfrac{N}{N^\circ} \\[2ex] h' = h - \dfrac{\tilde{\varphi}_x^\circ \varphi_y^\circ}{B^2 M^\circ}, & k' = k - \dfrac{\tilde{\psi}_y^\circ \psi_x^\circ}{A^2 N^\circ} \\[2ex] B' = B - \dfrac{\tilde{\varphi}^\circ \varphi^\circ}{M^\circ}, & A' = A - \dfrac{\tilde{\psi}^\circ \psi^\circ}{N^\circ} \end{cases} \tag{9.103}$$

with the definitions

$$\begin{aligned} \tilde{\varphi}^\circ &= B\varphi^\circ - 2\beta\frac{h_y}{Bh}\varphi_y^\circ - 2\frac{1}{\lambda_\circ}\frac{\beta}{Ak}\psi_x^\circ \\ \tilde{\psi}^\circ &= A\psi^\circ - 2\alpha\frac{k_x}{Ak}\psi_x^\circ - 2\lambda_\circ\frac{\alpha}{Bh}\varphi_y^\circ. \end{aligned} \tag{9.104}$$

9.5.2 *Invariance of the sl(4) Linear Representation*

Theorem 44 implies that the new eigenfunctions φ' and ψ' and their first derivatives constitute linear combinations of their unprimed counterparts with coefficients depending on φ°, ψ°, h, k, A, B and $\lambda_\circ$. Thus, the action of the Fundamental Transformation on the $so(3, 3)$ linear representation may be encoded in a transformation matrix P defined by

$$\Phi' = P\Phi \tag{9.105}$$

and satisfying

$$P^{\mathsf{T}} D P = D. \tag{9.106}$$

Accordingly, the transition from Φ to Φ' may be regarded as an (improper) rotation in $\mathbb{R}^6$ endowed with the metric D. However, it is readily shown that $\det P = -1$ so that $P \in O(3, 3)$ but $P \notin SO(3, 3)$. To exploit the $so(3, 3)$–$sl(4)$ isomorphism at the group level, that is the Plücker correspondence, we therefore need to conjugate the Fundamental Transformation $\mathbb{B}$ with the discrete invariance

$$\mathbb{D}: \quad (\varphi, \psi, h, k, A, B) \rightarrow (\varphi, -\psi, -h, -k, -A, -B) \tag{9.107}$$

represented by the transformation matrix $\hat{P} = \mathrm{diag}(-1, -1, -1, 1, 1, 1)$ with

$$\hat{\Phi} = \hat{P}\Phi, \quad \hat{P}^{\mathsf{T}} D \hat{P} = D, \quad \det \hat{P} = -1. \tag{9.108}$$

Indeed, the transformation matrix $\tilde{P}$ associated with

$$\tilde{\mathbb{B}} = \mathbb{D} \circ \mathbb{B}: \quad \tilde{\Phi} = \tilde{P}\Phi = \hat{P}P\Phi \tag{9.109}$$

enjoys the properties

$$\tilde{P}^{\mathsf{T}} D \tilde{P} = D, \quad \det \tilde{P} = 1 \tag{9.110}$$

and hence $\tilde{P} \in SO(3, 3)$.

If we now reformulate the relations (9.42), (9.43) and (9.46) as

$$\begin{aligned} \phi^1 &= \boldsymbol{r}_2 \wedge \boldsymbol{\eta}, \quad \phi^2 = \frac{1}{2}(\boldsymbol{r}_2 \wedge \boldsymbol{r}_1 + \boldsymbol{r} \wedge \boldsymbol{\eta}), \quad \phi^3 = \frac{1}{2}\boldsymbol{r}_1 \wedge \boldsymbol{r} \\ \phi^6 &= \boldsymbol{r}_1 \wedge \boldsymbol{\eta}, \quad \phi^5 = \frac{1}{2}(\boldsymbol{r}_1 \wedge \boldsymbol{r}_2 + \boldsymbol{r} \wedge \boldsymbol{\eta}), \quad \phi^4 = \frac{1}{2}\boldsymbol{r}_2 \wedge \boldsymbol{r} \end{aligned} \tag{9.111}$$

then the action of the transformation $\tilde{\mathbb{B}}$ on the Wilczynski tetrahedral $(\boldsymbol{r}_0, \boldsymbol{r}_1, \boldsymbol{r}_2, \boldsymbol{r}_3) = (\boldsymbol{r}, \boldsymbol{r}_1, \boldsymbol{r}_2, \boldsymbol{\eta})$ is obtained by solving the algebraic equations

$$\begin{aligned} \tilde{\phi}^1 &= \tilde{\boldsymbol{r}}_2 \wedge \tilde{\boldsymbol{\eta}}, \quad \tilde{\phi}^2 = \frac{1}{2}(\tilde{\boldsymbol{r}}_2 \wedge \tilde{\boldsymbol{r}}_1 + \tilde{\boldsymbol{r}} \wedge \tilde{\boldsymbol{\eta}}), \quad \tilde{\phi}^3 = \frac{1}{2}\tilde{\boldsymbol{r}}_1 \wedge \tilde{\boldsymbol{r}} \\ \tilde{\phi}^6 &= \tilde{\boldsymbol{r}}_1 \wedge \tilde{\boldsymbol{\eta}}, \quad \tilde{\phi}^5 = \frac{1}{2}(\tilde{\boldsymbol{r}}_1 \wedge \tilde{\boldsymbol{r}}_2 + \tilde{\boldsymbol{r}} \wedge \tilde{\boldsymbol{\eta}}), \quad \tilde{\phi}^4 = \frac{1}{2}\tilde{\boldsymbol{r}}_2 \wedge \tilde{\boldsymbol{r}}. \end{aligned} \tag{9.112}$$

Since proper rotations of the form (9.109), (9.110) are mapped via the Plücker correspondence to linear transformations

$$\tilde{\boldsymbol{r}}_i = \sum_{j=1}^{4} K_i{}^j \boldsymbol{r}_j, \quad K \in SL(4), \tag{9.113}$$

these algebraic equations are quadratic in K. At $\lambda = -1$, their solution is encapsulated in the following theorem:

Theorem 45 (A Bäcklund transformation for projective-minimal surfaces). *Let $\boldsymbol{r}$ be the position vector of a projective-minimal surface Σ and the eigenfunctions φ° and ψ° be defined as in Theorem 44. Then, a second projective-minimal surface $\tilde{\Sigma}$ is given by*

$$\tilde{\boldsymbol{r}} = f^0 \boldsymbol{r} + f^1 \boldsymbol{r}_1 + f^2 \boldsymbol{r}_2 \tag{9.114}$$

with coefficients

$$f^0 = -\frac{1}{2}\frac{\lambda_\circ \varphi^\circ + \psi^\circ}{\Omega}, \quad f^1 = \lambda_\circ \frac{\varphi^\circ_y}{Bh\Omega}, \quad f^2 = \frac{\psi^\circ_x}{Ak\Omega} \tag{9.115}$$

and

$$\Omega = \sqrt{\varphi^{\circ 2} - 2\frac{\varphi^\circ_y \varphi^\circ_x}{Bh} + 2\beta \frac{\varphi^{\circ 2}_y}{B^2}} = \sqrt{\psi^{\circ 2} - 2\frac{\psi^\circ_x \psi^\circ_y}{Ak} + 2\alpha \frac{\psi^{\circ 2}_x}{A^2}}. \tag{9.116}$$

The transformation $\tilde{\mathbb{B}}$ has the tangency property, that is the line segment $\tilde{\hat{\boldsymbol{r}}} - \hat{\boldsymbol{r}}$ which connects corresponding points on $\hat{\Sigma}$ and $\tilde{\hat{\Sigma}}$ is tangential to both surfaces.

Proof. The position vector of $\tilde{\Sigma}$ may be cast into the form

$$\tilde{\boldsymbol{r}} = g^0 \boldsymbol{r} + g^1 \boldsymbol{r}_x + g^2 \boldsymbol{r}_y. \tag{9.117}$$

Thus, if we make the usual identification of the surface $\tilde{\Sigma} \subset \mathbb{P}^3$ with a surface $\tilde{\hat{\Sigma}} \subset \mathbb{R}^3$, then the position vector of the latter reads

$$\tilde{\hat{\boldsymbol{r}}} = \hat{\boldsymbol{r}} + \frac{g^1 r^0 \hat{\boldsymbol{r}}_x + g^2 r^0 \hat{\boldsymbol{r}}_y}{g^0 r^0 + g^1 r^0_x + g^2 r^0_y}, \tag{9.118}$$

which implies that $\tilde{\hat{\boldsymbol{r}}} - \hat{\boldsymbol{r}}$ is tangent to $\hat{\Sigma}$. Moreover, the Fundamental Transformation is invertible and its inverse has the same form as $\tilde{\mathbb{B}}$, that is, $\hat{\boldsymbol{r}}$ constitutes a linear combination of $\tilde{\hat{\boldsymbol{r}}}$, $\tilde{\hat{\boldsymbol{r}}}_1$ and $\tilde{\hat{\boldsymbol{r}}}_2$. Consequently, the line segment $\tilde{\hat{\boldsymbol{r}}} - \hat{\boldsymbol{r}}$ is also tangential to $\tilde{\hat{\Sigma}}$. □

Exercises

1. Verify the first integral (9.76).
2. Show that the quantities $\tilde{\varphi}$ and $\tilde{\psi}$ as given by (9.88) constitute particular adjoint eigenfunctions and the corresponding bilinear potentials M° and N° assume the form (9.90).

9.6 One-Soliton Demoulin Surfaces

In the case of Demoulin surfaces, the Bäcklund transformation of the previous section may be readily related to the classical Moutard transformation. Thus, if $\alpha = \beta = 0$, then the bilinear potentials M° and N° become

$$M^\circ = \frac{1}{2}\varphi^{\circ 2}, \quad N^\circ = \frac{1}{2}\psi^{\circ 2} \tag{9.119}$$

so that the transformation formulae $(9.103)_{5,6}$ reduce to

$$A' = -A, \quad B' = -B. \tag{9.120}$$

This, in turn, implies that A and B are preserved by the transformation $\tilde{\mathbb{B}}$ and may therefore be normalised to $A = B = 1$. At the linear level, it has been shown that the $so(3,3)$ representation (9.63) for the Demoulin system (9.58) takes the form (9.60) and contains two Moutard equations. It is therefore natural to formulate the Fundamental Transformation (9.70) conjugated with the involution (9.107) as

$$\tilde{\varphi} = \frac{S}{\varphi^\circ}, \quad \tilde{\psi} = \frac{T}{\psi^\circ}, \tag{9.121}$$

where S and T are defined by

$$S = \varphi^\circ\varphi - 2M, \quad T = -\psi^\circ\psi + 2N. \tag{9.122}$$

The latter satisfy the skew-symmetric bilinear relations

$$\begin{aligned} S_x &= \varphi^\circ\varphi_x - \varphi^\circ_x\varphi, \quad S_y = \varphi^\circ_y\varphi - \varphi^\circ\varphi_y \\ T_x &= \psi^\circ\psi_x - \psi^\circ_x\psi, \quad T_y = \psi^\circ_y\psi - \psi^\circ\psi_y. \end{aligned} \tag{9.123}$$

Here, we have made use of the fact that the adjoint eigenfunctions $\tilde{\varphi}^\circ$ and $\tilde{\psi}^\circ$ coincide with the eigenfunctions φ° and ψ°, respectively, by virtue of the relations (9.104). Accordingly, the transformations (9.121) constitute two copies of the classical Moutard transformation and Theorem 44 reads:

Corollary 6 (A Bäcklund transformation for the Demoulin system). *The Demoulin system (9.58) and its $so(3,3)$ linear representation (9.60) are invariant under the Moutard-type transformation*

$$\begin{aligned} \varphi \to \tilde{\varphi} &= \frac{S}{\varphi^\circ}, \quad h \to \tilde{h} = h - 2(\ln \varphi^\circ)_{xy} \\ \psi \to \tilde{\psi} &= \frac{T}{\psi^\circ}, \quad k \to \tilde{k} = k - 2(\ln \psi^\circ)_{xy}, \end{aligned} \tag{9.124}$$

where $(\varphi^\circ, \psi^\circ)$ constitutes a solution of (9.60) with parameter $\lambda_\circ$ subject to the admissible constraint

$$\varphi^{\circ 2} - 2\frac{\varphi^\circ_x \varphi^\circ_y}{h} = \psi^{\circ 2} - 2\frac{\psi^\circ_x \psi^\circ_y}{k} \tag{9.125}$$

and the bilinear potentials S and T are given by

$$\begin{aligned} S &= \nu_3 \varphi^\circ \varphi - \nu_1 \psi^\circ \psi - \nu_2 \frac{\varphi^\circ_x \varphi_y}{h} - \nu_0 \frac{\varphi^\circ_y \varphi_x}{h} + \nu_1 \frac{\psi^\circ_x \psi_y}{k} + \nu_1 \frac{\psi^\circ_y \psi_x}{k} \\ T &= \nu_3 \psi^\circ \psi - \nu_1 \varphi^\circ \varphi - \nu_2 \frac{\psi^\circ_x \psi_y}{k} - \nu_0 \frac{\psi^\circ_y \psi_x}{k} + \nu_1 \frac{\varphi^\circ_x \varphi_y}{h} + \nu_1 \frac{\varphi^\circ_y \varphi_x}{h} \end{aligned} \tag{9.126}$$

with the constants

$$\begin{aligned} \nu_0 &= \frac{2\lambda_\circ^2}{\lambda^2 - \lambda_\circ^2}, \quad \nu_1 = \frac{2\lambda\lambda_\circ}{\lambda^2 - \lambda_\circ^2} \\ \nu_2 &= \frac{2\lambda^2}{\lambda^2 - \lambda_\circ^2}, \quad \nu_3 = \frac{\lambda^2 + \lambda_\circ^2}{\lambda^2 - \lambda_\circ^2}. \end{aligned} \tag{9.127}$$

If we start with the seed solution $h = k = 1$ of the Demoulin system, then the simplest non-trivial eigenfunctions which do not just generate solutions of the Tzitzeica equation are given by [133]

$$\varphi^\circ = e^{i\gamma_2} \cosh \gamma_1, \quad \psi^\circ = -e^{i\gamma_2} \sinh \gamma_1 \tag{9.128}$$

with

$$\gamma_1 = \frac{1}{2}(\kappa x + \kappa^{-1} y), \quad \gamma_2 = \frac{\sqrt{3}}{2}(\kappa x - \kappa^{-1} y), \quad \lambda_\circ = \kappa^3. \tag{9.129}$$

It is then readily shown that the constraint (9.125) is identically satisfied with

$$\Omega \sim e^{i\gamma_2} \Delta, \quad \Delta = \sqrt{\cosh 2\gamma_1} \tag{9.130}$$

and the new solution of the Demoulin system reads

$$\tilde{h} = 1 - \frac{1}{2\cosh^2\gamma_1}, \quad \tilde{k} = 1 + \frac{1}{2\sinh^2\gamma_1}. \tag{9.131}$$

The latter consists of a typical sech^2-shaped soliton ($\tilde{h}$) and a 'singular soliton' ($\tilde{k}$). However, the geometrically relevant quantities in the Gauss-Weingarten equations for the associated Demoulin surfaces prove to be non-singular. For instance, $\tilde{p} = -1/\tilde{k}$ and $\tilde{q} = -1/\tilde{h}$ take the simple form

$$\tilde{p} = -1 + \frac{1}{\cosh 2\gamma_1}, \quad \tilde{q} = -1 - \frac{1}{\cosh 2\gamma_1}. \tag{9.132}$$

These quantities are depicted as functions of γ_1 in Figure 9.1.

The Gauss-Weingarten equations (9.8) associated with the seed solution $p = q = -1,\ V = W = 0$ reduce to

$$\boldsymbol{r}_{xx} = -\boldsymbol{r}_y, \quad \boldsymbol{r}_{yy} = -\boldsymbol{r}_x \tag{9.133}$$

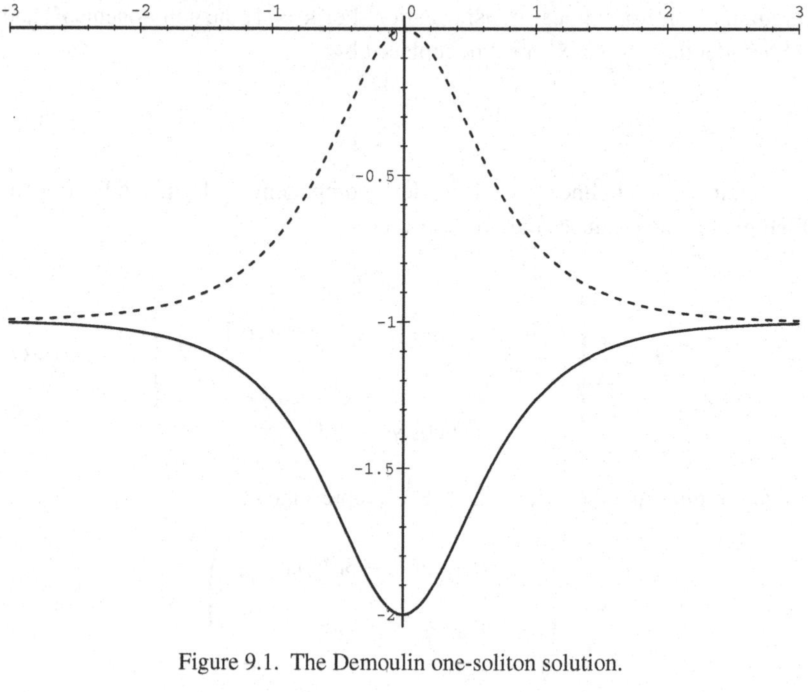

Figure 9.1. The Demoulin one-soliton solution.

and hence the position vector of the seed surface Σ is given by

$$\boldsymbol{r} = \begin{pmatrix} 1 \\ e^{-(x+y)} \\ e^{(x+y)/2}\cos\gamma_3 \\ e^{(x+y)/2}\sin\gamma_3 \end{pmatrix}, \quad \gamma_3 = \frac{\sqrt{3}}{2}(x-y) \tag{9.134}$$

up to a linear transformation generated by a non-singular but otherwise arbitrary constant matrix. Even though the eigenfunctions φ° and ψ° are complex, the reality of $\tilde{p}$ and $\tilde{q}$ guarantees that the Bäcklund transformation for projective-minimal surfaces delivers a class of real Demoulin surfaces via separation of the complex position vector $\tilde{\boldsymbol{r}}$ as given by (9.114) into its real and imaginary parts. Thus, the real position vector of the one-parameter class of Demoulin surfaces $\tilde{\Sigma}(\kappa)$, $\kappa \neq 1$ is readily shown to be

$$\tilde{\boldsymbol{r}} = \frac{1}{\Delta}\begin{pmatrix} \kappa^3\cosh\gamma_1 - \sinh\gamma_1 \\ e^{-(x+y)}s \\ e^{(x+y)/2}[s\sin\gamma_3 + \sqrt{3}\,t\cos\gamma_3] \\ e^{(x+y)/2}[s\cos\gamma_3 - \sqrt{3}\,t\sin\gamma_3] \end{pmatrix} \tag{9.135}$$

$$s = \kappa\cosh\gamma_1 + \sinh\gamma_1, \quad t = \kappa\cosh\gamma_1 - \sinh\gamma_1$$

modulo an arbitrary linear transformation. For $\kappa = 1$, the components $\tilde{r}^0$ and $\tilde{r}^1$ are identical so that $\tilde{r}^1$ must be replaced by

$$\lim_{\kappa\to 1}\frac{\tilde{r}^1 - \tilde{r}^0}{\kappa - 1} \tag{9.136}$$

to obtain a fourth linearly independent component. Indeed, application of l'Hôpital's rule produces the position vector

$$\tilde{\boldsymbol{r}} = \frac{1}{\Delta}\begin{pmatrix} e^{-\gamma_1} \\ e^{-\gamma_1}\left(\frac{2}{\sqrt{3}}\gamma_3 + (e^{-\gamma_1} - 3e^{\gamma_1})\cosh\gamma_1\right) \\ e^{2\gamma_1}\sin\gamma_3 + \sqrt{3}\cos\gamma_3 \\ e^{2\gamma_1}\cos\gamma_3 - \sqrt{3}\sin\gamma_3 \end{pmatrix}, \tag{9.137}$$

which implies that the surface $\tilde{\tilde{\Sigma}} \subset \mathbb{R}^3$ is represented by

$$\tilde{\tilde{\boldsymbol{r}}} = \begin{pmatrix} \frac{2}{\sqrt{3}}\gamma_3 + (e^{-\gamma_1} - 3e^{\gamma_1})\cosh\gamma_1 \\ e^{3\gamma_1}\sin\gamma_3 + \sqrt{3}\,e^{\gamma_1}\cos\gamma_3 \\ e^{3\gamma_1}\cos\gamma_3 - \sqrt{3}\,e^{\gamma_1}\sin\gamma_3 \end{pmatrix}. \tag{9.138}$$

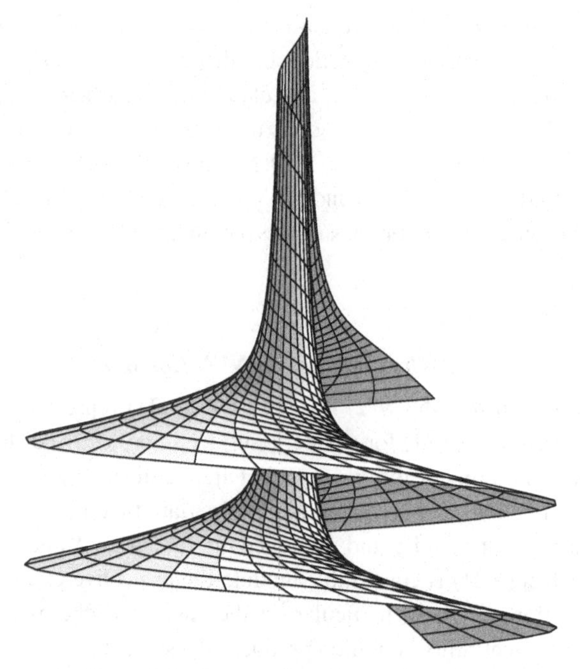

Figure 9.2. A one-soliton Demoulin surface.

The above parametrisation shows that the surface $\tilde{\tilde{\Sigma}}$ is helicoidal since it is generated by the curve $\tilde{\tilde{\boldsymbol{r}}} = \tilde{\tilde{\boldsymbol{r}}}(\gamma_1, \gamma_3 = 0)$ which is uniformly rotated and translated in space. The 'stationary one-soliton' Demoulin surface $\tilde{\tilde{\Sigma}}$ parametrised in terms of γ_1 and γ_3 is displayed in Figure 9.2.

9.7 Isothermal-Asymptotic Surfaces. The Stationary mNVN Equation

It has been shown in Section 9.2 that Tzitzeica surfaces constitute projective-minimal surfaces subject to the constraint

$$p = q. \tag{9.139}$$

The latter condition defines *isothermal-asymptotic* surfaces in projective geometry. This terminology is due to Fubini [143] and reflects the fact that the Darboux cubic form (9.15) associated with such surfaces simplifies to

$$p\,(dx^3 + dy^3). \tag{9.140}$$

Remarkably, the Gauss-Mainardi-Codazzi equations underlying isothermal-asymptotic surfaces are integrable [129]. We here present the connection between the Gauss-Weingarten equations for isothermal-asymptotic surfaces and the standard linear representation for the stationary *modified Nizhnik-Veselov-Novikov* (mNVN) equation and elaborate on the geometric interpretation of the Miura-type transformation to the *Nizhnik-Veselov-Novikov* (NVN) equation as discussed by Ferapontov [129]. Moreover, the well-known Darboux-type transformations for the NVN and mNVN equations are interpreted in terms of the Bäcklund transformation for surfaces parametrised in terms of asymptotic coordinates (cf. Chapter 1).

9.7.1 The Stationary mNVN Equation

It is readily shown that the $sl(2)$ linear system (9.31) represents the Gauss-Weingarten equations (9.31) for a generic surface in projective differential geometry written in terms of the Wilczynski tetrahedral $(\boldsymbol{r}, \boldsymbol{r}_1, \boldsymbol{r}_2, \boldsymbol{\eta})$. Indeed, if we regard the relations (9.21) as definitions of certain functions α and β which now depend on both x and y and introduce the functions A and B according to $(9.23)_{1,3}$, then (9.31) is compatible modulo the projective Gauss-Mainardi-Codazzi equations (9.9). In particular, in the case of isothermal-asymptotic surfaces, the compatibility condition produces the system

$$\boxed{\begin{aligned} p_{yyy} - 2p_y W - pW_y &= p_{xxx} - 2p_x V - pV_x \\ W_x &= 3pp_y \\ V_y &= 3pp_x \end{aligned}} \tag{9.141}$$

which, remarkably, is the stationary version of the mNVN equation [47]

$$\begin{aligned} p_t &= p_{xxx} - 2p_x V - pV_x - p_{yyy} + 2p_y W + pW_y \\ W_x &= 3pp_y, \quad V_y = 3pp_x. \end{aligned} \tag{9.142}$$

The mNVN equation constitutes an integrable 2+1-dimensional generalisation of the mKdV equation

$$p_t = 2p_{xxx} + 12p^2 p_x \tag{9.143}$$

which is obtained by setting $p = p(x - y, t)$.

In the general case, the $sl(4)$–$so(3, 3)$ isomorphism encapsulated in the relations (9.42), (9.43) implies that the $so(3, 3)$ linear representation (9.47) is also compatible modulo the projective Gauss-Mainardi-Codazzi equations (9.9). In the particular case $p = q$, the components ϕ^1, ϕ^2, ϕ^5 and ϕ^6 may conveniently be expressed in terms of

$$\phi^3 = \bar{\psi}, \quad \phi^4 = \bar{\varphi}, \tag{9.144}$$

that is

$$\phi^1 = -\bar{\psi}_{yy} + \frac{p_y}{p}\bar{\psi}_y + \left[W - \frac{1}{2}\left(\frac{p_y}{p}\right)^2\right]\bar{\psi}, \quad \phi^2 = -\bar{\psi}_y + \frac{p_y}{p}\bar{\psi}$$
$$\phi^6 = -\bar{\varphi}_{xx} + \frac{p_x}{p}\bar{\varphi}_x + \left[V - \frac{1}{2}\left(\frac{p_x}{p}\right)^2\right]\bar{\varphi}, \quad \phi^5 = -\bar{\varphi}_x + \frac{p_x}{p}\bar{\varphi}, \tag{9.145}$$

so that the $so(3,3)$ linear representation for the stationary mNVN equation reduces to

$$\begin{aligned} \bar{\varphi}_y &= p\bar{\psi}, \quad \bar{\varphi}_{xxx} = \bar{\varphi}_{yyy} + 2V\bar{\varphi}_x + V_x\bar{\varphi} - 3p_y\bar{\psi}_y - 2pW\bar{\psi} \\ \bar{\psi}_x &= p\bar{\varphi}, \quad \bar{\psi}_{yyy} = \bar{\psi}_{xxx} + 2W\bar{\psi}_y + W_y\bar{\psi} - 3p_x\bar{\varphi}_x - 2pV\bar{\varphi}. \end{aligned} \tag{9.146}$$

The above linear system constitutes nothing but the stationary reduction of the linear representation

$$\begin{aligned} \bar{\varphi}_y &= p\bar{\psi} \\ \bar{\psi}_x &= p\bar{\varphi} \\ \bar{\varphi}_t &= \bar{\varphi}_{xxx} - \bar{\varphi}_{yyy} - 2V\bar{\varphi}_x - V_x\bar{\varphi} + 3p_y\bar{\psi}_y + 2pW\bar{\psi} \\ -\bar{\psi}_t &= \bar{\psi}_{yyy} - \bar{\psi}_{xxx} - 2W\bar{\psi}_y - W_y\bar{\psi} + 3p_x\bar{\varphi}_x + 2pV\bar{\varphi} \end{aligned} \tag{9.147}$$

of the mNVN equation. A linear representation of the stationary mNVN equation which contains an arbitrary parameter may therefore be obtained by setting

$$\bar{\varphi}(x,y,t) = e^{\lambda t}\bar{\varphi}(x,y), \quad \bar{\psi}(x,y,t) = e^{\lambda t}\bar{\psi}(x,y), \tag{9.148}$$

leading to

$$\begin{aligned} \bar{\varphi}_y &= p\bar{\psi} \\ \bar{\psi}_x &= p\bar{\varphi} \\ \lambda\bar{\varphi} &= \bar{\varphi}_{xxx} - \bar{\varphi}_{yyy} - 2V\bar{\varphi}_x - V_x\bar{\varphi} + 3p_y\bar{\psi}_y + 2pW\bar{\psi} \\ -\lambda\bar{\psi} &= \bar{\psi}_{yyy} - \bar{\psi}_{xxx} - 2W\bar{\psi}_y - W_y\bar{\psi} + 3p_x\bar{\varphi}_x + 2pV\bar{\varphi}. \end{aligned} \tag{9.149}$$

The geometrically relevant case (9.146) is retrieved if $\lambda = 0$.

To summarise, the standard linear representation of the stationary mNVN equation admits a canonical geometric interpretation in terms of isothermal-asymptotic surfaces represented in projective space $\mathbb{P}^5$. However, a natural link may also be established by considering the realisation of isothermal-asymptotic surfaces Σ as surfaces $\hat{\Sigma}$ in Euclidean space $\mathbb{R}^3$. This is shown below.

9.7.2 The Stationary NVN Equation

The Gauss-Weingarten equations for isothermal-asymptotic surfaces Σ read

$$\boldsymbol{r}_{xx} = p\boldsymbol{r}_y + \frac{1}{2}(V - p_y)\boldsymbol{r}, \quad \boldsymbol{r}_{yy} = p\boldsymbol{r}_x + \frac{1}{2}(W - p_x)\boldsymbol{r}. \tag{9.150}$$

Accordingly, if r^0 constitutes a corresponding scalar solution, then the position vector $\hat{\boldsymbol{r}}$ of the 'isothermal-asymptotic' surface $\hat{\Sigma} \subset \mathbb{R}^3$ satisfies the linear system

$$\hat{\boldsymbol{r}}_{xx} = a\hat{\boldsymbol{r}}_x + p\hat{\boldsymbol{r}}_y, \quad \hat{\boldsymbol{r}}_{yy} = b\hat{\boldsymbol{r}}_y + p\hat{\boldsymbol{r}}_x, \tag{9.151}$$

where $a = -2(\ln r^0)_x$, $b = -2(\ln r^0)_y$. Conversely, any vector-valued solution $\hat{\boldsymbol{r}}$ of (9.151) for appropriate functions a, b and p defines an isothermal-asymptotic surface $\Sigma \subset \mathbb{P}^3$.

The compatibility condition for (9.151) is readily shown to yield the nonlinear system

$$\begin{aligned} &\left(p_x + ap + \frac{1}{2}b^2 - b_y\right)_x = 3pp_y, \\ &\left(p_y + bp + \frac{1}{2}a^2 - a_x\right)_y = 3pp_x, \end{aligned} \quad a_y = b_x. \tag{9.152}$$

For a given solution (a, b, p) of this system, it is natural to set

$$W = p_x + ap + \frac{1}{2}b^2 - b_y, \quad V = p_y + bp + \frac{1}{2}a^2 - a_x \tag{9.153}$$

so that $(9.152)_{1,3}$ produce the relations

$$W_x = 3pp_y, \quad V_y = 3pp_x. \tag{9.154}$$

The relation $(9.152)_2$ confirms that there exists a function r^0 such that

$$a = -2(\ln r^0)_x, \quad b = -2(\ln r^0)_y. \tag{9.155}$$

Insertion into (9.153) now leads to the linear system

$$r^0_{xx} = pr^0_y + \frac{1}{2}(V - p_y)r^0, \quad r^0_{yy} = pr^0_x + \frac{1}{2}(W - p_x)r^0 \tag{9.156}$$

which is of the same form as the Gauss-Weingarten equations for isothermal-asymptotic surfaces. Accordingly, its compatibility condition produces the remaining relation

$$p_{yyy} - 2p_y W - pW_y = p_{xxx} - 2p_x V - pV_x. \tag{9.157}$$

Hence, the Gauss-Mainardi-Codazzi equations for isothermal-asymptotic surfaces are retrieved, and it is justified to refer to $\hat{\boldsymbol{r}}$ as the position vector of an isothermal-asymptotic surface $\hat{\Sigma} \subset \mathbb{R}^3$.

Since the surfaces $\hat{\Sigma}$ are parametrised in terms of asymptotic coordinates, we may employ the Lelieuvre formulae

$$\hat{\boldsymbol{r}}_x = \boldsymbol{\nu} \times \boldsymbol{\nu}_x, \quad \hat{\boldsymbol{r}}_y = \boldsymbol{\nu}_y \times \boldsymbol{\nu} \tag{9.158}$$

which relate the position vector $\hat{\boldsymbol{r}}$ of $\hat{\Sigma}$ to its scaled normal (co-normal) $\boldsymbol{\nu}$. The co-normal satisfies a Moutard equation (cf. Chapter 1) and insertion of $\hat{\boldsymbol{r}}_x$ and $\hat{\boldsymbol{r}}_y$ as given by (9.158) into the Gauss-Weingarten equations (9.151) reveals that the second derivatives $\boldsymbol{\nu}_{xx}$ and $\boldsymbol{\nu}_{yy}$ are likewise determined up to their components in $\boldsymbol{\nu}$-direction. Accordingly, there exist functions f, g and u such that the co-normal obeys the triad

$$\begin{aligned} \boldsymbol{\nu}_{xx} &= a\boldsymbol{\nu}_x - p\boldsymbol{\nu}_y + f\boldsymbol{\nu} \\ \boldsymbol{\nu}_{xy} &= u\boldsymbol{\nu} \\ \boldsymbol{\nu}_{yy} &= b\boldsymbol{\nu}_y - p\boldsymbol{\nu}_x + g\boldsymbol{\nu}. \end{aligned} \tag{9.159}$$

The compatibility conditions for the latter reduce to

$$\begin{aligned} u &= p^2 + a_y, & u &= p^2 + b_x \\ f &= p_y + bp, & g &= p_x + ap \\ u_x &= f_y - pg + au, & u_y &= g_x - pf + bu. \end{aligned} \tag{9.160}$$

Either of the first two relations define u where $a_y = b_x$. The relations $(9.160)_{3,4}$ determine the functions f and g, and the remaining equations adopt the form of $(9.152)_{1,3}$. Thus, the linear triad (9.159) is compatible if and only if its coefficients are parametrised in terms of a solution of the Gauss-Mainardi-Codazzi equations of isothermal-asymptotic surfaces. Consequently, any vector-valued solution $\boldsymbol{\nu}$ of (9.159) defines uniquely an isothermal-asymptotic surface $\hat{\Sigma}$ via the Lelieuvre formulae (9.158).

It is readily verified that the pair

$$\begin{aligned} \boldsymbol{\nu}_{xxx} - 3v\boldsymbol{\nu}_x &= \boldsymbol{\nu}_{yyy} - 3w\boldsymbol{\nu}_y \\ \boldsymbol{\nu}_{xy} &= u\boldsymbol{\nu}, \end{aligned} \tag{9.161}$$

where the functions v and w are defined by

$$v = \frac{2}{3}V + a_x, \quad w = \frac{2}{3}W + b_y, \tag{9.162}$$

is a differential consequence of the linear triad (9.159). Its compatibility condition produces the stationary NVN equation [276, 371]

$$\boxed{\begin{aligned} u_{xxx} - 3(vu)_x &= u_{yyy} - 3(wu)_y \\ w_x &= u_y \\ v_y &= u_x \end{aligned}} \tag{9.163}$$

with

$$\begin{aligned} u_t = u_{xxx} - 3(vu)_x - u_{yyy} + 3(wu)_y \\ w_x = u_y, \quad v_y = u_x \end{aligned} \tag{9.164}$$

being the 2+1-dimensional counterpart. The NVN equation reduces for solutions of the form $u = u(x - y, t)$ to the KdV equation

$$u_t = 2u_{xxx} + 12uu_x. \tag{9.165}$$

The NVN equation admits coherent structure solutions (dromions) which, for fixed 'time' t, decay exponentially in any direction [20]. It is recalled that the existence of dromions was first established by Boiti et al. [49] for the Davey-Stewartson I equation via a binary Darboux transformation. The latter, with regard to both structure and form, is nothing but the classical Fundamental Transformation.

The standard Lax pair

$$\begin{aligned} \phi_t = \phi_{xxx} - 3v\phi_x - \phi_{yyy} + 3w\phi_y \\ \phi_{xy} = u\phi \end{aligned} \tag{9.166}$$

for the NVN equation reduces to that for the stationary NVN equation by setting

$$\phi(x, y, t) = e^{\lambda t}\phi(x, y) \tag{9.167}$$

so that (9.161) may be regarded as a vector-valued version of

$$\begin{aligned} \lambda\phi = \phi_{xxx} - 3v\phi_x - \phi_{yyy} + 3w\phi_y \\ \phi_{xy} = u\phi \end{aligned} \tag{9.168}$$

evaluated at $\lambda = 0$. Thus, the co-normal $\boldsymbol{\nu}$ to the isothermal-asymptotic surface $\hat{\Sigma}$ is naturally associated with the stationary NVN equation while the position vector $\hat{\boldsymbol{r}}$ is closely related to the stationary mNVN equation. This geometric observation is encapsulated in the following theorem [129] and may be regarded as a geometric interpretation of the analogue of the important Miura transformation

$$u = p^2 + p_x \tag{9.169}$$

between the mKdV and KdV equations

$$p_t = p_{xxx} - 6p^2 p_x, \quad u_t = u_{xxx} - 6uu_x. \tag{9.170}$$

Theorem 46 (A link between the stationary mNVN and NVN equations). *If (p, V, W) is a solution of the stationary NVN equation (9.141) and r^0 a corresponding solution of the linear system (9.156) then*

$$\begin{aligned} u &= p^2 - 2(\ln r^0)_{xy} \\ v &= \frac{2}{3}V - 2(\ln r^0)_{xx} \\ w &= \frac{2}{3}W - 2(\ln r^0)_{yy} \end{aligned} \tag{9.171}$$

satisfies the stationary NVN equation (9.163).

Proof. Insertion of the parametrisation (9.155) into the relations $(9.160)_1$ and (9.162) produces the transformation formulae (9.171). □

At the linear level, the above connection is made as follows. Let ϕ be an eigenfunction associated with the stationary NVN equation, that is a solution of the Lax pair (9.168). Even though ϕ is a solution of the Moutard equation $(9.159)_2$, the remaining two equations of the linear triad (9.159) are not satisfied by ϕ. It is therefore natural to introduce functions $\bar{\varphi}$ and $\bar{\psi}$ according to

$$\begin{aligned} \phi_{xx} &= a\phi_x - p\phi_y + f\phi + \bar{\varphi} \\ \phi_{xy} &= u\phi \\ \phi_{yy} &= b\phi_y - p\phi_x + g\phi + \bar{\psi}\,. \end{aligned} \tag{9.172}$$

Since the above system is compatible if we formally set $\bar{\varphi} = \bar{\psi} = 0$, it is evident that the compatibility conditions for (9.172) give rise to linear homogeneous equations for $\bar{\varphi}$ and $\bar{\psi}$. Indeed, collection of the terms involving $\bar{\varphi}$ and $\bar{\psi}$ yields

$$\bar{\varphi}_y = p\bar{\psi}, \quad \bar{\psi}_x = p\bar{\varphi} \tag{9.173}$$

which coincides with the 'scattering problem' $(9.149)_{1,2}$ for the stationary mNVN equation. The third-order equation $(9.168)_1$ may then be written as

$$\lambda\phi = \bar{\varphi}_x + a\bar{\varphi} - \bar{\psi}_y - b\bar{\psi}\,. \tag{9.174}$$

On solving the latter for ϕ, the Moutard equation $(9.172)_2$ is readily shown to be satisfied identically while $(9.172)_{1,3}$ may be cast into the form

$$\begin{aligned} \lambda\bar{\varphi} &= \bar{\varphi}_{xxx} - \bar{\varphi}_{yyy} - 2V\bar{\varphi}_x - V_x\bar{\varphi} + 3p_y\bar{\psi}_y + 2pW\bar{\psi} \\ -\lambda\bar{\psi} &= \bar{\psi}_{yyy} - \bar{\psi}_{xxx} - 2W\bar{\psi}_y - W_y\bar{\psi} + 3p_x\bar{\varphi}_x + 2pV\bar{\varphi}. \end{aligned} \tag{9.175}$$

Thus, the linear representation (9.149) of the mNVN equation is retrieved. This connection between the linear representations of the NVN and mNVN equations is exploited in the following section to simplify considerably the analytic formulation of a Bäcklund transformation for isothermal-asymptotic surfaces.

Exercises

1. Show that, on appropriate interpretation of the functions α, β and A, B, the $sl(2)$ and $so(3,3)$ linear representations (9.31) and (9.47) are compatible modulo the Gauss-Mainardi-Codazzi equations of projective differential geometry.
2. Verify that the quantities $\bar{\varphi}$ and $\bar{\psi}$ as defined by the linear triad (9.172) constitute eigenfunctions of the stationary mNVN equation, that is they satisfy the linear representation (9.149).
3. Prove that the projective Gauss-Weingarten equations (9.151) and the relations (9.155) imply that

$$|\hat{\boldsymbol{r}}_x, \hat{\boldsymbol{r}}_y, \hat{\boldsymbol{r}}_{xy}| = \frac{c}{(\boldsymbol{r}^0)^4}$$

and hence $\hat{\boldsymbol{r}}$ may be normalised in such a way that $c = 1$. In this case, show that

$$\boldsymbol{\nu} = (\boldsymbol{r}^0)^2\,\hat{\boldsymbol{r}}_x \times \hat{\boldsymbol{r}}_y$$

obeys the Lelieuvre formulae (9.158).

9.8 A Bäcklund Transformation for Isothermal-Asymptotic Surfaces

As a starting point for the construction of a Bäcklund transformation for isothermal-asymptotic surfaces, we first construct an invariance of the stationary mNVN equation based on the $so(3, 3)$ linear representation (9.149). The Plücker correspondence may then, in principle, be exploited to obtain, by purely algebraic means, a mapping between a seed isothermal-asymptotic surface Σ and a second isothermal-asymptotic surface Σ'. However, we choose here to relate the invariance of the $so(3, 3)$ linear representation directly to an invariance of the linear triad for the co-normal $\boldsymbol{\nu}$ via the procedure outlined in the previous section. Finally, the Lelieuvre formulae are used to formulate a Bäcklund transformation at the surface level.

9.8.1 An Invariance of the mNVN Equation

A Bäcklund transformation for a two-dimensional integrable system may sometimes be conveniently constructed via dimensional reduction of a Bäcklund transformation for a three-dimensional system in which it is embedded. This approach sheds light on the origin of bilinear potentials of the form (9.98). Here, it is illustrated by constructing a Bäcklund transformation for the stationary mNVN equation by means of a transformation for its non-stationary counterpart which is reminiscent of the classical Fundamental Transformation.

Let $(\bar{\varphi}, \bar{\psi})$ and $(\varphi^\circ, \psi^\circ)$ be two pairs of eigenfunctions of the mNVN equation satisfying the linear representation (9.147). Bilinear potentials M and M° may then be defined by the compatible equations

$$M_x = \varphi^\circ \bar{\varphi}, \quad M_y = \psi^\circ \bar{\psi} \tag{9.176}$$

and

$$M^\circ_x = \varphi^{\circ 2}, \quad M^\circ_y = \psi^{\circ 2}. \tag{9.177}$$

A short calculation now reveals that the quantities $\bar{\varphi}'$ and $\bar{\psi}'$ given by

$$\bar{\varphi}' = \bar{\varphi} - \varphi^\circ \frac{M}{M^\circ}, \quad \bar{\psi}' = \bar{\psi} - \psi^\circ \frac{M}{M^\circ}, \tag{9.178}$$

satisfy the primed version of the scattering problem $(9.147)_{1,2}$ with

$$p' = p - \frac{\varphi^\circ \psi^\circ}{M^\circ}. \tag{9.179}$$

To show that the entire linear representation (9.147) may be preserved by the above transformation, it is required to prescribe compatible time-evolutions of the bilinear potentials M and M°. These may be found in a purely algebraic manner for large classes of 2+1-dimensional integrable systems by employing a pseudo-differential operator formalism [281]. In the present case, one obtains

$$\begin{aligned} M_t = {} & \varphi^\circ_{xx}\bar{\varphi} + \varphi^\circ\bar{\varphi}_{xx} - \varphi^\circ_x\bar{\varphi}_x - 2V\varphi^\circ\bar{\varphi} \\ & - \psi^\circ_{yy}\bar{\psi} - \psi^\circ\bar{\psi}_{yy} + \psi^\circ_y\bar{\psi}_y + 2W\psi^\circ\bar{\psi} \end{aligned} \tag{9.180}$$

and, in particular,

$$M^\circ_t = 2\varphi^\circ_{xx}\varphi^\circ - \varphi^{\circ 2}_x - 2V\varphi^{\circ 2} - 2\psi^\circ_{yy}\psi^\circ + \psi^{\circ 2}_y + 2W\psi^{\circ 2}. \tag{9.181}$$

Indeed, the compatibility conditions for M and M° are readily shown to be satisfied modulo the linear representation for the mNVN equation and insertion of the transformation laws (9.178) into the primed analogue of $(9.147)_{3,4}$ delivers

$$W' = W - \frac{3}{2}(\ln M^\circ)_{yy}, \quad V' = V - \frac{3}{2}(\ln M^\circ)_{xx}. \tag{9.182}$$

In this connection, it is noted that (9.179) implies that

$$p'^2 = p^2 - (\ln M^\circ)_{xy}. \tag{9.183}$$

By construction, the quantities W', V' and p' constitute another solution of the mNVN equation.

A Bäcklund transformation for the stationary mNVN equation and its linear representation may now be obtained by specialising the eigenfunctions and bilinear potentials to

$$\begin{aligned} \bar{\varphi}(x, y, t) &= e^{\lambda t}\bar{\varphi}(x, y), & \varphi^\circ(x, y, t) &= e^{\lambda_\circ t}\varphi^\circ(x, y) \\ \bar{\psi}(x, y, t) &= e^{\lambda t}\bar{\psi}(x, y), & \psi^\circ(x, y, t) &= e^{\lambda_\circ t}\psi^\circ(x, y) \\ M(x, y, t) &= e^{(\lambda+\lambda_\circ)t}M(x, y) & M^\circ(x, y, t) &= e^{2\lambda_\circ t}M^\circ(x, y). \end{aligned} \tag{9.184}$$

As a consequence, the time-evolutions (9.180) and (9.181) provide explicit expressions for M and M°, respectively.

Theorem 47 (An invariance of the linear representation for the mNVN equation). *If (p, W, V) is a solution of the stationary mNVN equation and $(\bar{\varphi}, \bar{\psi})$, $(\varphi^\circ, \psi^\circ)$ are corresponding eigenfunctions which obey the linear representation*

(9.149) with λ *and* $\lambda_\circ$*, respectively, then*

$$
\begin{aligned}
p' &= p - \frac{\varphi^\circ \psi^\circ}{M^\circ} \\
W' &= W - \frac{3}{2}(\ln M^\circ)_{yy} \\
V' &= V - \frac{3}{2}(\ln M^\circ)_{xx}
\end{aligned}
\tag{9.185}
$$

constitutes a second solution of the stationary mNVN equation and an associated pair of eigenfunctions is given by

$$
\bar{\varphi}' = \bar{\varphi} - \varphi^\circ \frac{M}{M^\circ}, \quad \bar{\psi}' = \bar{\psi} - \psi^\circ \frac{M}{M^\circ} \tag{9.186}
$$

with bilinear potentials

$$
\begin{aligned}
(\lambda + \lambda_\circ) M &= \varphi^\circ_{xx}\bar{\varphi} + \varphi^\circ \bar{\varphi}_{xx} - \varphi^\circ_x \bar{\varphi}_x - 2V\varphi^\circ\bar{\varphi} \\
&\quad - \psi^\circ_{yy}\bar{\psi} - \psi^\circ \bar{\psi}_{yy} + \psi^\circ_y \bar{\psi}_y + 2W\psi^\circ\bar{\psi} \\
\lambda_\circ M^\circ &= \varphi^\circ_{xx}\varphi^\circ - \frac{1}{2}\varphi^{\circ 2}_x - V\varphi^{\circ 2} - \psi^\circ_{yy}\psi^\circ + \frac{1}{2}\psi^{\circ 2}_y + W\psi^{\circ 2}.
\end{aligned}
\tag{9.187}
$$

An invariance of the $so(3,3)$ *linear representation for isothermal-asymptotic surfaces is obtained by setting* $\lambda = 0$.

9.8.2 An Invariance of the NVN Equation and a Bäcklund Transformation for Isothermal-Asymptotic Surfaces

One may now solve the system of quadratic equations provided by the Plücker correspondence to obtain the action of the transformation given in Theorem 47 on isothermal-asymptotic surfaces. We here exploit the connection between the stationary mNVN equation and the stationary NVN equation to gain important insight into the geometric nature of the Bäcklund transformation for isothermal-asymptotic surfaces. To this end, it is recalled (cf. Chapter 1) that a mapping between two surfaces $\hat{\Sigma}$ and $\hat{\Sigma}'$ parametrised in terms of asymptotic coordinates is given by

$$
\hat{\boldsymbol{r}}' = \hat{\boldsymbol{r}} + \boldsymbol{\nu}' \times \boldsymbol{\nu}, \quad \boldsymbol{\nu}' = \frac{\boldsymbol{S}}{\phi^\circ}, \tag{9.188}
$$

where the co-normal $\boldsymbol{\nu}$ is related to the position vector $\hat{\boldsymbol{r}}$ by the Lelieuvre formulae (9.158) and the bilinear potential $\boldsymbol{S}$ is defined by

$$
\boldsymbol{S}_x = \phi^\circ \boldsymbol{\nu}_x - \phi^\circ_x \boldsymbol{\nu}, \quad \boldsymbol{S}_y = \phi^\circ_y \boldsymbol{\nu} - \phi^\circ \boldsymbol{\nu}_y \tag{9.189}
$$

with underlying Moutard equations

$$\boldsymbol{\nu}_{xy} = u\boldsymbol{\nu}, \quad \phi^\circ_{xy} = u\phi^\circ. \tag{9.190}$$

A notable property of the above transformation is that $\hat{\boldsymbol{r}}' - \hat{\boldsymbol{r}}$ is tangential to both surfaces $\hat{\Sigma}$ and $\hat{\Sigma}'$.

Since the Moutard equation (9.190) is part of the linear triad satisfied by the co-normal $\boldsymbol{\nu}$ associated with isothermal-asymptotic surfaces $\hat{\Sigma} \subset \mathbb{R}^3$, it is natural to investigate under what circumstances the triad (9.159) is mapped to itself. As a first step, we observe that if ϕ° depends parametrically on t and satisfies the 'time-evolution' (9.166), then the time-evolution

$$\begin{aligned}\boldsymbol{S}_t = {} & \phi^\circ\boldsymbol{\nu}_{xxx} - \phi^\circ_{xxx}\boldsymbol{\nu} + \phi^\circ\boldsymbol{\nu}_{yyy} - \phi^\circ_{yyy}\boldsymbol{\nu} \\ & + 2(\phi^\circ_{xx}\boldsymbol{\nu}_x - \phi^\circ_x\boldsymbol{\nu}_{xx}) + 2(\phi^\circ_{yy}\boldsymbol{\nu}_y - \phi^\circ_y\boldsymbol{\nu}_{yy}) \\ & + 3v(\phi^\circ_x\boldsymbol{\nu} - \phi^\circ\boldsymbol{\nu}_x) + 3w(\phi^\circ_y\boldsymbol{\nu} - \phi^\circ\boldsymbol{\nu}_y)\end{aligned} \tag{9.191}$$

is compatible with the defining relations (9.189) provided that (9.161) (or its time-dependent analogue) holds. Moreover, application of the transformation $(9.188)_2$ produces

$$\begin{aligned}\boldsymbol{\nu}'_t &= \boldsymbol{\nu}'_{xxx} - 3v'\boldsymbol{\nu}'_x - \boldsymbol{\nu}'_{yyy} + 3w'\boldsymbol{\nu}'_y \\ \boldsymbol{\nu}'_{xy} &= u'\boldsymbol{\nu}',\end{aligned} \tag{9.192}$$

where the primed coefficients are given by

$$\begin{aligned}u' &= u - 2(\ln\phi^\circ)_{xy} \\ v' &= v - 2(\ln\phi^\circ)_{xx} \\ w' &= w - 2(\ln\phi^\circ)_{yy}.\end{aligned} \tag{9.193}$$

The relations (9.192), (9.193) embody the standard Bäcklund transformation for the NVN equation [20] and its Lax pair. To ensure that $\boldsymbol{\nu}'$ does not depend on t, that is, that it satisfies the primed version of the geometrically relevant linear equation $(9.161)_1$, we set

$$\phi^\circ(x,y,t) = e^{\lambda_\circ t}\phi^\circ(x,y), \quad \boldsymbol{S}(x,y,t) = e^{\lambda_\circ t}\boldsymbol{S}(x,y) \tag{9.194}$$

so that ϕ° is a solution of the Lax pair (9.168) for the stationary NVN equation and $\boldsymbol{S}$ is explicitly given by

$$\begin{aligned}\lambda_\circ\boldsymbol{S} = {} & \phi^\circ\boldsymbol{\nu}_{xxx} - \phi^\circ_{xxx}\boldsymbol{\nu} + \phi^\circ\boldsymbol{\nu}_{yyy} - \phi^\circ_{yyy}\boldsymbol{\nu} \\ & + 2(\phi^\circ_{xx}\boldsymbol{\nu}_x - \phi^\circ_x\boldsymbol{\nu}_{xx}) + 2(\phi^\circ_{yy}\boldsymbol{\nu}_y - \phi^\circ_y\boldsymbol{\nu}_{yy}) \\ & + 3v(\phi^\circ_x\boldsymbol{\nu} - \phi^\circ\boldsymbol{\nu}_x) + 3w(\phi^\circ_y\boldsymbol{\nu} - \phi^\circ\boldsymbol{\nu}_y).\end{aligned} \tag{9.195}$$

It is emphasized that $\boldsymbol{S}$ does indeed satisfy the defining relations (9.189) and (u', v', w') now constitutes a second solution of the stationary NVN equation.

It may be verified directly that, if $\boldsymbol{S}$ is specialised to (9.195), not only the pair (9.161) but also the linear triad (9.159) is invariant under the transformation $(9.188)_2$. To simplify the expressions for the new coefficients p', a', b' (and f', g'), we compare the form of p' with that given in Theorem 47. One is led to introduce the quantity

$$M^\circ = \varphi^\circ \phi^\circ_x + \psi^\circ \phi^\circ_y - \frac{1}{2}(\varphi^\circ_x + a\varphi^\circ + \psi^\circ_y + b\psi^\circ)\phi^\circ, \tag{9.196}$$

where φ° and ψ° are eigenfunctions of the mNVN equation related to the eigenfunction ϕ° of the NVN equation via the triad (9.172). Moreover, differentiation and use of (9.174) produces the usual relations

$$M^\circ_x = \varphi^{\circ 2}, \quad M^\circ_y = \psi^{\circ 2}. \tag{9.197}$$

Thus, at the level of the mNVN equation, the transformations (9.186) and $(9.188)_2$ are identical up to the discrete invariance $p' \to -p'$ and we obtain the following:

Theorem 48 (A Bäcklund transformation for isothermal-asymptotic surfaces). *Let $\hat{\boldsymbol{r}}$ be the position vector of an isothermal-asymptotic surface $\hat{\Sigma}$ and $\boldsymbol{\nu}$ its associated co-normal. Then, the position vector $\hat{\boldsymbol{r}}'$ and co-normal $\boldsymbol{\nu}'$ of its Bäcklund transform $\hat{\Sigma}'$ are given by (9.188), where the bilinear potential $\boldsymbol{S}$ is defined by (9.195) and ϕ° is a solution of (9.168) with parameter $\lambda_\circ$. If the bilinear potential M° is given by (9.196) with the eigenfunctions φ° and ψ° being related to ϕ° by the triad (9.172) then the second solution of the (m)NVN equation and the nonlinear system (9.152) reads*

$$\begin{aligned}
u' &= u - 2(\ln \phi^\circ)_{xy}, & p' &= -p + \frac{\varphi^\circ \psi^\circ}{M^\circ} \\
v' &= v - 2(\ln \phi^\circ)_{xx}, & V' &= V - \frac{3}{2}(\ln M^\circ)_{xx} \\
w' &= w - 2(\ln \phi^\circ)_{yy}, & W' &= W - \frac{3}{2}(\ln M^\circ)_{yy} \\
a' &= a + \left(\ln \frac{M^\circ}{\phi^{\circ 2}}\right)_x, & b' &= b + \left(\ln \frac{M^\circ}{\phi^{\circ 2}}\right)_y.
\end{aligned} \tag{9.198}$$

The Bäcklund transformation $\mathbb{B} : \hat{\Sigma} \to \hat{\Sigma}'$ possesses the tangency property, that is, $\hat{\boldsymbol{r}}' - \hat{\boldsymbol{r}}$ is tangential to both $\hat{\Sigma}$ and $\hat{\Sigma}'$.

Large classes of isothermal-asymptotic surfaces may now be generated by means of iterative application of the Bäcklund transformation $\mathbb{B}$ to known (classical) isothermal-asymptotic surfaces such as quadrics, cubics, quartics of Kummer and projective transforms of surfaces of revolution [51, 135, 136, 143, 227]. The solutions of the stationary (modified) NVN equation underlying these seed surfaces have been discussed by Ferapontov [129]. Soliton surfaces may be obtained by choosing seed surfaces corresponding to the trivial solution $(a, b, p) = \text{const}$. Their associated Bäcklund transforms are akin to the one-soliton affine spheres discussed in Chapter 3 (cf. Exercise 3).

Exercises

1. Verify that $\boldsymbol{\nu}'$ as given by $(9.188)_2$, (9.189) and (9.191) satisfies the Lax pair (9.192) and therefore (9.193) constitutes an invariance of the NVN equation.
2. Show that (9.196) coincides with the expression for M° given in Theorem 47.
3. (a) Determine the isothermal-asymptotic surfaces associated with

$$u = p = 1, \quad f = g = b = a, \quad v = w = \frac{1}{3}a(a+2).$$

   (b) Specify the real constants α, β and μ in such a way that

$$\phi^\circ = e^{i\alpha(\nu x - y/\nu)} \cosh[\beta(\nu x + y/\nu)]$$

   is a particular solution of (9.168) with $\lambda_\circ = i\mu$ and show that the corresponding solution u' of the stationary NVN equation represents a real soliton.

   (c) Apply Theorem 48 to construct *real* one-soliton isothermal-asymptotic surfaces. Show that these constitute surfaces of revolution if $\nu = 1$.

Appendix A

The su(2)–so(3) Isomorphism

The transition between the 3×3 Gauss-Weingarten equations for integrable surfaces and 2×2 representations for the associated soliton equations encoded in the Gauss-Mainardi-Codazzi equations has been seen to be based on the $su(2)$-$so(3)$ isomorphism (cf. Chapter 2). The isomorphism between the Lie algebras $su(2)$ and $so(3)$ is reflected at the group level by the existence of a two-to-one homomorphic mapping of the Lie group $SU(2)$ onto the Lie group $SO(3)$ [87]. This may be exploited to map the Gauss-Weingarten equations with underlying $SO(3)$ structure to its $SU(2)$-valued counterpart. In particular, the triad $\{\boldsymbol{r}_x, \boldsymbol{r}_y, \boldsymbol{N}\}$ associated with a surface may be encoded in an $SU(2)$ matrix. The $SO(3)$–$SU(2)$ connection is obtained as follows.

Let e_k be the canonical trace-free and skew-symmetric generators of the $su(2)$ Lie algebra defined by

$$e_1 = \frac{1}{2i}\begin{pmatrix} 0 & 1 \\ 1 & 0 \end{pmatrix}, \quad e_2 = \frac{1}{2i}\begin{pmatrix} 0 & -i \\ i & 0 \end{pmatrix}, \quad e_3 = \frac{1}{2i}\begin{pmatrix} 1 & 0 \\ 0 & -1 \end{pmatrix}. \tag{A.1}$$

These obey the commutator relations

$$[e_k, e_l] = \epsilon_{kl}{}^m e_m, \tag{A.2}$$

where $\epsilon_{kl}{}^m$ denotes the standard alternating Levi-Civita symbol and Einstein's summation convention has been adopted. Accordingly, any linear differential equation of the form

$$\dot{\phi} = g^m e_m \phi \tag{A.3}$$

with some given functions g^m admits $SU(2)$-valued solutions, that is, $\phi^\dagger \phi = \mathbb{1}$, $\det \phi = 1$. Thus, if

$$\phi \in SU(2) \tag{A.4}$$

then we may define a 3×3-matrix Φ by

$$\Phi_{kl} = -2\,\mathrm{Tr}\,(\phi^{-1} e_k \phi e_l). \tag{A.5}$$

Differentiation and evaluation by means of (A.3) results in

$$\begin{aligned} \dot{\Phi}_{kl} &= -2\,\mathrm{Tr}\,(\phi^{-1}[e_k, \dot{\phi}\phi^{-1}]\phi e_l) \\ &= -2g^m\,\mathrm{Tr}\,(\phi^{-1}[e_k, e_m]\phi e_l) \\ &= -2g^m \epsilon_{km}{}^n\,\mathrm{Tr}\,(\phi^{-1} e_n \phi e_l) \\ &= g^m \epsilon_{km}{}^n \Phi_{nl} \end{aligned} \tag{A.6}$$

so that

$$\dot{\Phi} = g^m L_m \Phi, \tag{A.7}$$

where the matrices L_m are given by

$$(L_m)_k{}^n = \epsilon_{km}{}^n, \tag{A.8}$$

that is

$$L_1 = \begin{pmatrix} 0 & 0 & 0 \\ 0 & 0 & -1 \\ 0 & 1 & 0 \end{pmatrix}, \quad L_2 = \begin{pmatrix} 0 & 0 & 1 \\ 0 & 0 & 0 \\ -1 & 0 & 0 \end{pmatrix}, \quad L_3 = \begin{pmatrix} 0 & -1 & 0 \\ 1 & 0 & 0 \\ 0 & 0 & 0 \end{pmatrix}. \tag{A.9}$$

The latter constitute the adjoint representation of the $so(3)$ Lie algebra and satisfy the same commutator relations as the $su(2)$ matrices e_m. In fact, since

$$-2\,\mathrm{Tr}\,(e_k e_l) = \delta_{kl}, \tag{A.10}$$

where δ_{kl} designates the standard Kronecker symbol, the definition (A.5) of the matrix Φ is equivalent to

$$\phi^{-1} e_k \phi = \sum_m \Phi_{km} e_m \tag{A.11}$$

and hence

$$-2\,\mathrm{Tr}\,(\phi^{-1} e_k e_l \phi) = -2 \sum_{m,n} \Phi_{km} \Phi_{ln}\,\mathrm{Tr}\,(e_m e_n). \tag{A.12}$$

Evaluation of the latter produces

$$\delta_{kl} = \sum_m \Phi_{km} \Phi_{lm} \tag{A.13}$$

which implies that

$$\Phi^{\mathsf{T}}\Phi = \mathbb{1}. \tag{A.14}$$

In a similar manner, one may verify that

$$\det \Phi = 1, \tag{A.15}$$

whence

$$\Phi \in SO(3). \tag{A.16}$$

Accordingly, the $SU(2)$ matrix ϕ is mapped to an $SO(3)$ matrix Φ by the transformation (A.5). It is noted that Φ is invariant under $\phi \rightarrow -\phi$ which shows that the mapping $SU(2) \rightarrow SO(3)$ is, at least, two-to-one.

To prove that the $SU(2)$ Lie group indeed provides a 'double covering' of the $SO(3)$ Lie group, we now introduce the parametrisation

$$\phi = \begin{pmatrix} \phi_1 & -\bar{\phi}_2 \\ \phi_2 & \bar{\phi}_1 \end{pmatrix}, \quad |\phi_1|^2 + |\phi_2|^2 = 1 \tag{A.17}$$

of any matrix $\phi \in SU(2)$. It is then easy to show that the associated matrix $\Phi \in SO(3)$ reads

$$\Phi = \begin{pmatrix} \Re(\phi_1^2 - \phi_2^2) & \Im(\phi_1^2 - \phi_2^2) & 2\Re(\phi_1\bar{\phi}_2) \\ -\Im(\phi_1^2 + \phi_2^2) & \Re(\phi_1^2 + \phi_2^2) & -2\Im(\phi_1\bar{\phi}_2) \\ -2\Re(\phi_1\phi_2) & -2\Im(\phi_1\phi_2) & |\phi_1|^2 - |\phi_2|^2 \end{pmatrix}. \tag{A.18}$$

In particular, the relations

$$\frac{\Phi_{11} + i\Phi_{21}}{1 - \Phi_{31}} = \frac{\bar{\phi}_1 - \phi_2}{\phi_1 + \bar{\phi}_2}, \quad \frac{\Phi_{12} + i\Phi_{22}}{1 - \Phi_{32}} = \frac{\bar{\phi}_1 + i\phi_2}{\bar{\phi}_2 - i\phi_1} \tag{A.19}$$

obtain by virtue of the condition $(A.17)_2$. Conversely, for a given $SO(3)$ matrix Φ, the *linear* equations (A.19) define two complex functions ϕ_1 and ϕ_2 up to a real multiplicative factor, the absolute value of which is calculated using the normalisation condition $(A.17)_2$. In this way, a matrix $\phi \in SU(2)$ of the form (A.17) is uniquely constructed up to the transformation $\phi \rightarrow -\phi$.

Appendix B

CC-Ideals

Constant coefficient ideals (cc-ideals) have been utilised by Harrison [158] in connection with the construction of Bäcklund transformations for the Ernst equation of general relativity. The notion of cc-ideals has subsequently been used extensively in the context of 1+1-dimensional integrable systems [122, 169–171, 173, 323, 336, 339]. There exist strong connections with the prolongation technique developed by Wahlquist and Estabrook [123, 378].

We here consider an infinite-dimensional Lie algebra which admits a faithful matrix representation obeying the commutator relations

$$[X_i, X_j] = c^k{}_{ij} X_k, \quad c^k{}_{ij} = \text{const} \tag{B.1}$$

and the Jacobi identity

$$[[X_i, X_j], X_k] + [[X_j, X_k], X_i] + [[X_k, X_i], X_j] = 0. \tag{B.2}$$

In terms of the structure constants $c^k{}_{ij}$, the latter is given by

$$c^a{}_{ij} c^b{}_{ak} + c^a{}_{jk} c^b{}_{ai} + c^a{}_{ki} c^b{}_{aj} = 0, \tag{B.3}$$

where Einstein's summation convention has been adopted. The matrices (generators) X_i may be associated with *dual* one-forms ξ^i which, in turn, give rise to the two-forms

$$\omega^k = d\xi^k - \frac{1}{2} c^k{}_{ij} \xi^i \xi^j. \tag{B.4}$$

Here, we have suppressed the wedge between differential forms. The two-forms ω^k generate a closed differential ideal since

$$d\omega^k = 0 \bmod \omega^l. \tag{B.5}$$

Indeed, modulo ω^l, we obtain

$$\begin{aligned} d\omega^k &= d^2\xi^k - \frac{1}{2}c^k{}_{ij}(d\xi^i\xi^j - \xi^i d\xi^j) \\ &= \frac{1}{2}c^k{}_{ij}c^j{}_{ab}\xi^i\xi^a\xi^b \\ &= \frac{1}{6}\left(c^k{}_{ij}c^j{}_{ab} + c^k{}_{aj}c^j{}_{bi} + c^k{}_{bj}c^j{}_{ia}\right)\xi^i\xi^a\xi^b \\ &= 0 \end{aligned} \tag{B.6}$$

by virtue of the Jacobi identity (B.3) and $c^k{}_{ji} = -c^k{}_{ij}$. This guarantees that the equations $\omega^k = 0$ are integrable. The matrix-valued one-form

$$\Omega = -d\Phi + X_i\xi^i\Phi, \tag{B.7}$$

where Φ denotes a matrix-valued function, is also closed modulo Ω and ω^l, that is

$$\begin{aligned} d\Omega &= -d^2\Phi + X_i d\xi^i\Phi - X_i\xi^i X_j\xi^j\Phi \\ &= \left(X_k d\xi^k - \frac{1}{2}[X_i, X_j]\xi^i\xi^j\right)\Phi \\ &= \frac{1}{2}(X_k c^k{}_{ij} - [X_i, X_j])\xi^i\xi^j\Phi \end{aligned} \tag{B.8}$$

modulo (Ω, ω^l) and hence

$$d\Omega = 0 \bmod (\Omega, \omega^l). \tag{B.9}$$

Thus, the equation $\Omega = 0$ is likewise integrable and, accordingly, the one-form equation

$$d\Phi = X_i\xi^i\Phi \tag{B.10}$$

may be regarded as a linear representation of the 'cc-ideal'[1]

$$d\xi^k = \frac{1}{2}c^k{}_{ij}\xi^i\xi^j. \tag{B.11}$$

To obtain differential equations from the system of two-forms ω^k, it is now required to assume that only a finite number of one-forms ξ^i, $i \in \mathbb{J} \subset \mathbb{Z}$ are

[1] Strictly speaking, the cc-ideal is *generated* by the one-forms ω^k.

non-vanishing, that is

$$\xi^i = 0, \quad i \notin \mathbb{J}. \tag{B.12}$$

The set of two-forms ω^k then splits into a differential system and an algebraic system, namely

$$\begin{aligned} \rho^k &= d\xi^k - \frac{1}{2}c^k{}_{ij}\xi^i\xi^j, \quad k \in \mathbb{J} \\ \sigma^k &= \qquad -\frac{1}{2}c^k{}_{ij}\xi^i\xi^j, \quad k \notin \mathbb{J}. \end{aligned} \tag{B.13}$$

It is emphasized that the system (ρ^k, σ^k) remains closed under exterior differentiation, that is $d(\rho^k, \sigma^k) = 0 \bmod (\rho^l, \sigma^l)$. The maximum dimension of an integral manifold on which the two-forms ρ^k and σ^k vanish is Cartan's genus g. The one-forms ξ^i may thus be represented as linear combinations of g linearly independent one-forms η^i, $i = 1, \ldots, g$, that is

$$\xi^i = a^i{}_j\eta^j, \quad i \in \mathbb{J} \tag{B.14}$$

with functions $a^i{}_j$ depending on the coordinates which parametrise the integral manifold. These have to be chosen in such a way that the algebraic one-forms σ^k vanish identically. The remaining conditions $\rho^k = 0$ then reduce to a system of first-order differential equations which may be combined to one (or several) higher order equation(s). By construction, the g *linear* matrix differential equations (B.10) are compatible modulo this system of *nonlinear* differential equations.

The linearly independent one-forms η^i may be found by investigating the two-forms ρ^k with respect to *exact* one-forms to be used as coordinate differentials on the integral manifold. Thus, if there exist more than g exact one-forms, then we may choose any g linearly independent exact one-forms as coordinate differentials. The remaining exact one-forms constitute differentials of potentials. Interchange of coordinates and potentials may be exploited to derive various differential equations which admit the *same* cc-ideal.

As an illustration, we consider the loop algebra $so(3)\otimes\mathbb{R}(\lambda, \lambda^{-1})$ [186] of the $so(3)$ Lie algebra with associated commutator relations

$$\left[X_i^n, X_j^m\right] = \epsilon_{ij}{}^k X_k^{n+m}, \tag{B.15}$$

where $\epsilon_{ij}{}^k$ denotes the standard alternating Levi-Civita symbol. A matrix representation is given by

$$X_i^n = \lambda^n Y_i, \quad \lambda = \text{const}, \tag{B.16}$$

where the matrices Y_i satisfy the $so(3)$ commutator relations

$$[Y_i, Y_j] = \epsilon_{ij}{}^k Y_k. \tag{B.17}$$

Canonical representations of the $so(3)$ Lie algebra are the adjoint representation $Y_i = L_i$ or the $su(2)$ representation $Y_i = e_i$ as discussed in Appendix A.

To set up a particular cc-ideal, we now assume that the only non-vanishing one-forms $\xi^1, \xi^2, \xi^3, \xi^4$ are those dual to the generators

$$X_1 = X_2^0, \quad X_2 = X_3^1, \quad X_3 = X_1^{-1}, \quad X_4 = X_3^{-1}, \tag{B.18}$$

respectively. The corresponding cc-ideal then becomes

$$\begin{aligned} d\xi^1 &= \xi^2\xi^3, & \xi^1\xi^2 &= 0 \\ d\xi^2 &= 0, & \xi^3\xi^4 &= 0 \\ d\xi^3 &= \xi^1\xi^4 & & \\ d\xi^4 &= \xi^3\xi^1. & & \end{aligned} \tag{B.19}$$

Cartan's genus is $g = 2$ since the algebraic part of the above cc-ideal shows that

$$\xi^1 = a\xi^2, \quad \xi^3 = b\eta, \quad \xi^4 = c\eta \tag{B.20}$$

for as yet unspecified functions a, b, c and a one-form η. Its differential part implies that

$$d\xi^2 = 0, \quad d\left(\sqrt{b^2 + c^2}\,\eta\right) = 0, \quad d\left(\frac{1}{2}a^2\xi^2 + c\eta\right) = 0, \tag{B.21}$$

which, in turn, guarantees the existence of the differentials

$$dx = \xi^2, \quad dy = \sqrt{b^2 + c^2}\,\eta, \quad dz = \frac{1}{2}a^2\xi^2 + c\eta. \tag{B.22}$$

Thus, in terms of the coordinates x and y, the one-forms ξ^i may be parametrised according to

$$\xi^1 = a\,dx, \quad \xi^2 = dx, \quad \xi^3 = \sin\omega\,dy, \quad \xi^4 = -\cos\omega\,dy \tag{B.23}$$

and the cc-ideal reduces to the system of differential equations

$$a_y = -\sin\omega, \quad \omega_x = -a. \tag{B.24}$$

We therefore conclude that the cc-ideal (B.19) represents the classical sine-Gordon equation

$$\omega_{xy} = \sin\omega \tag{B.25}$$

and the potential z defined by

$$dz = \frac{1}{2}\omega_x^2 dx - \cos\omega\, dy \tag{B.26}$$

corresponds to the 'conservation law'

$$\left(\omega_x^2\right)_y + (2\cos\omega)_x = 0. \tag{B.27}$$

Alternatively, if we choose y and z as coordinates, then x becomes a potential and the one-forms ξ^i admit the parametrisation

$$\begin{aligned} \xi^1 &= A(\cos\omega\, dy + dz), & \xi^3 &= \sin\omega\, dy \\ \xi^2 &= \frac{1}{2}A^2(\cos\omega\, dy + dz), & \xi^4 &= -\cos\omega\, dy \end{aligned} \tag{B.28}$$

with $A = 2/a$. Insertion into the cc-ideal produces the system

$$A_y - A_z\cos\omega + \frac{1}{2}A\omega_z\sin\omega = 0, \quad \omega_z = -A \tag{B.29}$$

or, equivalently,

$$\omega_{yz} = \omega_{zz}\cos\omega - \frac{1}{2}\omega_z^2\sin\omega. \tag{B.30}$$

The latter equation therefore encapsulates the sine-Gordon equation (B.25) via the reciprocal transformation $(y, z) \to (x, y)$ with the variable x being defined by

$$dx = \frac{1}{2}\omega_z^2(\cos\omega\, dy + dz). \tag{B.31}$$

By construction, it is encoded in the same cc-ideal as the sine-Gordon equation. This avatar of the sine-Gordon equation was set down in [195].

The linear representation corresponding to any parametrisation of the two-forms ξ^i may be obtained from (B.10) by means of the matrix representation (B.16), where λ plays the role of the 'spectral' parameter. For instance, if we choose the $su(2)$ representation

$$Y_1 = \frac{1}{2i}\begin{pmatrix} 0 & 1 \\ 1 & 0 \end{pmatrix}, \quad Y_2 = \frac{1}{2i}\begin{pmatrix} 0 & -i \\ i & 0 \end{pmatrix}, \quad Y_3 = \frac{1}{2i}\begin{pmatrix} 1 & 0 \\ 0 & -1 \end{pmatrix} \tag{B.32}$$

then the linear representation of the sine-Gordon equation becomes

$$\begin{aligned}\Phi_x &= \frac{1}{2}\left[\begin{pmatrix} 0 & \omega_x \\ -\omega_x & 0 \end{pmatrix} - i\lambda \begin{pmatrix} 1 & 0 \\ 0 & -1 \end{pmatrix}\right]\Phi \\ \Phi_y &= -\frac{i}{2\lambda}\begin{pmatrix} -\cos\omega & \sin\omega \\ \sin\omega & \cos\omega \end{pmatrix}\Phi,\end{aligned} \tag{B.33}$$

which is nothing but the standard AKNS representation up to the transformation $\lambda \to -\lambda$ (cf. Chapter 2).

Appendix C

Biographies

Albert Victor Bäcklund

Born: *Väsby, Sweden 1845*
Died: *Lund, Sweden 1922*

Bäcklund received his tertiary education at the University of Lund. In 1864, he was appointed as an assistant at the Astronomical Observatory where he became a student of Professor Axel Möller. In 1868, Bäcklund received his Doktor Philosophiae for a thesis concerning the measurement of latitude from astronomical observations. Bäcklund thereafter turned to geometry and, in particular, to the work inspired by the Norwegian mathematician Sophus Lie.

In 1874, Bäcklund was awarded a government travel scholarship to pursue his studies on the Continent for six months. He spent most of his time in Leipzig and Erlangen where he met both Klein and Lindemann. Ideas which Bäcklund gained in this period led his later work on the geometry of what have come to be known as Bäcklund transformations.

In 1878, Bäcklund was appointed to the Extraordinary Chair in Mechanics and Mathematical Physics at Lund. He was elected Fellow of the Royal Swedish Academy of Sciences (KVA) in 1888. In 1897, Bäcklund was awarded the Chair in Physics at the University of Lund where he was later Rector. Bäcklund retired in 1910, but continued his scholarly research until his death.

Bäcklund is usually associated with the type of transformation of surfaces that bears his name, the extension of which have had major impact in soliton theory. Bäcklund's geometric work in this area was originally motivated by an attempt to extend Lie's theory of contact transformations. Bäcklund also made a significant incursion into the theory of characteristics which originated in the work of Monge and Ampére. Indeed, Bäcklund was regarded by both Goursat and Hadamard as the founder of the modern theory of characteristics.

Bibliography

C.W. Oseen, *Albert Victor Bäcklund*, Year-Book of the Royal Swedish Academy of Sciences [in Swedish] (1924). German translation in *Jahresber. Deutsch. Math.-Verein.* **38**, 113–152 (1929).

Gaston Darboux

Born: *Nîmes, France 1842*
Died: *Paris, France 1917*

Darboux entered the École Normale Supérieure in 1861 and completed his doctoral thesis entitled *Sur Les Surfaces Orthogonales* in 1866. Between 1873 and 1878, he was suppléant to Liouville in the Chair of Rational Mechanics at the Sorbonne and in 1880 succeeded to the Chair of Higher Geometry previously held by Chasles and retained it until his death. In 1884, he became a member of the Académie des Sciences. In 1902, he was elected a Fellow of the Royal Society of London and awarded its Sylvester Medal in 1916.

Darboux's primary contributions were to geometry, although he also made important incursions into analysis. He produced extensive works on orthogonal systems of surfaces, notably his *Leçons sur la théorie générale des surfaces, t. 1-4* (1887–1896) and the *Leçons sur les systèmes orthogonaux et les coordonnées curvilignes* (1898).

Bibliography

1. E. Lebon, *Gaston Darboux*, Paris (1910, 1913).
2. L.P. Eisenhart, Darboux's contribution to geometry, *Bull. Am. Math. Soc.* **24**, 227–237 (1918).
3. D. Hilbert, Gaston Darboux, *Acta Math.* **42**, 269–273 (1919).

Bibliography and Author Index

The numbers in bold following each entry give the pages of this book on which the entry is cited.

[1] M.J. Ablowitz, D.J. Kaup, A.C. Newell and H. Segur, Nonlinear evolution equations of physical significance, *Phys. Rev. Lett.* **31**, 125–127 (1973). **(64, 204)**

[2] M.J. Ablowitz, D.J. Kaup, A.C. Newell and H. Segur, The inverse scattering transform-Fourier analysis for nonlinear problems, *Stud. Appl. Math.* **53**, 249–134 (1974). **(210)**

[3] M.J. Ablowitz and H. Segur, *Solitons and the Inverse Scattering Transform*, SIAM, Philadelphia (1981). **(211)**

[4] J.E. Adkins, A reciprocal property of the finite plane strain equations, *J. Mech. Phys. Solids* **6**, 267–275 (1958). **(97)**

[5] S.I. Agonov and E.V. Ferapontov, Theory of congruences and systems of conservation laws, *J. Math. Sci.* **94**, 1748–1794 (1999). **(230)**

[6] M.A. Akivis and V.V. Goldberg, *Projective Differential Geometry of Submanifolds*, Math. Library **49**, North-Holland (1993). **(329)**

[7] M.S. Alber, R. Camassa, D.D. Holm and J.E. Marsden, The geometry of peaked solutions of a class of integrable pdes, *Lett. Math. Phys.* **32**, 137–151 (1994). **(239)**

[8] M.S. Alber, R. Camassa, D.D. Holm and J.E. Marsden, On the link between umbilic geodesics and soliton solutions of nonlinear ODEs, *Proc. R. Soc. Lond. A* **450**, 677–692 (1995). **(239)**

[9] G. Albrecht and W.L.F. Degen, Construction of Bézier rectangles and triangles on the symmetric Dupin horn cyclide by means of inversion, *Computer Aided Geometric Design* **14**, 349–357 (1997). **(198)**

[10] S. Allen and D. Dutta, Cyclides in pure blending I, *Computer Aided Geometric Design* **14**, 51–75 (1997). **(198)**

[11] S. Allen and D. Dutta, Cyclides in pure blending II, *Computer Aided Geometric Design* **14**, 77–102 (1997). **(198)**

[12] S. Allen and D. Dutta, Supercyclides and blending, *Computer Aided Geometric Design* **14**, 637–651 (1997). **(198)**

[13] L.K. Antanovskii, C. Rogers and W.K. Schief, A note on a capillarity model

and the nonlinear Schrödinger equation, *J. Phys. A: Math. Gen.* **30**, L555–L557 (1997). **(119)**

[14] M. Antonowicz, On the Bianchi-Bäcklund construction for affine minimal surfaces, *J. Phys. A: Math. Gen.* **20**, 1989–1996 (1987). **(88)**

[15] M. Antonowicz and A.P. Fordy, Factorisation of energy dependent Schrödinger operators: Miura maps and modified systems, *Commun. Math. Phys.* **124**, 465–486 (1989). **(217)**

[16] M. Antonowicz and A.P. Fordy, Hamiltonian structure of nonlinear evolution equations, in A.P. Fordy, ed, *Soliton Theory: A Survey of Results*, pp. 273–312, Manchester University Press (1990). **(217)**

[17] M. Antonowicz and A. Sym, New integrable nonlinearities from affine geometry, *Phys. Lett. A* **112**, 1–2 (1985). **(88)**

[18] N. Asano, T. Taniuti and N. Yajima, Perturbation method for nonlinear wave modulation: II, *J. Math. Phys.* **10**, 2020–2024 (1969). **(119)**

[19] C. Athorne, On the characterization of Moutard transformations, *Inverse Problems* **9**, 217–232 (1993). **(111, 113)**

[20] C. Athorne and J.J.C. Nimmo, On the Moutard transformation for integrable partial differential equations, *Inverse Problems*, **7**, 809–826 (1991). **(362, 368)**

[21] A.V. Bäcklund, Om ytor med konstant negativ krökning, *Lunds Universitets Årsskrift* **19**, 1–48 (1883). **(17)**

[22] J.A. Baker and C. Rogers, Invariance properties under a reciprocal Bäcklund transformation in gasdynamics, *J. Mécanique Théor. Appl.* **1**, 563–578 (1982). **(229)**

[23] T.W. Barnard, $2N\pi$ Ultrashort light pulses, *Phys. Rev. A* **7**, 373–376 (1973). **(22, 30)**

[24] V.I. Baspalov and V.I. Talanov, Filamentary structure of light beams in nonlinear liquids, *JETP Engl. Transl.* **3**, 307–310 (1966). **(119)**

[25] H. Bateman, The lift and drag functions for an elastic fluid in two-dimensional irrotational flow, *Proc. Natl. Acad. Sci. U.S.A.* **24**, 246–251 (1938). **(229)**

[26] R. Beals, M. Rabelo and K. Tenenblat, Bäcklund transformations and inverse scattering solutions for some pseudospherical surface equations, *Stud. Appl. Math.* **81**, 125–151 (1989). **(22)**

[27] V.A. Belinsky and V.E. Zakharov, Integration of Einstein's equations by means of the inverse scattering technique and construction of exact solutions, *Sov. Phys. JETP* **48**, 985–994 (1978). **(297, 305)**

[28] E. Beltrami, Saggio di interpretazione della geometria non-euclidea, *Giornale di Matematiche* **6**, 284–312 (1868). **(17)**

[29] D.J. Benney and G.J. Roskes, Wave instabilities, *Stud. Appl. Math.* **48**, 377–385 (1969). **(163)**

[30] R. Betchov, On the curvature and torsion of an isolated vortex filament, *J. Fluid. Mech.* **22**, 471–479 (1965). **(60)**

[31] L. Bianchi, Ricerche sulle superficie a curvatura constante e sulle elicoidi. Tesi di Abilitazione, *Ann. Scuola Norm. Sup. Pisa (1)* **2**, 285–304 (1879). **(17)**

[32] L. Bianchi, Sopra i sistemi tripli ortogonali di Weingarten, *Ann. Matem.* **13**, 177–234 (1885). **(17, 60, 72)**

[33] L. Bianchi, Sopra alcone nuove classi di superficie e di sistemi tripli ortogonali, *Ann. Matem.* **18**, 301–358 (1890). **(45, 50, 297)**

[34] L. Bianchi, Sulle deformazioni infinitesime delle superficie flessibili ed inestendibili, *Rend. Lincei* **1**, 41–48 (1892). **(299)**

[35] L. Bianchi, Sulla trasformazione di Bäcklund per le superficie pseudosferiche, *Rend. Lincei* **5**, 3–12 (1892). **(28)**

[36] L. Bianchi, Ricerche sulle superficie isoterme e sulla deformazione delle quadriche, *Ann. Matem.* **11**, 93–157 (1905). **(152, 171, 184)**

[37] L. Bianchi, *Lezioni di geometria differenziale* **1-4**, Zanichelli, Bologna (1923–1927). **(18, 21, 28, 152, 154, 182)**

[38] O. Bjørgum, On Beltrami vector fields and flows, Part I., *Universitet I. Bergen, Årbok Naturvitenskapelig* rekke n-1 (1951). **(139)**

[39] W. Blaschke, *Differentialgeometrie*, Chelsea Publishing Company, New York, Reprinted (1967). **(88, 91, 100, 127, 335)**

[40] A.I. Bobenko, Surfaces in terms of 2 by 2 matrices. Old and new integrable cases, in A. Fordy and J. Woods, eds, *Harmonic Maps and Integrable Systems*, Vieweg, pp. 83–128 (1994). **(40)**

[41] A. Bobenko and U. Eitner, Bonnet surfaces and Painlevé equations, *J. Reine Angew. Math.* **499**, 47–79 (1998). **(118)**

[42] A.I. Bobenko and U. Eitner, Painlevé equations in differential geometry of surfaces, *Lecture Notes in Mathematics* **1753** Springer Verlag, Berlin, Heidelberg (2000). **(118)**

[43] A. Bobenko, U. Eitner and A. Kitaev, Surfaces with harmonic inverse mean curvature and Painlevé equations, *Geom. Dedicata* **68**, 187–227 (1997). **(118)**

[44] A.I. Bobenko and A.V. Kitaev, On asymptotic cones of surfaces with constant curvature and the third Painlevé equation, *Manuscripta. Math.* **97**, 489–516 (1998). **(118)**

[45] A.I. Bobenko and R. Seiler, eds, *Discrete Integrable Geometry and Physics*, Clarendon Press, Oxford (1999). **(237)**

[46] W. Boem, On cyclides in geometric modelling, *Computer Aided Geometric Design* **7**, 243–255 (1990). **(198)**

[47] L.V. Bogdanov, Veselov-Novikov equation as a natural two-dimensional generalization of the Korteweg-de Vries equation, *Teoret. Mat. Fiz.* **70**, 309–314 (1987). **(330)**

[48] M. Boiti, C. Laddomada and F. Pempinelli, Multiple-kink-soliton solutions of the nonlinear Schrödinger equation, *Il Nuovo Cimento B* **65**, 248–258 (1981). **(149)**

[49] M. Boiti, J. Leon, L. Martina and F. Pempinelli, Scattering of localized solitons in the plane, *Phys. Lett. A* **132**, 432–439 (1988). **(196, 362)**

[50] M. Boiti, F. Pempinelli and P.C. Sabatier, First and second order nonlinear evolution equations, *Inverse Problems* **9**, 1–37 (1993). **(163)**

[51] G. Bol, *Projektive Differentialgeometrie*, Göttingen (1954). **(329, 330, 332, 336, 341, 370)**

[52] A. Yu. Boldin, S.S. Safin and R.A. Shapirov, On an old article of Tzitzeica and the inverse scattering method, *J. Math. Phys.* **34**, 5801–5809 (1993). **(91)**

[53] O. Bonnet, Mémoire sur la théorie des surfaces applicables sur une surface donnée, *J. l'École Polytech.* **41**, 201–230 (1865); *J. l'École Polytech.* **42**, 1–151 (1867). **(18)**

[54] E. Bour, Théorie de la déformation des surfaces, *J. l'École Imperiale Polytech.* **19**, Cahier 39, 1–48 (1862). **(17, 152)**

[55] C. Brezinski, A general extrapolation algorithm, *Numer. Math.* **35**, 175–187 (1980). **(237)**

[56] P. Broadbridge, J.H. Knight and C. Rogers, Constant rate rainfall infiltration in a bounded profile: solutions of a nonlinear model, *Soil. Soc. Am. J.* **52**, 1526–1533 (1988). **(229)**

[57] P. Broadbridge and C. Rogers, Exact solutions for vertical drainage and redistribution in soils, *J. Eng. Math.* **24**, 225–43 (1990). **(229)**

[58] P. Broadbridge and P. Tritscher, An integrable fourth order nonlinear evolution equation applied to the thermal grooving of metal surfaces, *IMA J. Appl. Math.* **53**, 249–265 (1994). **(232)**

[59] F. Burstall, Isothermic surfaces in arbitrary co-dimension, Atti del Congresso Internazionale in onore di Pasquale Calapso, Rendiconti del Sem. Mat. di. Messina, 57–68 (2001). **(163, 171)**

[60] F. Burstall, Isothermic surfaces: conformal geometry, Clifford algebras and integrable systems, *Math. DG/0003096* (2000). **(163, 171)**

[61] F. Burstall, U. Hertrich-Jeromin, F. Pedit and U. Pinkall, Curved flats and isothermic surfaces, *Math. Z.* **225**, 199–209 (1997). **(171, 189)**

[62] P. Calapso, Sulla superficie a linee di curvatura isoterme, *Rend. Circ. Mat. Palermo* **17**, 275–286 (1903). **(152, 154, 165)**

[63] F. Calogero and A. Degasperis, Coupled nonlinear evolution equations solvable via the inverse spectral transform, and solitons that come back: the boomeron, *Lett. Nuovo Cimento* **16**, 425–433 (1976). **(155, 161, 164)**

[64] F. Calogero and A. Degasperis, Bäcklund transformations, nonlinear superposition principle, multisoliton solutions and conserved quantities for the “boomeron” nonlinear evolution equation, *Lett. Nuovo Cimento* **16**, 434–438 (1976). **(155, 161, 164, 191)**

[65] F. Calogero and A. Degasparis, *Spectral Transform and Solitons*, North Holland Publishing Company, Amsterdam (1982). **(233, 266)**

[66] F. Calogero and A. Degasperis, A modified modified Korteweg-de Vries equation, *Inverse Problems* **1**, 57–66 (1985). **(243)**

[67] R. Camassa and D.D. Holm, An integrable shallow water equation with peaked solitons, *Phys. Rev. Lett.* **71**, 1661–1664 (1993). **(239)**

[68] E. Cartan, *Les systèmes différentielles extérieurs et leurs applications à métriques*, Hermann, Paris (1945). **(261)**

[69] P.J. Caudrey, J.C. Eilbeck, J.D. Gibbon and R.K. Bullough, Exact multisoliton solution of the inhomogeneously broadened self-induced transparency equations, *J. Phys. A: Math. Gen.* **6**, L53–L56 (1973). **(130)**

[70] P.J. Caudrey, J.D. Gibbon, J.C. Eilbeck and R.K. Bullough, Exact multi-soliton solutions of the self-induced transparency and sine-Gordon equations, *Phys. Rev. Lett.* **30**, 237–239 (1973). **(130)**

[71] A. Cayley, On the cyclide, *Q.J. Pure Appl. Math.* **12**, 148–165 (1873). **(198)**

[72] H.M. Cekirge and C. Rogers, On elastic-plastic wave propagation: transmission of elastic-plastic boundaries, *Arch. Mech.* **29**, 125–141 (1977). **(98)**

[73] H.M. Cekirge and E. Varley, Large amplitude waves in bounded media I: reflexion and transmission of large amplitude shockless pulses at an interface, *Philos. Trans. R. Soc. Lond. A* **273**, 261–313 (1973). **(98)**

[74] B. Cenkl, Geometric deformations of the evolution equations and Bäcklund transformations, *Physica D* **18**, 217–219 (1986). **(21)**

[75] Ö. Ceyhan, A.S. Fokas and M. Gürses, Deformations of surfaces associated with integrable Gauß-Mainardi-Codazzi equations, *J. Math. Phys.* **41**, 2251–2270 (2000). **(42)**

[76] H.H. Chen and C.S. Liu, Nonlinear wave and soliton propagation in media with arbitrary inhomogeneities, *Phys. Fluids* **21**, 377–380 (1978). **(119)**

[77] S.S. Chern, Surface theory with Darboux and Bianchi, *Miscellanea Mathematica*, pp. 59–69, Springer, Berlin (1991). **(17)**

[78] S.S. Chern and K. Tenenblat, Foliations on a surface of constant curvature and the modified Korteweg-de Vries equations, *J. Diff. Geom.* **16**, 347–349 (1981). **(22)**

[79] S.S. Chern and K. Tenenblat, Pseudospherical surfaces and evolution equations, *Stud. Appl. Math.* **74**, 55–83 (1986). **(22)**

[80] S.S. Chern and C.L. Terng, An analogue of Bäcklund's theorem in affine geometry, *Rocky Mountain J. Math.* **10**, 105–124 (1980). **(88)**

[81] F.J. Chinea, Vector Bäcklund transformations and associated superposition principle, in C. Hoenselaers and W. Dietz, eds, *Solutions of Einstein's Equations: Techniques and Results*, Lecture Notes in Physics, pp. 55–67, Springer-Verlag, Berlin (1984). **(298, 326)**

[82] J. Cieśliński, An algebraic method to construct the Darboux matrix, *J. Math. Phys.* **36**, 5670–5706 (1995). **(266, 270)**

[83] J. Cieśliński, The Darboux-Bianchi transformation for isothermic surfaces. Classical results versus the soliton approach, *Diff. Geom. Appl.* **7**, 1–28 (1997). **(171)**

[84] J. Cieśliński, A generalized formula for integrable classes of surfaces in Lie algebras, *J. Math. Phys.* **38**, 4255–4272 (1997). **(208)**

[85] J. Cieśliński, P.K.H. Gragert and A. Sym, Exact solutions to localised induction-approximation equation modelling smoke-ring motion, *Phys. Rev. Lett.* **57**, 1507–1510 (1986). **(150)**

[86] J. Cieśliński, P. Goldstein and A. Sym, Isothermic surfaces in E^3 as soliton surfaces, *Phys. Lett. A* **205**, 37–43 (1995). **(154, 192)**

[87] J.F. Cornwell, *Group Theory in Physics*, Vols. I, II, Academic Press, London (1984). **(371)**

[88] C.M. Cosgrove, Relationships between the group-theoretic and soliton-theoretic techniques for generating stationary axisymmetric gravitational solutions, *J. Math. Phys.* **21**, 2417–2447 (1980). **(297, 305)**

[89] E. Cosserat, Sur les systèmes conjugués et sur la déformation des surfaces, *C. R. Acad. Sci. Paris* **113**, 460–463 (1891); sur les systèmes cycliques et sur la déformation des surfaces, *C.R. Acad. Sci. Paris* **113**, 498–500. **(299)**

[90] J. Crank, *The Mathematics of Diffusion*, 2nd ed, Oxford University Press, (1975). **(232)**

[91] M.M. Crum, Associated Sturm-Liouville systems, *Q. J. Math. Oxford* **6**, 121–127 (1955). **(266)**

[92] G. Darboux, Sur une proposition relative aux equations linéaires, *C.R. Acad. Sci. Paris* **94**, 1456–1459 (1882). **(17, 152, 266)**

[93] G. Darboux, *Leçons sur la théorie générale des surfaces*, Gauthier-Villars, Paris (1887). **(109)**

[94] G. Darboux, Sur les surfaces dont la courbure totale est constante, *C.R. Acad. Sci. Paris* **97**, 848–850 (1883); sur les surfaces à courbure constante, *C.R. Acad. Sci. Paris* **97**, 892–894; sur l'équation aux dérivées partielles des surfaces à courbure constante, *C.R. Acad. Sci. Paris* **97**, 946–949. **(17)**

[95] G. Darboux, Sur les surfaces isothermiques, *C.R. Acad. Sci. Paris* **128**, 1299–1305 (1899). **(152, 154, 171, 175)**

[96] A. Davey, The propagation of a weak nonlinear wave, *J. Fluid. Mech.* **53**, 769–781 (1972). **(119)**

[97] L.S. Da Rios, Sul moto d'un liquido indefinito con un filetto vorticoso, *Rend. Circ. Mat. Palermo* **22**, 117–135 (1906). **(60, 119, 121)**

[98] A. Davey and K. Stewartson, On three-dimensional packets of surface waves, *Proc. R. Soc. Lond. A* **338**, 101–110 (1974). **(163)**

[99] P.G. deGennes, *Superconductivity of Metals and Alloys*, Benjamin, New York (1966). **(119)**

[100] A. Degasperis, C. Rogers and W.K. Schief, Isothermic surfaces generated via

Bäcklund and Moutard transformations. Boomeron and zoomeron connections, to appear in *Stud. Appl. Math.* (2002). **(164, 198)**

[101] A. Demoulin, Sur les systèmes et les congruences K, *C.R. Acad. Sci. Paris* **150**, 156–159 (1910). **(186)**

[102] A. Demoulin, Sur deux transformations des surfaces dont les quadriques de Lie n'ont que deux ou trois points charactéristiques, *Bull. l'Acad. Belgique* **19**, 479–502, 579–592, 1352–1363 (1933). **(329, 335)**

[103] J. de Pont, Essays on the cyclide patch, *PhD Thesis*, Cambridge University (1984). **(198)**

[104] W. Dietz and C. Hoenselaers, Two mass solutions of Einstein's vacuum equations: the double Kerr solution, *Ann. Phys.* **165**, 319–383 (1985). **(311)**

[105] L.A. Dmitrieva, Finite-gap solutions of the Harry Dym equation, *Phys. Lett. A* **182**, 65–70 (1993). **(234)**

[106] L.A. Dmitrieva, N-loop solitons and their link with the complex Harry Dym equation, *J. Phys. A: Math. Gen.* **27**, 8197–8205 (1994). **(226, 234)**

[107] L. Dmitrieva and M. Khlabystova, Multisoliton solutions of the (2+1)-dimensional Harry Dym equation, *Phys. Lett. A* **237**, 369–380 (1998). **(239)**

[108] M.P. do Carmo, *Differential Geometry of Curves and Surfaces*, Prentice-Hall, Inc., Englewood Cliffs, New Jersey (1976). **(18)**

[109] R.K. Dodd, General relativity, in A.P. Fordy, ed, *Soliton Theory: A Survey of Results*, pp. 174–207, Manchester University Press (1990). **(319)**

[110] R.K. Dodd, Soliton immersions, *Commun. Math. Phys.* **197**, 641–665 (1998). **(208)**

[111] R.K. Dodd and R.K. Bullough, Polynomial conserved densities for the sine-Gordon equations, *Proc. R. Soc. Lond. A* **352**, 481–503 (1977). **(88)**

[112] A. Doliwa and P. Santini, An elementary geometric characterisation of the integrable motions of a curve, *Phys. Lett. A* **185**, 373–384 (1994). **(60)**

[113] A. Doliwa, P.M. Santini and M. Mañas, Transformations of quadrilateral lattices, *J. Math. Phys.* **41**, 944–990 (2000). **(167)**

[114] C. Dupin, *Applications de Géometrie et de Mécanique*, Bachelier, Paris (1822). **(198, 200)**

[115] J. Ehlers, *Les théories relativistes de la gravitation*, CRNS, Paris (1959). **(310)**

[116] L.P. Eisenhart, *Riemannian Geometry*, Princeton University Press, Princeton, New Jersey (1950). **(157, 158, 161)**

[117] L.P. Eisenhart, *Non-Riemannian Geometry*, American Mathematical Society, New York (1958). **(309)**

[118] L.P. Eisenhart, *A Treatise on the Differential Geometry of Curves and Surfaces*, Dover, New York (1960). **(47, 60, 68, 127)**

[119] L.P. Eisenhart, *Transformations of Surfaces*, Chelsea, New York (1962). **(72, 89, 109, 117, 155, 157, 167, 175, 180)**

[120] F. Emde, Der Einfluß der Feldlinien auf Divergenz und Rotor, *Arch. Elektrotechn.* **39**, 2–8 (1948). **(139)**

[121] F. Ernst, New formulation of the axially symmetric gravitational field problem. I/II, *Phys. Rev.* **167**, 1175–1178; *Phys. Rev.* **168**, 1415–1417 (1968). **(49, 297, 304, 309)**

[122] F.B. Estabrook, Moving frames and prolongation algebras, *J. Math. Phys.* **23**, 2071–2076 (1982). **(374)**

[123] F.B. Estabrook and H.D. Wahlquist, Prolongation structures of nonlinear evolution equations. II, *J. Math. Phys.* **17**, 1293–1297 (1976). **(261, 312, 374)**

[124] E.V. Ferapontov, Reciprocal transformations and their invariants, *Diff. Uravnen* **25**, 1256–1265 (1989). **(230)**

[125] E.V. Ferapontov, Reciprocal transformations and hydrodynamic symmetries, *Diff. Uravnen* **27**, 1250–1263 (1993). **(230)**

[126] E.V. Ferapontov, Nonlocal Hamiltonian operators of hydrodynamic type: differential geometry and applications, *Trans. Am. Math. Soc.* **170**, 33–58 (1995). **(230)**

[127] E.V. Ferapontov, Dupin hypersurfaces and integrable Hamiltonian systems of hydrodynamic type which do not possess Riemann invariants, *Diff. Geom. Appl.* **5**, 121–152 (1995). **(198, 230)**

[128] E.V. Ferapontov, Surfaces in Lie sphere geometry and the stationary Davey-Stewartson hierarchy, *Sfb 288 Preprint* **287**, Technische Universität, Berlin (1997). **(163)**

[129] E.V. Ferapontov, Stationary Veselov-Novikov equation and isothermally asymptotic surfaces in projective-differential geometry, *Diff. Geom. Appl.* **11**, 117–128 (1999). **(330, 357, 358, 363, 370)**

[130] E.V. Ferapontov, Lie sphere geometry and integrable systems, *Tohoku Math. J.* **52**, 199–233 (2000). **(230)**

[131] E.V. Ferapontov, Integrable systems in projective differential geometry, *Kyushu J. Math.* **54**, 183–215 (2000). **(329, 330)**

[132] E.V. Ferapontov, C. Rogers and W.K. Schief, Reciprocal transformations of two-component hyperbolic systems and their invariants, *J. Math. Anal. Appl.* **228**, 365–376 (1998). **(230)**

[133] E.V. Ferapontov and W.K. Schief, Surfaces of Demoulin: differential geometry, Bäcklund transformation and integrability, *J. Geom. Phys.* **30**, 343–363 (1999). **(329, 354)**

[134] R.P. Feynman, R.B. Leighton and M. Sands, *The Feynman Lectures on Physics*, Vol. II, Addison-Wesley (1964). **(105)**

[135] S.P. Finikov, *Projective-Differential Geometry*, Moscow-Leningrad (1937). **(135, 329, 330, 370)**

[136] S.P. Finikov, *Theory of Congruences*, Moscow-Leningrad (1950). **(117, 330, 370)**

[137] A. Fokas, A symmetry approach to exactly solvable evolution equations, *J. Math. Phys.* **21**, 1318–1325 (1980). **(240, 243)**

[138] A.S. Fokas and I.M. Gelfand, Surfaces on Lie groups, on Lie algebras and their integrability, *Comm. Math. Phys.* **177**, 203–220 (1996). **(208)**

[139] A.S. Fokas, I.M. Gelfand, F. Finkel and Q.M. Liu, A formula for constructing infinitely many surfaces on Lie algebras and Lie groups, to appear in *Selecta Math.* **(208)**

[140] A.P. Fordy and J. Gibbons, Integrable nonlinear Klein Gordon equations, *Commun. Math. Phys.* **77**, 21–30 (1980). **(91, 113)**

[141] M.V. Foursov, P.J. Olver, and E.G. Reyes, On formal integrability of evolution equations and local geometry of surfaces, *Diff. Geom. Appl.* **15**, 183–199 (2001). **(22)**

[142] B.D. Fried and Y.H. Ichikawa, On the nonlinear Schrödinger equation for Langmuir waves, *J. Phys. Soc. Japan* **33**, 789–792 (1972). **(119)**

[143] G. Fubini and E. Ĉech, *Geometria Proiettiva Differenziale*, Zanichelli, Bologna (1926). **(329, 330, 370)**

[144] B. Gaffet, $SU(3)$ symmetry of the equations of uni-dimensional gas flow, with arbitrary entropy distribution, *J. Math. Phys.* **25**, 245–255 (1984). **(88, 95)**

[145] B. Gaffet, An infinite Lie group of symmetry of one-dimensional gas flow for a class of entropy distributions, *Physica D* **11**, 287–308 (1984). **(88, 95)**

[146] B. Gaffet, An $SL(3)$-Symmetrical F-Gordon Equation: $z_{\alpha\beta} = \frac{1}{3}(e^z - e^{-2z})$,

Lecture Notes in Physics **246**, pp. 301–319, Springer Verlag, Berlin (1986). **(88, 95)**

[147] B. Gaffet, The non-isentropic generalisation of the classical theory of Riemann invariants, *J. Phys. A: Math. Gen.* **20**, 2721–2731 (1987). **(88, 95)**

[148] B. Gaffet, A class of 1-d gas flows soluble by the inverse scattering transform, *Physica D* **26**, 123–139 (1987). **(88, 95)**

[149] R. Geroch, A method for generating solutions of Einstein's equations. I/II, *J. Math. Phys.* **12**, 918–924 (1971); *J. Math. Phys.* **13**, 394–404 (1972). **(310)**

[150] H.M. Gibbs and R.E. Slusher, Peak amplification and pulse breakup of a coherent optical pulse in a simple atomic absorber, *Phys. Rev. Lett.* **24**, 638–641 (1970). **(22, 31)**

[151] D. Gilbarg, On the flow patterns common to certain classes of plane fluid motions, *J. Math. and Phys.* **26**, 137–142 (1947). **(120)**

[152] L. Godeaux, *La théorie des surfaces et l'espace réglé (Géometrie projective differentielle)*, Actualités scientifiques et industrielles, N138, Hermann, Paris (1934). **(329, 334)**

[153] B. Grammaticos, V. Papageorgiu and A. Ramani, KdV equations and integrability detectors, *Acta Appl. Math.* **39**, 335–348 (1995). **(236)**

[154] R. Grimshaw, Slowly varying solitary waves: II, Nonlinear Schrödinger equation, *Proc. R. Soc. Lond. A* **368**, 377–388 (1979). **(119)**

[155] C. Gu, H. Hu and Z. Zhou, Darboux Transformation in Soliton Theory and its Geometric Applications, Shanghal Scientific & Technical Publishers (1999). **(266)**

[156] A. Haar, Über adjungierte Variationsprobleme und adjungierte Extremalflächen, *Math. Ann.* **100**, 481–502 (1928). **(229)**

[157] B.K. Harrison, Bäcklund transformation for the Ernst equation of general relativity, *Phys. Rev. Lett.* **41**, 1197–1200 (1978). **(297, 305, 311, 317)**

[158] B.K. Harrison, Unification of Ernst equation Bäcklund transformations using a modified Wahlquist-Estabrook technique, *J. Math. Phys.* **24**, 2178–2187 (1983). **(374)**

[159] I. Hauser and F.J. Ernst, A homogeneous Hilbert problem for the Kinnersley-Chitre transformations, *J. Math. Phys.* **21**, 1126–1140 (1980). **(305)**

[160] I. Hauser and F. Ernst, Proof of a Geroch conjecture, *J. Math. Phys.* **22**, 1051–1063 (1981). **(311)**

[161] A. Hasegawa and F. Tappert, Transmission of stationary nonlinear optical pulses in dispersive dielectric fibers: I. Anomolous dispersion, *Appl. Phys. Lett.* **23**, 142–144 (1973). **(119)**

[162] H. Hasimoto, A soliton on a vortex filament, *J. Fluid. Mech.* **51**, 477–485 (1972). **(60, 120)**

[163] H. Hasimoto and H. Ono, Nonlinear modulation of gravity waves, *J. Math. Soc. Japan* **33**, 805–811 (1972). **(119)**

[164] R. Hermann, *The Geometry of Nonlinear Differential Equations, Bäcklund Transformations and Solitons*, Part A, Math. Sci. Press, Brookline, Mass. (1976). **(111)**

[165] U. Hertrich-Jeromin and F. Pedit, Remarks on the Darboux transform of isothermic surfaces, *Doc. Math.* **2**, 313–333 (1997). **(171)**

[166] R. Hirota, Exact solution of the Korteweg-de Vries equation for multiple collisions of solitons, *Phys. Rev. Lett.* **27**, 1192–1194 (1971). **(198)**

[167] R. Hirota and J. Satsuma, A simple structure of superposition formula of the Bäcklund transformation, *J. Phys. Soc. Japan* **45**, 1741–1750 (1978). **(79)**

[168] C. Hoenselaers, HKX transformations. An introduction, in C. Hoenselaers and

W. Dietz, eds, *Solutions of Einstein's Equations: Techniques and Results*, Lecture Notes in Physics, pp. 68–84, Springer-Verlag, Berlin (1984). **(311)**

[169] C. Hoenselaers, The sine-Gordon prolongation algebra, *Progr. Theor. Phys.* **74**, 645–654 (1985). **(249, 374)**

[170] C. Hoenselaers, More prolongation structures, *Progr. Theor. Phys.* **75**, 1014–1029 (1986). **(249, 374)**

[171] C. Hoenselaers, Equations admitting $o(2, 1) \times R(t, t^{-1})$ as a prolongation algebra, *J. Phys. A: Math. Gen.* **21**, 17–31 (1988). **(249, 374)**

[172] C. Hoenselaers and W. Dietz, eds, *Solutions of Einstein's Equations: Techniques and Results*, Lecture Notes in Physics, Springer Verlag, Berlin (1984). **(297)**

[173] C. Hoenselaers and W.K. Schief, Prolongation structures for Harry Dym type equations and Bäcklund transformations of cc-ideals, *J. Phys. A: Math. Gen.* **25**, 601–622 (1992). **(249, 374)**

[174] A.N.W. Hone, The associated Camassa-Holm equation and the KdV equation, *J. Phys. A: Math. Gen.* **32**, L307–L314 (1999). **(230)**

[175] L.N. Howard, *Constant Speed Flows*, PhD Thesis, Princeton University (1953). **(120)**

[176] R.W.H.T. Hudson, *Kummer's Quartic Surface*, Cambridge University Press (1990). **(330)**

[177] N. Ibragimov, Sur l'équivalence des équations d'évolution qui admettent une algèbre de Lie-Bäcklund infinie, *C.R. Acad. Sci. Paris* **293**, 657–660 (1981). **(234)**

[178] Y.H. Ichikawa, T. Imamura and T. Tanuiti, Nonlinear wave modulation in collisionless plasma, *J. Phys. Soc. Japan* **33**, 189–197 (1972). **(119)**

[179] N. Jacobson, *Lie algebras*, Dover Publications, Inc., New York (1962). **(113)**

[180] A. Jeffrey, Equations of evolution and waves, in C. Rogers and T.B. Moodie, eds, *Wave Phenomena: Modern Theory and Applications*, North Holland, Amsterdam (1986). **(226)**

[181] M.E. Johnston, *Geometry and the Sine Gordon Equation*, M.Sc. Thesis, University of New South Wales (1994). **(40, 83)**

[182] M.E. Johnston, C. Rogers, W.K. Schief and M.L. Seiler, On moving pseudospherical surfaces: a generalised Weingarten system, *Lie Groups and Their Applications* **1**, 124–136 (1994). **(72)**

[183] H. Jonas, Über die Transformation der konjugierten Systeme und über den gemeinsamen Ursprung der Bianchischen Permutabilitätstheoreme, *Sitzungsberichte Berl. Math. Ges.* **14**, 96–118 (1915). **(89, 167, 180)**

[184] H. Jonas, Sopra una classe di transformazioni asintotiche, applicabili in particolare alle superficie la cui curvatura è proporzionale alla quarta potenza della distanza del piano tangente da un punto fisso, *Ann. Mat. Pura Appl. Bologna Ser. III* **30**, 223–255 (1921). **(88)**

[185] H. Jonas, Die Differentialgleichung der Affinsphären in einer neuen Gestalt, *Math. Nachr.* **10**, 331–361 (1953). **(88, 92, 93, 94, 100)**

[186] V.G. Kac, *Infinite Dimensional Lie Algebras*, Cambridge University Press (1985). **(113, 250)**

[187] L.P. Kadanoff, Exact solutions for the Saffman-Taylor problem with surface tension, *Phys. Rev. Lett.* **65**, 2986–2988 (1986). **(239)**

[188] T. Kakutani and H. Ono, Weak nonlinear hydromagnetic waves in cold collisionless plasma, *J. Phys. Soc. Japan* **26**, 1305–1318 (1969). **(71)**

[189] T. Kambe and T. Takao, Motion of distorted vortex rings, *J. Phys. Soc. Japan* **31**, 591–599 (1971). **(60)**

[190] N. Kamran and K. Tenenblat, On differential equations describing pseudo-spherical surfaces, *J. Diff. Eq.* **115**, 75–98 (1995). **(22)**

[191] V.I. Karpman and E.M. Kruskal, Modulated waves in nonlinear dispersive media, *Sov. Phys. JETP* **28**, 277–281 (1969). **(119)**

[192] D.J. Kaup, The method of solution for stimulated Raman scatttering and two-photon propagation, *Physica D* **6**, 143–154 (1983). **(130)**

[193] J.P. Keener and J.J. Tyson, The dynamics of scroll waves in excitable media, *SIAM Rev.* **38**, 1–39 (1992). **(120)**

[194] P.L. Kelley, Self focussing of optic beams, *Phys. Rev. Lett.* **15**, 1005–1008 (1965). **(119)**

[195] J.G. Kingston and C. Rogers, Reciprocal Bäcklund transformations of conservation laws, *Phys. Lett. A* **92**, 261–264 (1982). **(230, 378)**

[196] J.G. Kingston, C. Rogers and D. Woodall, Reciprocal auto-Bäcklund transformations, *J. Phys. A: Math. Gen.* **17**, L35–L38 (1984). **(230, 243)**

[197] W. Kinnersley, Symmetries of the stationary Einstein-Maxwell field equations I, *J. Math. Phys.* **18**, 1529–1537 (1977). **(311)**

[198] W. Kinnersley and D.M. Chitre, Symmetries of the stationary Einstein-Maxwell field equations II, *J. Math. Phys.* **18**, 1538–1542 (1978). **(311)**

[199] W. Kinnersley and D.M. Chitre, Symmetries of the Einstein-Maxwell field equations III, *J. Math. Phys.* **19**, 1926–1931 (1978). **(305)**

[200] P. Klimczewski, M. Nieszporski and A. Sym, Luigi Bianchi, Pasquale Calapso and solitons, *Preprint Instytut Fizyki Teoretycznej*, Uniwersytet Warszawski (2000). **(152)**

[201] A. Kochendörfer and A. Seeger, Theorie der Versetzungen in eindimensionalen Atomreihen I. Periodisch angeordnete Versetzungen, *Z. Phys.* **127**, 533–550 (1950). **(21)**

[202] K. Konno and A. Jeffrey, Some remarkable properties of two loop soliton solutions, *J. Phys. Soc. Japan* **52**, 1–3 (1983). **(226)**

[203] K. Konno, W. Kameyama and H. Sanuki, Effect of weak dislocation potential on nonlinear wave equation in an anharmonic crystal, *J. Phys. Soc. Japan* **37**, 171–176 (1974). **(71)**

[204] K. Konno and H. Sanuki, Bäcklund transformation for equation of motion for nonlinear lattice under weak dislocation potential, *J. Phys. Soc. Japan* **39**, 22–24 (1975). **(78)**

[205] B.G. Konopelchenko, Elementary Bäcklund transformations, nonlinear superposition principles and solutions of the integrable equations, *Phys. Lett. A* **87**, 445–448 (1982). **(237)**

[206] B.G. Konopelchenko, Soliton eigenfunction equations: the IST integrability and some properties, *Rev. Math. Phys.* **2**, 399–440 (1990). **(204, 217)**

[207] B.G. Konopelchenko, The non-Abelian (1+1)-dimensional Toda lattice as the periodic fixed point of the Laplace transform for (2+1)-dimensional integrable systems, *Phys. Lett. A* **156**, 221–222 (1991). **(118)**

[208] B.G. Konopelchenko, Induced surfaces and their integrable dynamics, *Stud. Appl. Math.* **96**, 9–51 (1996). **(208)**

[209] B.G. Konopelchenko and U. Pinkall, Integrable deformations of affine surfaces via the Nizhnik-Veselov-Novikov equation, *Phys. Lett. A* **245**, 239–245 (1998). **(88)**

[210] B.G. Konopelchenko and C. Rogers, On a 2+1-dimensional nonlinear system of Loewner-type, *Phys. Lett. A* **152**, 391–397 (1991). **(64, 99)**

[211] B.G. Konopelchenko and C. Rogers, On generalised Loewner systems: novel integrable equations in 2+1 dimensions, *J. Math. Phys.* **34**, 214–242 (1993). **(64, 99)**

[212] B.G. Konopelchenko and W.K. Schief, Lamé and Zakharov-Manakov systems: Combescure, Darboux and Bäcklund transformations, *Preprint AM 93/9 Department of Applied Mathematics*, The University of New South Wales (1993). **(167)**

[213] B.G. Konopelchenko and W.K. Schief, Three-dimensional integrable lattices in Euclidean spaces: conjugacy and orthogonality, *Proc. R. Soc. Lond. A* **454**, 3075–3104 (1998). **(167)**

[214] B.G. Konopelchenko, W. Schief and C. Rogers, A 2+1-dimensional sine-Gordon system: its auto-Bäcklund transformation, *Phys. Lett. A* **172**, 39–48 (1992). **(110)**

[215] D. Kramer, *GR9 Abstracts* **1**, 42 (1980). **(319)**

[216] D. Kramer, Equivalence of various pseudopotential approaches for Einstein-Maxwell fields, *J. Phys. A: Math. Gen.* **15**, 2201–2207 (1982). **(305)**

[217] D. Kramer and G. Neugebauer, Zu axialsymmetrischen stationären Lösungen der Einsteinschen Feldgleichungen für das Vakuum, *Comm. Math. Phys.* **10**, 132–139 (1968). **(297, 308)**

[218] D. Kramer and G. Neugebauer, The superposition of two Kerr solutions, *Phys. Lett. A* **75**, 259–261 (1980). **(319)**

[219] D. Kramer and G. Neugebauer, Bäcklund transformations in general relativity, in C. Hoenselaers and W. Dietz, eds, *Solutions of Einstein's Equations: Techniques and Results*, Lecture Notes in Physics, pp. 1–25, Springer-Verlag, Berlin (1984). **(319)**

[220] D. Kramer, G. Neugebauer and T. Matos, Bäcklund transforms of chiral fields, *J. Math. Phys.* **32**, 2727–2730 (1991). **(305)**

[221] D. Kramer, H. Stephani, H. Herlt and M. MacCallum, *Exact Solutions of Einstein's Equations*, Cambridge University Press (1980). **(297, 309)**

[222] M. Lakshmanan, Th.W. Ruijgrok and C.J. Thompson, On the dynamics of a continuum spin system, *Physica A* **84**, 577–590 (1976). **(60, 128)**

[223] G.L. Lamb Jr., Analytical descriptions of ultrashort optical pulse propagation in a resonant medium, *Rev. Mod. Phys.* **43**, 99–124 (1971). **(22, 30)**

[224] G.L. Lamb Jr., Solitons on moving space curves, *J. Math. Phys.* **18**, 1654–1661 (1977). **(60, 61)**

[225] G.L. Lamb, *Elements of Soliton Theory*, John Wiley, New York (1980). **(148)**

[226] G. Lamé, *Leçons sur les coordonnées curvilignes et leurs diverses applications*, Mallet-Bechelier, Paris (1859). **(60)**

[227] E.P. Lane, *Projective Differential Geometry of Curves and Surfaces*, University of Chicago Press, Chicago (1932). **(109, 329, 330, 370)**

[228] P.D. Lax, Integrals of nonlinear equations of evolution and solitary waves, *Comm. Pure Appl. Math.* **21**, 467–490 (1968). **(217)**

[229] D. Levi, Nonlinear differential difference equations as Bäcklund transformations, *J. Phys. A: Math. Gen.* **14**, 1082–1098 (1981). **(237)**

[230] D. Levi and R. Benguria, Bäcklund transformations and nonlinear differential difference equations, *Proc. Natl. Acad. Sci. U.S.A.* **77**, 5025–5027 (1980). **(237)**

[231] D. Levi and O. Ragnisco, Bäcklund transformations for chiral field equations, *Phys. Lett. A* **87**, 381–384 (1982). **(270)**

[232] D. Levi, O. Ragnisco and A. Sym, Bäcklund transformation vs. the dressing method, *Lett. Nuovo Cimento* **33**, 401–406 (1982). **(266, 270)**

[233] D. Levi, O. Ragnisco and A. Sym, Dressing method vs. classical Darboux transformation, *Il Nuovo Cimento B* **83**, 34–42 (1984). **(266, 270)**

[234] D. Levi and A. Sym, Integrable systems describing surfaces of non-constant curvature, *Phys. Lett. A* **149**, 381–387 (1990). **(21, 54, 299)**

[235] T. Levi-Civita, Attrazione Newtoniana dei Tubi Sottili e Vortici Filiformi, *Ann. R. Scuola Norm. Sup. Pisa*, Zanichelli, Bologna (1932). **(60)**

[236] T. Lewis, Some special solutions of the equations of axially symmetric gravitational fields, *Proc. R. Soc. Lond. A* **136**, 176–192 (1932). **(309)**

[237] C. Loewner, A transformation theory of partial differential equations of gasdynamics, *NACA Technical Note* **2065**, 1–56 (1950). **(98, 229)**

[238] C. Loewner, Generation of solutions of systems of partial differential equations by composition of infinitesimal Bäcklund transformations, *J. Anal. Math.* **2**, 219–242 (1952). **(64)**

[239] L.G. Loitsyanskii, *Mechanics of Liquids and Gases*, International Series of Monographs in Aeronautics and Astronautics, Pergamon Press, New York (1966) (Translation Editor K. Stewartson). **(105)**

[240] F. Lund and T. Regge, Unified approach to strings and vortices with soliton solutions, *Phys. Rev. D* **14**, 1524–1535 (1976). **(120, 129, 204)**

[241] D. Maison, Are the stationary, axially symmetric Einstein equations completely integrable? *Phys. Rev. Lett.* **41**, 521–522 (1978). **(297, 305, 311)**

[242] M.H. Martin, A new approach to problems in two-dimensional flow, *Q. Appl. Math.* **8**, 137–350 (1951). **(96, 229)**

[243] M.H. Martin, The propagation of a plane shock into a quiet atmosphere, *Can. J. Math.* **5**, 37–39 (1953). **(96)**

[244] R.R. Martin, *Principal patches for computational geometry*, PhD Thesis, Cambridge University (1982). **(198)**

[245] R.R. Martin, J. de Pont and T.J. Sharrock, Cyclide surfaces in computer aided design, in J.A. Gregory, ed, *The Mathematics of Surfaces*, Oxford University Press (1986). **(198)**

[246] A.W. Marris, On motions with constant speed and streamline parameters, *Arch. Rat. Mech. Anal.* **90**, 1–14 (1985). **(120)**

[247] A.W. Marris and S.L. Passman, Vector fields and flows on developable surfaces, *Arch. Rat. Mech. Anal.* **32**, 29–86 (1969). **(120, 137, 138, 142, 144)**

[248] A.W. Marris and C.C. Wang, Solenoidal screw fields of constant magnitude, *Arch. Rat. Mech. Anal.* **39**, 227–244 (1970). **(140)**

[249] A. Masotti, Decomposizione intrinseca del vortice a sue applicazioni, *Instituto Lombardo di Scienze a Lettere Rendiconti (2)* **60**, 869–874 (1927). **(139)**

[250] Y. Matsuno, *Bilinear Transformation Method*, Academic Press (1984). **(198)**

[251] V.B. Matveev and M.A. Salle, *Darboux Transformations and Solitons*, Springer-Verlag, Berlin (1991). **(266, 270)**

[252] R.A. Matzner and C.W. Misner, Gravitational field equations for sources with axial symmetry and angular momentum, *Phys. Rev.* **154**, 1229–1232 (1967). **(310)**

[253] J.C. Maxwell, On the cyclide, *Q.J. Pure Appl. Math.* **9**, 111–126 (1868). **(198)**

[254] O. Mayer, Contribution à l'étude des surfaces minima projectives, *Bull. Sci. Math. Ser. 2* **56**, 146–168, 188–200 (1932). **(329)**

[255] S.L. McCall and E.L. Hahn, Self-induced transparency by pulsed coherent light, *Phys. Rev. Lett.* **18**, 908–911 (1967). **(130)**

[256] D. McLean, A method of generating surfaces as a composite of cyclide patches, *Comput. J.* **4**, 433–438 (1985). **(198)**

[257] R.I. McLachlan and H. Segur, A note on the motion of surfaces, *Phys. Lett. A* **194**, 165–172 (1994). **(68)**

[258] A.M. Meirmanov, V.V. Pukhnachov and S.I. Shmarev, *Evolution Equations and Lagrangian Coordinates*, de Gruyter, Berlin (1997). **(230)**

[259] A.V. Michailov, The reduction problem and the inverse scattering method, *Physica D* **3**, 73–117 (1981). **(88, 91, 105, 329)**

[260] L.M. Milne-Thomson, *Theoretical Hydrodynamics*, Macmillan & Company Ltd, London (1962). **(151)**

[261] F. Minding, Wie sich entscheiden lässt, ob zwei gegebene krumme Flächen aufeinander abwickelbar sind order nicht; nebst Bemerkungen über die Flächen von unverändlichem Krümmungsmasse, *J. für die reine und angewandte Mathematik* **18**, 297–302 (1838). **(17)**

[262] R.M. Miura, Korteweg-de Vries equation and generalizations: I. A remarkable explicit nonlinear transformation, *J. Math. Phys.* **9**, 1202–1204 (1968). **(217)**

[263] H. Motz, V.P. Pavlenko and J. Weiland, Acceleration and slowing down of nonlinear packets in a weakly nonuniform plasma, *Phys. Lett. A* **76**, 131–133 (1980). **(119)**

[264] Th. Moutard, Sur la construction des équations de la forme $\frac{1}{z}\frac{\partial^2 z}{\partial x \partial y} = \lambda(x, y)$ qui admettent une intégrale générale explicite, *J. l'Ecole Polytechn.*, Cahier **45**, 1–11 (1878). **(103, 266)**

[265] W.W. Mullins, Theory of thermal grooving, *J. Appl. Phys.* **28**, 333–339 (1957). **(232)**

[266] M.F. Natale and D.A. Tarzia, Explicit solutions to the two-phase Stefan problem for Storm-type materials, *J. Phys. A: Math. Gen.* **33**, 395–404 (2000). **(229)**

[267] G. Neugebauer, Bäcklund transformations of axially symmetric stationary gravitational fields, *J. Phys. A: Math. Gen.* **12**, L67–L70 (1979). **(297, 305)**

[268] G. Neugebauer, A general integral of the axially symmetric stationary Einstein equations, *J. Phys. A: Math. Gen.* **13**, L19–L21 (1980). **(319, 325)**

[269] G. Neugebauer and D. Kramer, Einstein-Maxwell solitons, *J. Phys. A: Math. Gen.* **16**, 1927–1936 (1983). **(270, 277)**

[270] G. Neugebauer and R. Meinel, General N-soliton solution of the AKNS class on arbitrary background, *Phys. Lett. A* **100**, 467–470 (1984). **(266, 270, 277, 279)**

[271] A.C. Newell, *Solitons in Mathematics and Physics*, SIAM, Philadelphia (1985). **(214)**

[272] F.W. Nijhoff, H.W. Capel, G.L. Wiersma and G.R.W. Quispel, Bäcklund transformations and three-dimensional lattice equations, *Phys. Lett. A* **105**, 267–272 (1984). **(237)**

[273] J.J.C. Nimmo and W.K. Schief, Superposition principles associated with the Moutard transformation: an integrable discretization of a 2+1-dimensional sine-Gordon system, *Proc. R. Soc. Lond. A* **453**, 255–279 (1997). **(105, 237)**

[274] J.J.C. Nimmo and W.K. Schief, An integrable discretization of a 2+1-dimensional sine-Gordon equation, *Stud. Appl. Math.* **100**, 295–309 (1998). **(105, 237)**

[275] J.J.C. Nimmo, W.K. Schief and C. Rogers, Termination of Bergman series. Connection to the B_n Toda system, *J. Eng. Math.* **36**, 137–148 (1999). **(98)**

[276] L.P. Nizhnik, Integration of multidimensional nonlinear equations by the inverse problem method, *Dokl. Akad. Nauk SSSR* **254**, 332–335 (1980). **(362)**

[277] K. Nomizu and T. Sasaki, *Affine Differential Geometry*, Cambridge University Press (1994). **(88)**

[278] A.W. Nutbourne and R.R. Martin, *Differential Geometry Applied to the Design of Curves and Surfaces*, Ellis Horwood, Chichester (1988). **(198, 245)**

[279] J. Nycander, D.G. Dritschel and G.G. Sutyrin, The dynamics of long frontal waves in the shallow-water equations, *Phys. Fluids A* **5**, 1089–1091 (1993). **(231)**

[280] W. Oevel and C. Rogers, Gauge transformations and reciprocal links in 2 + 1-dimensions, *Rev. Math. Phys.* **5**, 299–330 (1993). **(217, 239)**

[281] W. Oevel and W. Schief, Darboux theorems and the KP hierarchy, in

P.A. Clarkson, ed, *Applications of Analytic and Geometric Methods to Nonlinear Differential Equations*, pp. 192–206, Kluwer Academic Publishers, Dordrecht (1993). **(366)**

[282] F. Pempinelli, Localized soliton solutions for the Davey-Stewartson I and Davey-Stewartson III equations, in P.A. Clarkson, ed, *Applications of Analytic and Geometric Methods to Nonlinear Differential Equations*, pp. 207–215, Kluwer, Dordrecht (1993). **(196)**

[283] R. Perline, Localized induction equation and pseudospherical surfaces, *J. Phys. A: Math. Gen.* **27**, 5335–5344 (1994). **(87)**

[284] R. Perline, Localized induction hierarchy and Weingarten systems, *Phys. Lett. A* **220**, 70–74 (1996). **(87)**

[285] K. Pohlmeyer, Integrable Hamiltonian systems and iteractions through quadratic constraints, *Comm. Math. Phys.* **46**, 207–221 (1976). **(120, 129, 204)**

[286] G. Power and P. Smith, Reciprocal properties of plane gas flows, *J. Math. Mech.* **10**, 349–361 (1961). **(229)**

[287] M.J. Pratt, Cyclides in computer aided geometric design, *Computer Aided Geometric Design* **7**, 221–242 (1990). **(198)**

[288] R. Prim, On the uniqueness of flows with given streamlines, *J. Math. and Phys.* **28**, 50–53 (1949). **(120)**

[289] R.C. Prim, Steady rotational flow of ideal gases, *Arch. Rat. Mech. Anal.* **1**, 425–497 (1952). **(95)**

[290] R. Prus, *Geometry of Bianchi surfaces in E^3*, Master Thesis, Warsaw University (1995). **(57)**

[291] G.R.W. Quispel, F.W. Nijhoff, H.W. Capel and J. van der Linden, Linear integral equations and nonlinear difference-difference equations, *Physica A* **125**, 344–380 (1984). **(237)**

[292] M.L. Rabelo, On equations which describe pseudospherical surfaces, *Stud. Appl. Math.* **81**, 221–248 (1989). **(22)**

[293] A. Razzaboni, Delle superficie nelle quali un sistema di geodetiche sono del Bertrand, Bologna Mem (5) **10**, 539–548 (1903). **(245)**

[294] E.G. Reyes, Conservation laws and Calapso-Guichard deformations of equations describing pseudo-spherical surfaces, *J. Math. Phys.* **41**, 2968–2989 (2000). **(22)**

[295] C. Rogers, Reciprocal relations in non-steady one-dimensional gasdynamics, *Z. Angew. Math. Phys.* **19**, 58–63 (1968). **(223, 229)**

[296] C. Rogers, Invariant transformations in non-steady gasdynamics and magneto-gasdynamics, *Z. Angew. Math. Phys.* **20**, 370–382 (1969). **(229)**

[297] C. Rogers, The construction of invariant transformations in plane rotational gasdynamics, *Arch. Rat. Mech. Anal.* **47**, 36–46 (1972). **(229)**

[298] C. Rogers, Application of a reciprocal transformation to a two-phase Stefan problem, *J. Phys. A: Math. Gen.* **18**, L105–L109 (1985). **(229)**

[299] C. Rogers, On a class of moving boundary problems in nonlinear heat conduction: application of a Bäcklund transformation, *Int. J. Nonlinear Mech.* **21**, 249–256 (1986). **(229)**

[300] C. Rogers, On the Heisenberg spin equation in hydrodynamics, *Research Report, Inst. Pure Appl. Math.*, Rio de Janeiro, Brazil (2000). **(120, 151)**

[301] C. Rogers and P. Broadbridge, On a nonlinear moving boundary problem with heterogeneity: application of a Bäcklund transformation, *Z. Angew. Math. Phys.* **39**, 122–128 (1988). **(229)**

[302] C. Rogers and P. Broadbridge, On sedimentation in a bounded column, *Int. J. Nonlinear Mech.* **27**, 661–667 (1992). **(229)**

[303] C. Rogers and S. Carillo, On reciprocal properties of the Caudrey-Dodd-Gibbon and Kaup-Kuperschmidt hierarchies, *Physica Scripta* **36**, 865–869 (1987). **(239)**

[304] C. Rogers, S.P. Castell and J.G. Kingston, On invariance properties of conservation laws in non-dissipative planar magneto-gasdynamics, *J. de Mécanique* **13**, 243–354 (1974). **(229)**

[305] C. Rogers and J.G. Kingston, Non-dissipative magneto-hydrodynamic flows with magnetic and velocity field lines orthogonal geodesics, *Soc. Ind. Appl. Math. J. Appl. Math.* **26**, 183–195 (1974). **(137, 142)**

[306] C. Rogers and J.G. Kingston, Reciprocal properties in quasi one-dimensional non-steady oblique field magneto-gasdynamics, *J. de Mécanique* **15**, 185–192 (1976). **(229)**

[307] C. Rogers, J.G. Kingston and W.F. Shadwick, On reciprocal-type invariant transformations in magneto-gasdynamics, *J. Math. Phys.* **21**, 395–397 (1980). **(229)**

[308] C. Rogers and M.C. Nucci, On reciprocal Bäcklund transformations and the Korteweg-de Vries hierarchy, *Physica Scripta* **33**, 289–292 (1986). **(233)**

[309] C. Rogers, M.C. Nucci and J.G. Kingston, On reciprocal auto-Bäcklund transformations: application to a new nonlinear hierarchy, *Il Nuovo Cimento* **96**, 55–63 (1986). **(238)**

[310] C. Rogers and T. Ruggeri, A reciprocal Bäcklund transformation: application to a nonlinear hyperbolic model in heat conduction, *Lett. Nuovo Cimento* **44**, 289–296 (1985). **(229)**

[311] C. Rogers and W.F. Shadwick, *Bäcklund Transformations and Their Applications*, Academic Press, New York (1982). **(21, 31, 99, 198, 205)**

[312] C. Rogers and W.K. Schief, Intrinsic geometry of the NLS equation and its auto-Bäcklund transformation, *Stud. Appl. Math.* **26**, 267–287 (1998). **(137, 142, 146)**

[313] C. Rogers and W.K. Schief, On geodesic hydrodynamic motions, Heisenberg spin connections. *J. Math. Anal. Appl.* **251**, 855–870 (2000). **(120, 151)**

[314] C. Rogers, W.K. Schief and M.E. Johnston, Bäcklund and his works: applications in soliton theory, in *Geometric Approaches to Differential Equations*, P.J. Vassiliou and I.G. Lisle, eds, *Australian Mathematical Society Lecture Series* **15**, pp. 16–55, Cambridge University Press (2000). **(124)**

[315] C. Rogers, M.P. Stallybrass and D.L. Clements, On two-phase filtration under gravity and with boundary infiltration: application of a Bäcklund transformation, *J. Nonlinear Analysis, Theory, Methods and Applications* **7**, 785–799 (1983). **(229)**

[316] C. Rogers and P. Wong, On reciprocal Bäcklund transformations of inverse scattering schemes, *Physica Scripta* **30**, 10–14 (1984). **(224, 233)**

[317] C. Rogers and B. Yu Guo, A note on the onset of melting in a class of simple metals. Condition on the applied boundary flux, *Acta Math. Sci.* **8**, 425–430 (1988). **(229)**

[318] O. Rozet, Sur certaines congruences W attachée aux surfaces dont les quadriques de Lie n'ont que deux points characteristiques, *Bull. Sci. Math. II* **58**, 141–151 (1934). **(329, 334)**

[319] M.A. Salle, Darboux transformations for non-abelian and nonlocal equations of the Toda chain type, *Teoret. Mat. Fiz.* **53**, 227–237 (1982). **(270)**

[320] P.M. Santini and A.S. Fokas, Recursion operators and bi-Hamiltonian structures in multidimensions. I, *Comm. Math. Phys.* **115**, 375–419 (1988). **(163)**

[321] R. Sasaki, Soliton equations and pseudospherical surfaces, *Nucl. Phys. B* **154**, 343–357 (1979). **(22)**
[322] T. Sasaki, On a projectively minimal hypersurface in the unimodular affine space, *Geom. Dedicata* **23**, 237–251 (1987). **(329)**
[323] W.K. Schief, Bäcklund transformations for the (un)pumped Maxwell-Bloch system and the fifth Painlevé equation, *J. Phys. A: Math. Gen.* **27**, 547–557 (1994). **(249, 374)**
[324] W.K. Schief, On a 2+1-dimensional integrable Ernst-type equation, *Proc. R. Soc. Lond. A* **446**, 381–398 (1994). **(49)**
[325] W.K. Schief, Self-dual Einstein spaces via a permutability theorem for the Tzitzeica equation, *Phys. Lett. A* **223**, 55–62 (1996). **(91, 105, 237)**
[326] W.K. Schief, On the geometry of an integrable (2+1)-dimensional sine-Gordon system, *Proc. R. Soc. Lond. A* **453**, 1671–1688 (1997). **(86)**
[327] W.K. Schief, Self-dual Einstein spaces and a discrete Tzitzeica equation. A permutability theorem link, in P.A. Clarkson and F.W. Nijhoff, eds, *Symmetries and Integrability of Difference Equations*, London Mathematical Society, Lecture Note Series *255*, pp. 137–148, Cambridge University Press (1999). **(91, 105, 237)**
[328] W.K. Schief, Integrable discretization of geodesics of constant torsion and pseudospherical surfaces, *in preparation* (2002). **(263)**
[329] W.K. Schief, The Painlevé III, V and VI transcendents as solutions of the Einstein-Weyl equations, *Phys. Lett. A* **267**, 265–275 (2000). **(45)**
[330] W.K. Schief, Hyperbolic surfaces in centro-affine geometry. Integrability and discretization, *Chaos, Solitons and Fractals* **11**, 97–106 (2000). **(88, 105)**
[331] W.K. Schief, Isothermic surfaces in spaces of arbitrary dimension: integrability, discretization and Bäcklund transformations. A discrete Calapso equation, *Stud. Appl. Math.* **106**, 85–137 (2001). **(163, 171, 172, 176, 183, 184, 188, 190, 237)**
[332] W.K. Schief, On Laplace-Darboux-type sequences of generalized Weingarten surfaces, *J. Math. Phys.* **41**, 6566–6599 (2000). **(45, 118)**
[333] W.K. Schief, On the geometry of the Painlevé V equation and a Bäcklund transformation, to appear in *The ANZIAM J. (J. Austral. Math. Soc.)* (2002). **(45, 118)**
[334] W.K. Schief, On the integrability of geodesic Bertrand curves, *in preparation* (2002). **(245)**
[335] W.K. Schief, Nested toroidal surfaces in magnetohydrostatics. Generation via soliton theory, *in preparation* (2002). **(120)**
[336] W.K. Schief and C. Rogers, The affinsphären equation. Moutard and Bäcklund transformations, *Inverse Problems* **10**, 711–731 (1994). **(88, 91, 95, 98, 249, 374)**
[337] W.K. Schief and C. Rogers, On a Laplace sequence of nonlinear integrable Ernst-type equations, *Prog. Nonlinear Diff. Eq.* **26**, 315–321 (1996). **(118)**
[338] W.K. Schief and C. Rogers, Loewner transformations: adjoint and binary Darboux connections, *Stud. Appl. Math.* **100**, 391–422 (1998).**(98)**
[339] W.K. Schief and C. Rogers, Binormal motion of curves of constant curvature and torsion. Generation of soliton surfaces, *Proc. R. Soc. Lond. A* **455**, 3163–3188 (1999). **(240, 242, 249, 253, 260, 261, 374)**
[340] W.K. Schief, C. Rogers and S.P. Tsarev, On a 2+1-dimensional Darboux system: integrable and geometric connections, *Chaos, Solitons and Fractals* **5**, 2357–2366 (1995). **(110)**
[341] B.G. Schmidt, The Geroch group is a Banach Lie group, in C. Hoenselaers and

W. Dietz, eds, *Solutions of Einstein's Equations: Techniques and Results*, Lecture Notes in Physics, pp. 113–127, Springer-Verlag, Berlin (1984). **(311)**

[342] E.I. Schulman, On the integrability of equations of Davey-Stewartson type, *Math. Theor. Phys.* **56**, 720–724 (1984). **(163)**

[343] B.F. Schutz, *Geometric Methods of Mathematical Physics*, Cambridge University Press, Cambridge (1980). **(65)**

[344] A.C. Scott, Propagation of magnetic flux on a long Josephson junction, *Il Nuovo Cimento B* **69**, 241–261 (1970). **(22)**

[345] A. Seeger, H. Donth and A. Kochendörfer, Theorie der Versetzungen in eindimensionalen Atomreihen III. Versetzungen, Eigenbewegungen und ihre Wechselwirkung, *Z. Phys.* **134**, 173–193 (1953). **(21, 22, 30)**

[346] A. Seeger and A. Kochendörfer, Theorie der Versetzungen in eindimensionalen Atomreihen II. Beliebig angeordnete und beschleunigte Versetzungen, *Z. Phys.* **130**, 321–336 (1951). **(21)**

[347] T.J. Sharrock, Surface design with cyclide patches, *PhD Thesis*, Cambridge University (1985). **(198)**

[348] K. Shimuzu and Y.H. Ichikawa, Automodulation of ion oscillation modes in plasma, *J. Phys. Soc. Japan* **33**, 789–792 (1972). **(119)**

[349] H. Steudel, Space-time symmetry of self-induced transparency and of stimulated Raman scattering, *Phys. Lett. A* **156**, 491–492 (1991). **(120, 130, 134)**

[350] H. Steudel, Solitons in stimulated Raman scattering and resonant two-photon propagation, *Physica D* **6**, 155–178 (1983). **(120, 130)**

[351] R. Steuerwald, Über die Enneper'sche Flächen und Bäcklund'sche Transformation, *Abh. Bayer. Akad. Wiss.* **40**, 1–105 (1936). **(40)**

[352] D.J. Struick, *Lectures on Classical Differential Geometry*, 2nd ed, Addison-Wesley Publishing Company, Inc., Reading, Mass. (1961). **(18, 32, 63)**

[353] A. Sym, Soliton surfaces, *Lett. Nuovo Cimento* **33**, 394–400 (1982). **(204, 210)**

[354] A. Sym, Soliton surfaces II. Geometric unification of solvable nonlinearities, *Lett. Nuovo Cimento* **36**, 307–312 (1983). **(286, 292)**

[355] A. Sym, Soliton surfaces V. Geometric theory of loop solitons, *Lett. Nuovo Cimento* **41**, 33–40 (1984). **(222, 227)**

[356] A. Sym, Soliton surfaces and their applications, in R. Martini, ed, *Geometric Aspects of the Einstein Equations and Integrable Systems*, Springer, Berlin (1985). **(124, 204, 208, 210, 304)**

[357] M. Tabor, Painlevé property for partial differential equations, in A.P. Fordy, ed, *Soliton Theory: A Survey of Results*, pp. 427–446, Manchester University Press (1990). **(234)**

[358] J. Tafel, Surfaces in $\mathbb{R}^3$ with prescribed curvature, *J. Geom. Phys.* **294**, 1–10 (1995). **(210, 304)**

[359] L.A. Takhtajan, Integration of the continuous Heisenberg spin chain through the inverse scattering method, *Phys. Lett. A* **64**, 235–237 (1977). **(128)**

[360] V.I. Talanov, Self focussing of wave beams in nonlinear media, *JETP Lett. Engl. Transl.* **2**, 138–141 (1965). **(119)**

[361] T. Taniuki and H. Washimi, Self trapping and instability of hydromagnetic waves along the magnetic field in a cold plasma, *Phys. Rev. Lett.* **21**, 209–212 (1968). **(119)**

[362] B. Temple, Systems of conservation laws with invariant submanifolds, *Trans. Am. Math. Soc.* **280**, 781–795 (1983). **(230)**

[363] K. Tenenblat, *Transformations of Manifolds and Applications to Differential*

Equations, Pitman Monographs and Surveys in Pure and Applied Mathematics **93**, Longman, Harlow (1998). **(22)**

[364] G. Thomsen, Über eine liniengeometrische Behandlungsweise der projektiven Flächentheorie und die projektive Geometrie der Systeme von Flächen zweiter Ordnung, *Abhandl. Math. Sem. Hamburg* **4**, 232–266 (1926). **(329)**

[365] G. Thomsen, Sulle superficie minime proiettive, *Ann. Math.* **5**, 169–184 (1928). **(329)**

[366] P. Tritscher and P. Broadbridge, Grain boundary grooving by surface diffusion: an analytic nonlinear model for a symmetric groove, *Proc. R. Soc. Lond. A* **450**, 569–587 (1995). **(232)**

[367] H.S. Tsien, Two dimensional subsonic flow of compressible fluids, *J. Aeronaut. Sci.* **6**, 399–407 (1939). **(229)**

[368] T. Tsuzuki, Nonlinear waves in the Pitaevsky-Gross equation, *J. Low Temp. Phys.* **4**, 441–457 (1971). **(119)**

[369] G. Tzitzeica, Sur une nouvelle classe de surfaces, *C.R. Acad. Sci. Paris* **144**, 1257–1259 (1907); sur une classe de surfaces, *C.R. Acad. Sci. Paris* **146**, 165–166 (1908). **(88)**

[370] G. Tzitzeica, Sur une nouvelle classe de surfaces, *C.R. Acad. Sci. Paris* **150**, 955–956, 1227–1229 (1910). **(88)**

[371] A.P. Veselov and S.P. Novikov, Finite-gap two-dimensional potential Schrödinger operators. Explicit formulas and evolution equations, *Dokl. Akad. Nauk SSSR* **279**, 20–24 (1984). **(362)**

[372] A. Voss, Encyclopädie der mathematischen Wissenschaften, Bd. III, DGa, Leipzig (1902). **(152)**

[373] M.G. Vranceanu, Les éspaces non-holonomes et leurs applications mécaniques, *Mém. Sci. Mathém.* **76**, 1–70 (1936). **(140)**

[374] M. Wadati, Wave propagation in nonlinear lattice: I, *J. Phys. Soc. Japan* **38**, 673–680 (1975). **(79)**

[375] M. Wadati, H. Sanuki and K. Konno, Relationships among inverse method, Bäcklund transformation and an infinite number of conservation laws, *Progr. Theor. Phys.* **53**, 419–436 (1975). **(266)**

[376] M. Wadati, K. Konno and Y.H. Ichikawa, New integrable nonlinear evolution equations, *J. Phys. Soc. Japan*, **47**, 1698–1700 (1979). **(205, 224, 225)**

[377] H.D. Wahlquist and F.B. Estabrook, Bäcklund transformations for solutions of the Korteweg-de Vries equation, *Phys. Rev. Lett.* **31**, 1386–1390 (1973). **(236)**

[378] H.D. Wahlquist and F.B. Estabrook, Prolongation structures of nonlinear evolution equations, *J. Math. Phys.* **16**, 1–7 (1975). **(261, 312, 374)**

[379] R.H. Wasserman, On a class of three-dimensional compressible fluid flows, *J. Math. Anal. Appl.* **5**, 119–135 (1962). **(120)**

[380] C.E. Weatherburn, *Differential Geometry of Three Dimensions*, Vol. I, Cambridge University Press (1927). **(127, 142, 143, 199, 245)**

[381] C.E. Weatherburn, *Differential Geometry of Three Dimensions*, Vol. II, Cambridge University Press (1930). **(128, 141, 142)**

[382] J. Weiss, On classes of integrable systems and the Painlevé property, *J. Math. Phys.* **25**, 13–24 (1984). **(239)**

[383] H. Weyl, Zur Gravitationstheorie, *Ann. Phys.* **54**, 117–145 (1917). **(319)**

[384] E.I. Wilczynski, Projective-differential geometry of curved surfaces, *Trans. Am. Math. Soc.* **8**, 233–260 (1907); *Trans. Am. Math. Soc.* **9**, 79–120, 293–315 (1908); *Trans. Am. Math. Soc.* **10**, 176–200, 279–296 (1909). **(329, 331)**

[385] D. Wójcik and J. Cieśliński, eds, *Nonlinearity & Geometry*, Polish Scientific Publishers PWN, Warsaw (1998). **(17)**

[386] W.L. Yin and A.C. Pipkin, Kinematics of viscometric flow, *Arch. Rat. Mech. Anal.* **37**, 111–135 (1970). **(141)**

[387] H.C. Yuen and B.M. Lake, Nonlinear wave concepts applied to deep water waves, in K. Lonngren and A. Scott, eds, *Solitons in Action*, Academic Press, New York (1978). **(119)**

[388] N.J. Zabusky, A synergetic approach to problems of nonlinear dispersive wave propagation, in W.F. Ames, ed, *Nonlinear Partial Differential Equations*, Academic Press, New York (1967). **(71)**

[389] N.J. Zabusky and M.D. Kruskal, Interaction of 'solitons' in a collisionless plasma and recurrence of initial states, *Phys. Rev. Lett.* **15**, 240–243 (1965). **(22)**

[390] V.E. Zakharov, Stability of periodic waves of finite amplitude on the surface of a deep fluid, *J. Appl. Mech. Tech. Phys.* **9**, 86–94 (1968). **(119)**

[391] V.E. Zakharov, Description of the n-orthogonal curvilinear coordinate systems and Hamiltonian integrable systems of hydrodynamic type. I. Integration of the Lamé equations, *Duke Math. J.* **94**, 103–139 (1998). **(61)**

[392] V.E. Zakharov and S.V. Manakov, Construction of multidimensional nonlinear integrable systems and their solutions, *Funct. Anal. Pril.* **19**, 11–25 (1985). **(110)**

[393] V.E. Zakharov, S.V. Manakov, S.P. Novikov and L.P. Pitaevskii, *The Theory of Solitons: The Inverse Problem Method* [in Russian], Nauka, Moscow (1980). **(266, 270)**

[394] V.E. Zakharov and A.V. Mikhailov, Relativistically invariant two-dimensional models of field theory which are integrable by means of the inverse scattering method, *Sov. Phys. JETP* **47**, 1017–1027 (1978). **(48)**

[395] V.E. Zakharov and A.B. Shabat, A scheme for integrating the nonlinear equations of mathematical physics by the method of the inverse scattering transform, *Funct. Anal. Appl.* **8**, 226–235 (1974). **(266, 270)**

Subject Index

abnormality, 139, 140
  vanishing, 142
adjoint eigenfunctions, 172, 177, 348, 353
  and parallel nets, 173, 174, 178, 179
adjoint relations, 229
adjoint triad, 91
Adkins reciprocal principle, 97
affine geometry, 8, 88
  and Tzitzeica surfaces, 91
affine Lie algebra
  and the Toda lattice, 343
affine normal, 91, 99, 100
affinsphäre, 96, 101, 370
  and the Tzitzeica equation, 8, 108
  and Tzitzeica surfaces, 88, 91
  the Lelieuvre formulae, 93
  in projective geometry, 335
affinsphären equation, 8, 88, 93
  in gasdynamics, 89, 95–97, 101
affine transformation, 88
AKNS hierarchy, 288, 293, 294
AKNS systems, 14, 204, 210, 211, 267, 275, 281, 286, 292
  linked WKI system, 224, 233
  matrix Darboux transformation, 276, 277, 279
  mKdV hierarchy, 261
  NLS eigenfunction hierarchy, 12
  NLS equation, 10, 136, 137, 213, 214
  NLS hierarchy, 13, 210–216, 221
  scattering problem, 210, 211, 218, 220, 229, 261, 277, 281, 282
  sine-Gordon equation, 7, 11, 64–68, 205–207, 209, 211, 267, 379
  ZS-AKNS system, 4, 5
Alfvén wave, 7, 71
anharmonic lattice, 7
  and the KdV equation, 2
  and motion of a pseudospherical surface, 68, 71
  Bäcklund transformation, qv
asymptotic coordinates, 58, 100, 330, 331, 358
  in Lelieuvre formulae, 51
  on Bianchi surfaces, 55
  on hyperbolic surfaces, 5, 45, 46, 47, 49, 52, 53, 89
  on isothermal-asymptotic surfaces, 367
  on pseudospherical surfaces, 3, 17, 20, 23, 25, 63, 65, 68, 205, 206, 208, 209
  on Tzitzeica surfaces, 90, 91
asymptotic lines
  and the Plücker correspondence, 341
  and the Wilczynski tetrahedral, 336
  on hyperbolic surfaces, 47
  on pseudospherical surfaces, 27, 63, 68, 215

Bäcklund parameter, 17, 22, 30, 185, 262
  and breathers, 38, 107
  in ϵ-algorithm, 237
  in Lie group invariance, 22, 26, 27, 235
  in gauge transformation, 75
  in iterated matrix Darboux transformations, 287
  time-dependent, 76, 78, 80, 85
Bäcklund transformations (BTs)
  AKNS hierarchy, 294
  and reciprocal invariance, 222
  and the dressing method, 266
  anharmonic lattice, 73, 77, 79, 80
  as conjugation of Bianchi and Lie transformations, 22, 26
  generation of multi-solitons by, 28–30
  generation of N-loop solitons by, 226, 277

Bäcklund transformations (BTs) (*contd.*)
  nonlinear superposition principles
    (*see* permutability theorems)
  Bianchi permutability theorem, 28–30
  Bianchi surfaces, 7, 14, 46, 53–55, 297, 300, 303, 311
  Bianchi system, 7, 15, 45, 53, 327
  classical, 5, 13, 17
  conjugate nets, 109
  connection with matrix Darboux transformations, 13, 267–276
  Demoulin surfaces, 329, 353
  Demoulin system, 16, 353–354
  Dupin cyclides, 198, 199, 202
  Einstein's equations, 297
  Ernst equation, 14, 15, 297, 311, 374
  extended Dym equation, 13, 243, 245, 252–255
  extended sine-Gordon system, 13, 262–265
  generation of multi-solitons by, 28–30
  generation of N-loop solitons by, 226, 277
  Godeaux-Rozet surfaces, 329
  Harrison transformation, 15, 297, 311, 318, 319, 322, 327, 374
  hyperbolic surfaces, 7, 49–53
  in continuum mechanics, 5, 99
  in crystal dislocation theory, 3, 21, 30
  in gasdynamics, 98, 99, 229
  in ultrashort optical pulse propagation, 3, 22, 30, 31
  isothermal-asymptotic surfaces, 16, 358, 364, 365, 367, 369, 370
  isothermic surfaces, 11, 171–178, 184, 186, 187
  isothermic system, 154
  KdV hierarchy, 12, 235
  Lamé system, 61
  mKdV equation, 79
  m^2KdV equation, 243, 254
  Neugebauer transformation, qv
  NLS equation, 13, 120, 124, 137, 146–151, 284, 286
  NLS hierarchy, 13, 282–284
  NLS surfaces, 13, 146–151, 267, 271–276, 277
  permutability theorems, qv
  potential mKdV equation, 79
  potential KdV equation, 12, 236
  potential mKdV hierarchy, 284, 285
  projective-minimal surfaces, 16, 341, 343–352, 356
  pseudospherical surfaces, 2, 3, 5, 13, 17, 22, 37–41, 46, 73, 75, 258, 259, 262, 267–271, 277
  Razzaboni surfaces, 245
  sine-Gordon equation, 17, 22–41, 77, 78, 285
  stationary mNVN equation, 365–367
  Toda lattice equation, 114
  Tzitzeica equation, 8, 88, 91, 102
  Tzitzeica surfaces, 8, 46, 101–103, 109
  vector Calapso system, 11, 166, 188–191
  Weingarten surfaces, 43–44
  Weingarten system, 73, 75, 76

Beltrami-Enneper theorem, 63
Bergman series, 98
Bertrand curves, 245
Bianchi diagram, 28
  associated with the ϵ-algorithm, 237
  for Bianchi surfaces, 326
  for isothermic surfaces, 185
  for the Ernst equation, 325
  for the Fundamental Transformation, 180, 181
  for the sine-Gordon equation, 28, 29, 292
  generic, 292
Bianchi lattice, 30, 31
Bianchi quadrilateral, 11
  constant cross-ratio property, 181, 185, 186
  planarity and cyclicity properties, 181
Bianchi surfaces, 46, 53, 298–300, 306
  Bäcklund transformation, qv
  generation of, 55–57
  nonlinear superposition of unit normals, 327
  normal equation, 14, 326
  Sym-Tafel formula, qv
Bianchi system, 5, 6, 48, 53–57, 299, 302
  and the Ernst equation, 327
  Bäcklund transformation, qv
  linear representation, qv
  2+1-dimensional version, 49
Bianchi transformation, 3, 13, 22, 26, 27, 326–327
  and extended sine-Gordon surfaces, 263–265
bilinear form, 198
  for the Calapso system, 198, 203
bilinear operator formalism, 198
bilinear potential, 104, 166
  and invariance of NVN equation, 365–367
  and parallel nets, 345, 349
  and projective-minimal surfaces, 350
  in BT for the Demoulin system, 354
  in BT for isothermal-asymptotic surfaces, 369
  in the Fundamental Transformation, 172, 173, 179
  in the Moutard transformation, 104, 187, 188, 353
  in the permutability theorem for the vector Calapso system, 189
binary Darboux transformation
  as a Fundamental Transformation, 362

binormal motion
  generation of the NLS equation, 120–122, 220, 240
  of curves of constant curvature, 205, 240–258
  of curves of constant torsion, 205, 258–265
Bonnet surfaces, 10, 153
Bonnet's theorem, 18, 246
  concerning parallel surfaces, 43
boomeron equation, 155, 161, 164, 191
breathers, 38, 279
  and the NLS equation, 9, 120, 124
  and the sine-Gordon equation, 3, 6, 22, 30, 38–41
  and the Tzitzeica equation, 8, 107, 108
breather surfaces
  NLS, 124, 126
  pseudospherical, 39–41
  Tzitzeica, 108

Calapso equation, 11, 155, 162, 163, 199
  and constant mean curvature surfaces, 166, 187
  and Dupin cyclides, 200–202
  and the Enneper surface, 202, 203
  Christoffel transform, 156
  Darboux-Ribaucour transformation, 11, 191, 192
  Moutard transformation, 192–198
Camassa-Holm equation, 230, 239
capillarity model, 119
Cartan matrix, 113
Caudrey-Dodd-Gibbon hierarchy, 239
cc-ideals, 249, 374–379
  for the extended Dym and m^2KdV equations, 250–253
Christoffel symbols, 19, 47
Christoffel transform, 155, 160, 199
  and BT for isothermic surfaces, 176, 177
  and parallel nets, 184
  and the Enneper surface, 195, 201, 202
  and the vector Calapso system, 164
  as degenerate Ribaucour transform, 189
  of a minimal surface, 196
circles
  as lines of curvature on Dupin cyclides, 199, 200, 203
Clairin's method, 4
Combescure transformation, 168
  and the Fundamental Transformation, 4, 169, 170
commutation theorem, 317, 318
commuting flows, 214
compatible integrable systems, 10, 137
  and motion of soliton surfaces, 68, 130
  and NLS hierarchy, 212
computer-aided design, 198, 245
  and Dupin cyclides, 198
  and dual Bertrand curves, 245
conformal coordinates, 45, 152, 159
congruences, 230
  and systems of conservation laws, 230
conjugate coordinates, 157
  and the classical Darboux system, 110
  and the Fundamental Transformation, 171, 172
  and Laplace-Darboux transformations, 117
  on isothermic surfaces, 159
conjugate lines
  on Tzitzeica surfaces, 92
conjugate net equation, 167
  and the Combescure transformation, 167
  and the Fundamental Transformation, 11, 167, 170
  and the radial transformation, 168
  in projective space, 341, 345
conjugate nets, 8, 89, 109
  and Laplace-Darboux invariants, 342
  and Laplace-Darboux transformations, 9, 109, 110, 117, 118
  and the Combescure transformation, 167
  and the Fundamental transformation, 169
  and the Levy transform, 170, 171
  and the radial transformation, 168, 171
  and the Ribaucour transformation, 11, 175
  isometric deformation, 299
  parallel, 172, 175
  permutability theorem, qv
conservation laws, 129, 131, 132, 217, 218, 219, 244, 261, 309, 378
  and the Crum transformation, 266
  reciprocal transformations, 223, 230, 232, 239
constant length property, 14
  and BT for extended Dym surfaces, 205, 256
  and BT for extended sine-Gordon surfaces, 262
  and BT for NLS surfaces, 146, 272
  and BT for Weingarten surfaces, 44
  and the classical BT, 13, 23
  for matrix Darboux transformations, 285–286, 288, 291
constant curvature curves, 61
  motion of, 63, 64, 205, 240–258
constant mean curvature (CMC) surfaces, 42–43, 45, 156, 160
constant torsion curves, 61
  motion of, 62, 63, 68, 80, 81, 205, 258–265
constitutive laws
  in gasdynamics, 8, 96, 98
contact transformation
  and the Ernst equation, 15, 308

continuum mechanics
  BTs in, 99
  reciprocal transformations in, 229, 230
cross-ratio, 184
  constant property, 11, 181, 185, 186
  reality, 183, 187
Crum transformation, 5, 266
crystal dislocations, 3, 21, 22, 30
cubic surfaces, 330
curvature coordinates, 31, 36, 152
  and Dupin cyclides, 200
  and pseudospherical surfaces, 31, 32, 60
  the classical BT in, 37
  the Weingarten system in, 72
curvature nets, 11, 175

Darboux cubic form, 332, 357
Darboux matrix, 253, 254, 256, 266, 270, 288, 292, 296
  and the Ernst equation, 321, 322
  as gauge matrix, 270
Darboux system, 110
Darboux-Ribaucour transformation, 11, 191, 192
Darboux transformation, 115, 116
  and the Toda lattice equation, 116
  classical, 5, 13, 17, 171, 191, 266, 270
  conjugation with a Laplace-Darboux transformation, 115, 118
  matrix, qv
Da Rios equations, 60, 119, 120, 121, 145
Davey-Stewartson equation
  I 362, II 11, 163, III 11, 163, 191, 196, 203
Demoulin surfaces, 15, 16, 335, 339
  and periodic Godeaux sequences, 341, 342
  one-soliton, 353, 355–357
Demoulin system, 16, 335, 336
  as a periodic Toda lattice, 341–343
  BT, qv
developables, 142, 144
Dini surface, 35–36
  as a one-soliton surface, 6, 36
  motion of, 8, 80, 82–87
distance property, 14, 302–303
dressing method, 266
dromions
  and the Davey-Stewartson I equation, 362
  and the Davey-Stewartson III equation, 191, 196
dual affinsphären equation, 96
dual Bertrand curves, 245
dual extended Dym surface, 245, 246
dual extended sine-Gordon surface
  Bianchi-type transformation, 264, 265
dual Ernst equation, 297, 306–310, 312, 313, 319, 323–325
dual isothermic surfaces, 155, 159, 166
dual one-form, 374
dual (adjoint) triad, 91
Dupin cyclides, 11, 189, 191, 198–203
Dupin hypersurfaces
  and systems of hydrodynamic type 198, 230
Dupin's theorem, 141
Dym equation, 12, 239, 242
  invariance under reciprocal transformation, 12, 233, 234
Dym hierarchy, 12
  invariance under reciprocal transformation, 222, 228, 233, 234
  link with mKdV and KdV hierarchies, 205, 234, 235
  2+1-dimensional, 239

Ehlers transformation, 310
  and the Neugebauer transformation, 15, 313
  and the Weyl class, 319
eigenfunction equations, 204, 217–220, 222, 228
eigenfunction matrix, 207–209
eigenfunctions, 172, 269
  adjoint, qv
  and BT for the Demoulin system, 354
  and elementary matrix Darboux transformations, 277–279, 287, 288–292, 301
  and extended Dym surfaces, 246
  and generic permutability theorems, 292–294
  and isothermal-asymptotic surfaces, 369
  and NLS surfaces, 272, 275, 276
  and parallel conjugate nets, 178, 179
  and pseudospherical surfaces, 271
  and the classical Moutard transformation, 353
  and the Ernst equation, 306
  and the KdV hierarchy, 235
  and the mNVN equation, 365–367
  and the vector Calapso system, 165, 166, 188
Einstein's equations, 14, 297, 309, 319
Enneper surface, 40, 156, 195, 202, 203
Ernst equation, 6, 14, 15, 49, 297, 303–329
  adjoint, qv
  and cc-ideals, 374
  and Laplace-Darboux transformations, 118
  Ehlers transformation, qv
  Geroch transformation, qv
  Harrison transformation, qv
  Hoenselaers-Kinnersley-Xanthopoulos (HKX) transformation, qv
  Kinnersley-Chitre potentials, qv
  Kramer-Neugebauer transformations, qv
  Matzner-Misner transformation, qv
  Möbius invariance, 15, 313, 327

Möbius transformations, 309, 310, 322
Neugebauer transformation, qv
potential, 304, 306, 308, 319, 324, 325, 326
soliton surfaces, 210
Ernst-type systems, 318, 319, 324
Euler-Lagrange equations, 15, 329, 333
extended Dym (ED) equation, 13, 205, 242–259
and binormal motion, 242
and m^2KdV equation, 242, 243
BT, qv
cc-ideal formulation, 249–252
reciprocal invariance, 243, 245, 246
extended Dym surfaces, 13, 243, 244, 255–258
dual, qv
linear representation, qv
extended sine-Gordon equation, 264
extended sine-Gordon surfaces, 245, 259–265
dual, qv
linear representation, qv
extended sine-Gordon system, 13, 205, 260
BT, qv

Fermi-Pasta-Ulam problem, 1, 2
Fordy-Gibbons system, 113
Fundamental Transformation (FT), 11, 16, 89, 105, 167, 169, 170, 362, 365
and BT for projective-minimal surfaces, 343, 345–353
and conjugate nets, 171–173, 178
and the Ribaucour transformation, 173–175, 182
as conjugation of Combescure and radial transformations, 170
permutability theorem, qv

Galilean transformation, 12, 235
gasdynamics
affinsphären equation in, 8, 88, 89, 95–99
Loewner transformations in, 98, 99
Monge-Ampère equations in, 101, 229
reciprocal transformations in, 205, 229
gauge matrix, 270, 279, 292
gauge transformations, 7, 215, 222, 228, 270, 331, 343
acting on linear representations, 7, 12, 67, 73–75, 136, 148, 216, 221, 273, 276
and Ernst potential, 306
and generation of NLS eigenfunction hierarchy, 12, 216, 221
composition with reciprocal transformations, 224, 234, 239
conjugation with matrix Darboux transformation, 279
induction of BTs, 7, 75
Miura transformations as, 217
of Laplace-Darboux invariants, 110, 111
of projective Gauss-Weingarten equations, 332
Gaussian (total) curvature, 2, 6, 8, 10, 14, 19, 20, 89, 131, 143
in motion of pseudospherical surfaces, 68, 71, 80
in terms of principal curvatures, 10, 41
in the Beltrami-Enneper theorem, 63
of Bianchi surfaces, 45, 53, 303, 327
of Hasimoto (NLS) surfaces, 126, 214
of hyperbolic surfaces, 46, 51
of pseudospherical surfaces, 21, 33, 205, 268
of Tzitzeica surfaces, 90, 100
of Weingarten surfaces, 42, 43, 44
Gauss equations, 18, 144
and conjugate nets, 117
and Pohlmeyer-Lund-Regge model, 130
and the Tzitzeica transformation, 103, 105
compatibility, 19
connection with soliton theory, 1
for hyperbolic surfaces, 89
for pseudospherical surfaces, 17, 21, 27
for Tzitzeica surfaces, 90, 91, 101–104, 109
Gauss-Mainardi-Codazzi equations, 3, 5, 19, 20, 143, 145, 158, 162, 241, 267, 371
and involution, 176
and the AKNS system, 207, 210
and the classical Bianchi system, 6, 20, 21, 48, 53
and the Da Rios system, 145
and the Fundamental Transformation, 175
and the Pohlmeyer-Lund-Regge model, 131, 204
and the Ribaucour transformation, 175
for extended sine-Gordon surfaces, 259
for hyperbolic surfaces, 3, 5, 46, 50, 53, 56, 89
for isothermal-asymptotic surfaces, 16, 357, 361
for isothermic surfaces, 10, 154
for NLS surfaces, 372
for projective-minimal surfaces, 335, 339, 349
for pseudospherical surfaces, 3, 21
projective, 15, 331, 333, 334, 358, 364
Gauss-Weingarten equations, 5, 18, 158, 159, 162, 172, 267, 371
and the AKNS system, 205
and the Fundamental Transformation, 175
and the Pohlmeyer-Lund-Regge model, 204
and the Ribaucour transformation, 173, 175
and the vector Calapso system, 164, 165
for Bianchi surfaces, 298, 299
for Demoulin surfaces, 355
for Dupin cyclides, 200

Gauss-Weingarten equations (*contd.*)
for extended Dym surfaces, 246–248
for extended sine-Gordon surfaces, 261, 262
for hyperbolic surfaces, 6, 47, 50, 54, 56
for isothermal-asymptotic surfaces, 358, 361
for isothermic surfaces, 153, 155, 161, 177, 191, 199
for projective-minimal surfaces, 336
for pseudospherical surfaces, 5, 7, 23, 32, 64, 65, 206, 209
projective, 331, 332, 333, 364
general relativity, 297, 309
geodesic curvature, 127, 142, 143
geodesic form, 144
geodesics, 142, 151, 241
on extended Dym surfaces, 245
on NLS surfaces, 127, 215
Geroch transformation, 15, 310, 311
Ginzburg-Landau equation, 119
Godeaux-Rozet surfaces, 15, 329, 334, 335
Godeaux sequences, 16
periodic, 341, 342
gravity waves
and the NLS equation, 119
great wave of translation, 1

Hamiltonian systems
of hydrodynamic type, 198, 230
Harrison transformation, 15, 297, 311, 317–319, 324, 325
and BT for the classical Bianchi system, 327
and cc-ideals, 374
decomposition, 322, 323, 328
Hasimoto surfaces (*see* NLS surfaces)
Hasimoto transformation, 121, 145, 272
Heat conduction, 10
reciprocal transformations in, 229
Heisenberg spin equation, 9, 60, 128, 129, 204, 229
as base member of the NLS eigenfunction hierarchy, 204, 219, 227
in hydrodynamics, 120, 151
reciprocal transformation, 227, 230
helicoid
Dini surface, 6, 35
one-soliton Demoulin surface, 357
helix
binormal evolution, 256
motion of a soliton on, 257, 258, 263
hexenhut, 105, 106
hodograph transformation, 97
Hoenselaers-Kinnersley-Xanthopoulos (HKX) transformation, 311
Huygens' tractrix, 6, 34, 35, 36
hydrodynamics, 120, 137, 142, 151
hyperbolic paraboloid, 55, 59
hyperbolic surfaces, 5, 7, 18, 20, 89
Bianchi system, 6, 20, 45
Beltrami-Enneper theorem, 63
BT, qv
NLS, 123, 126
spherical representation, 46

integrable discretisation, 14, 105, 237, 273
inverse harmonic mean curvature surfaces, 45
inverse scattering transform, 4, 61, 211
inversion
of isothermic surfaces, 156
involution
and the isothermic system, 155, 156
composition with the Ribaucour transformation, 176, 177, 184
isometric surfaces, 17
isometry
and Bonnet surfaces, 10
isomorphism, 7, 215, 268, 269, 358, 371, 373
between 3×3 and 2×2 linear representations, 7, 67, 371
and Plücker correspondence, 335, 340, 341
isothermal-asymptotic surfaces, 15, 16, 329, 330, 335, 357–370
and the stationary mNVN equation, 358–369
BT, qv
isothermic surfaces, 10, 11, 152–203
and the Calapso equation, qv
and the stationary Davey-Stewartson II equation, 11, 163
and the stationary Davey-Stewartson III equation, 11, 196, 203
BT, qv
constant mean curvature surfaces as, 156
Dupin cyclides as, 198–203
Gauss-Mainardi-Codazzi system, qv
Gauss-Weingarten system, qv
in $\mathbb{R}^{n+2}$, 156–162
minimal surface of Enneper as, 156
one-soliton, 191–193
vector Calapso system, qv
isothermic system, 10, 11, 154, 187, 191, 196, 199, 202
involutory invariance, 155
isotherm transformation, 232

Jacobi identity, 66
Josephson junctions, 22

Kac-Moody algebra, 113
Kadomtsev-Petviashivili (KP) hierarchy, 239
Kaup-Kuperschmidt hierarchy, 239
kelch, 107
Kerr black hole, 15, 319
Kerr-NUT fields, 15, 319

Killing-Cartan metric, 210
kink, 3, 22, 30
Kinnersley-Chitre potentials, 311
Korteweg-de Vries (KdV) equation, 2, 4, 22, 204, 210, 211
  and the Dym equation, 234
  as NVN reduction, 362
  Miura transformation, 7, 217, 363
Korteweg-deVries (KdV) hierarchy, 12, 205, 235
  BT, qv
  connection with mKdV and Dym hierarchies, 234, 235
Kramer-Neugebauer transformation, 308, 309, 314–316, 327
  in decomposition of Harrison transformations, 323, 328
  inverse, 310

Lagrangian description
  in 1+1-dimensional gasdynamics, 8, 95
Lamé system, 61, 68
Laplace – Beltrami operator, 99
Laplace – Darboux invariants, 110, 111, 118
  and periodicity, 112
  and the Toda lattice, 112, 114, 342
  under a Darboux transformation, 115
Laplace-Darboux transformations, 9, 89, 109–118
  and conjugate nets, 89, 109, 110, 117–118
  and the Toda lattice, 110, 111–116, 341–343
Lax pairs
  and the cc-ideal formulation, 249
  and the Sym-Tafel formula, 285, 298, 303
  elementary matrix Darboux transformations for, 285, 286
  for the projective Gauss-Mainardi-Codazzi equations, 333
  for the Bianchi system, 54
  for the Ernst equation, 297, 305–308
  for the extended Dym equation, 247
  for the extended sine-Gordon system, 261
  for the KdV equation, 217
  for the NLS equation, 147, 148, 149, 271, 284
  for the NVN equation, 362, 363, 368
  for the vector Calapso system, 11, 164–166, 188, 189, 193
  non-isospectral, 14, 54, 298, 300–303
Lax triad, 91
Legendre transformation, 98
Lelieuvre formulae, 47, 51
  and isothermal-asymptotic surfaces, 16, 361, 364, 365, 367
  and Tzitzeica surfaces, 92, 93
Levy sequences, 5
Levy transformation, 167, 170, 171
Lewis – Papapetrou metric, 309, 310
Lie algebras, 66, 208, 248, 336, 340, 371, 372, 374, 376, 377
  and soliton surfaces, 210, 211, 304
  and the Toda lattice, 112
Lie-Bäcklund transformation, 234
Lie groups, 48, 65, 66, 208, 371–373
Lie point symmetry
  and the Bianchi system, 6, 299
  and the Davey-Stewartson III equation, 11, 191, 196
  and the KdV hierarchy, 235
  and the NLS equation, 273, 276
  and the projective Gauss-Mainardi-Codazzi equations, 333
  and the sine-Gordon equation, 3, 22, 26, 27
Lie quadrics, 329
Lie sphere transformations, 230
linear representations (*see also* AKNS systems and Lax pairs), 11, 26, 154, 206, 209, 215, 217, 267, 277
  and cc-ideals, 252–254
  and the Sym-Tafel procedure, 11, 204, 262
  for extended Dym surfaces, 246–249, 256
  for extended sine-Gordon surfaces, 260, 261
  for projective-minimal surfaces, 335–341, 345, 350
  for the Bianchi system, 298, 299
  for the Demoulin system, 243, 354
  for the Ernst equation, 305, 306, 319–324
  for the mKdV equation, 82, 87
  for the NLS eigenfunction hierarchy, 12, 221
  for the NLS equation, 122, 124, 129, 148, 214, 272, 273, 276
  for the NLS hierarchy, 212, 213
  for the projective Gauss-Mainardi-Codazzi equations, 333, 337
  for the sine-Gordon equation, 26, 27, 64–68, 73–75, 210, 270, 378, 379
  for the SIT system, 130
  for the stationary mNVN equation, 358, 359, 364, 365–367
  for the Toda lattice, 113, 114
  for the Tzitzeica equation, 88, 89–91, 100, 101, 343
  for the vector isothermic system, 177, 188
  Gauss-Weingarten connection, 5
  non-isospectral, 14, 298, 300–302
  the Plücker correspondence, 16, 337-341, 344
lines of curvature
  and Dupin's theorem, 141
  and parallel surfaces, 41, 42
  and the Fundamental Transformation, 175
  circular, 198, 199, 203
  on a surface of revolution, 33, 153

lines of curvature (*contd.*)
  on hyperbolic surfaces, 49
  on the pseudosphere, 35
  on Weingarten surfaces, 43
  planar, 40, 156
LKR system, 49
Lobachevski geometry, 17
localised induction hierarchy, 87
Loewner transformations, 64, 98, 99, 229
loop soliton equation, 12, 225–226, 228
  link with the mKdV equation, 233, 238
loop soliton hierarchy
  invariance under reciprocal transformation, 230, 232
loop solitons, 12, 63, 192, 205, 215, 222–229
lump solution
  of the Calapso equation, 11, 195, 202

magnetohydrodynamics, 137, 142
magnetohydrostatics, 120
Martin formulation, 96, 97, 229
Masotti relation, 139, 142
matrix Darboux transformations, 13, 14, 171, 255, 262, 267, 270
  and cc-ideals, 249–254
  and non-isospectral linear representations, 300–303
  and superposition principles for the AKNS hierarchy, 292–294
  and the constant length property, 285, 286, 298
  and the Harrison transformation, 324, 325, 328
  for pseudospherical surfaces, 75, 267–271, 276, 281
  for the Bianchi system, 54, 326
  for the Ernst equation, 15, 297, 311, 319–325
  for the mKdV hierarchy, 284
  for the NLS hierarchy, 282, 283
  iteration, 287–292, 295
matrix Laplace-Darboux transformations, 118
matrix Schrödinger equation, 191
Matzner-Misner transformation, 15, 310, 313
Maxwell-Bloch system, 10, 120, 130
  compatibility with the NLS equation, 135–137
  the SIT equations, 132, 133
mean curvature, 143, 144
  CMC surfaces, 42–43, 153, 156, 161, 166
  in terms of principal curvatures, 10, 41
  of Hasimoto (NLS) surfaces, 126, 127
  of Weingarten surfaces, 42, 43, 44
Minding's theorem, 17
minimal surfaces, 156, 161, 166, 195, 196, 201
Miura transformation, 7, 12, 216, 217, 221, 229, 235, 358, 363
Minkowski space, 210, 292, 304, 326
  isothermic surfaces in, 11, 161, 163
mKdV equation, 7, 8, 10, 12, 204, 210, 211, 261, 330
  and motion of a curve, 60, 80–81, 237
  and motion of a pseudospherical surface, 68, 71, 130, 137
  as base member of the mKdV hierarchy, 214
  BT, qv
  connection to the loop soliton equation, 224, 225, 233, 238
  linear representation, qv
  Miura transformation, 217, 363
  $2+1$-dimensional version (mNVN equation), 330, 358
mKdV hierarchy
  and motion of a curve, 12, 205, 225, 237–238, 240
  as a specialisation of the NLS hierarchy, 214, 234
  link with Dym and KdV hierarchies, 205, 234, 235
  Miura transformation, 235
mKP hierarchy, 239
m^2KdV equation, 243
  BT, qv
  cc-ideal formalism, 250–254
  reciprocal link to the extended Dym equation, 242, 243, 246
  soliton on helix, 257
mNVN equation, 362
  and isothermal-asymptotic surfaces, 16, 330, 357–367
  Miura transformation, 363
Monge-Ampère equations
  in gasdynamics, 96, 97, 101, 229
  related to the Tzitzeica equation, 96, 97, 98, 101
Moutard equation, 52, 104, 363, 364
  and bilinear potential, 165
  and the Demoulin system, 353
  and the Lelieuvre formulae, 361, 363, 368
  in the vector Calapso system, 163, 164, 188
Moutard transformation, 5, 8, 9, 11, 52, 187–188, 266
  and the construction of isothermic surfaces, 191–198
  and the construction of Tzitzeica surfaces, 101–109
  and the Demoulin system, 353–354

Neugebauer transformation, 15, 297, 299, 311–318, 327
  commutation theorem, 317, 318
  iteration, 314, 316, 317
nonlinear optics
  and the NLS equation, 4, 119

nonlinear Schrödinger (NLS) equation, 4, 9, 10, 12, 13, 119, 120, 121, 142–146, 204, 210, 211, 216, 229, 241, 272, 283
  AKNS system, qv
  and binormal motion, 120–129, 215, 219, 220, 240
  and the Da Rios equations, 60, 119, 145
  and the Heisenberg spin equation, 128, 219
  and the unpumped Maxwell-Bloch system, 130, 135–137
  as base member of the NLS hierarchy, 212
  BT, qv
  Lie invariance, 273, 276
  single soliton solution, 122–123, 227
NLS eigenfunction hierarchy, 12, 216, 219, 221, 228, 229, 279
  and soliton surfaces of the NLS hierarchy, 222
  and the WKI system, 224
  potential, 12, 216, 220
  reciprocal transformation, 222–224, 227
NLS hierarchy, 11, 12, 13, 204, 212–214, 222–224
  AKNS system, qv
  and the localised induction hierarchy, 87
  and the mKdV hierarchy, 214, 234
  BT, qv
  Miura transformation, 216, 217, 221
  soliton surfaces, 205, 217–222
NLS (Hasimoto) surfaces, 10, 12, 13, 120, 149, 204, 215, 216, 263, 281
  and binormal motion of a curve, 122
  and loop solitons, 228
  and unpumped Maxwell-Bloch system, 130, 137
  compatible motion of, 130, 137
  corresponding to breathers, 126, 150, 151
  BT, qv
  geometric properties, 124–129, 137–146
  mean curvature, 127
  total (Gaussian) curvature, 126
  single soliton, 122–125, 228
  smoke rings, qv
NVN equation, 16, 358, 362
  and isothermal-asymptotic surfaces, 367–370
  Miura transformation, 363
nonlinear sigma model, 6, 48, 49, 129
numerical analysis
  ϵ-algorithm, 12, 205, 236–237

offset curves
  and dual Bertrand curves, 245

parallel conjugate nets, 172–175, 178, 179, 184
parallel curves, 128, 142, 215, 241
parallel surfaces, 6, 13, 41–45, 245
Pauli matrices, 48, 66, 205
permutability theorems
  and integrable discretisation, 14, 105
  for an anharmonic lattice model, 78, 79
  for an AKNS hierarchy, 14, 288, 292–296
  for Bianchi surfaces, 298, 326–327
  for conjugate nets, 178–183
  for Demoulin surfaces, 329
  for Godeau-Rozet surfaces, 329
  for isothermic surfaces, 154, 170, 184–186
  for the Ernst equation, 15, 297, 324–325
  for the Fundamental Transformation, 11, 178–183
  for the potential KdV equation, 4, 12, 205, 235–237, 239
  for the Ribaucour transformation, 182
  for the sine-Gordon equation, 3, 5, 28–31, 190, 288, 292
  for the vector Calapso system, 187–191
  generation of solitons and breathers via, 6, 37–41
Plücker correspondence, 16, 335, 336–341, 344, 351, 352, 365, 367
Pohlmeyer-Lund-Regge model, 9, 120, 129–132, 134, 204
polar reciprocal affinsphären, 92
potential KdV equation
  permutability theorem, qv
potential mKdV equation, 79, 224, 226, 230, 233, 240
  and motion of a Dini surface, 82–85
potential mKdV hierarchy, 284, 285
Prim gas, 95
principal curvatures, 10, 41, 42, 153
  vector analogues, 158
projective area functional, 15, 329, 333
projective metric, 332
projective minimal surfaces, 15, 16, 329, 333–341
  BT, qv
  linear representations, qv
projective transformations, 330
pseudo-differential operator formalism, 366
pseudopotentials, 15, 262, 326
  and Neugebauer's BT, 15, 312, 315–319, 327
  and a matrix Darboux transformation, 321–324
  and a permutability theorem, 324–325, 328
pseudosphere of Beltrami, 6, 32, 34–35
pseudospherical surfaces, 2, 3, 5, 6, 7, 10, 11, 13, 17–41, 46, 190, 204, 215, 216, 222, 258, 259, 263, 299
  and motion of constant torsion curves, 61–63
  BT, qv

pseudospherical surfaces (*contd.*)
  Dini, qv
  Gauss-Mainardi-Codazzi equations, qv
  Gauss-Weingarten equations, qv
  matrix Darboux transformations, qv
  motion of, 60, 68–80, 82–87
  position vector, 205–209

quadric, 153, 198, 330, 370
quartics of Kummer, 330, 370
quasi-geostrophic flow, 231

radial transformation, 167, 168, 171
  and decomposition of the Fundamental Transformation, 11, 169, 170
reciprocal transformations, 12, 205, 215, 222–235, 238, 239
  and dual extended Dym surfaces, 13, 245, 246
  and generation of N-gap solutions, 234
  and loop solitons, 222–228
  and the extended Dym equation, 242–246
  and the NLS eigenfunction hierarchy, 223, 224
  in link between the AKNS and WKI systems, 224, 233
  in link between Dym, mKdV and KdV hierarchies, 205, 229, 234, 235
  in physical applications, 205, 229, 230
  invariance of the Dym hierarchy under, 12, 228, 233, 239
recursion operator
  and the NLS hierarchy, 11, 12, 204, 212, 282
relativistic vortices
  and the Pohlmeyer-Lund-Regge model, 129
Ribaucour sphere congruence, 175
Ribaucour transformation, 11, 173–175, 187, 189
  Bianchi quadrilateral for, 183
  composition with an involution, 176
  permutability theorem, qv
Riccati equation, 58, 283, 312, 313
  and Lax pairs, 54, 149, 270–271, 284
  and the Kramer-Neugebauer transformation, 315
  and the Neugebauer BT, 316
Riemann invariants, 95

Schwarzschild solution, 15, 319
Schwarzian derivative, 332
Serret-Frenet relations, 61, 80, 121, 129, 138, 146, 149, 220, 221, 243, 244
sine-Gordon equation, 2, 3, 4, 5, 7, 11, 13, 14, 17, 21–41, 60–68, 75, 78, 130, 204, 215, 222, 235, 295
  and motion of curves, 60–64, 80
  as reduction of the Pohlmeyer-Lund-Regge model, 129
  BT, qv
  linear representation, qv
  permutability theorem, qv
  solitons, qv
  2 + 1-dimensional, 110
sinh-Gordon equation, 43, 167
  as Toda lattice reduction, 109
  and CMC surfaces, 156, 160
SIT system, 9, 110, 120
  and the NLS system, 129, 130, 132–134
smoke rings, 10, 120, 137, 150
solitons
  and an anharmonic lattice system, 78–79
  and the AKNS hierarchy, 279
  and the Ernst equation, 311
  and the extended Dym equation, 257
  and the KdV equation, 4, 236, 239
  and the mKdV equation, 79
  and the m^2KdV equation, 79
  and the NLS equation, 9, 122–124, 215
  and the sine-Gordon equation, 5, 6, 32, 34, 36–39, 41, 83, 84, 85
  and the Tzitzeica equation, 8, 105–108
  and the Weingarten equations, 85–86
  generation via BTs, 28–31
  history, 1–5
  in crystal dislocation theory, 3, 21, 22
  in gasdynamics, 95
  in stimulated Raman scattering, 130
  in ultrashort optical pulse propagation, 3, 4, 22, 30, 31
soliton surfaces, 209, 210, 217–220, 290
  Bianchi, qv
  Demoulin, qv
  Ernst, qv
  dual, 13, 94, 205
  extended Dym, qv
  extended sine-Gordon, qv
  isothermal-asymptotic, qv
  isothermic, qv
  loop soliton generation, 222–228
  NLS (Hasimoto), qv
  potential mKdV, 82–85
  pseudospherical, qv
  Sym-Tafel formula, qv
  Tzitzeica, qv
  Weingarten, qv
spectral parameter, 6, 54, 68, 88, 135, 154, 165, 204, 205, 221, 225, 299, 378
spectral transform, 266
spherical (Gauss) representation, 6, 46–49
  and Bianchi orthonormal basis, 50
SRS system, 10, 120, 134, 135
stereographic projection, 58
surfaces of revolution
  as isothermic surfaces, 153
  extended Dym, 258, 259
  isothermal-asymptotic, 153

projective transforms, 330
pseudospherical, 32–35
Tzitzeica, 94, 100, 105, 106
Sym-Tafel formula, 13, 14, 204, 208, 243, 262, 277, 285, 290, 303
and Bianchi surfaces, 298–300
and extended Dym surfaces, 267, 269, 270
and NLS surfaces, 215, 275
and pseudospherical surfaces, 267, 269, 270

tau function, 197, 198
Theorema egregium, 19, 33, 143
Toda lattice model, 8, 9, 16, 329
and Demoulin system, 16, 336, 341–343
and Laplace-Darboux transformations, 89, 109–116
and termination of Bergman series, 98
toroids
in magnetohydrostatics, 120
torus
as a Dupin cyclide, 198, 200, 202
total curvature (*see* Gaussian curvature)
total moments, 141
triply orthogonal systems of surfaces
Dupin's theorem, 141
Lamé system, 61
Weingarten system, 7, 8, 60, 68, 71–73, 80, 85–87
Tzitzeica condition, 8, 9, 88–91
Tzitzeica equation, 8, 9, 88–91, 93, 103, 105, 343
and the Demoulin system, 343, 354
as Toda lattice reduction, 113
BT, qv
linear representation, qv
in gasdynamics, 95, 96
Tzitzeica surfaces, 8, 46, 52, 88, 343
as isothermal-asymptotic surfaces, 335, 357
as projective-minimal surfaces, 15, 335
BT, qv
construction, 101–108
in gasdynamics, 89–99
of revolution, 100
Tzitzeica transformation, 103, 104–109

ultrashort optical pulses, 3, 22
and the classical permutability theorem, 30, 31
propagation of $2N\pi$ pulses, 3, 31

vector Calapso system, 162–166, 187–190
and the matrix boomeron equation, 164
and the stationary Davey-Stewartson II equation, 163
BT, qv
Lax pair, qv
vector isothermic system, 159, 165, 166, 189
and BT for isothermic surfaces, 177
vortex filament
and generation of Hasimoto surfaces, 122
and the Da Rios equations, 60, 199
and the NLS equation, 119, 120

Wahlquist-Estabrook procedure, 130, 261
and BT for the Ernst equation, 297, 311, 374
Weingarten equations, 18, 90, 130
for hyperbolic surfaces, 46, 51
Weingarten surfaces, 41–45
BT, qv
generalised, 118
Weingarten system, 7, 60, 71–76, 78, 80
and motion of pseudospherical surfaces, 10, 60, 71, 72
BT, qv
single soliton solution, 85–87
Weyl class, 319
Whitham's averaged variational principle, 119
Wilczynski moving tetrahedral, 16, 335, 336, 341, 351, 358
WKI equation, 227
loop soliton solution, 228, 229
WKI system, 224, 227, 230, 233
WTC procedure, 234
zoomeron equation, 11, 155, 191
and the isothermic system, 161
dromion-type solutions, 191, 196